Human Factors in Lighting

Human Factors in Lighting
Second edition

Peter R. Boyce
Lighting Research Center

Taylor & Francis
Taylor & Francis Group

LONDON AND NEW YORK

First published 2003
by Taylor & Francis
11 New Fetter Lane, London EC4P 4EE

Simultaneously published in the USA and Canada
by Taylor & Francis Inc
29 West 35th Street, New York, NY 10001

Taylor & Francis is an imprint of the Taylor & Francis Group

© 2003 Peter R. Boyce

Typeset in Sabon by
Newgen Imaging Systems (P) Ltd, Chennai, India
Printed and bound in Great Britain by
TJ International Ltd, Padstow, Cornwall

Every effort has been made to ensure that the advice and information
in this book is true and accurate at the time of going to press.
However, neither the publisher nor the authors can accept any legal
responsibility or liability for any errors or omissions that may be
made. In the case of drug administration, any medical procedure or
the use of technical equipment mentioned within this book, you are
strongly advised to consult the manufacturer's guidelines.

British Library Cataloguing in Publication Data
A catalogue record for this book is available from the British Library

Library of Congress Cataloging in Publication Data
Boyce, P. R.
    Human factors in lighting / P.R. Boyce. – [2nd ed.]
      p. cm.
    Includes bibliographical references and index.
    1. Light – Physiological effect. 2. Lighting – Physiological aspects.
3. Visual perception. I. Title.

QP82.2.L5 B69 2003
621.32′2–dc21   2002040936

ISBN 0–7484–0949–1 (hbk)
ISBN 0–7484–0950–5 (pbk)

To my parents, Robert James Boyce and Kathleen Mary Boyce, for their unstinting support and encouragement.

To my wife, Susan Boyce, for her enduring love and endless patience.

To my daughter, Anna Jane Boyce, for being a daughter of whom any father would be proud.

Contents

Preface

This book is the culmination of a long journey, a journey that started in 1966 when I joined the Electricity Council Research Centre, at Capenhurst, in England, and was directed to take an interest in lighting. So began what has become a lifetime's study of the uses and abuses of light, and the impacts it has on people. To me, lighting is attractive because it is a "crossroads" subject, one that requires the integration of knowledge from the fields of physics, physiology, psychology, and ergonomics. To me, the study of lighting is valuable because of the potential it has to make a difference in people's lives. This combination of integration and influence has been irresistible.

For 24 years I worked at Capenhurst, steadily building my understanding of light and lighting, but essentially working alone. During that time, the first edition of *Human Factors in Lighting* was published. Then, in 1990, pushed by the lack of interest in lighting following privatization of the electricity industry, and pulled by the opportunity to have day-to-day contact with people with similar interests, I moved to the Lighting Research Center at Rensselaer Polytechnic Institute, in New York State. It has turned out to be a wonderful move. At the LRC, I have been able to work with many bright and insightful individuals and to enjoy the delights of teaching small classes of interested students from all over the world. The result has been a widening of my horizons and a deepening of my understanding. One outcome is this second edition of *Human Factors in Lighting*.

Like Gaul, this book is divided into three parts. Part I is devoted to the fundamentals of light, vision, and the circadian system. Part II covers the generalities of how lighting affects the ability to work, causes visual discomfort, and influences the perception of spaces and objects. Part III is devoted to the specifics of the use of light in different contexts and for different purposes. Those new to the subject are advised to read Part I before proceeding further. Those familiar with the fundamentals can read wherever their curiosity takes them.

During this journey I have had the good fortune to work with many gifted individuals. Of particular note are Donald McIntyre, the late Ian Griffiths, Geoffrey Brundrett, Kit Cuttle, David Loe, Warren Julian,

Jennifer Veitch, Guy Newsham, Russ Leslie, Andrew Bierman, Neil Eklund, John Bullough, Yukio Akashi, Claudia Hunter, and Mark Rea. I have learned something from all of them but particularly from Mark Rea. He has been an unending source of inspiration and challenges for many years. I thank them all for their willingness to share the excitement of ideas and the joy of science.

Of course, to turn the understanding generated by research into practice requires contributions from the fields of manufacture and design. During my journey I have had the pleasure of working with many talented individuals active in these fields. Of particular note are John Baker, Terry McGowan, Howard Brandston, Naomi Miller, Bill Blitzer, Ton Begmann, Kate Conway, and Peter Bleasby. I thank them all for their willingness to show me the view outside the laboratory door.

Peter R. Boyce
Troy, New York

Acknowledgments

This book could not have been completed without the help and support of many people. It is a pleasure to acknowledge the contributions of Suzanne Hayes, for obtaining many varied and obscure publications; Julia Cerotti, for checking many of the references; Yimin Gu and Susan Sechrist, for the preparation of many of the figures; and Sandra Vasconez for her review of Chapter 1.

I am also pleased to acknowledge the roles of Alan Balfour, Dean of the School of Architecture, and of Mark Rea, Director of the Lighting Research Center, both of whom supported my request for sabbatical leave to finish this book; and of Rensselaer Polytechnic Institute, who granted it.

Finally, the cooperation of the following authors and publishers in giving permission for the reproduction of copyright material is gratefully acknowledged.

American Journal of Physiology, for Figure 3.11
Andrew Bierman, for Figures 1.11 and 1.15
Arnold Wilkins, for Figure 13.9
Eugenio van Someren, for Figure 13.12
Holophane, A Division of Acuity Lighting Group Inc, for Figure 8.13
Claudia Hunter, for Figure 4.1
Lei Deng, for Figures 1.14 and 1.17
Lighting Research Center, for Figure 9.2
McGraw-Hill Inc, for Figures 2.5 and 2.10
National Eye Institute, for Figure 12.4
Oxford University Press, for Figure 12.2
Ross De Alessi Lighting Design, for Figure 14.5
R.A.Weale, for Figures 12.2 and 12.6
Society of Automotive Engineers, for Figure 10.7
The Chartered Institution of Building Services Engineers, for Figures 1.10, 7.11, 7.16, 9.13, 10.6, and 10.8
The Illuminating Engineering Society of North America, for Figures 1.4, 1.5, 1.6, 1.7, 1.9, 1.12, 1.16, 2.9, 2.20, 2.25, 2.30, 2.31, 3.1, 4.10, 4.11, 4.13, 7.6, 7.8, and 9.15

Part I
Fundamentals

1   Light

1.1   Introduction

This book is concerned with the interaction of people and light. To fully understand this interaction, it is first necessary to understand what light is, how its characteristics can be quantified, and how it is produced and controlled. These topics are the subject of this chapter.

1.2   Light and radiation

To the physicist, light is simply part of the electromagnetic spectrum that stretches from cosmic rays with wavelengths of the order of femtometers to radio waves with wavelengths of the order of kilometers (Figure 1.1). What distinguishes the wavelength region between 380 and 780 nm from the rest of the electromagnetic spectrum is the response of the human visual system. Photoreceptors in the human eye absorb energy in this wavelength range and thereby initiate the process of seeing. Other creatures are sensitive to different parts of the electromagnetic spectrum but light is defined by the visual response of humans.

Unfortunately for simplicity, the response of the human visual system is not the same at all wavelengths in the range 380–780 nm. This makes it impossible to adopt the radiometric quantities conventionally used to measure the characteristics of the electromagnetic spectrum for quantifying light. Rather, a special set of quantities have to be derived from the radiometric quantities by weighting them by the spectral sensitivity of the human visual system.

The principle used for the measurement of the human spectral sensitivity is the equivalence of visual effect, the effect in question being the perception of brightness. Radiation consisting of a single wavelength somewhere between 380 and 780 nm will be seen as having both a brightness and a color. An observer viewing two equal-size visual fields presented for the same time, and with the same single wavelength and the same radiance will consider the two fields indistinguishable, i.e. equal in all respects, so they have the same visual effect. If the two fields have the same wavelength but different radiances, the field with the higher radiance will be perceived to

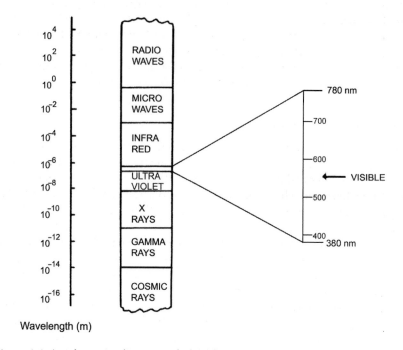

Wavelength (m)

Figure 1.1 A schematic diagram of the electromagnetic spectrum showing the location of the visible spectrum. The divisions between the different types of electromagnetic radiation are indicative only.

be brighter. When both the wavelength and the radiance of the two fields are different, the two fields will be seen to differ in brightness and color. In this situation, it is possible to achieve brightness equivalence by altering the radiance of one field until the two fields look equally bright. If R_1 and R_2 are, respectively, the radiances of the two fields at wavelengths λ_1 and λ_2, brightness equivalence can be represented by the equation:

$$V_1 R_1 = V_2 R_2$$

where V_1 and V_2 are the weighting factors necessary to make the equation correct for the measured radiances. Since the only measured values are radiances, each brightness equivalence match produces a ratio V_1/V_2. By establishing brightness equivalences for many pairs of wavelengths and using the transitive principle of mathematics, i.e. $V_1/V_3 = (V_1/V_2) \times (V_2/V_3)$, it is possible to express the sensitivity of the visual system at each wavelength relative to its sensitivity at an arbitrarily chosen standard wavelength, i.e. $V_\lambda/V_{standard}$. The standard wavelength usually chosen is the one for which human visual sensitivity is a maximum, i.e. the wavelength at which, for a constant radiance, the brightness is the greatest. Then, by giving $V_{standard}$ a value of unity and plotting the resulting V_λ against wavelength, a curve can be produced which quantifies the relative efficiency of different wavelengths in producing the same perception of brightness. Such a curve is the relative spectral sensitivity curve of the human visual system. It contains the information necessary to convert the fundamental radiometric quantities into quantities suitable for measuring light.

1.3 The CIE standard observers

Unfortunately, a unique spectral sensitivity curve applicable to all people in all conditions does not and cannot exist. Different relative spectral sensitivity curves are obtained depending on the method used for measuring brightness equivalence, on what visual photoreceptors are stimulated, and on what channel of the visual system is being accessed (Kaiser, 1981). Further details of these matters are given in Chapter 2. For the moment it is sufficient to know that the human retina has two classes of visual photoreceptors, one class operating when light is plentiful, in what are called photopic conditions (cone photoreceptors) and the other operating when light is very limited, in what are called scotopic conditions (rod photoreceptors). These two photoreceptor types have very different relative spectral sensitivities. What these spectral sensitivities are has been the subject of international agreement. The body which organizes these agreements is the Commission Internationale de l'Eclairage (CIE). In 1924, the CIE adopted the CIE Standard Photopic Observer, based on the work of Gibson and Tyndell (1923), who took data from several experiments and proposed a smooth and symmetric spectral sensitivity curve. The experiments from which the data were taken used small test fields, usually less than 2° in diameter, and the amount of light was sufficient to put the visual system into the photopic state. Later work by Judd (1951) showed that the CIE Standard Photopic Observer was too insensitive at short wavelengths, a result which eventually lead the CIE to formally recognize a modified photopic spectral sensitivity curve (CIE, 1990a) with greater sensitivity than the CIE Standard Photopic Observer at wavelengths below 460 nm. This CIE Modified Photopic Observer was stated to be a supplement to the CIE Standard Photopic Observer, not a replacement for it. As a result, the CIE Standard Photopic Observer has continued to be widely used by the lighting industry. This is acceptable because the modified sensitivity at wavelengths below 460 nm has been shown to make little difference to the photometric properties of nominally white light sources that emit radiation over a wide range of wavelengths. It is only for light sources that emit significant amounts of radiation below 460 nm that changing from the CIE Standard Photopic Observer to the CIE Modified Photopic Observer can be expected to make a significant difference to measured photometric properties (CIE, 1978). Some colored signals, colored displays, and narrow band light sources, such as blue light emitting diodes, fall into this category.

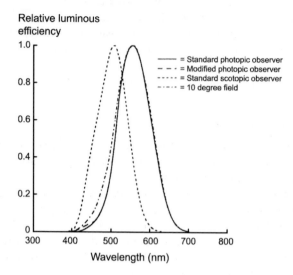

Figure 1.2 The relative luminous efficiency functions for the CIE Standard Photopic Observer, the CIE Modified Photopic Observer, the CIE Standard Scotopic Observer, and the relative luminous efficiency function for a 10° field of view in photopic conditions.

In 1951, the CIE adopted the CIE Standard Scotopic Observer, based on measurements by Wald (1945) and Crawford (1949) using an area covering the central 20° of the visual field with a photopic luminance of approximately 0.00003 cd/m^2. While this is scientifically interesting because it represents the spectral response of the rod photoreceptors, it is rarely used by the lighting industry because the provision of almost any lighting installation worthy of the name will take the human visual system out of the scotopic state.

The CIE Standard and Modified Photopic Observers and the CIE Standard Scotopic Observer are shown in Figure 1.2, the Standard and Modified Photopic Observers having maximum sensitivities at 555 nm and the Standard Scotopic Observer having a maximum sensitivity at 507 nm (CIE, 1983a, 1990a). These relative spectral sensitivity curves are formally known as the 1924 CIE Spectral Luminous Efficiency Function for Photopic Vision, the CIE 1988 Modified Two Degree Spectral Luminous Efficiency Function for Photopic Vision, and the 1951 CIE Spectral Luminous Efficiency Function for Scotopic Vision, respectively. More commonly, they are known as the CIE $V(\lambda)$, CIE $V_M(\lambda)$, and the CIE $V'(\lambda)$ curves. These curves are the basis of the conversion from radiometric quantities to photometric quantities, the quantities used to characterize light.

1.4   Photometric quantities

The most fundamental measure of the electromagnetic radiation emitted by a source is its radiant flux. This is a measure of the rate of flow of energy emitted and is measured in watts. The most fundamental quantity used to measure light is luminous flux. Luminous flux is radiant flux multiplied, wavelength by wavelength, by the relative spectral sensitivity of the human visual system, over the wavelength range 380–780 nm. This process can be represented by the equation:

$$\Phi = K_m \sum \Psi_\lambda V_\lambda \Delta\lambda$$

where Φ is the luminous flux (lumens), Ψ_λ the radiant flux in a small wavelength interval $\Delta\lambda$ (watts), V_λ the relative luminous efficiency function for the conditions, and K_m a constant (lumens/watt).

In System Internationale (SI) units, the radiant flux is measured in watts and the luminous flux in lumens. The value of K_m is 683 lm/W for the CIE Standard and Modified Photopic Observers and 1,699 lm/W for the CIE Standard Scotopic Observer. These numbers arise from the decision of the CIE that 1 W of radiant flux at 555 nm should produce 683 lm, for both photopic and scotopic conditions. As 555 nm is the maximum sensitivity of the CIE Standard and Modified Photopic Observers, the constant is unchanged for the photopic condition. But for the CIE Standard Scotopic Observer, the relative spectral sensitivity is only 0.402 at 555 nm. Therefore, the constant for scotopic conditions is 683/0.402 = 1,699 lm/W. It is always important to identify which of the standard observers is being used in any particular measurement or calculation. This requirement has lead the CIE to recommend that whenever the Standard Scotopic Observer is being used, the word scotopic should precede the measured quantity, i.e. scotopic luminous flux.

Luminous flux is used to quantify the total light output of a light source in all directions. While this is important, for lighting practice it is also important to be able to quantify the luminous flux emitted in a given direction. The measure that quantifies this concept is luminous intensity. Luminous intensity is the luminous flux emitted per unit solid angle, in a specified direction. The unit of measurement is the candela, which is equivalent to a lumen per steradian. Luminous intensity is used to quantify the distribution of light from a luminaire.

Both luminous flux and luminous intensity have area measures associated with them. The luminous flux falling on unit area of a surface is called the illuminance. The unit of measurement of illuminance is the lm/m^2 or lux. The luminous intensity emitted per unit projected area of a source in a given direction is the luminance. The unit of measurement of luminance is cd/m^2. The illuminance incident on a surface is the most widely used electric lighting design criterion. The luminance of a surface is a correlate of its brightness. Table 1.1 summarizes these photometric quantities.

As might be expected, there is a relationship between the amount of light incident on a surface and the amount of light reflected from the same surface.

Table 1.1 The photometric quantities

Measure	Definition	Units
Luminous flux	That quantity of radiant flux which expresses its capacity to produce visual sensation	Lumens (lm)
Luminous intensity	The luminous flux emitted in a very narrow cone containing the given direction divided by the solid angle of the cone, i.e. luminous flux/unit solid angle	Candela (cd)
Illuminance	The luminous flux/unit area at a point on a surface	Lumen/ meter^2
Luminance	The luminous flux emitted in a given direction divided by the product of the projected area of the source element perpendicular to the direction and the solid angle containing that direction, i.e. luminous intensity/unit area	Candela/ meter^2
Reflectance	The ratio of the luminous flux reflected from a surface to the luminous flux incident on it	
For a diffuse surface	Luminance = (illuminance × reflectance)/π	
Luminance factor	The ratio of the luminance of a reflecting surface viewed from a given direction to that of a perfect white uniform diffusing surface identically illuminated	
For a non-diffuse surface, for a specific direction and lighting geometry	Luminance = (illuminance × luminance factor)/π	

For a perfectly diffusely reflecting surface, the relationship is give by the equation:

$$\text{luminance} = (\text{illuminance} \times \text{reflectance})/\pi$$

where luminance is expressed in cd/m^2 and illuminance is expressed in lm/m^2.

For a diffusely reflecting surface, reflectance is defined as the ratio of reflected luminous flux to incident luminous flux.

For a non-diffusely reflecting surface, i.e. a surface with some specularity, the same equation between luminance and illuminance applies but reflectance is replaced with luminance factor. Luminance factor is defined as the ratio of the luminance of the surface viewed from a specific position and lit in a specified way to the luminance of a diffusely reflecting white

Table 1.2 Some photometric units of measurement for illuminance and luminance and the multiplying factors necessary to change them to SI units

Quantity	Unit	Dimensions	Multiplying factor
Illuminance	Lux	Lumen/meter^2	1.00
	Meter candle	Lumen/meter^2	1.00
	Phot	Lumen/centimeter^2	10,000
	Footcandle	Lumen/foot^2	10.76
Luminance	Nit	Candela/meter^2	1.00
	Stilb	Candela/centimeter^2	10,000
		Candela/inch^2	1,550
		Candela/foot^2	10.76
	Apostilb[a]	Lumen/meter^2	0.32
	Blondel[a]	Lumen/meter^2	0.32
	Lambert[a]	Lumen/centimeter^2	3,183
	Footlambert[a]	Lumen/foot^2	3.43

Note

a These four units are based on an alternative definition of luminance. This definition is that if a surface can be considered as perfectly diffusely reflecting, its luminance in any direction is the product of the illuminance on the surface and the reflectance of the surface. Thus the luminance has dimensions of lumens/unit area. This definition is deprecated in the SI system.

Table 1.3 Typical illuminance and luminance values

Situation	Illuminance (lm/m^2)	Typical surface	Luminance (cd/m^2)
Clear sky in summer in temperate zones	150,000	Grass	2,900
Overcast sky in summer in temperate zones	16,000	Grass	300
Textile inspection	1,500	Light gray cloth	140
Office work	500	White paper	120
Heavy engineering	300	Steel	20
Good road lighting	10	Concrete road surface	1.0
Moonlight	0.5	Asphalt road surface	0.01

surface viewed from the same direction and lit in the same way. It should be clear from this definition, that a non-diffusely reflecting surface can have many different values of the luminance factor. Table 1.1 summarizes these definitions.

Unfortunately for consistency, photometry has a long history which has generated a number of different units of measurement for illuminance and luminance. Table 1.2 lists some of the alternative units, together with the multiplying factors necessary to convert from the alternative unit to the SI units of lm/m^2 for illuminance and cd/m^2 for luminance. The SI units will be used throughout this book.

Both illuminance and luminance are widely used in lighting practice to quantify the end result of installing a lighting system and the stimulus to the visual system. Being able to define these quantities is useful but, in addition, it is always helpful to have an idea of what are representative magnitudes for these quantities in different situations. Table 1.3 shows some illuminances and luminances typical of commonly occurring situations, all measured using the CIE Standard Photopic Observer.

1.5   Some limitations

Although the photopic photometric quantities defined above can be calculated or measured precisely, it is important to appreciate that they only represent the visual effect of light in a particular state. Specifically, they represent the brightness response of the central 2° of the retina, i.e. the fovea, in high light level conditions. Changing either field size or light level can change the spectral sensitivity of the visual system.

The effect of field size was recognized by the CIE in 1964 when a provisional relative spectral sensitivity curve for the central 10° of the visual field in photopic conditions was approved (CIE, 1986; see Figure 1.2). This curve shows greater sensitivity to short wavelength light than the CIE Standard Photopic Observer because the visual field extends beyond the macula, an area covering the central 5° of the retina and containing a pigment that attenuates short-wavelength light, and into the area where short-wavelength cone photoreceptors are found.

As for the effect of changing light level, it should be appreciated that there is a large gap in the luminance range between photopic and scotopic conditions for which there are no CIE recommendations for relative spectral sensitivity, namely between approximately 0.001 and 3 cd/m². This range is called the mesopic condition. For the fovea, the CIE Standard Photopic Observer still applies in the mesopic range because there are only medium- and long-wavelength cones present in the fovea, which is what the CIE Standard Photopic Observer is based on. In the rest of the visual field the spectral sensitivity is in a state of continual change as the balance between rod and cone photoreceptors changes with light level, until either rods dominate, as in scotopic vision, or cones dominate as in photopic vision. Mesopic vision is important because much exterior lighting, such as on roads, actually provides conditions that are in the mesopic range. Nonetheless, all the photometric quantities that are used to characterize road lighting, and other exterior applications, are based on the CIE Standard Photopic Observer. This practice can lead to situations where the photometric measurements bear little relation to the visual effect of the light source for off-axis vision (see Chapter 10). The absence of a CIE mesopic photometry system is not for want of trying (CIE, 1989). Indeed several different systems have been suggested, most based on some weighted combination of photopic and scotopic measurements. Others have abandoned the perception of brightness as the metric of visual effect and, using reaction time, have developed a comprehensive system of photometry that covers photopic, mesopic, and scotopic light levels (He *et al.*, 1998). This latter proposal may be rather too radical to be adopted by the CIE but it does indicate the gathering strength of interest in developing a system of mesopic photometry.

Two other systematic effects that lead to different relative spectral sensitivities from the values represented by the CIE Standard Photopic Observer occur with age or with defective color vision. As discussed in Chapter 12, as the eye ages the transmittance of the lens decreases, particularly at the short-wavelength end of the visible spectrum. This will lead to a reduced sensitivity in this wavelength region for older people (Sagawa and Takahashi, 2001). For people with defective color vision, either there are missing photopigments or the photopigments are different from the normal. In either case, the relative spectral sensitivity of such people is likely to depart from that of the CIE Standard Photopic Observer.

In addition to these systematic effects, there are the inevitable individual differences between people. Figure 1.3 shows the range of relative spectral sensitivity for 52 observers, taken from the data of Gibson and Tyndell (1923), from which the CIE Standard Photopic Observer was derived. Clearly, there are wide individual differences in spectral sensitivity. This implies that the fact that the photometric quantities can be calculated

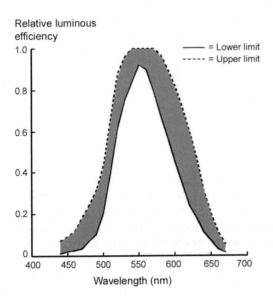

Figure 1.3 Range of relative luminous efficiency functions for 52 observers. The results for all the observers fall in the shaded area (after Judd and Wyszecki, 1963).

and/or measured precisely is no guarantee that they will be closely related to the visual effects produced. Despite this limitation, the CIE Standard Observers have a definite value. They provide a globally agreed means for the lighting industry to quantify the performance of its products, in terms of luminous flux and luminous intensity distributions; and for designers to quantify what their lighting systems deliver, in terms of illuminance and luminance. Despite the utility of such measures, whenever considering the photometric quantities for a given lighting situation it is always important to ask whether whatever CIE Standard Observer is being used in the calculation of the photometric quantities is appropriate to the situation. If it is not, then the apparent precision of the measurement may be misleading.

1.6   Colorimetric quantities

The photometric quantities described above do not take into account the wavelength combination of the light received at the eye. Thus it is possible for two luminous fields to have the same luminance but to be made up of totally different combinations of wavelengths. In this situation, and provided either photopic or mesopic conditions prevail, the two fields may look different in color. Exactly what color will be seen depends not only on the spectral distribution of the radiation incident on the retina, but also on several other factors, such as the luminance and color of the surroundings and the state of adaptation of the observer. As the saying goes, color is a pigment of our imagination. It is a perception developed in the brain from past experience and the information contained in the retinal image. Light itself is not colored. Nonetheless, to have a means of characterizing the color perception associated with different light sources and other stimuli to the visual system, some way had to be found to provide a quantitative measure of color. The CIE colorimetry system provides such a measure.

1.6.1   *The CIE colorimetry system*

The basis of the CIE colorimetry system is color matching. Color matching measurements are another example of visual equivalence in the sense that the observer is simply asked to determine whether two fields are the same color. From extensive color matching measurements, the CIE Color Matching Functions have been determined. These functions are essentially the relative spectral sensitivity curves of the human observers with normal color vision and can be considered as another form of standard observer. There are three color matching functions, as might be expected from the fact that humans with normal color vision can match any color of light with a combination of not more than three wavelengths of light from the long-, medium-, and short-wavelength regions. Although the existence of three color matching functions is analogous to the existence of the three cone photoreceptor types involved in color vision (see Chapter 2), it must be emphasized that the CIE color matching functions are not based on physiology. They are mathematical constructs that reflect the relative spectral sensitivities required to ensure that all the spectral distributions that are seen as the same color have the same position in the CIE colorimetry system and that every spectral distribution that is seen as a different color occupies a different position. Figure 1.4 shows two sets of color matching functions, the 1931 Standard Observer for a 2° field and the 1964 Standard Observer for a 10° field (CIE, 1986). The CIE 1931 Standard Observer is used for colors occupying visual fields up to 4° of angular subtense. The CIE 1964 Standard Observer is used for colors covering visual fields greater than 4° in angular subtense. The values of the color matching functions at different wavelengths are known as the spectral tristimulus values.

The color of a light source can be represented mathematically by multiplying the spectral power distribution of the light source, wavelength by wavelength, by each of the three color matching functions $x(\lambda)$, $y(\lambda)$, and $z(\lambda)$, the outcome being the amounts of three imaginary primary colors X, Y, and Z required to match the light source color. In the form of equations, X, Y, and Z are given by

$$X = h\Sigma S(\lambda) \cdot x(\lambda) \cdot \Delta\lambda$$
$$Y = h\Sigma S(\lambda) \cdot y(\lambda) \cdot \Delta\lambda$$
$$Z = h\Sigma S(\lambda) \cdot z(\lambda) \cdot \Delta\lambda$$

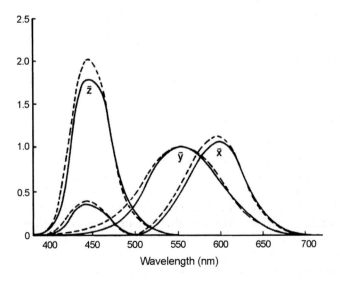

Figure 1.4 Two sets of color matching functions: the CIE 1931 Standard Observer (2°) (solid line) and the CIE 1964 Standard Observer (10°) (dashed line).

where $S(\lambda)$ is the spectral radiant flux of the light source (W/nm), $x(\lambda)$, $y(\lambda)$, $z(\lambda)$ the spectral tristimulus values from the appropriate color matching function, $\Delta\lambda$ the wavelength interval (nm), and h an arbitrary constant.

If only relative values of the X, Y, and Z are required, an appropriate value of h is one that makes $Y = 100$. If absolute values of the X, Y, and Z are required it is convenient to take $h = 683$ since then the value of Y is the luminous flux in lm.

If the color being calculated is for light reflected from a surface or transmitted through a material, the spectral reflectance or spectral transmittance is included as a multiplier in the above equations. For a reflecting surface, an appropriate value of h is one that makes $Y = 100$ for a reference white because then the actual value of Y is the percentage reflectance of the surface.

Having obtained the X, Y, and Z values, the next step is to express their individual values as proportions of their sum, i.e.

$$x = X/(X + Y + Z) \quad y = Y/(X + Y + Z) \quad z = Z/(X + Y + Z)$$

The values x, y, and z are known as the CIE chromaticity coordinates. As $x + y + z = 1$, only two of the coordinates are required to define the chromaticity of a color. By convention, the x and y coordinates are used. Given that a color can be represented by two coordinates, then all colors can be represented on a two-dimensional surface. Figure 1.5 shows the CIE 1931 chromaticity diagram, the two axes being the x and y chromaticity coordinates. It is possible to identify a number of interesting features on the CIE 1931 chromaticity diagram. The outer curved boundary is called the spectrum locus. All pure colors, i.e. those that consist of a single wavelength, lie on this curve. The straight line joining the ends of the spectrum locus is the purple boundary and is the locus of the most saturated purples obtainable. At the center of the diagram is a point called the equal energy point. This is the point where a colorless surface will be located. Close to the equal energy point is a curve called the Planckian locus. This curve passes through the chromaticity coordinates of objects that operate as a black body, i.e. the spectral power distribution of the light source is determined solely by its temperature.

The CIE 1931 chromaticity diagram can be considered as a map of the relative location of colors. The saturation of a color increases as the chromaticity coordinates get closer to the spectrum locus and further from the equal energy point. The hue of the color is determined by the direction in which the chromaticity coordinates move. These characteristics have been formalized as dominant wavelength and excitation purity. To determine the dominant wavelength of a surface lit by a known light source, a line is drawn through the two points represented by the chromaticity coordinates of the light source alone and the surface when lit by the light source, and

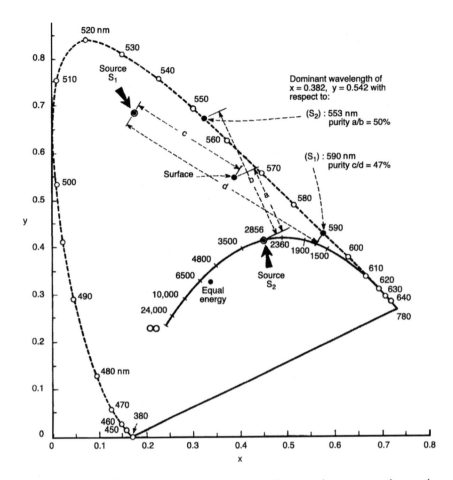

Figure 1.5 The CIE 1931 chromaticity diagram showing the spectrum locus, the Planckian locus, the equal energy point, and the method of calculating dominant wavelength and excitation purity for a surface and two different light sources (from IESNA, 2000a).

extended to the spectrum locus. The wavelength at which the extended line intersects the spectrum locus is the dominant wavelength. As for excitation purity, this is the ratio of the distance from the chromaticity coordinates of the light source to the chromaticity coordinates of the lit surface, divided by the total distance from the light source alone to the intersection of the line with the spectrum locus. Two examples of such calculations are shown in Figure 1.5.

Strictly, any discussion as to how a specific combination of wavelengths will appear, based on the chromaticity diagram, is nonsense. The only thing that a set of chromaticity coordinates tells us about a color is that colors with the same chromaticity coordinates will match. They tell us nothing about the appearance of the matched colors. But this is an argument for color vision zealots. The fact is a red surface lit by a nominally white light source will always plot in one part of diagram and a green in another part and so on. Thus, although the CIE 1931 chromaticity diagram is not theoretically pure, it is useful for indicating approximately how a color will appear, a value recognized by the CIE when it specified chromaticity coordinate limits for signal lights and surfaces so that they will be recognized as red, green, yellow, and blue (CIE, 1994).

Given that different colors plot at different positions on the CIE 1931 chromaticity diagram, it would seem reasonable to expect that the distance between two sets of chromaticity coordinates would be correlated to how different the two colors represented by the chromaticity coordinates appear. While this is approximately true, the correlation is very low. This is because the CIE 1931 chromaticity diagram is perceptually non-uniform. Green colors cover a large area while red colors are compressed in the bottom right corner. This perceptual non-uniformity makes any attempt to quantify large color differences using the CIE 1931 chromaticity diagram futile. In an attempt to improve this situation, the CIE first introduced the CIE 1960 Uniform Chromaticity Scale (UCS) diagram and then, in 1976, recommended the use of the CIE 1976 UCS diagram. Both diagrams are simply linear transformations of the CIE 1931 chromaticity diagram. The axes for the CIE 1976 UCS diagram are

$$u' = 4x/(-2x + 12y + 3) \quad v' = 9y/(-2x + 12y + 3)$$

where x and y are the CIE 1931 chromaticity coordinates. Figure 1.6 shows the CIE 1976 UCS diagram.

While the 1976 UCS diagram is more perceptually uniform that the CIE 1931 chromaticity diagram, it is of limited value for determining color differences. This is because it is two dimensional, considering only the hue and saturation of the color. To completely describe a color, a third dimension is needed, that of brightness for a self-luminous object and lightness for a reflecting object (Wyszecki, 1981). In 1964, the CIE introduced the U^*, V^*, W^* three-dimensional color space for use with surface colors, where

$$U^* = 13\,W^*(u - u_n)$$
$$V^* = 13\,W^*(v - v_n)$$
$$W^* = 25\,Y^{0.33} - 17 \quad \text{(where Y has a range from 1 to 100)}$$

W^* is called a lightness index and approximates the Munsell value of a surface color (see Section 1.6.2). The coordinates u, v, refer to the chromaticity coordinates of the surface color in the CIE 1960 UCS diagram

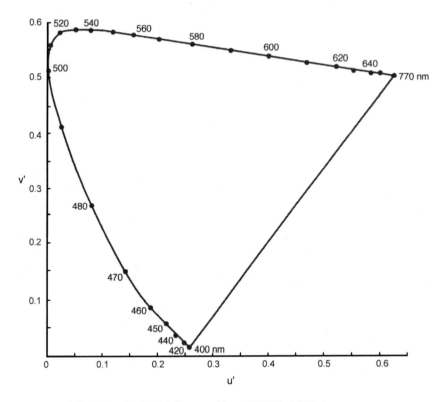

Figure 1.6 The CIE 1976 UCS diagram (from IESNA, 2000a).

while the chromaticity coordinates u_n, v_n refer to a spectrally neutral color lit by the source, that is placed at the origin of the U^*, V^* system. This U^*, V^*, W^* system is little used now, about the only purpose for which it is routinely used is the calculation of the CIE Color Rendering Indices discussed later.

The U^*, V^*, W^* color space is little used now because it has been superceded by two other color spaces introduced by the CIE in 1976 (Robertson, 1977; CIE, 1986). These two color spaces are known by the initialisms CIELUV and CIELAB. The three coordinates of the CIELUV color space are given by the expressions:

$$L^* = [116\,(Y/Y_n)^{0.33} - 16] \quad \text{for } Y/Y_n > 0.008856$$
$$L^* = 903.29\,(Y/Y_n) \quad \text{for } Y/Y_n \leq 0.008856$$
$$u^* = 13\,L^*(u' - u'_n)$$
$$v^* = 13\,L^*(v' - v'_n)$$

where u' and v' are the chromaticity coordinates from the CIE 1976 UCS diagram and u'_n, v'_n, Y_n are values for a nominally achromatic color, usually the surface with 100 percent reflectance ($Y = 100$) lit by the light source.

The three coordinates of the CIELAB color space are given by the expressions:

$$L^* = 116\, f(Y/Y_n) - 16$$
$$a^* = 500\, [f(X/X_n) - f(Y/Y_n)]$$
$$b^* = 200\, [f(Y/Y_n) - f(Z/Z_n)]$$

where $f(q) = q^{0.33}$ for $q > 0.008856$, $f(q) = 7.787q + 0.1379$ for $q \leq 0.008856$, and $q = X/X_n$, or Y/Y_n, or Z/Z_n.

Again, X_n, Y_n, Z_n are, respectively, the values of the X, Y, and Z for a nominally achromatic surface, usually that of the light source with $Y_n = 100$.

Each of these color spaces have a color difference formula associated with them. For the CIELUV color space, the color difference is given by

$$\Delta E^*_{uv} = [(\Delta L^*)^2 + (\Delta u^*)^2 + (\Delta v^*)^2]^{0.5}$$

For the CIELAB color space, the color difference is given by

$$\Delta E^*_{ab} = [(\Delta L^*)^2 + (\Delta a^*)^2 + (\Delta b^*)^2]^{0.5}$$

These two color spaces are now widely used to set color tolerances for manufacture in many industries. But why two color spaces? When introduced, there was insufficient experimental evidence to indicate which would be most satisfactory for a wide range of industrial applications, but it was believed that either would be better than the then prevalent U^*, V^*, W^* system. As an indication of the perceptual uniformity of the CIELUV and CIELAB systems, Figure 1.7 shows loci of constant Munsell hue and chroma for a value of 5 (see Section 1.6.2), plotted on u^*, v^* and a^*, b^* planes through the CIELUV and CIELAB color spaces (Anon, 1977). If the CIELUV and CIELAB color spaces were perceptually uniform, these loci should form equally spaced concentric circles for saturation and equally spaced radial lines for hue. As can be seen in Figure 1.7, neither CIELUV nor CIELAB is perfectly perceptually uniform but they are both a lot better than the alternative U^*, V^*, W^* color difference system or the more primitive CIE UCS diagrams. Further, both CIELUV and CIELAB can be used as the basis for developing metrics correlated with lightness, chroma, and hue. The link between these color spaces and color appearance will be discussed in Chapter 6.

1.6.2   Color order systems

While the CIE colorimetric system is valuable for quantifying colors, it does lack a physical presence. This need is met by a variety of color

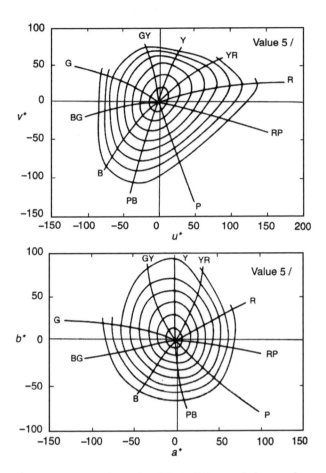

Figure 1.7 Loci of constant Munsell hue and chroma for a value of 5, plotted on planes through the CIELUV and CIELAB color spaces (from IESNA, 2000a).

ordering systems. A color ordering system is a physical, three-dimensional representation of color space. In a sense it is an atlas of colors and like an atlas, the separation between adjacent colors is intended to be uniform in all directions. There are several different color ordering systems used in different parts of the world (Billmeyer, 1987). One of the most widely used is the Munsell system. Figure 1.8 shows the organization of the Munsell system. The azimuthal hue dimension consists of 100 steps arranged around a circle, with five principal hues (red, yellow, green, blue, and purple) and five intermediate hues (yellow-red, green-yellow, blue-green, purple-blue, and red-purple). The vertical value scale contains 10 steps from black to white. The horizontal chroma scale contains up to 20 steps from gray to highly saturated. Each of the three scales is designed to provide equal steps of perception for an observer with normal color vision looking at the samples lit by daylight, with a gray or white surround. The position of any color in the Munsell system is identified by an alphanumeric reference made up of three terms, hue, value, and chroma, e.g. a strong red is given the alphanumeric 7.5R/4/12. Achromatic surfaces, i.e. colors that lie along the vertical Value axis and hence that have no hue or chroma, are coded as Neutral 1, Neutral 2, etc. depending on their reflectance. To a first approximation, the percentage reflectance of a surface is given by the product of V and $(V-1)$ of the surface, where V is the Munsell value of the surface.

The utility of a color ordering system is that it makes colors manifest and hence makes it easy to communicate about color in a more precise way than words permit. For example, rather than someone in New York telling someone in London that the required color is lightish, yellowish green, it is much better to say that the color required is Munsell reference 5YG/8/2 because then, provided both parties have access to a Munsell system publication, they can physically see what the required color is. While communicating through the Munsell system, or any other color ordering system, is more precise than words, it is not as precise as using the numbers generated by the CIE color spaces. However, sometimes precision has to give way to convenience. Building materials, such as paints, plastic, and ceramics are commonly classified in terms of a color ordering system.

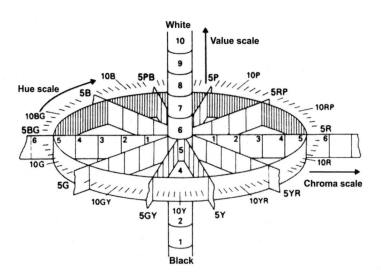

Figure 1.8 The organization of the Munsell color order system. The hue letters are B = blue, PB = purple/blue, P = purple, RP = red/purple, R = red, YR = yellow/red, Y = yellow, GY = green/yellow, G = green, BG = blue/green.

The existence of several different color atlas systems used in different parts of the world, as well as the quantitative CIE colorimetry system, would seem to be a recipe for confusion. Fortunately, this is usually avoided by the fact that conversions are available between many of the color ordering systems and the CIE colorimetry system. For example, the German DIN system provides both Munsell and CIE equivalents of its components (Richter and Witt, 1986). The name categories of the Inter-Society Color Council – National Bureau of Standards Method (Kelly and Judd, 1965) are given in terms of the Munsell system (National Bureau of Standards, 1976). Conversions between the Munsell system and the CIE colorimetry system are given in the American Society for Testing and Materials Test Method D1535 (ASTM, 1996c), based on Nickerson (1957).

1.6.3 Application metrics

While the CIE colorimetry system is the most complete and most widely accepted means of quantifying color, it is undeniably complex. Therefore, the lighting industry has used the CIE colorimetry system to derive two single-number metrics to characterize the color properties of light sources; correlated color temperature and the CIE General Color Rendering Index (CRI). You will find these two metrics given in most lamp manufacturers' catalogs. Correlated color temperature is a metric for the color appearance of the light emitted by a light source. The CIE General CRI is a metric of the effect a light source has on the appearance of surface colors.

1.6.3.1 Correlated color temperature

In principle, the color of the light emitted by a light source can be characterized by its chromaticity coordinates. In practice, this is rarely done. Rather, the correlated color temperature is used. The basis of this measure is the fact that the spectral emission of a black body is defined by Planck's radiation law and hence is a function of its temperature only. Figure 1.9 shows a section of the CIE 1931 chromaticity diagram with the Planckian locus shown. The locus is the curved line joining the chromaticity coordinates of black bodies at different temperatures. The lines running across the Planckian locus are iso-temperature lines. When the chromaticity coordinates of a light source lie directly on the Planckian locus, the color appearance of that light source is expressed by the color temperature, i.e. the temperature of the black body that has the same chromaticity coordinates. For light sources that have chromaticity coordinates close to the Planckian locus but not on it, their color appearance is quantified as the correlated color temperature, i.e. the temperature of the iso-temperature line that is closest to the actual chromaticity coordinates of the light source. The temperatures are usually given in Kelvin (K). An alternative metric, namely reciprocal color temperature is sometimes used, this being measured

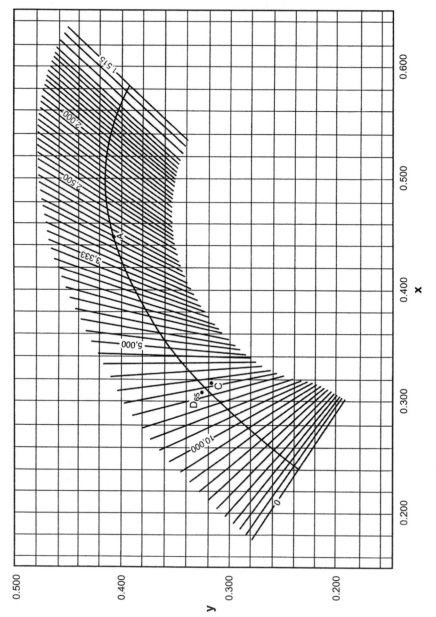

Figure 1.9 The Planckian locus and lines of constant correlated color temperature plotted on the CIE 1931 (x, y) chromaticity diagram. Also shown are the chromaticity coordinates of CIE Standard Illuminants, A, C, and D65 (from IESNA, 2000a).

as 1,000,000 divided by the correlated color temperature measured in Kelvin and expressed as reciprocal megaKelvin (MK^{-1}). The advantage of this metric is that a difference of 1 MK^{-1} indicates approximately the same color difference at any color temperature above 1,800 K.

Correlated color temperature is a very convenient and easily understandable metric of light source color appearance, applicable to nominally white light sources. As a rough guide, such light sources have correlated color temperatures ranging from 2,700 to 7,500 K. A 2,700 K light source, such as an incandescent lamp, will have a yellowish color appearance and be described as "warm," while a 7,500 K lamp, such as some types of fluorescent lamp, will have a bluish appearance and be described as "cool." The most commonly used fluorescent lamps are in the correlated color temperature range of 3,000–4,100 K. It is important to appreciate that light sources that have chromaticity coordinates that lie beyond the range of the iso-temperature lines shown in Figure 1.9 should not be given a correlated color temperature. Such lamps will appear greenish when the chromaticity coordinates lie above the Planckian locus or purplish if they lie below it.

1.6.3.2 CIE color rendering index

As for the effect a given light source will have on the appearance of surface colors, in principle, this can be given by calculating the chromaticity coordinates of each color in one of the CIE color spaces. Differences between different surface colors can then be estimated by calculating their separation in color space. This is reasonable if a specific set of surface colors is of interest but for most lighting applications, where many different but unspecified colors are used, more general advice is desirable. This is where the CIE CRI comes in. The CIE CRI measures how well a given light source renders a set of standard test colors relative to their rendering under a reference light source of the same correlated color temperature as the light source of interest (CIE, 1995a). The reference light source used is an incandescent light source for light sources with a correlated color temperature below 5,000 K and some form of daylight for light sources with correlated color temperature above 5,000 K. The actual calculation involves obtaining the positions of a surface color in the CIE 1964, U^*, V^*, W^* color space under the reference light source and under the light source of interest and expressing the difference between the two positions on a scale that give perfect agreement between the two positions a value of 100. The CIE has 14 standard test colors. The first eight form a set of pastel colors arranged around the hue circle. Test colors 9–14 represent colors of special significance, such as skin tones and vegetation. The result of the calculation for any single color is called the CIE Special CRI, for that color. The average of the Special CRIs for the first eight test colors is called the CIE General CRI. It is this latter index that is usually presented in light source manufacturers' catalogs.

The CIE General CRI has its limitations. First, it should be appreciated that just because two light sources have the same General CRI, it does not mean that they render colors the same way. The General CRI is an average and there are many combinations of Special CRI values that give the same average. Second, different light sources are being compared with different reference light sources. This makes the meaning of comparisons between different light sources uncertain, yet comparing light sources is what the General CRI is most widely used to do. Third, there is considerable argument about the method used to correct for chromatic adaptation. These limitations should be borne in mind when evaluating the CIE General CRIs for different light sources.

1.6.3.3   *Color vector maps*

The great attraction of the CIE General CRI is that it reduces the complexity of the rendering of colors to a single number. But this reduction leads to a considerable loss of information. An alternative but similar approach to quantifying the color properties of light sources that preserves the complexity of color rendering has been developed by Philips Lighting BV. Figure 1.10 shows a plot of the difference in position in color space for 215 test colors (Opstelten, 1983) when lit by the light source of interest and a reference light source of the same correlated color temperature, plotted on the a^*, b^* plane of the CIELAB color space (van Kemenade and van der Burgt, 1988). The origin of each arrow on the map is the chromaticity of the color under the reference light source and the head of the arrow is the chromaticity of the color when lit by the light source of interest. Obviously, the shorter the arrows, the more similarly the light source of interest renders colors relative to their rendering under the reference light source. Further, the direction of the arrow gives the direction of any change in color rendering. Arrows that point toward the origin of the figure indicate a reduction in chroma under the light source of interest, while arrows that point across radial lines from the origin indicate a shift in hue. A common feature of Figure 1.10 is that greater color shifts occur in some hue/chroma areas and smaller shifts occur in others. Clearly, such a method of displaying the color rendering properties of light sources gives much more information than the single number of the CIE General CRI but understanding the diagram requires some thought which has reduced its popularity.

1.6.3.4   *Color gamut*

While the color vector map overcomes one of the problems inherent in the CIE General CRI, namely the fact that light sources that render colors in different ways can have the same General CRI, it still relies on the comparison with a reference light source and light sources with different correlated

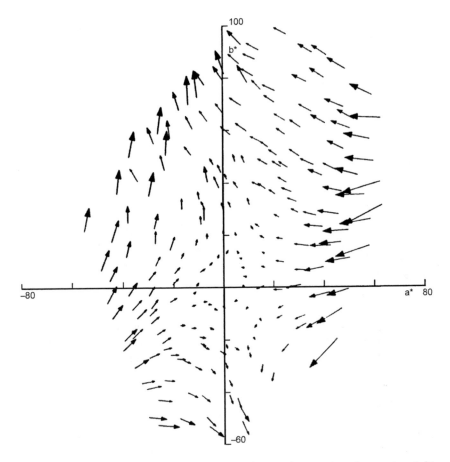

Figure 1.10 Color mismatch vectors for 215 object colors projected onto the *a**, *b** plane of the CIELAB color space for a metal halide lamp (after van Kemenade and van der Burgt, 1988).

color temperatures will have different reference light sources. The color gamut is a way to overcome this limitation. The color gamut is obtained by calculating the position of the first eight CIE standard test colors under the light source of interest and plotting them on the CIE 1976 UCS diagram. When the plotted positions are joined together, the color gamut is formed. Figure 1.11 shows the color gamuts for a number of different light sources. A great deal can be learnt from the color gamut. From a consideration of its shape and the spacing between the positions of the individual test colors, the extent to which the different parts of the hue circle can be discriminated is apparent. From its location on the CIE 1976 UCS diagram, the appearance

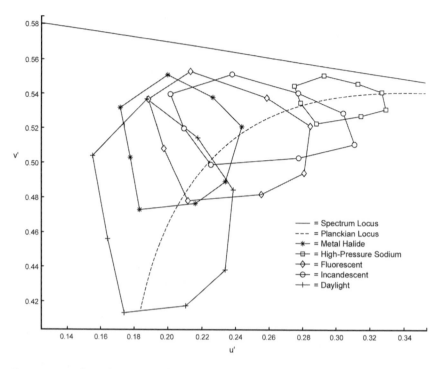

Figure 1.11 The color gamuts for high-pressure sodium, incandescent, fluorescent, and metal halide light sources, and for the CIE Standard Illuminant D65, simulating daylight, all plotted on the CIE 1976 UCS diagram. The dotted curve is the Planckian locus.

of colors can be appreciated to some degree. By plotting different light sources on the same diagram it is easy to make comparisons between light sources. Further, by including the gamut area of an ideal light source, such as daylight, it is possible to evaluate how close to the ideal light source is the light source of interest, as far as color rendering is concerned.

The color gamut does share one limitation of the CIE General CRI in that it uses the same eight standard test colors but that is more a matter of convenience than anything else. There is no reason why a wider range of colors should not be used. Of course, using more colors makes any comparison more difficult but that is a reflection of the reality of color rendering. Failure to recognize this fact has lead to the proposal to develop another single number index, the gamut area, as a metric of color rendering. The gamut area is the area enclosed by the boundary of the color gamut. Gamut area suffers from the defects of any single number index, in that light sources with the same gamut area can render colors differently, but at least it is free from having different reference light sources.

Table 1.4 Scotopic/photopic ratios for a number of widely used electric light sources

Light source	Photopic efficacy (lm/W)	Scotopic efficacy (lm/W)	Scotopic/ photopic ratio
Incandescent	14.7	20.3	1.38
Fluorescent (3,500 K)	84.9	115.9	1.36
Mercury vapor	52.3	66.8	1.28
Metal halide	107.4	181.7	1.69
High-pressure sodium	126.9	80.5	0.63
Low-pressure sodium	180.0	40.8	0.23

Source: He *et al.* (1997).

1.6.3.5 Scotopic/photopic ratio

One other measure of light source color characteristics that has been gaining interest in recent years is the scotopic/photopic ratio (Berman, 1992). This is calculated by taking the relative spectral power distribution, in radiometric units, of the light source and weighting it by the CIE Standard Scotopic and Photopic Observers and expressing the resulting scotopic lumens and photopic lumens as a ratio. The value of scotopic/photopic ratios is that they express the relative effectiveness of different light sources in stimulating the rod and cone photoreceptors in the human visual system. A light source with a higher scotopic/photopic ratio will stimulate the rods more than a light source with a lower scotopic/photopic ratio when both produce the same photopic luminous flux. This information is useful when considering light sources for applications where the operation of both rod and cone photoreceptors is likely. Table 1.4 gives scotopic/photopic ratios for a number of commonly used light sources.

1.7 Sources of light

Illumination is produced naturally, by the sun, and artificially, by oil and gas flames and electric light sources. The development and growth in use of artificial sources of light over the last century has fundamentally changed the pattern of life for millions of people on Earth.

1.7.1 Natural light

Natural light is light received on Earth from the Sun, either directly or after reflection from the Moon. The prime characteristic of natural light is its variability. Natural light varies in magnitude, spectral content, and distribution with different meteorological conditions, at different times of day and year, at different latitudes. Moonlight is of little interest as a source of illumination but daylight is used, and strongly desired, for the lighting of buildings. Daylight can be divided into two components, sunlight and skylight. Sunlight is light received at the Earth's surface, directly from the sun. Sunlight produces strong, sharp-edged shadows. Skylight is light from the sun received at the Earth's surface after scattering in the atmosphere. It is this scattered light that gives the sky its blue appearance, as compared to the blackness of space. Skylight produces only weak, diffuse shadows. The balance between sunlight and skylight is determined by the nature of the atmosphere and the distance which the light passes through it. The greater the amount of water vapor and the longer the distance, the higher is the proportion of skylight.

The illuminances on the Earth's surface produced by daylight can cover a large range, from 150,000 lx on a sunny summer's day to 1,000 lx on a heavily overcast day in winter. Several models exist for predicting the daylight incident on a plane, at different locations, for different atmospheric conditions (Robbins, 1986). These models can be used to predict the contribution of daylight to the lighting of interiors.

Figure 1.12 shows a spectral power distribution of daylight. It is clear that daylight contain significant amounts of ultraviolet and infrared radiation and that, over the visible wavelengths, daylight is a continuous spectrum. The correlated color temperature of daylight can range from 4,000 K for an overcast day to 40,000 K for a clear blue sky. For calculating the appearance of objects under natural light, the CIE recommends the use of one of three different spectral distributions corresponding to correlated color temperatures of 5,500, 6,500, and 7,500 K (Wyszecki and Stiles, 1982). Figure 1.13 shows the relative spectral irradiance distributions for several different phases of daylight.

In addition to the amount of daylight and its spectral power distribution, daylight varies in distribution over the sky. The luminance distribution of natural light in the sky has been standardized by the CIE for a completely overcast sky in 1955 and for a clear sky in 1967. The luminance distribution for a completely overcast sky is given by the expression:

$$L = L_z(1 + 2\cos\eta)/3$$

where L is the luminance of an element of the sky in kcd/m^2, L_z the luminance of the sky at the zenith in kcd/m^2, and η the angle between the element of sky and the zenith in radians.

The maximum luminance occurs at the zenith, the luminance distribution being symmetrical about the zenith and independent of the actual altitude of the sun.

The luminance of an element of clear sky (L) is given by the expression (CIE, 1973):

$$L = L_z(1 - e^{-0.32/\cos\gamma})(0.91 + 10\,e^{-3\eta} + 0.45\cos^2\eta)$$
$$/0.274\,(0.91 + 10\,e^{-3z} + 0.45\cos^2 z)$$

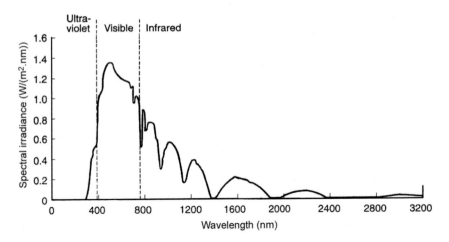

Figure 1.12 The spectral irradiance of daylight over the ultraviolet, visible, and infrared regions of the electromagnetic spectrum (from IESNA, 2000a).

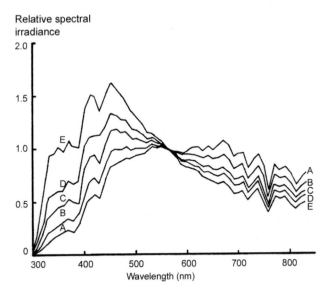

Figure 1.13 The relative spectral irradiance distributions of the phases of daylight. The distributions are normalized at 560 nm. The standard phases of daylight have correlated color temperatures of A = 4,800 K, B = 5,500 K, C = 6,500 K, D = 7,500 K, and E = 10,000 K (after Judd *et al.*, 1964).

where L_z is the luminance of the sky at the zenith in kcd/m^2, γ the angle of the sky element from the zenith in radians, η the angle between the sky element and the sun in radians, and z the angle of the sun from the zenith in radians.

In practice, the luminance distribution of the sky varies between these two standards depending on the degree of cloud cover. To estimate the amount of daylight admitted to a building and its consequences for the use of electric lighting and air-conditioning, it is necessary to make some assumptions about the manner of this variation. Quantitative data for use under different sky conditions is given in Krochmann and Seidl (1974). Daylight availability data are available for different climate types and there are also average models, the average being taken over the complete year (Littlefair, 1981). These databases can be used to estimate the cost benefits of using a lighting control system linked to daylight as well as the impact of daylight on the annual energy consumption of a building. A simple worst-case approach is still sometimes used, namely to assume a CIE standard overcast sky producing an illuminance on the ground of 5,000 lx. As for sunlight, there are standard skypath diagrams which can be used to predict the position of a patch of sunlight in a building for any specific window or skylight size, form, and orientation. Details of these approaches can be found in Robbins (1986) and the IESNA *Lighting Handbook* (IESNA, 2000a). Daylight is a feature of many buildings. It is highly regarded by people, at least in climates where daylight is limited for part of the year. It has a marked effect on the design and control of buildings in all climates.

1.7.2 *Artificial light – flame sources*

The first form of artificial lighting used by humans was firelight, created by the combustion of wood. Developments in basic technology lead to the creation of the oil lamp, the candle and, ultimately, the gas lamp, all of which depend on combustion of a fuel. Oil lamps, candles, and gas lamps are sometimes used today, either through necessity or for the atmosphere they evoke. However, they are rarely used for functional lighting where an electricity supply is available. This is for three reasons. The first is the fire hazard posed by open flames in buildings. The second is the level of air pollution produced by combustion of fuel in confined spaces (the tradition of spring cleaning originated with the amount of carbon deposited on room surfaces during the winter months from gas lighting). The third, and the most important, is the low luminous efficacy of these flame sources. Luminous efficacy is the ratio of the amount of light emitted by the light source to the power supplied to it and is measured in lm/W. Typical luminous efficacy values for candles, oil lamps, and gas flames are 0.1, 0.3, and 1 lm/W, respectively. These values are two orders of magnitude lower than the ubiquitous fluorescent lamp widely used for functional lighting in commercial buildings and one order of magnitude less than the common incandescent lamp used in homes.

1.7.3 Artificial light sources – electric/general illumination

The lighting industry makes several thousand different types of electric lamps. Those used for providing general illumination can be divided into two classes; incandescent lamps and discharge lamps. Incandescent lamps produce light by heating a tungsten filament to incandescence. Discharge lamps produce light by an electric discharge in a gas. Incandescent lamps operate directly from the electricity supply. Discharge lamps require control gear between the lamp and the electricity supply, because different electrical conditions are required to initiate the discharge and to sustain it.

1.7.3.1 The incandescent lamp

The most common form of incandescent lamp is known to many as the household bulb. This produces light by heating a thin tungsten filament to incandescence in an inert gas atmosphere. The spectral emission of the incandescent lamp is a continuum over the visible spectrum (Figure 1.14) although the exact spectrum is determined by the temperature of the filament. This is easily seen when an incandescent lamp is dimmed. Reducing the voltage reduces the current through the filament and hence the temperature of the filament. The result is that the color appearance of the light emitted by the lamp becomes more yellow and then red until, at very low voltages, no light can be seen at all, although the lamp may still be emitting infrared radiation. The design of an incandescent lamp is a matter of balancing luminous efficacy against life. A high light output and hence a higher luminous efficacy can be achieved by heating the filament to just below its melting point but then the life is short. When a long lamp life is desirable, as in traffic signals, the filament is heated to a lower than usual temperature. For incandescent lamps used in households around the world, the luminous efficacy is around 12 lm/W and the life is about 1,000 h. The incandescent lamp has been commercially available since the 1880s and is probably still the most widely used lamp in the world. It is small, cheap, simple to operate, has good color properties, and is easy to dim. This list is enough to suggest why the incandescent lamp has withstood the test of time so well.

1.7.3.2 The tungsten halogen lamp

The tungsten halogen lamp is essentially an incandescent lamp with a halogen in the gas filling. The inclusion of the halogen allows the filament to be run at a higher temperature because, although the tungsten is evaporated off the filament faster at the higher temperature, the halogen chemically reacts with the evaporated tungsten to form a tungsten halogen compound

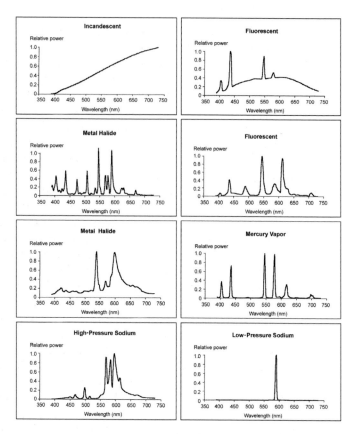

Figure 1.14 Relative spectral power distributions for incandescent, two forms of metal halide, high-pressure sodium, two forms of fluorescent, mercury vapor, and low-pressure sodium light sources. All the spectral power distributions are normalized to unity for the wavelength with the maximum output. The two forms of metal halide differ in the chemicals used in the arc tube. The two forms of fluorescent differ in the mixture of phosphors used.

which diffuses back to the filament where the higher temperature causes it to separate into tungsten and halogen, depositing the tungsten back on the filament. This cycle ensures the light output is maintained at a higher level for longer than would be the case without the halogen. The higher filament temperature also implies a higher luminous efficacy, around 20 lm/W. The spectral emission of the tungsten halogen lamp is a continuum across the visible spectrum, as would be expected given its fundamental incandescent nature. The tungsten halogen lamp has been commercially available since the 1960s. Its small size, in combination with an appropriate reflector, have made it a favorite for accent lighting in retail applications.

1.7.3.3   *The fluorescent lamp*

The fluorescent lamp is a discharge lamp in that the physical means for producing light is the excitation of a gaseous discharge. The fluorescent lamp, in either its linear or compact form, consists of a glass tube containing a mercury atmosphere. Heating the electrodes produces a stream of negatively charged electrons. These electrons are accelerated through the mercury gas by the potential difference between the electrodes. The accelerated electrons collide with the gas atoms producing two effects. The first is the ionization of the atom into electrons and a positively charged particle called an ion. This increases the electron concentration and hence maintains the discharge. The second possible outcome is that the atom absorbs most of the energy of the colliding electron and thereby raises the energy state of its own captive electrons to a higher level. These energy levels are discrete and when the captive electrons shortly afterwards decay back to their resting level, energy is radiated at a wavelength determined by the energy level structure of the atom. For mercury at a low pressure, which is what fills a fluorescent lamp, most of the radiation emitted by the discharge is in the ultraviolet region of the electromagnetic spectrum. To produce radiation in the visible spectrum, the inner surface of the glass tube is coated with a phosphor. This absorbs the ultraviolet radiation from the discharge and emits radiation in the visible spectrum. This two-step process is evident in the spectral emission of the fluorescent lamp (see Figure 1.14), which usually consists of a series of strong emission lines, from the discharge, superimposed on a continuous emission spectrum, from the phosphor. By changing the phosphor mix, different spectral emissions can be created, so fluorescent lamps are available with a wide range of color properties.

The fluorescent lamp is a discharge lamp and therefore needs to have a control system to alter the electrical conditions from those required to start the discharge to those required to maintain it. This control system, which is usually called a ballast, can be electromagnetic or electronic. Ballasts are available that make it possible to dim fluorescent lamps over a wide range with little change in color properties. The fluorescent lamp has been commercially available since the late 1930s and today is widely used in commercial applications, mainly because of its high luminous efficacy (in the range 60–110 lm/W).

1.7.3.4   *The mercury vapor lamp*

The mercury vapor lamp is similar to the fluorescent lamp in that it is a discharge lamp based on a mercury atmosphere in an arc tube. The difference is that the mercury vapor lamp is a high-pressure lamp. The result is that the spectral emission of the gas discharge is moved into the visible region, although it still consists of a series of intense spectral lines (see Figure 1.14). The mercury vapor lamp is also available with a phosphor coating on the inside of the envelope, the phosphor coating being used to improve the color properties of the lamp. The mercury vapor lamp has been commercially available since the early 1930s and is now fading into disuse. It has a lower luminous efficacy than competing high-pressure discharge lamps and poor color properties. Its only saving grace is its low cost.

1.7.3.5 The metal halide lamp

The metal halide lamp is also a high-pressure gas discharge lamp based on a mercury discharge, but it is different from the mercury vapor lamp in that it has metal halides, such as scandium and sodium iodides, in the arc tube. When the arc tube reaches operating temperature, the metal halides are vaporized. At the core of the discharge the metal halides separate into the metals and halogen, the metals emitting radiation in the visible région. At the cooler edge of the arc tube, the metals and halogen recombine and then repeat the process. The result is a spectrum consisting of many discrete spectral lines (see Figure 1.14). As may be imagined, the chemistry of the metal halide lamp is very complex. The result is that early metal halide lamps gained a reputation for showing shifts in color properties over life and even between different lamps from the same manufacturer when new. Recent developments in arc tube materials and design have gone a long way to alleviate this problem (van Lierop *et al.*, 2000). The metal halide lamp was introduced to the market in the late 1960s and is now the lamp of choice where a high lumen output light source with high luminous efficacy and good color properties is required.

1.7.3.6 The low-pressure sodium lamp

The other broad class of discharge lamp is based around sodium. Electrically, the low-pressure sodium lamp operates in the same manner as the fluorescent lamp but in this case a phosphor is unnecessary because the spectral emission from the sodium discharge is concentrated in two spectral lines, which are both close to 589 nm. Because this wavelength is near to the peak spectral sensitivity of the human visual system at 555 nm, the low-pressure sodium lamp has the highest luminous efficacy of all the artificial light sources (up to 180 lm/W). Unfortunately, its color properties are what might be expected from a monochromatic source, non-existent. For this reason, its use is restricted to applications where color is of little consequence, such as road lighting away from inhabited areas, and then only in countries where luminous efficacy is valued above all else.

1.7.3.7 The high-pressure sodium lamp

Conceptually, the high-pressure sodium lamp is the same as the low-pressure sodium lamp but the much higher pressure has an effect on the spectral emission. The increased pressure in the discharge leads to self-absorption of radiation within the discharge and interactions between the closely packed atoms. The combined effect of these phenomena is to reduce the power at 589 nm and to spread the spectral emission over a much wider range of wavelengths (see Figure 1.14). The result is a combination of high luminous efficacy and modest color properties. Exactly what the balance is between luminous efficacy and color properties depends on the pressure in the arc tube. Two levels of pressure are commercially produced. One produces light with an orange color appearance but has a high luminous efficacy. The other produces light with a white color appearance but with a lower luminous efficacy. The former is commonly used for street lighting and in industrial applications. The latter is used for display lighting. The high-pressure sodium lamp in its high luminous efficacy form has been commercially available since the early 1960s. It soon replaced the mercury vapor lamp to become the most widely used light source for exterior lighting and for much industrial lighting, although both these positions are now being challenged by the metal halide lamp.

1.7.3.8 Electrodeless lamps

All the discharge light sources discussed above create a discharge by applying a voltage across two electrodes placed in the arc tube. In the last decade, a number of electrodeless lamps have been introduced into the market. These use an electromagnetic field, at either radio or microwave frequencies, to create a discharge. The lamps using radio frequencies are called induction lamps and are essentially a fluorescent lamp. The electromagnetic field excites the mercury in an enclosure that then emits radiation mainly in the ultraviolet region. This is then absorbed by a phosphor and re-radiated in the visible region. The luminous efficacy and color properties of induction lamps are similar to those of fluorescent lamps, their main advantage being the longer life produced by not having any electrodes to fail.

The electrodeless lamp using microwave frequencies produces an electromagnetic field in a cavity in which is rotating a small sphere filled with sulfur. The excited sulfur becomes a plasma that emits a radiation with a continuous spectrum. The sulfur lamp, as this is called, is not currently in production but it offers an intriguing look into the possibilities for new light sources that combine high luminous efficacy with good color properties (Turner *et al.*, 1997).

1.7.3.9 Others

It should not be thought that the above represent all the light sources available. There are many forms of each class of lamp available together with some lamps developed for special applications. Among the latter are cold cathode fluorescent lamps used for advertising and decorative effect and short-arc xenon and metal halide lamps used in searchlights, and for television.

Details of these light sources can be found in the IESNA *Lighting Handbook* (IESNA, 2000a).

1.7.4   Light source characteristics

Electric light sources can be characterized on several different dimensions. They are:

- Luminous efficacy = the ratio of luminous flux produced to power supplied (lm/W). If the lamp needs a ballast to operate, the watts supplied should include the power demand of the ballast.
- Spectral power distribution = the radiant flux (W) emitted at different wavelengths.
- Correlated color temperature (see Section 1.6.3.1).
- CIE General CRI (see Section 1.6.3.2).
- Lamp life = the number of burning hours until either lamp failure or a stated percentage reduction in light output occurs. Lamp life can vary widely with switching cycle.
- Warm-up time = the time from switch-on to full light output.
- Re-strike time = the time delay between the lamp being switched off before it will re-ignite.

Figure 1.14 shows the spectral power distribution of the lamp types discussed above. Table 1.5 summarizes the other characteristics for many of the same lamp types. The values in the Table 1.5 should be treated as indicative only. Details of the characteristics of any specific lamp should always be obtained from the manufacturer. A more detailed discussion of the construction, operation, and properties of the light sources listed in Table 1.5 is available in Coaton and Marsden (1996).

1.7.5   Artificial light sources – electric-signs and -signals

Many of the lamp types discussed above are also used for internally and externally illuminated signs and signals. For example, incandescent and compact fluorescent lamps have been used in exit signs; metal halide lamps are used to externally illuminate road signs; tubular fluorescent lamps are used for externally illuminated billboards and internally illuminated advertising signs; miniature incandescent lamps are used for brake lights and direction indicators on vehicles, as well as for lighting instrument panels. However, there are other light sources that are used primarily for signs and signals alone. The three that will be discussed here are the light emitting diode (LED), the electroluminescent panel, and the nuclear light source.

1.7.5.1   Light emitting diodes

The LED is a semiconductor that emits light when a current is passed through it. The spectral emission of the LED depends on the materials used

Table 1.5 Summary of the properties of some widely used electric light sources

Light source	Luminous efficacy (lm/W)	Correlated color temperature (K)	CIE General CRI	Lamp life (h)	Warm-up time (min)	Re-strike time (min)
Incandescent	8–19	2,700	100	750–2,000	Instant	Instant
Tungsten halogen	8–20	2,900	100	2,000–6,000	Instant	Instant
Tubular fluorescent	60–110	3,000–5,000	50–95	9,000–20,000	10	Instant
Compact fluorescent	50–70	2,700–4,100	80–85	9,000–20,000	10	Instant
Mercury vapor	30–60	3,200–7,000	15–50	16,000–24,000	4	3–10
Metal halide	50–110	3,000–6,500	65–95	3,000–20,000	6	5–20
High-pressure sodium	60–140	2,100–2,500	20–70	10,000–24,000	10–12	0–1
Low-pressure sodium	100–180	1,800	n.a.	16,000–18,000	4–6	1

to form the semiconductor. For light, the most common LED material combinations are now aluminum indium gallium phosphide (AlInGaP) and indium gallium nitride (InGaN). These combinations are replacing older combinations of gallium arsenide phosphide (GaAsP), gallium phosphide (GaP), and aluminum gallium arsenide (AlGaAs). LEDs typically produce narrow band radiation (Figure 1.15), the spectral emission being characterized by the wavelength at which the maximum emission occurs (peak wavelength) and the half bandwidth, this being the half the difference in wavelengths at which the radiant flux is half the maximum radiant flux. Figure 1.16 shows the chromaticity coordinates of some current LEDs. AlInGaP LEDs have peak wavelengths 626, 615, 605, and 590 nm, corresponding, respectively, to the perceptions of red, red-orange, orange, and amber. InGaN LEDs have peak wavelengths of 525, 505, 498, and 450 nm, corresponding to the perception of green, blue-green, blue-green, and blue. The half-bandwidth for AlInGaP LEDs is about 17 nm while that for InGaN LEDs is 35 nm.

The light output of LEDs is determined by the current through the semiconductor and its temperature. Basically, the higher the current and the lower the temperature, the higher the light output. Care should be taken when using LEDs not to exceed the maximum current recommended by the manufacturer. LEDs made from different materials can vary widely in life, from a few thousand hours to a projected 100,000 h, depending on the current used (Narendran *et al.*, 2001). The fact that LEDs are solid state devices and hence are less likely than incandescent lamps to fail due to mechanical damage caused by vibration, have made them an attractive option for vehicle signal lights. As for luminous efficacy, the latest LEDs typically have luminous efficacies in the range 20–30 lm/W. This luminous efficacy, together with the long life and the fact that most LEDs are configured to produce a highly directional beam of light, has made them an attractive option for use in exit signs and traffic signals to replace incandescent lamps.

It might be thought that the fact that the LED is a narrow band source of light would preclude its use for general lighting, apart from the entertainment industry, but this is not the case. One manufacturer has developed a system that combines red, green, and blue LEDs in one unit. Then, by applying different currents to the different LEDs, any color within the triangle formed on the CIE chromaticity diagram by the lines connecting the chromaticity coordinates of the individual LEDs can be produced. Another manufacturer has produced a nominally white LED by adding a phosphor to the epoxy that encapsulates an InGaN LED with a peak wavelength of 470 nm. Figure 1.17 shows the spectral emission of this white LED. Perceptually, the light emitted by this LED appears bluish-white. It has been used in a prototype post-top lantern for use in rural areas, where the ambient light levels are low (Brandston *et al.*, 2000), and for localized lighting applications, such as an aircraft reading light. There can be no doubt that

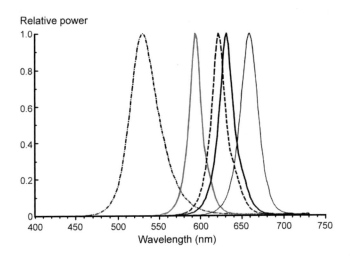

Figure 1.15 Relative spectral power distributions of green, yellow, and red LEDs.

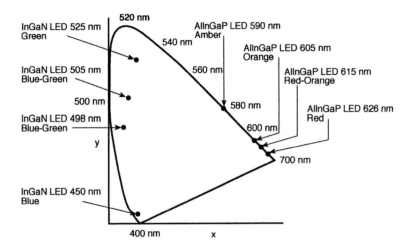

Figure 1.16 The chromaticity coordinates of different types of LEDs plotted on the CIE 1931 (*x*, *y*) chromaticity diagram. Different LEDs use different materials. In this figure, AlInGaP = aluminum indium gallium phosphide, InGaN = indium gallium nitride (from IESNA, 2000a).

the race is on to transfer LEDs from the current field of application in signs and signals to general lighting (Narendran and Bullough, 2001). The LED is a rapidly improving light source with great potential. It will be interesting to see how it develops.

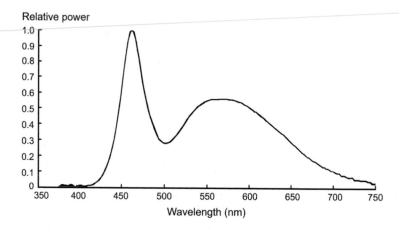

Figure 1.17 The relative spectral power distribution of a white LED. This LED is basically a InGaN LED with an integral phosphor.

1.7.5.2   Electroluminescent lamps

Electroluminescent lamps are a sandwich made up of a flat area conductor, a layer of dielectric–phosphor mixture, and another area conductor that is transparent. When a high, alternating voltage is applied across the two area conductors, the phosphor is excited and light is emitted. The color of the light emitted depends on the dielectric–phosphor combination used and the frequency of the applied voltage. Spectral emissions that are perceived as blue, yellow, green, and pink are available. Electroluminescent lamps have luminous efficacies less than incandescent lamps. However, the fact that they have a long life, low power requirements, and can be formed as either rigid ceramic or flexible plastic sheets or tapes have made them an attractive option for instrument panels and for backlighting liquid crystal displays. As with LEDs there has been a recent development that suggests an interesting future for the electroluminescent lamp. This is the light emitting polymer (Kwong *et al.*, 2001). Light emitting polymers are organic semiconducting materials that possess similar light emitting properties to the inorganic LEDs but also have the mechanical properties of plastics. Again, it will be interesting to see if light emitting polymers can establish a place in the panoply of widely used light sources.

1.7.5.3   Radioluminescent lamps

These light sources consist of a sealed glass tube filled with tritium gas and coated with a phosphor. Low-energy beta particles from the tritium are absorbed by the phosphor, which in turn emits light, the spectrum emitted depending on the phosphor used. These lamps require no power supply, so they have an infinitely high luminous efficacy. Unfortunately, they also emit very little light, the luminance of the glass tube being about $2\,cd/m^2$ (a T8 fluorescent tube has a luminance of about $10,000\,cd/m^2$). This low light output and the fact that their disposal is closely regulated have limited their use, one common application being for exit signs in situations where maintenance is difficult or where the atmosphere is hazardous, such as on an oil rig.

1.8   Control of light distribution

Being able to produce light is only part of what is necessary to produce illumination. The other part is to control the distribution of light from the light source. For daylight, this is done by means of window or skylight shape, placement, and glass transmittance properties (Robbins, 1986; Heschong *et al.*, 1998). For electric light sources, it is done by placing the light source in a luminaire. The luminaire provides electrical and mechanical support for the light source and controls the light distribution. The light distribution is controlled by using reflection, refraction, or diffusion, individually or in combination (Simons and Bean, 2000). One factor in the choice of which method of light control to adopt in a luminaire is the balance desired between the reduction in the luminance of the light source and the precision required in light distribution. Highly specular reflectors can provide precise control of light distribution, but do little to reduce the maximum luminance of the luminaire. Conversely, diffusers make precise control of light distribution impossible but do reduce the maximum luminance of the luminaire. Refractors are an intermediate case. The light distribution provided by a specific luminaire is quantified by the luminous intensity distribution. All reputable luminaire manufacturers provide luminous intensity distributions for their luminaires. Further details on the optical principles of luminaire design and the types of luminaires available can be obtained from the IESNA *Lighting Handbook* (IESNA, 2000a).

1.9   Control of light output

The control of daylight admitted through a window or skylight is achieved by mechanical structures, such as light shelves, or by adjustable blinds (Littlefair, 1990; CIBSE, 1999b). Whenever the sun, or a very bright sky, is likely to be directly visible through a window, some form of blind will be required. Blinds take various forms; horizontal, venetian, vertical, and roller being the most common. Blinds can also be manually operated or motorized, either under manual control or under photocell control. Probably the most important feature to consider when selecting a blind is the extent to which it preserves a view of the outside. Roller blinds which can be drawn down to a position where the sun and/or sky is hidden but the lower part of the window is still open are an attractive option. Roller blinds made of a mesh material can preserve a view through the whole window while reducing the luminance of the view out. Such blinds are an attractive option where the problem is an over-bright sky but will be of limited value when a direct view of the sun is the problem. The same applies to low transmission glass.

For electric light sources, control of light output is provided by switching or dimming systems. Switching systems can vary from the conventional manual switch to sophisticated daylight control systems which switch lamps near to windows off when there is sufficient daylight. Time switches are used to switch off all or parts of a lighting installation at the end of the working day. Occupancy sensors are used to switch off lighting when there is nobody in the space. Such switching systems can reduce electricity waste but they will be irritating if they switch lighting off when light is required and they may shorten lamp life if switching occurs frequently. The factors to be considered when selecting a switching system are whether to rely on a manual or an automatic system, and if it is automatic, how to match the switching to the activities in the space. If your interest is primarily in reducing electricity consumption, a good principle is to use automatic switch off and manual switch on. This principle uses human inertia for the benefit of reducing energy consumption. If you wish to rely on voluntary manual switching of lighting, care should be taken to make the lighting being switched visible from the control panel and to label the switches so that the operator knows which lamps are being switched.

As for dimming systems, these all reduce light output and energy consumption but a different system is required for each lamp type, and some lamp types cannot be dimmed. The factors to consider when evaluating a dimming system are, whether the light source can be dimmed, the range over which dimming can be achieved without flicker or the lamp extinguishing, the extent to which the color properties of the lamp change as the light output is reduced and any effect dimming has on lamp life and energy consumption.

Sophisticated lighting control systems are available for some light sources which allow the user to have a number of preset scenes. These systems use dimming and switching to alter the lighting of a space. They are commonly used in rooms with multiple functions, such as conference rooms.

1.10   Summary

This chapter is concerned with what light is, how it can be measured and how it is produced and controlled. Light is part of the electromagnetic spectrum between 380 and 780 nm. What differentiates this wavelength range from the rest of the electromagnetic spectrum is that the human visual system responds to it. The actual human spectral response has been standardized in an internationally agreed form represented by the CIE Standard Photopic and Scotopic Observers. Using the appropriate spectral sensitivity curve, the four basic photometric quantities can be derived – luminous flux, luminous intensity, illuminance, and luminance.

These measures are all concerned with the overall amount of light and not with its color. To deal with color, the two approaches of color models and the color atlas are considered. This leads to a description of the CIE colorimetric system, including the two-dimensional chromaticity diagrams and the three-dimensional color spaces. By using these measures, colors can be quantified and the color appearance and color rendering properties of light sources characterized, using such measures as correlated color temperature and CIE General CRI.

Having considered how light can be measured, the physical principles and properties of natural and artificial light sources are considered. Natural light is characterized by its variability in both quantity and spectral emission. Artificial light sources are more stable but differ considerably in their properties, particularly spectral content and the efficiency with which they convert electricity to light.

It is important to appreciate that while there are numerous metrics used to characterize a lighting situation or a light source, these metrics are simultaneously precise and inaccurate. The precision arises because the metrics can often be measured or calculated exactly. If they are regarded as simple physical measures they can be considered accurate. But they are not simple physical measures. The whole reason for having photometric and colorimetric quantities is to quantify the visual effect of light. Because of the complexity and flexibility of the human visual system and the differences between individuals, any one standardized metric of visual effect is inevitably an approximation. The photometric and colorimetric quantities discussed are the best approximations so far devised, but it should always be remembered that they are approximations and their apparent precision can be deceptive.

2 The visual system

2.1 Introduction

Light is necessary for the human visual system to operate. With light, we can see; without light we cannot. This chapter describes the structural, operational, and perceptual characteristics of the human visual system.

2.2 The structure of the visual system

The first thing to appreciate about the visual system is that it is not the eye alone. Rather, it is the eye and brain working together. The visual system is often likened to a camera but this analogy is misleading. The only parts of the visual system that resemble a camera are the optical components of the eye. Once the optical components have formed an image of the world on the retina, the camera analogy fails. All the rest of the visual system is an image-processing system that extracts specific aspects of the retinal image for interpretation by the brain. Despite this fact, the obvious starting point for a consideration of the visual system is the eye.

2.2.1   The visual field

Humans have two eyes, mounted frontally. This is the classic position of the eyes for a predator, the two frontally mounted eyes providing considerable overlap between the two visual fields and hence the good depth perception necessary to stalk and capture prey. Animals that are prey typically have their eyes mounted laterally so that their visual fields cover a larger portion of the world around them.

Figure 2.1 shows the approximate extent of the visual field of the two eyes in humans and the overlap between them. Given the limited field of view imposed by the frontal mounting of the two eyes, it is necessary for the two eyes to be able to move. There are two ways this can be done; by moving the head and by moving the eyes in the head. Most animals do both, although some creatures show a bias to one extreme or the other. Owls, for example, have very limited ability to move their eyes but can

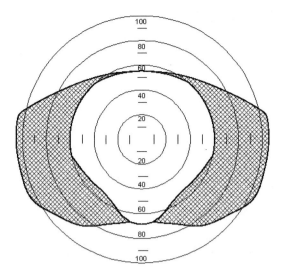

Figure 2.1 The binocular visual field expressed in degrees deviation from the point of fixation. The shaded areas are visible to only one eye (after Boff and Lincoln, 1988).

move their heads over a wide range. Humans have a more limited range of head movements but a wide range of eye movements.

2.2.2 Eye movements

Figure 2.2 shows the muscles used to adjust the position of the eye in its socket. There are six extra-ocular muscles arranged in opposing pairs. Each muscle is attached to the wall of the eye cavity in the skull, so that when opposing pairs of muscles are contracted and relaxed the eye moves. There are several different types of eye movements. When trying to stare directly at a target, without moving the eyes, a process called fixation, three types of eye movement occur. Tremor, a small oscillation in the eye position, is always present. It might be though that tremor is the outcome of noise in the eye position control system and has no other significance but when tremor is eliminated by an optical feedback system, vision rapidly fails, a structured visual field degrading into a uniform field (Pritchard *et al.*, 1960). Therefore, tremor of the retinal image is essential for the visual system to operate.

During fixation, the eye tends to drift slowly away from the fixation point but eventually fixation is restored by a rapid jump movement called a saccade. Saccades are very fast, velocities ranging up to 1,000°/s depending upon the distance moved. Saccadic eye movements have a latency of

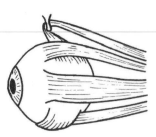

Figure 2.2 The arrangement of the muscles used to move the eye.

about 200 ms, which limits how frequently the line of sight can be moved to about five movements per second. Visual functions are substantially limited during saccadic movements. Figure 2.3 shows a pattern of fixations for people reading text. Movement between the fixation points is made by saccades.

About the only situation in which saccades rarely occur is in smooth pursuit eye movements. Such movements are relatively slow, up to 40°/s, and occur when trying to track a smoothly moving object, e.g. an aircraft in flight. Given a smoothly moving stimulus, the visual system can produce a smooth pursuit eye movement. The pursuit system cannot follow smoothly moving targets at high velocities, nor slow but erratically moving targets.

These eye movements all occur in a single eye, but movements in the two eyes are not independent. Rather, they are coordinated so that the lines of sight of the two eyes are both pointed at the same target at the same time. If the line of sight of the two eyes are not aimed at the same target at the same time, the target may be seen as double (diplopia). Movements of the two eyes which keep the primary lines of sight converged on a target, or which may be used to switch fixation from a target at one distance to a new target in the same direction but at a different distance, are called vergence movements. These movements are slow, up to 10°/s, but can occur as a jump movement or can smoothly follow a target moving in a fore-and-aft direction. Both types of movement involve a change in the angle between the two eyes.

2.2.3 Optics of the eye

Figure 2.4 shows a section through the eye, the upper and lower halves being adjusted for focus at near and far distances, respectively. The eye is basically spherical with a diameter of about 24 mm. The sphere is formed from three concentric layers. The outermost layer, called the sclera, protects the contents of the eye and maintains its shape under pressure. Over most of the eye's surface, the sclera looks white but at the front of the eye the

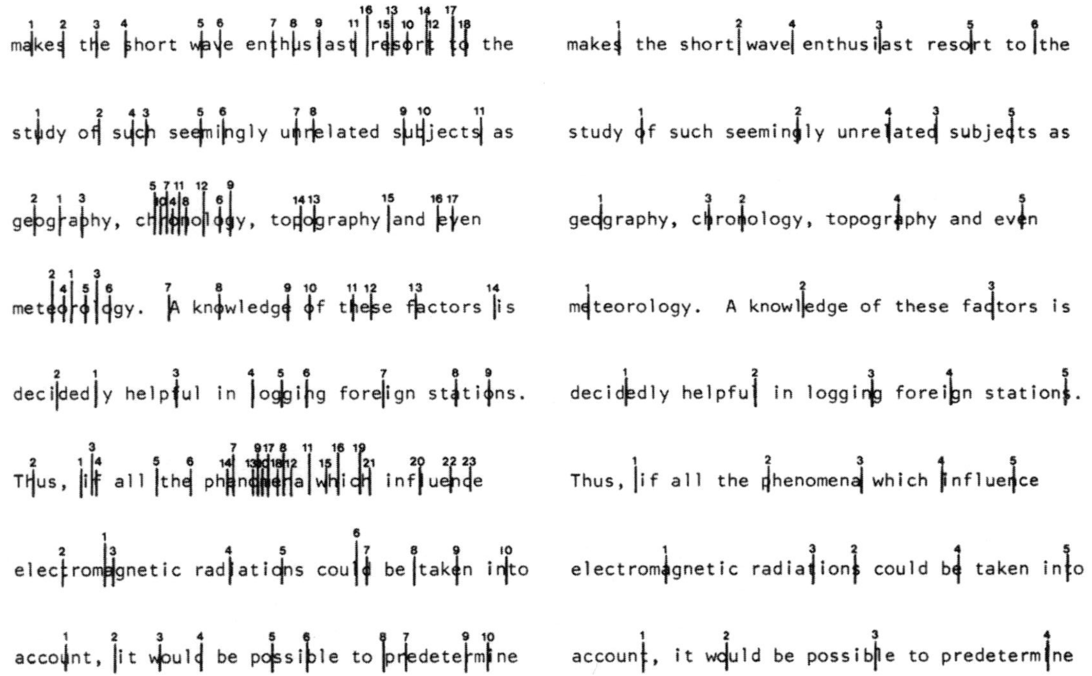

Figure 2.3 The patterns of fixations made by two readers reading the same passage. The intersection of a vertical line with the line of print indicates a fixation point. The numbers attached to the vertical lines give the order in which the fixations occurred for each line of print. The reader on the right has a more extensive vocabulary than the reader on the left (after Buswell, 1937).

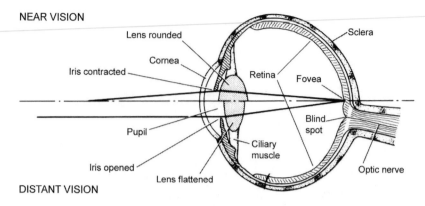

NEAR VISION

Lens rounded

Cornea

Iris contracted

Sclera

Retina

Fovea

Pupil

Blind spot

Ciliary muscle

Iris opened

Lens flattened

Optic nerve

DISTANT VISION

Figure 2.4 A section through the eye adjusted for near and distant vision.

sclera bulges up and becomes transparent. It is through this area, called the cornea, that light enters the eye. The next layer is the vascular tunic, or choroid. This layer contains a dense network of small blood vessels that provide oxygen and nutrients to the next layer, the retina. Without these supplies the retina would die. As the choroid approaches the front of the eye it separates from the sclera and forms the ciliary body. This element produces the watery fluid that lies between the cornea and the lens, called the aqueous humor. The aqueous humor provides oxygen and nutrients to the cornea and the lens, and takes away their waste products. Elsewhere in the eye this is done by blood but on the optical pathway through the eye, a transparent medium is necessary.

As the ciliary body extends further away from the sclera, it becomes the iris. The iris consists of two layers, an outer layer containing pigment and an inner layer containing blood vessels. The color of the iris is determined by the extent to which the outer layer is pigmented. If the outer layer is heavily pigmented the iris will appear brown but if it is lightly pigmented, the iris will appear to be a color formed by a combination of the outer and inner layers, usually blue, green, or gray. If there is no or very little pigment in the outer layer, as is the case with albinos, the color of the iris is determined by the inner layer and hence will appear pink.

The iris forms a circular opening, called the pupil, that admits light into the eye. The pupil can be changed in size by the operation of the two sets of muscles, one set that lie around the pupil and another that are directed radially out from the pupil. When the first set of muscles contract, the pupil is decreased in size. When the second set contract, the pupil expands. Pupil size varies with the amount of light reaching the retina but it is also influenced by the distance of the object from the eye, the age of the observer, and by emotional factors such as fear, excitement, and anger (Duke-Elder, 1944).

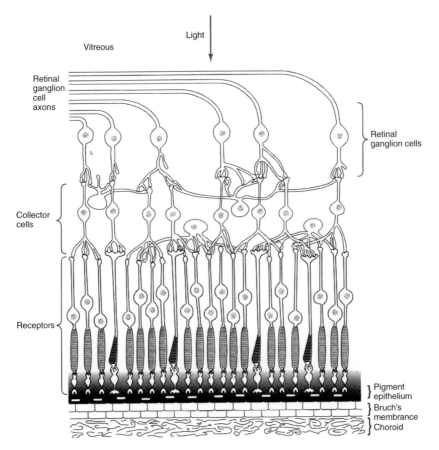

Figure 2.5 A section through the retina (after Sekular and Blake, 1994).

After passing through the pupil, light reaches the crystalline lens. The lens is fixed in position, but varies its focal length by changing its shape. The change in shape is achieved by contracting or relaxing the ciliary muscles. For objects close to the eye, the lens is fattened. For objects far away, the lens is flattened.

The space between the lens and the retina is filled with another transparent material, the jelly-like vitreous humor. After passing though the vitreous humor, light reaches the retina, the location where light is absorbed and converted to electrical signals. The retina is a complex structure, as can be seen from Figure 2.5, which shows a section through it. It can be considered as having three layers: a layer of photoreceptors, which can be divided into four types; a layer of collector cells which provide links between multiple photoreceptors, and a layer of ganglion cells. The axons of the ganglion cells form the optic nerve which produces the blind spot where it passes through the retina out of the eye. Light reaching the retina, passes through the ganglion and collector cell layers before reaching the photoreceptors, where it is absorbed. Any light that gets through the photoreceptor layer is absorbed by the pigment epithelium.

2.2.4   *The structure of the retina*

The retina is an extension of the brain. It derives from the same tissue as the brain, and like the brain, damaged cells are not replaced. The visual system has four photoreceptor types in the retina, each containing a different photopigment. These four types are conventionally grouped into two classes, rods and cones, these names being derived from their appearance under a microscope. All the rod photoreceptors are the same, containing the same photopigment and hence having the same spectral sensitivity. The relative spectral sensitivity of the rod photoreceptors is shown in Figure 2.6. The other three photoreceptor types are all cones, each with a different photopigment. Figure 2.7 shows the relative spectral sensitivity functions of

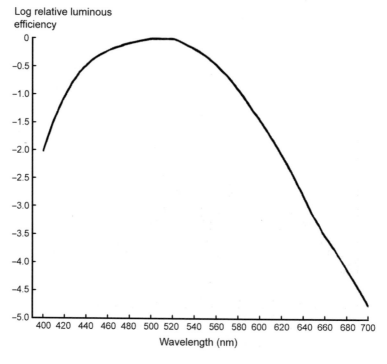

Figure 2.6 Log relative luminous efficiency of the rod photoreceptor (after CIE, 1990a).

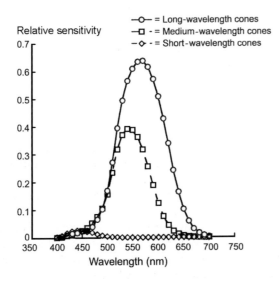

Figure 2.7 The relative spectral sensitivities of long- (L), medium- (M), and short-wavelength (S) cone photoreceptors (after Kaiser and Boynton, 1996).

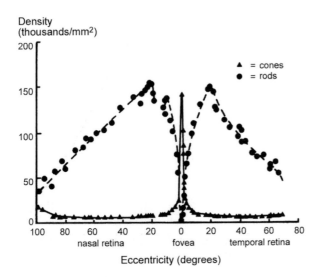

Figure 2.8 The distribution of rod and cone photoreceptors across the retina. The 0° indicates the position of the fovea.

the three cone photoreceptor types, called short-, medium-, and long-wavelength cones (S-, M-, and L-cones, respectively) after the wavelength region where they have the greatest sensitivity.

Rods and cones are distributed differently across the retina (Figure 2.8). The cones are concentrated in one small area that lies on the visual axis of the eye, called the fovea, although there is a low density of cones across the rest of the retina.

The three cone types are also not distributed equally across the retina. The L- and M-cones are concentrated in the fovea, their density declining gradually with increasing eccentricity. The S-cones are largely absent from the fovea, reach a maximum concentration just outside the fovea and then decline gradually in density with increasing eccentricity. The ratio of the L-, M-, and S-cone types in and around the fovea is approximately $32:16:1$ (Walraven, 1974).

Over the whole retina there are many more rods than cones, approximately 120 million rods and 8 million cones. The fact that there are many more rod than cone photoreceptors should not be taken to indicate that human vision is dominated by the rods. It is the fovea that allows resolution of detail and other fine discriminations and the fovea is entirely inhabited by cones. There are three other anatomical features that emphasize the importance of the fovea. The first is the absence of blood vessels. For most of the retina, light passes through a network of blood vessels before reaching the photoreceptors but blood vessels avoid crossing over the central area of the retina, called the macula, at the center of which is the fovea. The second is the fact that, even the collector and ganglion layers of the retina are pulled away over the fovea. This helps reduce the absorption and scattering of light in the region of the fovea and hence enhances the resolution of detail. The third is the fact that the outer limb of the cone photoreceptor can act as a waveguide, making cones most sensitive to light rays passing through the center of the lens (Crawford, 1972). This characteristic, known as the Stiles–Crawford effect, compensates to some extent for the poor quality of the eye's optics by making the fovea less sensitive to light passing through the edge of the lens or scattered in the optic media. The fovea is populated only with cones. Rod photoreceptors, which dominate the population of the rest of the retina, do not show a Stiles–Crawford effect.

2.2.5   The functioning of the retina

The retina is where the processing of the retinal image begins. Recordings of electrical output from single ganglion cells in the retinas of monkeys and cats, creatures that have visual systems similar to those of humans, have shown a number of important characteristics of the operation of the retina. The first is the fact that the electrical discharge is a series of voltage spikes of equal amplitude. Variations in the amount of light falling on the photoreceptors supplying signals to the ganglion cell through the network of collector cells, produce changes in the frequency with which these voltage spikes occur but not in their amplitude. The second is that there is a level of electrical discharge present even when there is no light falling on the photoreceptors, called the spontaneous discharge. The third is that illuminating photoreceptors with a spot of light, can produce either an increase or a decrease in the frequency of electrical discharges, relative to the level of frequency of discharges present when light is absent.

Further studies of the pattern of electrical discharges from a single ganglion cell have revealed two other important aspects of the operation of the retina. The first is the existence of receptive fields. A receptive field is the area of the retina that determines the output from a single ganglion cell. The size of a receptive field is measured by exploring the retina with a very small spot of light while measuring the electrical discharges from the ganglion cell. The boundary of the receptive field is determined by the point beyond which applying the spot of light fails to alter the spontaneous electrical discharge from the ganglion cell.

A given receptive field always represents the activity of a number of receptors, and often reflects input from different cone types as well as from rods. The sizes of receptive fields vary systematically with retinal location. Receptive fields around the fovea are very small. As eccentricity from the fovea increases, so does receptive field size.

The sensitivity of a receptive field to light is primarily determined by its size. Because all ganglion cells require some finite minimum electrical input to be stimulated, a receptive field that receives input from a large number of photoreceptors can be stimulated by a lower retinal illuminance than can a receptive field which receives input from only a few photoreceptors. Hence, the sensitivity to light of small receptive fields is usually significantly less than that of larger fields. The rod photoreceptors, which are concentrated outside the fovea, are organized into relatively large receptive fields. This combination of large receptive fields, relatively low spontaneous discharge levels and longer integration times makes the rod photoreceptor system significantly more sensitive to light than the cone photoreceptor system.

Within each ganglion cell receptive field there is a specific structure. Again, by recording the electrical discharges from a ganglion cell and exploring within a receptive field with a very small spot of light, it has been found that retinal receptive fields consist of a central circular area and a surrounding annular area. These two areas have opposing effects on the ganglion cell's electrical discharge. Either the central area increases and the annular surround decreases the rate of electrical discharge, or, in other receptive fields, the reverse occurs. These types of receptive fields are known as on-center/off-surround and off-center/on-surround fields, respectively. If either of these two types of retinal receptive fields is illuminated uniformly, the two types of effect on electrical discharge cancel each other, a process called lateral inhibition. However, if the illumination is not uniform across the two parts of the receptive field, a net effect on the ganglion cell discharge is evident. This pattern of response makes the retinal fields well suited to detect boundaries in the retinal image. There are an approximately equal number of on-center/off-surround and off-center/on-surround receptive fields in the retina. The electrical signals from the two types of receptive field do not cancel each other. Rather, the signals from the two types of receptive field are kept separate, indicating that they serve different aspects of vision.

While every retinal ganglion cell has a receptive field, not every ganglion cell is the same. In fact, there are two types of ganglion cell, called magnocellular (M) cells and parvocellular (P) cells. There are a number of important differences between the M- and P-cells. First, the axons of the M-cells are thicker than the axons of the P-cells, indicating that signals are transmitted more rapidly from the M-cells than from the P-cells. Second, there are many more P- than M-cells and they are distributed differently across the retina. The P-cells dominate in the fovea and parafovea and the M-cells dominate in the periphery. Third, for a given eccentricity, the P-cells have smaller receptive fields than the M-cells, explaining some of the local variation in receptive field size with eccentricity. Fourth, the M- and P-cells are sensitive to different aspects of the retinal image. The M-cells are more sensitive to rapidly varying stimuli and to small differences in illumination but are insensitive to differences in color. The P-cells are more sensitive to small areas of light and to color.

Overall, this brief description of the retina should have demonstrated that the retina is really the first stage of an image-processing system. The retina extracts information on boundaries in the retinal image and then extracts specific aspects of the stimulus within the boundaries, such as color. These aspects are then transmitted up the optic nerve, formed from the axons of the retinal ganglion cells, along different channels.

2.2.6 *The central visual pathways*

Figure 2.9 shows the pathways over which signals from the retina are transmitted. The optic nerves leaving the two eyes are brought together at the optic chiasm, rearranged and then extended to the lateral geniculate nuclei. Somewhere between leaving the eyes and arriving at the lateral geniculate nuclei, some optic nerve fibers are diverted to the superior colliculus, located at the top of the brain stem and responsible for controlling eye movements and to the suprachiasmatic nucleus in the hypothalamus and concerned with entraining circadian rhythms (see Chapter 3). As for vision, after the lateral geniculate nuclei, the two optic nerves spread out to supply information to various parts of the visual cortex, the part of the brain where vision occurs.

At the optic chiasm, the optic nerve from each eye is split and then parts of the optic nerves from the same side of the two eyes are combined.

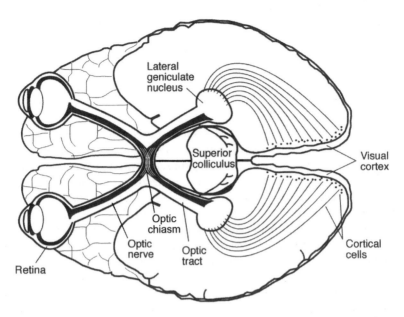

Figure 2.9 A schematic diagram of the pathways from the eyes to the visual cortex (from IESNA, 2000a).

This arrangement ensures that the signals from the same side of the two eyes are received together on the same side of the visual cortex.

The signals from the same side of the two eyes pass from the optic chiasm to a lateral geniculate nucleus. Anatomically, a lateral geniculate nucleus shows six distinct layers. Two of these layers receive signals from the M-ganglion cells of the retina while the other four layers receive signals from the P-ganglion cells. Each layer is arranged so that the location of the M- and P-cells on the retina is preserved. In other words, each layer preserves a map of the retina. As in the retina, electrophysiological recording of discharges from individual lateral geniculate nucleus cells have shown the existence of receptive fields, with either on-center/off-surround or off-center/on-surround. The division of function found in the retina is also present in the lateral geniculate nuclei. The magnocellular layers do not respond to color differences but the parvocellular layers do. Indeed, some receptive fields show strong responses when the center is illuminated by one color and the surround by another. The specific color combinations being red and green or yellow and blue, this being the basis of human color vision (see Section 2.2.7). The receptive fields in the parvocellular layers are smaller than in the magnocellular layer, so the parvocellular layer will be better at resolving detail; but the magnocellular layer will respond faster to a change in the amount of light.

From the above description it might seem that the lateral geniculate nuclei are just relay stations between the retinas of the two eyes and the visual cortex. However, there are reasons to suspect that they are more than this. Specifically, the lateral geniculate nuclei also receive signals from the reticular activating system, a part of the brain stem that determines the general level of arousal. It has been shown that low levels of arousal lead to an attenuated visual signal being transmitted from the lateral geniculate nuclei to the visual cortex (Livingston and Hubel, 1981).

The visual cortex is located at the back of cerebral hemispheres. It consists of another layered array, containing about one million or so cells. Apart from its amazing complexity, what is remarkable about it is the similarity of its organization to the organization of the retina and the lateral geniculate nuclei. For example, the magnocellular and parvocellular channels remain separated, signals from the magnocellular layers of the lateral geniculate nuclei are received in one layer of the visual cortex, while signals from the parvocellular layers go elsewhere. Further, each cortical cell reacts only to signals from a limited area of the retina, and the arrangement of the cortical cells replicates the arrangement of ganglion cells on the retina. Moreover, the number of cortical cells allocated to each part of the retina enhances the importance of the fovea. About 80 percent of the cortical cells are devoted to the central 10° of the visual field (Drasdo, 1977), the center of which is the fovea, a phenomenon called cortical magnification. As for how the cortical cells respond to light stimulation, again, on-center/off-surround and off-center/on-surround cells are found, but now they show sensitivity to the orientation of a boundary. Other cells do not show a clear on/off structure but are still sensitive to the orientation of a boundary and will respond strongly to a moving boundary of the appropriate orientation. There are also cortical cells grouped together that show no sensitivity to boundary orientation but are very sensitive to color differences. Yet, other cells respond more to signals from the left eye and others to the right eye while some respond equally to signals from both eyes. All this cellular diversity occurs at the entry level of the visual cortex. There is a much more complex structure beyond this in the higher areas of the visual cortex. Investigation of these areas have shown that different parts of the visual cortex are dedicated to specific discriminations. For example, areas have been identified in the visual cortex that are concerned with analyzing color, motion, and even human faces viewed from particular angles (Desimone, 1991).

2.2.7   *Color vision*

So far, this consideration of the structure of the human visual system has not considered the perception of color. Human color vision is trichromatic, i.e. it is based on the three different cone photoreceptors. These photoreceptors are characterized by having different wavelengths for peak sensitivity, but all having broadband spectral sensitivity with considerable overlap (Figure 2.7). The number of photoreceptor types used to form a color system is a matter of compromise. A single photoreceptor type containing a single photopigment is unable to discriminate differences in wavelength from differences in irradiance and so does not support color vision, e.g. rod photoreceptors. A system with many different photoreceptors each containing a different photopigment would be able to make many discriminations between wavelengths but at the cost of taking up more of the neural capacity of the visual system. Studies of the spectral emission of typical light sources have shown that trichromacy provides an accurate description of surface colors under most lighting conditions (Lennie and D'Zmura, 1988).

Figure 2.10 shows how the outputs from the three cone photoreceptor types are believed to be arranged into one non-opponent achromatic system and two opponent chromatic systems. The achromatic channel receives inputs from the M- and L-cones only. The red–green opponent channel produces the difference between the output of the M-cones and the sum of the outputs of the L- and S-cones. The other opponent channel, the yellow–blue channel, produces the difference between the S-cones and the sum of the M- and L-cones. This opponent structure for color vision influences

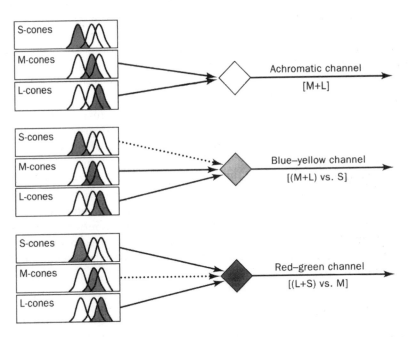

Figure 2.10 The organization of the human color system showing how the three cone photoreceptor types are believed to feed into one achromatic, non-opponent channel and two chromatic, opponent channels (after Sekular and Blake, 1994).

the perception of colors. This was shown in an experiment by Boynton and Gordon (1965). They presented monochromatic lights of different wavelengths and asked subjects to describe the appearance of each stimulus using only the color names, red, green, yellow, and blue. Either one or two names could be used for each color presented. The interesting result was that people very rarely described a color as red–green or yellow–blue, while yellow–red and green–blue were frequently used. Even 4-month-old infants divide incident light into four categories, corresponding to what adults call red, yellow, green, or blue (Bornstein *et al.*, 1976).

Physiologically, the outputs from the three different cone types are organized into opponent and non-opponent classes in the retina. Outputs in the non-opponent class always give an increase in activity with increasing retinal irradiance, although the magnitude of that increase will vary with the wavelength of the incident light. Outputs in the opponent class can show either an increase or a decrease in activity depending on the wavelength of the incident light. Cells of the non-opponent type are believed to constitute the achromatic channel shown in Figure 2.10, while cells of the opponent type form the opponent channels. The achromatic information is transmitted to the visual cortex by the magnocellular channel, while the chromatic information proceeds via the parvocellular channel.

The ability to discriminate the wavelength content of incident light makes a dramatic difference to the information that can be extracted from a scene. Creatures with only one type of photopigment, i.e. creatures without color vision, can only discriminate shades of gray, from black to white. Approximately 100 such discriminations can be made. Having two photopigment types increases the number of different combinations of irradiance and spectral content that can be discriminated to about 10,000. Having three types of photopigment increases the number of discriminations to approximately 1,000,000 (Neitz *et al.*, 2001). Thus, color vision is a valuable part of the visual system, and not a luxury which adds little to utility.

Unfortunately, a significant proportion of people have defective color vision, a condition characterized by abnormal color matching or color confusions. People with defective color vision are classified into three categories: monochromats, dichromats, and anomalous trichromats, according to the number of photoreceptors present and the nature of the photopigments present in the photoreceptors. Monochromats, although very rare, occur in two forms: rod monochromats, where there are no cone photoreceptors, only rod photoreceptors; and cone monochromats, where there are rod photoreceptors and only one type of cone photoreceptor, usually the S-cone. Rod monochromats are truly color-blind and see only differences in brightness. Cone monochromats have a very limited form of color vision in the luminance range where both rod and S-cones are operating. Both dichromats and anomalous trichromats have some perception of color, although not the same perception as people with normal color vision (for illustrations of how dichromats perceive colors, see McIntyre, 2002). Dichromats have two cone photoreceptors. They see a more limited range of colors than people with normal color vision and have a different spectral sensitivity, depending on which cone photoreceptor is missing (Wyszecki and Stiles, 1982). Dichromats with the L-cones missing are called protanopes. Dichromats with the M-cones missing are called deuteranopes, while dichromats with S-cones missing are called tritanopes. Anomalous trichromats have all three cone photopigments present, but one of the cones contains a photopigment that does not have the usual spectral sensitivity. Anomalous trichromats who have a defective long-wavelength photopigment are called protanomalous. Anomalous trichromats who have a defective medium-wavelength photopigment are called deuteranomalous, while anomalous trichromats who have a defective short-wavelength photopigment are called tritanomalous. The color vision of anomalous trichromats can vary widely from almost as bad as a dichromat to little different from someone with normal color vision.

Overall, about 8 percent of males and 0.4 percent of females have some form of defective color vision; about half being deuteranomalous. Steward and Cole (1989) surveyed people with defective color vision and found that many such people have some trouble with everyday tasks, such as selecting colored merchandise and judging the ripeness of fruit (see Table 2.1). Defective color vision is usually inherited, although it can also be acquired through age, disease, injury, or exposure to some chemicals There is little

Table 2.1 Percentage of people with different types of color vision reporting difficulties with everyday tasks

Activity	Dichromats (%)	Anomalous trichromats (%)	Normal (%)
Selecting clothes, cosmetics, etc.	86	66	0
Distinguishing the colors of wires, paints, etc.	68	23	0
Identifying plants and flowers	57	18	0
Determining when fruits and vegetables are ripe, by color	41	22	0
Determining when meat is cooked, by color	35	17	0
Difficulties in participating in or watching sports, because of color	32	18	0
Adjusting the color balance of a television satisfactorily	27	18	2
Recognizing skin conditions such as a rash or sunburn	27	11	0
Taking the wrong medication because of difficulties with color	0	3	0

Source: Steward and Cole (1989).

that can be done to overcome the limitations of defective color vision, although filters can sometimes be used to enhance specific color differences (McIntyre, 2002).

2.2.8 Conclusions

Much remains to be done before the visual system will be completely understood, but what is clear is that the visual system consists of two parts; an optical system that produces an image on the retina of the eye and an image-processing system that extracts different aspects of that image at various stages of its progress up the optic nerve, while preserving the location of the information. It is also clear that the visual system devotes most of its resources to analyzing the central area of the retina, particularly the fovea. This implies that peripheral vision is mainly devoted to identifying something that should be examined in detail by turning the head and eyes so the image of whatever it is falls on the fovea. One other point that should be appreciated is that the visual system is capable of making long-term adjustments to changed circumstances, both mechanically and neurally (Hofner and Williams, 2002). Specifically, following the removal of a cataract, the foveal cone photoreceptors have been shown to realign their main axis from pointing at the edge of the pupil to pointing at the center; and prolonged exposure to a colored environment has been shown to produce a shift in the perception of colors so as to compensate for the chromatically altered environment. Of course, these adjustments take place over many days and occur under rather extreme circumstances, but the visual system also makes continuous adjustments over much shorter times under normal conditions. These will be discussed next.

2.3 Continuous adjustments of the visual system

2.3.1 Adaptation

The human visual system can process information over an enormous range of luminances (about 12 log units), but not all at once. It continually adjusts itself to the prevailing conditions, aiming at reduced sensitivity and finer discrimination when there is plenty of light available and enhanced sensitivity and coarser discrimination when light is in short supply. When the visual system is adapted to a given luminance, much higher luminances appear as glaringly bright while much lower luminances are seen as black shadows. Figure 2.11 indicates the approximate limits within which differences in luminance can be discriminated for different adaptation luminances. An everyday example of this change in perception is the appearance of a vehicle headlight by day and night. The headlight has the same luminance under both conditions, but as the adaptation luminance decreases as night falls the brightness of the headlight increases until if viewed directly it is glaringly bright.

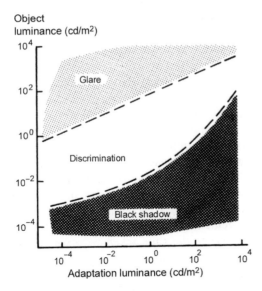

Figure 2.11 A schematic illustration of the range of object luminances within which discrimination is possible for different adaptation luminances. The boundaries are approximate (after Hopkinson and Collins, 1970).

To cope with the wide range of retinal illuminations to which it might be exposed, from a very dark night (0.000001 cd/m²) to a sunlit beach (100,000 cd/m²), the visual system changes its sensitivity through a process called adaptation. Adaptation is a continuous process involving three distinct changes:

Change in pupil size – The iris constricts and dilates in response to increased and decreased levels of retinal illumination. For young people, the diameter of the pupil can range from about 2 to 8 mm. For older people, the range is less (see Chapter 12). The amount of light transmitted through the pupil is proportional to its area so a range of diameters from 2 to 8 mm implies a maximum effect of pupil changes of 16–1. As the visual system can operate over a range of about 1,000,000,000,000–1, this indicates that the pupil plays only a minor role in the adaptation of the visual system. Iris constriction is faster (about 0.3 s) than dilation (about 1.5 s).

Neural adaptation – This is a fast (less than 200 ms) change in sensitivity produced by synaptic interactions in the retina. Neural processes account for virtually all the transitory changes in sensitivity of the eye at luminance values commonly encountered in electrically lighted environments, i.e. below luminances of about 600 cd/m². The facts that neural adaptation is fast, is operative at moderate light levels, and is effective over a luminance range of 2–3 log units explain why it is possible to look around most lit interiors without being conscious of being misadapted.

Photochemical adaptation – The four types of retinal photoreceptors contain four different pigments. When light is absorbed, the pigments break down into an unstable aldehyde of vitamin A and a protein (opsin). In the dark, the pigment is regenerated and is again available to absorb light. The sensitivity of the eye to light is largely a function of the percentage of unbleached pigment. Under conditions of steady retinal irradiance, the concentration of photopigment produced by the competing processes of bleaching and regeneration is in equilibrium. When the retinal irradiance is changed, pigment is bleached and regenerated so as to re-establish equilibrium. Because the time required to accomplish the photochemical reactions is of the order of minutes, changes in the sensitivity can lag behind the irradiance changes. The cone photoreceptors adapt much more rapidly than do the rod photoreceptors; even after exposure to high irradiances, the cones will achieve their maximum sensitivity in 10–12 min, while the rods will require 60 min (or longer) to achieve their maximum sensitivity. This is evident in Figure 2.12, which shows the time taken to reach maximum sensitivity, also known as complete dark adaptation.

Exactly how long it takes to adapt to a change in retinal illumination depends on the magnitude of the change, the extent to which it involves

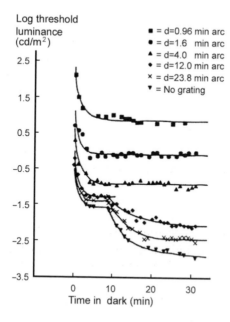

Figure 2.12 Log threshold luminances for the resolution of square wave gratings of bar widths from 0.96 to 23.8 min arc and for detecting a uniform target (no grating), plotted against time in the dark. The luminance to which the subjects were initially adapted was 5,011 cd/m² (after Brown *et al.*, 1953).

different photoreceptors and the direction of the change. For changes in retinal illumination of about 2–3 log units, neural adaptation is sufficient so adaptation should be complete in less than a second. For larger changes photochemical adaptation is necessary. If the change in retinal illumination lies completely within the range of operation of the cone photoreceptors, a few minutes will be sufficient for adaptation to occur. If the change in retinal illumination covers from cone photoreceptor operation to rod photoreceptor operation, tens of minutes may be necessary for adaptation to be completed. As for the direction of change, once the photochemical processes are involved, changes to a higher retinal illuminance can be achieved much more rapidly than changes to a lower retinal illuminance.

When the visual system is not completely adapted to the prevailing retinal illumination, its capabilities are limited. This state of changing adaptation is called transient adaptation. Transient adaptation is unlikely to be noticeable in interiors in normal conditions but can be significant where sudden changes from high to low retinal illumination occur, such as on entering a long road tunnel on a sunny day or in the event of a power failure in a windowless building.

The usual way of describing the state of adaptation is as the luminance of the visual field to which the observer is adapted. In the laboratory this is perfectly acceptable. The experimenter can determine the visual field and ensure that it is uniform in a luminance. In this situation, there is little doubt about what the adaptation luminance is. In the real world, determining the adaptation luminance is not so easy. If the observer has one main line of sight, such as might be the case of a driver approaching a tunnel entrance, then the average luminance within about 20° of the fixation point is a reasonable estimate of the adaptation luminance. If the observer has many fixation points, i.e. the observer is moving his/her eyes around a lot, then the average luminance of the whole scene is a good estimate. There are no clear rules for determining the adaptation luminance. The best that can be done is to look at the pattern of fixation points and the time spent at each to get a crude estimate for the adaptation luminance

2.3.2 *Photopic, scotopic, and mesopic vision*

This process of adaptation can change the spectral sensitivity of the visual system because at different retinal illuminances, different combinations of retinal photoreceptors are operating. The three states of sensitivity are conventionally identified as follows:

Photopic vision – This state of the visual system occurs at adaptation luminances higher than approximately 3 cd/m^2. For these luminances, the retinal response is dominated by the cone photoreceptors. This means that both color vision and fine resolution of detail are available.

Scotopic vision – This operating state of the visual system occurs at adaptation luminances less than approximately 0.001 cd/m^2. For these luminances only the rod photoreceptors respond to stimulation, the cone photoreceptors being insufficiently sensitive to respond to the low level of retinal irradiance. This means that color is not perceived, only shades of gray, and the fovea of the retina is blind. Therefore, in scotopic conditions, what limited resolution of detail there is occurs in the parafovea within a few degrees of the fovea.

Mesopic vision – This operating state of the visual system is intermediate between the photopic and scotopic states, i.e. between about 0.001 and 3 cd/m². In the mesopic state, both cones and rod photoreceptors are active. As luminance declines through the mesopic region, the fovea, which contains only cone photoreceptors, slowly declines in absolute sensitivity without significant change in spectral sensitivity, until vision fails altogether as the scotopic state is reached. In the periphery, the rod photoreceptors gradually come to dominate the cone photoreceptors, resulting in gradual deterioration in color vision and resolution and a shift in spectral sensitivity to shorter wavelengths.

Figure 2.13 shows the relative sensitivity of rod photoreceptors, and the cone photoreceptors in the fovea and outside the fovea, i.e. with and without the S-cones. It is clear that the two photoreceptor types differ in sensitivity. The rod photoreceptors are much more sensitive to light than

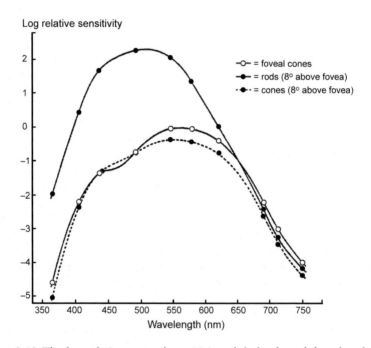

Figure 2.13 The log relative spectral sensitivity of dark-adapted foveal and peripheral cone, and rod photoreceptors. The spectral sensitivities are all normalized to the maximum sensitivity of the foveal cones (after Wald, 1945).

the cone photoreceptors, particularly for short-wavelength radiation. As for the two cone photoreceptors, the absence of the S-cones in the fovea reduces sensitivity at shorter wavelengths (Wald, 1945).

The relevance of the different operating states for lighting practice varies. Scotopic vision is largely irrelevant. Any lighting installation worthy of the name provides enough light to at least move the visual system into the mesopic state. Most interior lighting ensures that the visual system is operating in the photopic state. Current practice in exterior lighting ensures that the visual system operates near the boundary of the photopic and mesopic states.

The spectral sensitivity of the visual system is different in the photopic, mesopic, and scotopic states because different photoreceptors are dominant in each state. In the photopic state, cone photoreceptors are dominant everywhere. In the mesopic state, cone photoreceptors are dominant in the fovea, because that is all there is in the fovea, but in the peripheral retina, both rod and cone photoreceptors are active, the balance between them shifting as the retinal irradiance changes. In scotopic conditions, only rod photoreceptors are active and the fovea is blind. As if this pattern of changing spectral sensitivities were not enough, different spectral sensitivities occur within the photopic state, depending on the method used to make the measurements and hence the extent to which the achromatic, non-opponent and chromatic, opponent channels of the human visual system are stimulated (Figure 2.14).

To bring some order to this potential chaos, the Commission Internationale de l'Eclairage (CIE) has recognized three different spectral sensitivities known as the CIE Standard Photopic Observer, the CIE Modified Photopic Observer, and the CIE Standard Scotopic Observer (see Chapter 1). These relative luminous efficiency functions, shown in Figure 1.2, are used in the fundamental definition of light to convert from radiometric quantities to photometric quantities. There is no officially recognized spectral sensitivity for the mesopic state, despite extensive study (CIE, 1989). Because scotopic vision is irrelevant to lighting practice, the mesopic state has not officially been defined, and the CIE Modified Photopic Observer only makes a difference for light sources with a lot of power at short wavelengths, virtually all photometric quantities used in lighting practice are still measured using the CIE Standard Photopic Observer. Consequently, it should not come as a surprise when the visual effects of light sources with different spectral content are not the same when the two light sources are matched photometrically. The fact is that the CIE Standard Observers are primarily designed to facilitate the measurement of light rather than to describe the operation of the visual system precisely. The variability of the spectral sensitivity of the human visual system, depending on which photoreceptors are stimulated and which of the vision channels are active, implies that conditions in which the actual spectral sensitivity of the visual system is different from the CIE Standard Photopic Observer are likely to occur quite frequently.

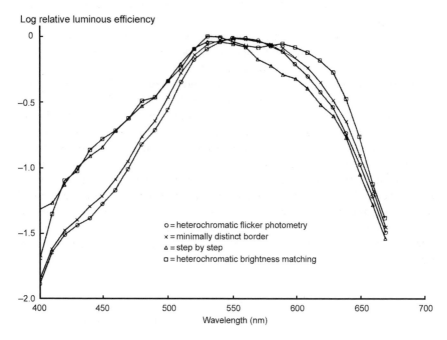

Log relative luminous efficiency

o = heterochromatic flicker photometry
× = minimally distinct border
▲ = step by step
□ = heterochromatic brightness matching

Wavelength (nm)

Figure 2.14 Log relative luminous efficiency measured by the heterochromatic flicker and the minimally distinct border methods, in which only the achromatic non-opponent channel is active, and by the heterochromatic brightness matching and step-by-step brightness matching methods, in which both the achromatic non-opponent channel and the chromatic opponent channels are active. The relative spectral sensitivity functions measured by the heterochromatic flicker and minimally distinct border methods closely match the CIE Modified Photopic Observer (after Comerford and Kaiser, 1975).

2.3.3 Accommodation

There are three optical components involved in the ability of the eye to focus an image on the retina. The first is the thin film of tears on the cornea. This film is important because it cleans the surface of the eye, starts the optical refraction process necessary for focusing objects, and smoothes out small imperfections in the surface of the second optical component, the cornea. The cornea covers the transparent anterior one-fifth of the eyeball (Figure 2.4). With the tear layer, it forms the major refracting component of the eye and gives the eye about 70 percent of its optical power. The crystalline lens provides most of the remaining 30 percent of the optical power. The ciliary muscles have the ability to change the curvature of the lens and thereby adjust the power of the eye's optical system in response to changing target distances; this change in optical power is called accommodation.

Accommodation is a continuous process, even when fixating, and is always a response to an image of the target located on or near the fovea rather than in the periphery of the retina. It is used to bring a defocused image into focus. It may be changed rapidly, so as to shift focus from one location to another, or gradually, so as to keep a target which is moving in a fore-and-aft direction in focus. Any condition, either physical or physiological, that handicaps the fovea, such as a low light level, will adversely affect accommodative ability. As adaptation luminance decreases below $0.03 \, \text{cd/m}^2$, the range of accommodation narrows so that it becomes increasingly difficult to focus objects near and far from the observer (Leibowitz and Owens, 1975). Blurred vision and eyestrain can be consequences of limited accommodative ability. When there is no stimulus for accommodation, as in complete darkness or in a uniform luminance visual field such as occurs in a dense fog, the visual system typically accommodates to approximately 70 cm away.

2.4 Capabilities of the visual system

The human visual system, like every other physiological system, has a limited range of capabilities. A convenient way to describe these limits is to set out what are called the thresholds of vision. Qualitatively, a visual threshold is the value of a stimulus to the visual system that can just be seen under a specified condition. A common experience of a threshold measurement occurs during a visit to an optician. To measure the smallest print size that can be read, the patient is shown a series of letters printed at the same luminance contrast, but in decreasing sizes. The patient starts with large-sized letters that can be read correctly every time they are presented. As the print size decreases, the letters become more difficult to read, until the patient starts to make mistakes, i.e. responds with some wrong letters. As the print size continues to decrease, more mistakes are made until the patient is essentially guessing, i.e. the percentage of correct responses is at the level of chance. Exactly what percentage of correct responses is taken as representing threshold is a matter of convention, the usual level being 50 percent, after correction for guessing.

Threshold measurements come in many different forms and depend on many different variables, most of which interact. Threshold measurements provide well-defined and sensitive metrics to explore the operation of the visual system and so have been extensively used in the field of vision science but for the practice of lighting, threshold measurements are mainly of interest for determining what will not be seen rather than how well something will be seen. Knowing what will not be seen is sometimes useful. For example, for a light source manufacturer it is useful to know what differences in light source spectrum are allowable before the same nominal light sources will be seen to differ in color; and how much light output fluctuation can be produced before the lamp will be seen to flicker. For the lighting designer, it is useful to know how much variation in luminance can be allowed on a wall while maintaining a perception of uniform illumination. The intention here is to summarize the thresholds of relevance to the practice of lighting and how they are effected by the characteristics of the human visual system. For these threshold measurements, it can be assumed that the observers were all fully adapted, that the target was presented on a field of uniform luminance, and, unless otherwise stated, that the observers' accommodation was correct.

2.4.1 Threshold measures

The threshold capabilities of the human visual system can conveniently be divided into spatial, temporal, and color classes.

2.4.1.1 Spatial threshold measures

Spatial threshold measures relate to the ability to detect a target from its background or to resolve detail within a target. For spatial threshold measures, it is usually assumed that the target does not vary with time. Common spatial threshold measures are threshold luminance contrast and visual acuity.

The luminance contrast of a target quantifies its visibility relative to its immediate background. The higher is the luminance contrast, the easier it is to detect the target. There are three different forms of luminance contrast commonly used for uniform luminance targets seen against a uniform luminance background. There is no agreement on how to measure luminance contrast for complex objects when contrast can occur within the target (Peli, 1990). For uniform targets seen against a uniform background, luminance contrast is defined as

$$C = (L_t - L_b)/L_b$$

where C is the luminance contrast, L_b the luminance of the background, and L_t the luminance of the target.

This formula gives luminance contrasts which range from 0 to 1 for targets which have details darker than the background and from 0 to infinity for targets which have details brighter than the background. It is widely used for the former, e.g. printed text on white paper

Another form of luminance contrast for a uniform targets seen against a uniform background is defined as

$$C = L_t/L_b$$

where C, L_b, and L_t are as defined in the previous equation.

This formula gives luminance contrasts that can vary from 0, when the target has zero luminance, to infinity, when the background has zero luminance. It is often used for self-luminous displays, e.g. computer monitors.

For targets that have a periodic luminance pattern, e.g. a grating, the luminance contrast is given by

$$C = (L_{max} - L_{min})/(L_{max} + L_{min})$$

where C is the luminance contrast, L_{max} the maximum luminance, and L_{min} the minimum luminance.

This formula gives luminance contrasts that range from 0 to 1, regardless of the relative luminances of the target and background. It is sometimes called the luminance modulation.

Given the different forms of luminance contrast measure, it is always important to understand which is being used.

Visual acuity is a measure of the ability to resolve detail for a target with a fixed luminance contrast. Many different targets can be used in the measurement of visual acuity, from spots, through standard optometric letters and Landolt rings, to gratings. Visual acuity is most meaningfully quantified as the angle subtended at the eye by the detail that can be resolved on 50 percent of the occasions the target is presented. This angle is usually expressed in minutes of arc. Using this measure, the visual acuity corresponding to "normal" vision is taken to be 1 min arc.

Unfortunately for simplicity, there are several other measures used to quantify visual acuity. One is the reciprocal of the angle subtended at the eye by the detail that can be resolved on 50 percent of the occasions the target is presented. A relative measure is used by the medical profession. This is the distance at which a patient can read a given size of letter or symbol relative to the distance an average member of the population with normal vision could read the same letter or symbol. For example, if the patient is said to have 20/200 vision it means that the patient can only read a given letter at 20 ft that an average member of the population with normal vision can read from 200 ft. Further, for a grating, visual acuity is sometimes expressed as spatial frequency, measured in cycles per degree. This is the number of cycles of the grating that subtend one degree from the observer's viewing position when the grating can be identified as a grating on 50 percent of the occasions it is presented.

Again, given the different forms of visual acuity that are used by different professions, it is important to be sure which metric is being used.

2.4.1.2 *Temporal threshold measures*

Temporal threshold measures relate to the speed of the response of the human visual system and its ability to detect fluctuations in luminance.

For temporal threshold measures, it is usually assumed that the target is fixed in position.

The ability of the human visual system to detect fluctuations in luminance can be measured as the frequency of the fluctuation, in hertz, and the amplitude of the fluctuation, for the stimulus that can be detected on 50 percent of the occasions it is presented. The amplitude is expressed as

$$M = (L_{max} - L_{min})/(L_{max} + L_{min})$$

where M is the modulation, L_{max} the maximum luminance, and L_{min} the minimum luminance.

This formula gives modulations that range from 0 to 1. Sometimes, modulation is expressed as a percentage modulation, calculated by multiplying the modulation by 100.

2.4.1.3   Color threshold measures

Color threshold measures are based on the separation in color space of two colors that can just be discriminated. In principle, the separation can be measured in any of the color spaces described in Chapter 1, but by far the most widely used has been the CIE 1931 (x, y) chromaticity diagram and the related Uniform Chromaticity Scale (UCS) diagrams.

2.4.2   Factors determining visual threshold

There are three distinct groups of factors that influence the measured threshold, using any of the above metrics. These groups are visual system factors, target characteristics, and the background against which the target appears.

Important visual system factors are the luminance to which the visual system is adapted, the position in the visual field where the target appears, and the extent to which the eye is correctly accommodated. The luminance to which the visual system is adapted determines which photoreceptor types are operating. The position in the visual field in which the target appears determines the size of the receptive field available to the visual system and the type of photoreceptors available. The state of accommodation determines the retinal image quality. As a general rule, the lower the luminance to which the visual system is adapted, the further the target is from the fovea, and the more mismatched the accommodation of the eye is to the viewing distance, the larger will be the threshold values.

Important target characteristics are the size and luminance contrast of the target and the color difference between the target and the immediate background. Any one of these three task characteristics can be the threshold measure of interest but the others will interact with it. This means that the visual acuity of a target will be different for targets of different luminance contrast and color difference. As a general rule, the closer the other target characteristics are to their own threshold, the greater will be the threshold of the measured variable. For example, the visual acuity for a low luminance contrast, achromatic target will be much larger than for a high luminance contrast, achromatic target.

As for the effect of the background against which the target appears, the important factors are the area, luminance, and color of the background. These factors are important because they determine the luminance and color adaptation state of the visual system and the potential for interacting with the image processing of the target. As a general rule, the larger the area around the target that is of a similar luminance to the target and neutral in color, the smaller will be the threshold measure.

2.4.3 Spatial thresholds

About the simplest possible visual task is the detection of a spot of light presented continuously against a uniform luminance background. For such a target, the visual system demonstrates spatial summation, i.e. the product of target luminance and target area is a constant. This relationship between target luminance and target area is known as Ricco's law. It implies that the total amount of energy required to stimulate the visual system so that the target can be detected is the same, regardless of whether it is concentrated in a small spot or distributed over a larger area. Spatial summation breaks down when the target is above a given size, called the critical size. The critical size varies with the angular deviation from the fovea. The critical size is about 0.5° at 5° from the fovea, and about 2° at 35° from the fovea (Hallet, 1963). There is very little spatial summation in the fovea, the critical size being about 6 min arc.

Given that the size of the target is above the critical size, the detection of the presence of a spot of light is determined simply by the luminance contrast. For the luminance of the surround greater than about 1–10 cd/m², i.e. in the photopic range, there is a constant relationship between the luminance difference of the target and the background luminance known as Weber's law. This relationship takes the form

$$(L_t - L_b)/L_b = k$$

where L_t is the luminance of the target, L_b the luminance of the background, and k a constant.

A more general picture of the effect of adaptation luminance on threshold contrast for targets of different size is shown in Figure 2.15. The increase in threshold contrast as adaptation luminance decreases is obvious, as is the increase in threshold contrast with decreasing target size (Blackwell, 1959). These data were obtained using a disc of different sizes presented for 1 s. Decreasing the presentation time, increases the threshold contrast, for all sizes, particularly at lower adaptation luminances.

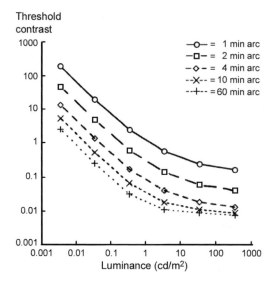

Figure 2.15 Threshold contrast plotted against background luminance for disk targets of various diameters, viewed foveally. The disks were presented for 1 s (after Blackwell, 1959).

Figure 2.16 shows the threshold contrast measured for circular targets of different sizes, occurring at different eccentricities from the fovea (Blackwell and Moldauer, 1958). It can be seen that the threshold luminance contrast is a minimum at the fovea and it increases as eccentricity from the fovea increases. Also apparent is the interaction between the size of the disk and the eccentricity of its locations. Specifically, higher threshold luminance contrasts are associated with smaller target sizes, and the increase in threshold luminance contrast with increasing eccentricity is greater the smaller the size of the disk. Another interaction with size occurs with the extent to which the target is focused on the retina. For disks less than 40 s of arc in diameter threshold contrast is rapidly reduced by blur but for disks greater than 20 min of arc in diameter there is no effect of blur on the detection of presence (Ogle, 1961).

Threshold luminance contrast is relevant to the detection of targets on a background. Targets with a luminance contrast close to or below the threshold value are unlikely to be seen and targets with a luminance contrast more than twice the threshold value are likely to be seen every time, provided the conditions are similar to those in which the threshold measurements were made.

Turning now to visual acuity, Figure 2.17 shows the variation in visual acuity with adaptation luminance for foveal viewing of the target. As adaptation luminance increases, visual acuity, measured as the reciprocal of the

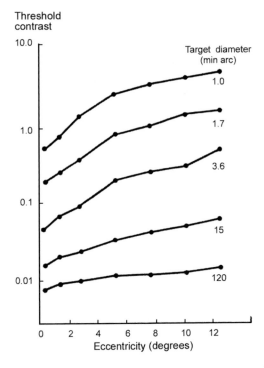

Figure 2.16 Threshold contrast for disk targets of various diameters, presented for 330 ms, at various degrees of eccentricity, at a background luminance of 257 cd/m² (after Blackwell and Moldauer, 1958).

minimum gap size, increases, approaching an asymptote at very high luminances corresponding to about 0.45 min arc (Shlaer, 1937).

Figure 2.18 shows the variation in visual acuity, measured as minimum gap size, with eccentricity from the fovea. These results show the expected deterioration in visual acuity with increasing eccentricity, the rate of deterioration being enhanced beyond about 20° eccentricity (Mandlebaum and Sloan, 1947).

Of course, these results are for photopic conditions. In mesopic and scotopic conditions, the variation of visual acuity with eccentricity is different. Figure 2.19 shows visual acuity, measured as the reciprocal of the minimum gap size, plotted against eccentricity, for a range of adaptation luminances. For the adaptation luminance of 3.2 cd/m², i.e. for photopic conditions, acuity is at about 1 min arc in the fovea and declines rapidly to about 10 min arc as eccentricity increases. For adaptation luminances below 0.006 cd/m², i.e. near the scotopic state where the fovea is blind and only the rod photoreceptors are active, visual acuity is best at about 10 min arc, 4–8° off-axis (Mandlebaum and Sloan, 1947).

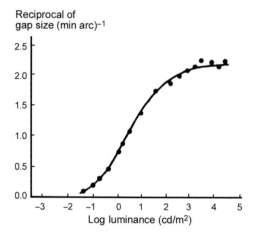

Figure 2.17 Visual acuity, expressed as the reciprocal of the minimum gap size, for a Landolt ring, plotted against log background luminance (after Shlaer, 1937).

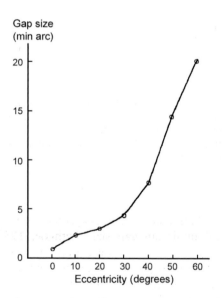

Figure 2.18 Visual acuity, expressed as the minimum gap size, for a Landolt ring target presented at different degrees of eccentricity. The target was presented for 220 ms on a background luminance of 245 cd/m² (after Mandlebaum and Sloan, 1947).

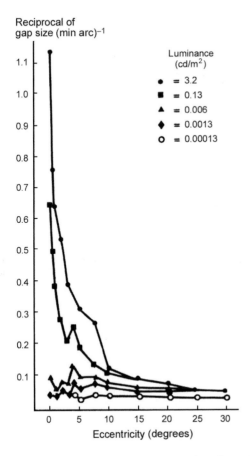

Figure 2.19 Visual acuity, expressed as the reciprocal of minimum gap size, for a Landolt ring target presented at different degrees of eccentricity, over a range of background luminances covering the photopic, mesopic, and scotopic states of the visual system (after Mandlebaum and Sloan, 1947).

Figure 2.20 shows the effect of the luminance of the background on visual acuity, measured as the reciprocal minimum gap size (Lythgoe, 1932). Visual acuity is measured using a Landolt ring, seen on a small rectangular area which is itself surrounded by a much larger area. When the luminances of the immediate background and the extensive surround are the same, visual acuity continues to improve monotonically as background luminance increases. When the luminance of the surround is very low relative to that of the immediate background, there is an optimum background luminance for visual acuity, above which visual acuity declines.

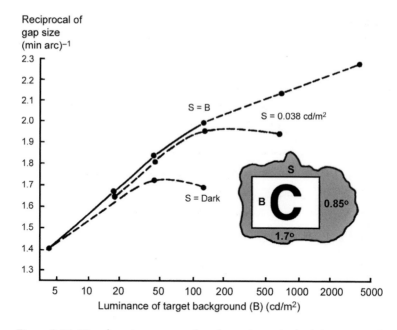

Figure 2.20 Visual acuity, expressed as the reciprocal of minimum gap size for a Landolt ring, plotted against background luminance, for different levels of surround luminance. The background luminance is the luminance of a rectangular area subtending 1.7° by 0.85° with the Landolt ring at its center. The surround luminance is the luminance of the area surrounding the background rectangle (after Lythgoe, 1932, from IESNA, 2000a).

In the above discussion, threshold contrast and visual acuity have been considered separately, because threshold contrast is usually measured with large size targets, without detail, and visual acuity is measured with high luminance contrast targets, with detail. But many things of practical interest vary in both luminance contrast and size of detail and these two target characteristics can be expected to interact. The threshold capabilities of the visual system to such targets can be expressed as the contrast sensitivity function. This is a rather grand name for what is essentially a simple piece of information, the frequency response of the visual system to spatial variations in luminance. The contrast sensitivity function of the visual system is measured using sine-wave grating targets of different spatial frequencies and adjustable modulation. The spatial frequency of the grating consists of the number of cycles of the grating that lie within a 1°-wide field of view for the observer, and hence is expressed in cycles per degree. The threshold contrast condition is usually measured as modulation but it is often displayed as contrast sensitivity which is the reciprocal of modulation.

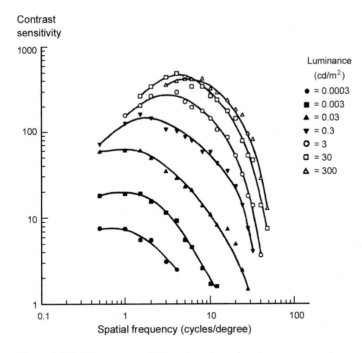

Figure 2.21 Contrast sensitivity functions for sine wave gratings at different levels of background luminance, covering the photopic, mesopic, and scotopic states of the visual system (after Van Nes and Bouman, 1967).

Figure 2.21 shows the contrast sensitivity functions for different adaptation luminances (Van Nes and Bouman, 1967).

The value of this apparently esoteric piece of information is that any variation in luminance across a surface can be represented as a waveform, and any waveform can be represented as a series of sine waves of different amplitudes and frequencies. The response of the visual system to sinewaves of different amplitudes and frequencies is given by the contrast sensitivity function. Thus, the contrast sensitivity function can be used to determine if a complex variation in luminance will be seen. If the luminance pattern has contrast sensitivities at all spatial frequencies that are greater than the threshold contrast sensitivities, the luminance pattern will be invisible. It is only when at least one spatial frequency has a contrast sensitivity below the threshold contrast sensitivity that the luminance pattern will be visible. The extent to which the luminance pattern will be seen in its entirety depends on the number of spatial frequencies for which the contrast sensitivity lies below the threshold contrast sensitivity, the more spatial frequencies for which this occurs, the more complete is the perception of the luminance pattern.

Contrast sensitivity functions can be used for many practical purposes. For example, they can be used to determine if the luminance variation of a wall washing installation will be noticed from a given distance and what size a road sign needs to be to be read from a given distance. The distance from which the observer views the luminance pattern is important because changing the viewing distance changes the spatial frequency of the pattern. As viewing distance increases, the spatial frequency of a fixed grating increases.

Returning now to Figure 2.21, it is apparent that increasing adaptation luminance increases both the contrast sensitivity and the maximum spatial frequency detectable, i.e. it produces a lower threshold contrast and a finer visual acuity. Also clear is that the change in contrast sensitivity function is slight for high luminances but it changes rapidly below adaptation luminance of about $30 \, \text{cd/m}^2$. The deterioration takes the form of reduced contrast sensitivities at all spatial frequencies and a shift in the spatial frequency at which maximum contrast sensitivity occurs. Another interesting feature of the contrast sensitivity function is the fact that it shows a maximum in contrast sensitivity. Both very high and very low spatial

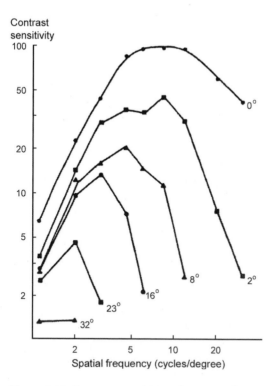

Figure 2.22 Contrast sensitivity functions for a 2.5° stimulus presented at different degrees of eccentricity (after Hilz and Cavonius, 1974).

frequencies show reduced contrast sensitivity and so are less likely to be seen than are intermediate spatial frequencies.

The effect of eccentricity on the contrast sensitivity function is shown in Figure 2.22. As might be expected from the increase in receptive field size with increasing eccentricity, the contrast sensitivity function shows a dramatic reduction in the highest spatial frequency visible as deviation from the fovea increases, as well as a reduction in peak contrast sensitivity. What this means is that it is not possible to see fine detail more than a few degrees away from the fovea.

As for the effect of incorrect accommodation, Campbell and Green (1965) showed that the effect of defocus is large at the high spatial frequencies but limited at low spatial frequencies.

2.4.4 *Temporal thresholds*

The simplest possible form of temporal visual task is the detection of a spot of light briefly presented against a uniform luminance background, i.e. a flash of light. For such a target the visual system demonstrates temporal summation, i.e. the product of target luminance and the duration of the flash is a constant. This relationship between target luminance and duration is known as Bloch's law. It implies that the total amount of energy required to stimulate the visual system so that the target can be detected is the same, regardless of the time for which the target is presented. Temporal summation breaks down above a fixed duration, called the critical duration. The critical duration varies with adaptation luminance, ranging from 0.1 s for scotopic luminances to 0.03 s for photopic luminances. For presentation times longer than the critical duration, presentation time has no effect, the ability to detect the flash being determined by the difference in luminance between the flash and the background.

While the ability to detect a single flash is of interest for signaling purposes, an aspect of temporal thresholds of wider relevance to lighting is the ability to detect flicker. All light sources operating from an alternating current electrical supply produce some fluctuation in light output, the waveform depending on the physical properties of the light source and the characteristics of the electrical supply to the light source. A light source is said to be flickering when the fluctuation in light output is visible. Figure 2.23 shows the maximum frequency of a sine wave fluctuation at 100 percent modulation that is visible at different retinal illuminations, for visual fields of different sizes (Hecht and Smith, 1936). This maximum frequency is called the critical fusion frequency (CFF). It is apparent from Figure 2.23 that the CFF increases with increasing retinal illumination and with area, although the increase is not a simple linear function. Rather, for large field sizes, such as might occur when using indirect lighting, the CFF increases linearly with retinal illumination in the scotopic state, shows little change

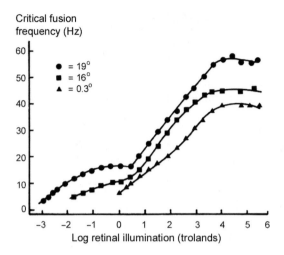

Figure 2.23 Critical fusion frequency plotted against log retinal illumination, for three different test field sizes (after Hecht and Smith, 1936).

in the mesopic state, and increases linearly in the photopic state until saturation occurs.

While the CFF is a useful metric of flicker detection, it only tells part of the story. Its limitation is that it is based on a stimulus with 100 percent modulation. Figure 2.24 shows a more general way of treating the temporal characteristics of the visual system; the temporal modulation transfer function (Kelly, 1961). The left panel of Figure 2.24 shows the percentage modulation amplitude, plotted against frequency of the oscillation at different levels of retinal illumination. These data were collected from a 60°-diameter field, uniformly illuminated, the flicker waveform being sinusoidal. This panel shows that increasing the retinal illumination increases the sensitivity to modulation and shifts the frequency for peak sensitivity from about 5 to 20 Hz. The other important point is that apart from the lowest retinal illuminance, the results for all the other retinal illuminances come to a common curve at low frequencies but have different curves at high frequencies. This implies that at low frequencies, the ability to detect flicker is determined by the percentage modulation but at high frequencies it is not. The right panel of Figure 2.24 shows the same data but now the vertical axis is plotted against the absolute modulation amplitude. The shift in frequency for peak sensitivity with increasing retinal illumination is again apparent but now the high-frequency end of the response for different retinal illuminations forms a common envelope. This means that the high-frequency response of the visual system is consistently related to the absolute modulation of the fluctuation not the percentage modulation. Flicker in lighting installations usually involves high frequencies.

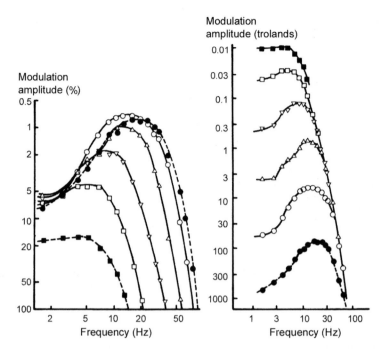

Figure 2.24 Temporal modulation transfer functions for a large visual field at different retinal illuminations. The modulation is expressed as percentage modulation in the left diagram and absolute modulation in the right diagram. The retinal illuminations are: • = 9,300 trolands; ○ = 850 trolands; △ = 77 trolands; ▽ = 7.1 trolands; □ = 0.65 trolands; ■ = 0.06 trolands (after Kelly, 1961).

Figure 2.24 can be used to determine if a light fluctuation will be visible for a large area fluctuation. For a sine wave oscillation, if the modulation at the given frequency is above the curve for the appropriate retinal illuminance, the flicker will not be visible. If it is below the curve, it will be visible. But what can be done if the waveform is not sinusoidal? The left panel of Figure 2.24 is the temporal equivalent of the contrast sensitivity function and can be used in an analogous way. To predict whether a given fluctuation waveform will be visible, the waveform should be represented by a Fourier series of different frequencies and amplitudes. If the modulations of all the components of the series lie above the appropriate temporal modulation transfer function curve, then the fluctuation will not be visible. If any of the components are below the curve, the fluctuation will be visible in some form. While these statements are true in principle, it should always be remembered that there are considerable individual differences between people in their sensitivity to flicker so to be sure that a flicker will not be seen, it is a good idea to use waveforms that have amplitudes and frequencies well clear of the threshold region represented by the temporal modulation transfer function.

2.4.5 Color thresholds

Both the spatial and temporal thresholds discussed above have been measured using achromatic targets lit by nominally white light, but, in the photopic state, the human visual system has a well-developed ability to discriminate colors. Figure 2.25 shows what are called the MacAdam ellipses, 10 times enlarged, plotted in the CIE (x, y) chromaticity diagram (MacAdam, 1942). Each ellipse represents the standard deviation in the chromaticity coordinates for color matches made between two, small visual fields with the reference field having the chromaticity of the center point of the ellipse. The lighting industry uses four-step MacAdam ellipses as its tolerance limits for quality control in lamp manufacture. Given that a four-step MacAdam ellipse represents four standard deviations and that four standard deviations should include the color matches made by more than 99.5 percent of the population, it may seem that this tolerance is too

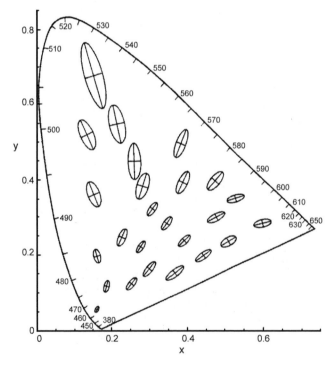

Figure 2.25 The CIE 1931 (x, y) chromaticity diagram with the MacAdam ellipses displayed, multiplied 10 times (after MacAdam, 1942, from IESNA, 2000a).

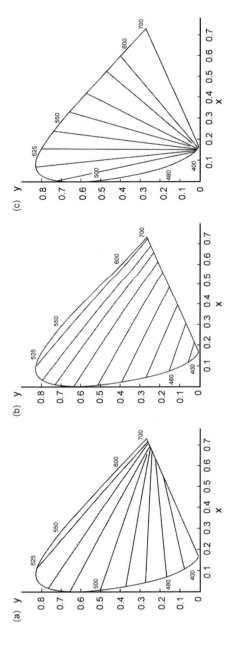

Figure 2.26 Isochromatic lines for dichromats, a = protanopes, b = deuteranopes, c = tritanopes. Surfaces represented by chomaticities at any point along a line will look the same color to a person with the given form of defective color vision, although they may differ in lightness.

lax. In practice, it has not been a problem, probably because the MacAdam ellipses were obtained in conditions ideal for comparison (simultaneous viewing of adjacent small fields by a highly practiced observer). Color discrimination between targets presented successively or between targets in which there are a wide range of colors and patterns present is more difficult (Narendran *et al.*, 2000).

Figure 2.25 is for people with normal color vision. People with defective color vision are unable to make such fine discriminations in color. Figure 2.26 shows what are called isochromatic lines on the CIE 1931 (x, y) chromaticity diagram for the three types of dichromat. All colors along a line will appear the same in hue and saturation to the dichromat although the may vary in lightness. The directions of the lines in Figure 2.26 demonstrate that protanopes and deuteranopes will have similar problems in discriminating amongst reds and greens but deuteranopes will find discriminating amongst purples much easier than will protanopes. As for tritanopes, these will have little difficulty discriminating amongst reds and greens but will have a problem discriminating between blues.

2.5   Interactions

The information given above represents a minute portion of the data available on visual thresholds for different conditions. Further, it is based on a restricted range of variables. The spatial thresholds all use an achromatic target seen on a field of uniform luminance. The temporal thresholds use fluctuations in luminance without a change in color. The color thresholds use side-by-side comparisons between small uniform visual fields with the same luminance. Nonetheless, the data given is enough to demonstrate the effects of the major factors, adaptation luminance, position in the visual field, and state of accommodation. Other factors, such as light spectrum and movement of the target interact with these major factors to determine threshold values. Specifically, Figure 2.27 shows the effect of light spectrum at the same adaptation luminance on visual acuity (Shlaer *et al.*, 1941). It can be seen that visual acuity is only slightly influenced by light spectrum, but what effect there is occurs at the long-wavelength end of the visible spectrum.

Figure 2.28 shows the visual acuity for smooth relative movement of target and observer at different velocities (Miller and Ludvigh, 1962). There is a slow deterioration in visual acuity with increasing velocity up to about 50°/s, but at higher velocities, visual acuity deteriorates rapidly. This result is understandable if it is assumed that for velocities below 40°/s, it is possible to use smooth pursuit eye movements to keep the target close to the fovea. Of course, this will not be possible if the target is moving in an unexpected manner, involving sudden changes of course and velocity.

There are many other factors that interact to determine a specific threshold condition. Further, there are large differences between individuals in threshold measures, particularly with age. Figure 2.29 shows the threshold

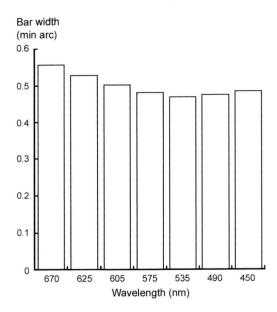

Figure 2.27 Visual acuity of a grating, measured as the width of a bar of the smallest resolvable grating, for light transmitted through different narrow-band filters. The peak wavelengths of the filters are 670, 625, 605, 575, 535, 490, and 450 nm (after Shlaer *et al.*, 1941).

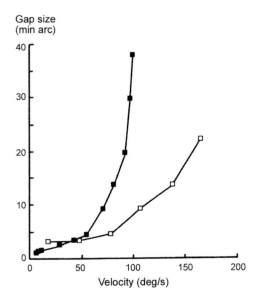

Figure 2.28 Visual acuity for Landolt rings, expressed as gap size in minutes of arc, for different angular velocities. The filled symbols are for the target moving. The open symbols are for the observer moving (after Miller and Ludvigh, 1962).

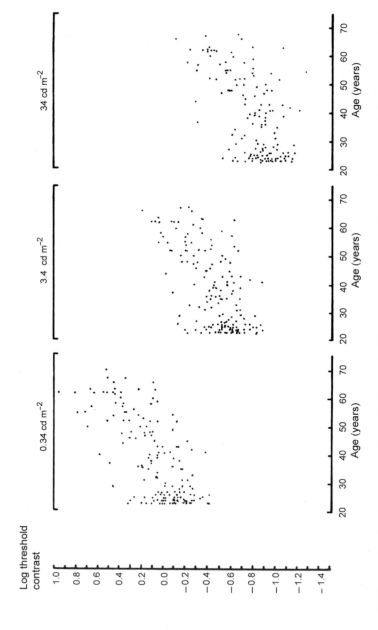

Figure 2.29 Log threshold contrasts for individuals of different ages at three different background luminances (after Blackwell and Blackwell, 1971).

contrast measurements for people of different ages at three different adaptation luminances. The trend in threshold luminance contrast with increasing adaptation luminance is obvious as is the general trend of increasing threshold contrast with increasing age. However, what is really impressive are the large differences in threshold contrast between individuals, different enough to ensure there is some overlap in threshold contrast between the people of 20 and 60 years of age (Blackwell and Blackwell, 1971). Really, if you want to know how a specific combination of factors will affect a specific threshold measure for a specific population, there is little alternative other than to make a direct measurement. However, if all you want is to ensure that a target presented will be clearly visible or definitely invisible, i.e. you want your target to be definitely above or below the relevant threshold then you may be able to use the data derived from the simplified conditions given above. Details of many different threshold measurements in a wide range of conditions can be found in Wyszecki and Stiles (1982) and Boff and Lincoln (1988).

2.6   Perception through the visual system

While thresholds define the limits of the capabilities of the human visual system, most of our life is spent looking at things that are well above threshold and hence clearly visible. The topic here is how we perceive these myriad of stimuli. The perception of the visual world is not solely determined by the physical stimuli presented to the visual system as the retinal image, nor by the characteristics of the visual system described above. Rather, the stimuli to the visual system are broken into different elements in the retina, different elements are then transmitted up the different visual channels to the visual cortex where the real world is re-assembled based on past experience and coincident information. As an example of the power of past experience, Figure 2.30 shows a surface with dents and dings in it. If this page is inverted the dents become dings and vice-versa, because it is unconsciously assumed that the light which is casting the shadows always comes from above. Clearly, there is a gap between our understanding of the visual system and its eventual output, perception of the visual world. The existence of this gap can be understood by an analogy. Consider the output of an orchestra. It consists of rhythm, melody, and tonality arranged in complex and subtle patterns, that can, in the right circumstances, generate an emotional response. However, our knowledge of how this is achieved is limited to how each instrument generates sound. In the world of vision, we have some idea of how each "instrument" behaves but not how they all fit together to generate the whole, the perception of the visual world.

When considering how we perceive the world, the overwhelming impression is one of stability in the face of continuous variation. As the eyes move in the head and the head itself moves about, the retinal images of objects move across the retina and change their shape and size according to the

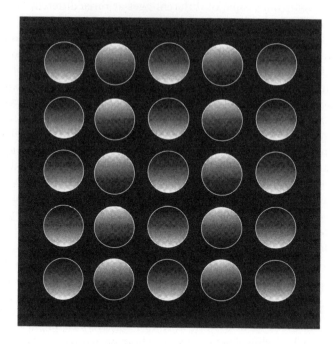

Figure 2.30 A surface with circular dents and dings. The distribution of light within each circular area determines whether it is seen as a dent or a ding (after IESNA, 2000a).

laws of physical optics. Further, throughout the day, the spectral content and distribution of daylight changes as the sun moves across the sky and the meteorological conditions vary. Despite these variations, our perception of objects rarely changes. This invariance of perception is called perceptual constancy. The evolutionary advantage in being able to recognize a tiger as a tiger over a wide range of lighting conditions is obvious.

2.6.1 The perceptual constancies

There are four fundamental attributes of an object which are maintained constant over a wide range of lighting conditions. They are described below.

Lightness – Lightness is the perceptual attribute related to the physical quantity, reflectance. In most lighting situations, it is possible to distinguish between the illuminance on a surface and its reflectance, i.e. to perceive the difference between a low-reflectance surface receiving a high illuminance and a high-reflectance surface receiving a low illuminance, even when both surfaces have the same luminance. It is this ability to perceptually separate the luminance of the retinal image into its components of illuminance and reflectance which ensures that a piece of coal placed near a window is always seen as black while a piece of paper far from the window is always seen as white, even when the luminance of the coal is higher than the luminance of the paper. This ability to separate illuminance from reflectance under most lighting conditions makes the use of luminance as the basis of lighting design criteria problematical (Jay, 1967, 1971).

Color – Physically, the stimulus a surface presents to the visual system depends on the spectral content of the light illuminating the surface and the spectral reflectance of the surface. However, quite large changes in the spectral content of the illuminant can be made without causing any changes in the perceived color of the surface, i.e. color constancy occurs. Color constancy is similar in many ways to lightness constancy. There are two factors that need to be separated; the spectral distribution of the incident light and the spectral reflectance of the surface. As long as the spectral content of the incident light can be identified the spectral reflectance of the surface, and hence its color, will be stable.

Size – As an object gets further away, the size of its retinal image gets smaller but the object itself is not seen as getting smaller. This is because by using clues such as texture and masking, it is usually possible to estimate the distance and then to compensate unconsciously for the increase in distance. Figure 2.31 shows an illustration of a room, called the Ames room after the inventor, where the cues to distance have been deliberately designed to be misleading when viewed from a specific position. The distortion in perceived size of the people standing in the two corners of the room is apparent.

Shape – As an object changes its orientation in space, its retinal image changes. Nonetheless, in most lighting conditions the distribution of light and shade across the object makes it possible to determine its orientation in space. This means that in most lighting conditions a circular plate that is tilted will continue to be seen as a tilted circular plate even though its retinal image is elliptical.

These constancies represent the application of everyday experience and the integration of all the information about the lighting available in the whole retinal image to the interpretation of a part of the retinal image which bears several alternative interpretations. Given this process, it should not be too surprising that the constancies can be broken by restricting the information available coincident with the object being viewed. For example, viewing a uniform luminance surface through an aperture which restricts the view to a limited part of the surface, will often eliminate lightness constancy, i.e. make it impossible to accurately judge the reflectance. Likewise, eliminating cues to distance, such as gradients in texture, motion parallax, and overlapping of objects will destroy size constancy; changing cues to the plane in which an object is lying will reduce shape constancy and eliminating information on the spectral content of the illuminant will reduce color constancy. In general, constancy is likely to break down whenever there is insufficient or misleading information available from the

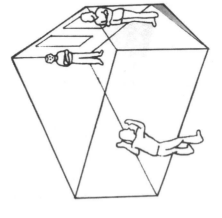

Figure 2.31 The Ames room: a demonstration that providing misleading clues to distance will break size constancy (after IESNA, 2000a).

surrounding parts of the visual field. The constancies are most likely to be maintained when there is enough light for the observer to see the object and the surfaces around it clearly, the light being provided by an obvious but not necessarily visible light source. It is also desirable that the light source has a spectral power distribution that covers all the visible spectrum and is delivered without disability glare. In addition, the constancies are most likely to be maintained when there are a variety of surface colors, including some small white surfaces and there are no large glossy areas, both factors that help with the identification of the spectral content of the light source (Lynes, 1971). Lighting conditions used in display lighting sometimes set out to break the constancies, particularly lightness constancy, in order to give the display some drama.

It is important to appreciate that even when the lighting conditions are such as to support it, perceptual constancy is not perfect. For example, lightness constancy will break down if large changes in illuminance occur. Figure 2.32 shows the apparent Munsell value of spectrally neutral surfaces (see Section 1.6.2), plotted against the illuminance on the surfaces. It shows that as the illuminance is decreased, the apparent Munsell value, i.e. the lightness, is reduced for all the Munsell samples, until at very low illuminances all the Munsell values are in the range of dark gray to black (Munsell value < 2). It should also be pointed out that this gradual breakdown in lightness constancy requires very large changes in illuminance

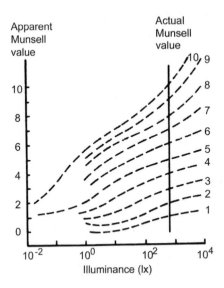

Figure 2.32 Apparent Munsell values at different illuminances for surfaces seen against a background of reflectance 0.2. The vertical line at an illuminance of 786 lx indicates the reference condition. At this illuminance, the apparent Munsell values of the surfaces have been normalized to their actual Munsell values (after Jay, 1971).

relative to those that occur in interior lighting, which typically lie between 100 and 1,000 lx.

2.6.2   Modes of appearance

While lighting has an important role in preserving or eliminating constancy, it also has a role in determining the perceived visual attributes of objects. Objects can have five different attributes: brightness, lightness, hue, saturation, transparency, and glossiness, depending on their nature and the way they are lit. These attributes are defined as follows:

Brightness – an attribute based on the extent to which an object is judged to be emitting more or less light.

Lightness – an attribute based on the extent to which an object is judged to be reflecting a greater or lesser fraction of the incident light.

Hue – an attribute based on the classification of a color as reddish, yellowish, greenish, bluish, or their intermediaries or as having no color.

Saturation – an attribute based on the extent to which a color is different from no color of the same brightness or lightness.

Transparency – an attribute based on the extent to which colors are seen behind or within an object.

Glossiness – an attribute based on the extent to which a surface is different from a matte surface with the same lightness, hue, saturation, and transparency.

Not all these attributes occur in every situation. Rather, different combinations of attributes occur in different modes of appearance. The four modes of appearance are:

Aperture mode – this occurs when an object or surface has no definite location in space, as occurs when a surface is viewed through an aperture.

Illuminant mode – this occurs when an object or surface is seen to be emitting light.

Object mode: volume – this occurs when a three-dimensional object has a definite location in space with defined boundaries.

Object mode: surface – this occurs when a two-dimensional surface has a definite location in space with defined boundaries.

Table 2.2 shows which of the attributes can be associated with each mode of appearance. Of particular interest to the perception of lighting is the shift between the attributes of brightness and lightness in different modes of appearance. An object which appears in the self-luminous mode, such as a video display terminal (VDT) screen or a light source, is perceived to have a brightness but not a lightness. In this mode of appearance, the concept of

Table 2.2 The visual attributes that can occur with each mode of
appearance

Attribute	Aperture	Illuminant	Volume	Surface
Brightness	*	*		
Lightness			*	*
Hue	*	*	*	*
Saturation	*	*	*	*
Transparency		*	*	*
Glossiness				*

reflectance is perceptually meaningless. However, an object which appears in the volume mode, such as a VDT screen or a light source which is turned off, does not have an attribute of brightness but does have a lightness in that its reflectance can be estimated.

A similar transformation occurs between the volume or surface modes of appearance and the aperture mode. Even non-self-luminous objects seen in the aperture mode are perceived as having a brightness but not a lightness. When seen in the object mode they have a lightness but not a brightness. This is important because lighting can be used to change the mode of appearance. For example, a painting hung on a wall has a lightness attribute when lighted so that both it and the wall appear in the object mode: surface. However, if the painting is illuminated solely with a carefully aimed framing spot so that the edge of the beam coincides with the edges of the painting, the painting is seen in the aperture mode and takes on a self-luminous quality with a brightness attribute. Adjusting the modes of appearance is an important technique in display lighting, both indoors and outdoors.

2.7 Summary

There is much about the visual system that remains a mystery but what is clear is that it involves both the eye and brain working together. The visual system consists of two parts; an optical system that produces an image on the retina of the eye and an image processing system that extracts different aspects of that image at various stages of its progress up the optic nerve to the visual cortex, while preserving the location where the information came from. It is also clear that the visual system devotes most of its resources to analyzing the central area of the retina, particularly the fovea. This implies that peripheral vision is mainly devoted to identifying something that should be examined in detail by turning the head and eyes so the image of whatever it is falls on the fovea.

The visual system can operate over a wide range of luminances, from sunlight to starlight. To do this it continually adjusts its sensitivity to light, increasing its sensitivity as the amount of light available falls. Decreasing the amount of light from daylight to darkness, takes the visual system through three distinct operating states, the photopic, the mesopic, and the scotopic. In the photopic condition, fine discriminations of size and color can be made. In the mesopic, the ability to make these discriminations deteriorates so that by the time the scotopic is reached, color can no longer be seen, detail is impossible to discriminate and the fovea is blind. Interior lighting usually allows the visual system to operate in the photopic state, while exterior lighting usually ensures the visual system is operating at the photopic/mesopic border. No lighting installation worthy of the name produces so little light that the visual system is in the scotopic state.

Like every other physiological system, the visual system has a limited range of capabilities. These limits are expressed by the thresholds of vision. A threshold is a stimulus that is detected a specified percentage of the times it is presented, usually 50 percent. There are many different thresholds, one of the most common being visual acuity, i.e. the smallest size that can be resolved. Others quantify the smallest luminance contrast that can be detected, the smallest color difference that can be detected, and the lowest flicker frequency that can be detected. Different thresholds occur under different conditions of lighting and stimulus presentation but, in general, vision becomes more limited as the amount of light decreases, the stimulus occurs further away from the fovea, and the degree of defocus increases. Threshold measurements provide well-defined and sensitive metrics to explore the operation of the visual system and so have been extensively used in the field of vision science but for the practice of lighting, threshold measurements are mainly of interest for determining what will not be seen rather than how well something will be seen.

Given that the details of a scene are clearly visible, i.e. they are well above their threshold values, the dominant characteristic of the visual system is the stability of perception in the face of continuous variation in the retinal image. Given lighting conditions that provide enough light with a wide spectral distribution in such a way that how the space is lit can be easily understood, the lightness, color, size, and shape of objects in the space remain constant no matter how they are viewed. It is only when the information about the space and the way it is lit is restricted or misleading, that these perceptual constancies will break down. Lighting can be used to reinforce or to undermine the perceptual constancies

This chapter is not intended to be an exhaustive review of the visual system. There are many other books that explore this topic in much greater detail. If you are interested in doing so, you are recommended to see Sekular and Blake (1994) and Kaiser and Boynton (1996).

3 The circadian system

3.1 Introduction

The lives of living things are characterized by changes in behavior that occur regularly over a 24-h cycle. As examples, consider the sleep–wake cycle present in all animals and the fact that many plants raise and lower their leaves from day to night. These rhythms, and the many others that occur, are called circarhythms. The physiology that controls such circarhythms is called the circadian system, from the Latin, *circa*, for "about," and *dies*, for "day," about a day. Circadian systems exist in a wide range of life, from unicellular organisms to humans, in insects, plants, fish, birds, and mammals. Further, there are cyclic patterns that occur over the seasons, called circannual rhythms, such as the seasonal breeding of mammals and seed germination in plants. These are believed to be controlled by the gradual change in light/dark ratio that occurs over the seasons, as signaled by the circadian system.

The reason for including information on the circadian system in humans in this book is that light entering the eye is a potent means for modifying the phase and amplitude of the system that synchronizes circarhythms in humans. The nature of the role of light in modifying circarhythms was first suggested by Bunning (1936), in relation to diurnal variations in plants. The results obtained suggested that the diurnal variations were not passive responses to the change from light to darkness but rather were driven by an endogenous (internal) clock, that was entrained by an exogenous (external) signal, in this case, the alternation of light and darkness (Kleinhoonte, 1929; Bunning and Stern, 1930). Bunning (1936) also hypothesized that this arrangement meant that interruptions of the normal light–dark cycle would have an effect in shifting the phase of the clock, the magnitude and direction of the phase shift being determined by when the interruption occurred. Specifically, exposing the circadian system to bright light early in the night would lead to a phase delay, while bright light presented late in the night would lead to a phase advance. Much research over the next 30 years showed that this hypothesis was correct and that the basic endogenous/ exogenous model of the circadian system occurred in many different forms of life (Gwinner, 1975; Pittendrigh, 1981). Among these life forms was humans (Sharp, 1960; Lobban, 1961; Aschoff, 1969). These findings lead to a series of studies undertaken in a temporal isolation facility (Wever, 1979). These experiments raised the question of how potent light was as an entraining stimulus, relative to other cues to time, such as social contact. Indeed, the relative phase shift that could be achieved by light–dark cycles alone compared to what could be done when social cues were added led the researchers to conclude that the effectiveness of light–dark cycles as an entraining mechanism was small compared to social cues. However, the experiments that lead to this conclusion were limited in two respects. The first was that the maximum illuminance that could be provided was about 1000 lx. The second was that the dark phase was not completely dark, the subjects being allowed to use table and bedside lamps. More recent work has demonstrated that both these limitations are likely to diminish the impact of the light–dark cycle as an entrainment cue. The view that the human circadian system had a reduced sensitivity to light relative to other creatures was challenged by Czeisler *et al.* (1981) who maintained that the light–dark cycle was a potent entrainment cue, when light was provided at higher illuminances and the dark cycle was truly dark. Since then, many studies have shown that the exposure to light is the principal exogenous stimulus to the human circadian system (Dijk *et al.*, 1995) although it is not the only one. Some studies have shown that social cues (Aschoff *et al.*, 1971), night-time activity (Van Reeth *et al.*, 1994), and fitness training (Van Someren *et al.*, 1997b) may also impact the circadian system. Conversely, some experiments with blind people have failed to show entrainment of the circadian system, even when the people lived in a conventional 24 h light–dark environment (Miles *et al.*, 1977; Klein *et al.*, 1993). How effective these other stimuli are for entrainment remains an open question but there can be little doubt that light is a major exogenous stimulus to the circadian system.

3.2 The structure of the circadian system

Like the visual system, the circadian system starts with the eye but unlike the visual system it does not transmit information directly to the visual cortex. Rather, after leaving the eye, the circadian system proceeds up the retinalhypothalamic tract (RHT) to the suprachiasmatic nuclei (SCN) and then by way of the paraventricular nucleus (PVN) and the superior cervical ganglion to the pineal gland (Figure 3.1). In dark conditions, the pineal gland synthesizes the hormone melatonin, which is then circulated throughout the body by the bloodstream. The anatomical linkage between the eye and the pineal gland is called the retinohypothalamic–pineal (RHP) axis. The location of the RHP axis in the midbrain and brainstem supports the idea that it developed very early in the evolution of humans and has features in common with many much simpler life forms (Menaker and Tosini, 1996).

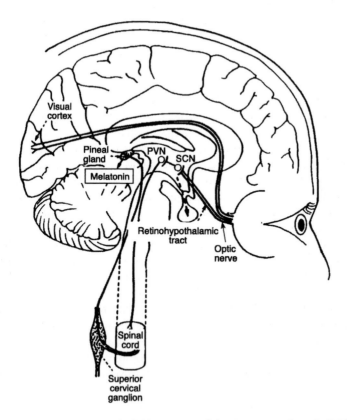

Figure 3.1 A simplified illustration of the RHP axis (from IESNA, 2000a).

The main difference between these simpler life forms and mammals, particularly humans, is that in the simpler life forms, the photoreceptors for light can be located in many different parts of the body, including directly in the pineal. In humans, the photoreceptors for light that stimulate the circadian system are located in the retina of the eye. This is shown by the fact that removal of the eyes eliminates all circadian responses to light in all the mammalian species examined to date, although it must be admitted that these are mainly various types of rodents (Roenneberg and Foster, 1997). Each of the components of the RHP axis will be considered in turn.

3.2.1   The retina

A detailed description of the structure of the retina of the eye is given in Chapter 2. Light reaching the retina provides signals to both the visual system and the circadian system, but through different neural connections. For the visual system, there are four kinds of photoreceptors; three cone types, each with a different photopigment, and one rod type, with a different photopigment to any of the cones. The photoreceptor, or photoreceptors, used to influence the human circadian system have not yet been identified. Using the night-time suppression of the hormone melatonin as a marker of the circadian system, Ruberg *et al.* (1996) examined the effect of light exposure on people with normal color vision, i.e. with all three cone types, and on protanopes and deutoranopes, i.e. people without the long-wavelength cone and the middle-wavelength cone, respectively. There was no difference in the level of melatonin suppression for the same illuminance for the different types of visual system. This demonstrates that all three cone types are not necessary for the human circadian system to work. Brainard *et al.* (2001a) found that radiation at 505 nm was four times as effective in suppressing melatonin as radiation at 555 nm, again suggesting that the cone photoreceptors used for photopic vision are not the primary means of influencing the circadian system. Even more surprising, measurements on blind people without any conscious perception of light, no pupil response to light and no electroretinogram (a measure of the gross level of electrical activity in the retina), showed melatonin suppression when the people were exposed to 6,000 lx (Czeisler *et al.*, 1995). This result suggests that none of the photoreceptors involved in vision need to be involved in the circadian system. Similar results have been found in retinally degenerate mice, in which all the rod photoreceptors are destroyed after about 90 days and all electrophysiological and behavioral responses to bright light have disappeared after about 90–150 days. Despite this apparent destruction of the mouse's visual system, these mice continued to show circadian responses to light that were identical to normally sighted mice (Foster *et al.*, 1991). Further, support for a unique photosensor for the circadian system comes from the discovery of photoreceptors based on vitamin B_2 in mice (Miyamoto and Sancar, 1998). The photoreceptors used for vision in mice are based on vitamin A. Even more surprising is the finding that these photoreceptors are not located at the same level of the retina as the rod and cone photoreceptors used in vision but rather at the collector and ganglion cell level. For the retinally degenerate mice results discussed above, the ganglion and collector levels of the retina are unaffected by the genetic modification used to destroy the rods and cone photoreceptors.

While identifying the photoreceptor or photoreceptors involved in signaling light levels to the circadian system is of great interest to physiologists, what matters to those concerned with lighting is the spectral sensitivity. Two recent studies (Brainard *et al.*, 2001b; Thapan *et al.*, 2001) have independently measured the spectral sensitivity of the human circadian system using melatonin suppression as a marker. Figure 3.2 shows the relative spectral sensitivity measured in these two studies. It is clear that melatonin suppression, and presumably all the other aspects of the circadian system, are primarily influenced by radiation at the short-wavelength end of the visible spectrum. A spectral sensitivity template, based on vitamin A

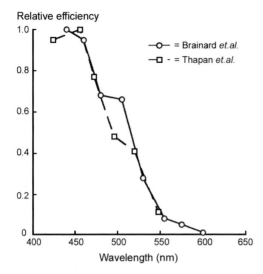

Figure 3.2 Measured relative efficiency of electromagnetic radiation at different wavelengths in stimulating the human circadian system, using melatonin suppression as a marker (after Brainard *et al.*, 2001a; Thapan *et al.*, 2001).

retinaldehyde photopigments, fitted to the data suggests a peak sensitivity at about 464 nm. This again suggests a new type of photoreceptor in the retina. However, Rea *et al.* (2002), while confirming the dominance of the short-wavelength end of the visible spectrum for the melatonin suppression, have pointed out that a good match to the measured spectral sensitivities could be achieved by a combination of short-wavelength cone and rod photoreceptors. Further, the axons of short-wavelength cones extend deep into the collector cell level and intertwine with the axons from the rod photoreceptors. Clearly, the argument over the photoreceptors that provide the stimulus to the circadian system still has some way to run, but there can be little doubt about the associated spectral sensitivity.

As for the characteristics of the channels that convey the electrical information from the retina to the SCN, these are well understood. Measurements in mouse and cat of the pathways from the retinal ganglion cells to the SCN have shown that ganglion cells that project to the SCN are infrequent, spread evenly across the retina and have extensive dendritic arbors (Moore *et al.*, 1995). Further, there is no attempt to preserve the location of each ganglion cell in its projection to the SCN. This structure emphasizes that as far as the circadian system is concerned the retina is a "photocell" designed to collect irradiance information. This concept is reinforced by measurements of the electrical output of cat retinal ganglion cells that project to the SCN. These ganglion cells were found to have 2–5° receptive field sizes, to respond most strongly to still or very slowly moving stimuli and to give sustained responses (Pu, 2000). All these characteristics, suggest a system designed to be insensitive to rapid fluctuations in light pattern.

3.2.2   The suprachiasmatic nuclei

The SCN are located in the hypothalamus of the brain. The SCN are recognized as the endogenous oscillator, i.e. the master clock, in mammals, including humans (Klein *et al.*, 1991). Measurements of the response of SCN neurons in rats to light indicate that they have very large receptive fields (20–40°) with no on/off structure (Groos and Mason, 1980). The lack of on/off structure is markedly different from the receptive field properties of the rat visual system but it must be remembered that an antagonistic surround is useful to identify edges in the visual scene, something that is essential for an efficient imaging system but unnecessary for a "photocell." It has also been shown that SCN neurons in rats respond to continuous light stimulation with a sustained response that can be as long as 30–60 min (Meijer and Rietveld, 1989). Other measurements have shown that the output of the SCN in rats is characterized by a high threshold and a limited dynamic range, features that serve to convert the differences between night and day into a simple square wave (Groos and Meijer, 1985). Assuming that the basic properties of the SCN in humans are the same as in rats, the overall picture presented by these results is a simplification of the signals received from the retinal ganglion cells, in terms of a more limited dynamic range and greater spatial and temporal summation.

The SCN send neural signals to many parts of the nervous system to coordinate physiological and behavioral activities. One that has been extensively studied is the path from the SCN to the pineal gland (Klein *et al.*, 1991). Signals on this path give rise to another form of oscillating output

3.2.3   The pineal gland

The pineal gland synthesizes and secretes the hormone melatonin during the dark phase of the 24 h light–dark cycle, regardless of whether the creature is diurnal or nocturnal in its activity pattern. Melatonin is easily absorbed into the bloodstream and hence serves as a chemical messenger throughout the body (Menaker, 1997). Melatonin detectors, which act as receivers of the message, have been found in many parts of the body. The message carried by melatonin is that of time, as determined by the SCN, the master clock. The essential role of melatonin is to synchronize the activation of many other physiological functions, not to the same time, but rather to the times in the 24-h cycle when they should occur (Cagnacci *et al.*, 1997b). Normally, high levels of melatonin are secreted at night and low levels are secreted during the day (Klein *et al.*, 1991; Wetterberg, 1993). However, the presence of light at night suppresses the synthesis of melatonin, the amount of suppression being determined by the retinal irradiance and the duration

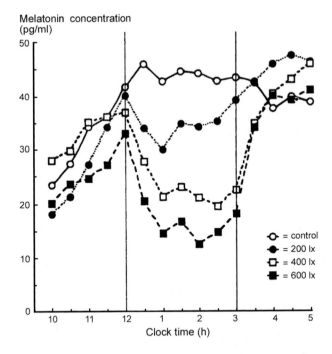

Figure 3.3 Melatonin concentrations at different times from 10 p.m. to 5 a.m. For the control condition, the subjects spent the entire period in a room where the illuminance was less than 10 lx. For the other conditions, the subjects spent the hours between 10 p.m. and midnight and between 3 a.m. and 5 a.m. in the room lit to less than 10 lx but between midnight and 3 a.m. they were exposed to either 200, 400, or 600 lx at the eye (after McIntyre *et al.*, 1989).

of exposure, for up to about 1 h. Exposure durations longer than 1 h fail to further reduce melatonin concentration, and recovery after exposure is complete within about 1 h. Figure 3.3 shows the mean melatonin concentrations measured for six subjects at half-hourly intervals between 10 p.m. and 5 a.m. Between midnight and 3 a.m., they were exposed to an illuminance at the eye of either 200, 400, or 600 lx (McIntyre *et al.*, 1989). The reduction in melatonin concentration with exposure to light is clear as is the recovery when the light is removed.

3.3   Characteristics of the circadian system

The RHP axis has a number of important characteristics. Probably the most notable is the fact that it continues to oscillate even the absence of any external cues to time. The period of this oscillation in humans is longer than 24 h. When these longer periods occur over a number of days, the circadian

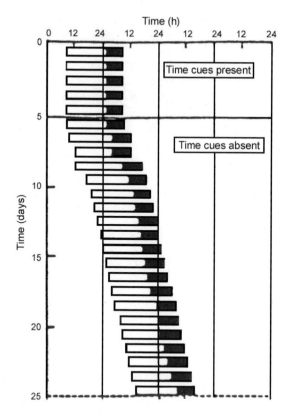

Figure 3.4 Sleep–wake cycles for an individual over 25 days, sleep periods being indicated by black shading. For the first 5 days, a constant light–dark cycle was present. At day 6 and for all the following days, a constant dim light level is provided throughout the 24 h. For the first 5 days the period of the circadian cycle is 24 h but after day 6, the period increases to more than 25 h and the circadian system starts to free-run resulting in a steady drift in sleep period.

system is said to be free-running. Figure 3.4 shows the sleep–wake cycle measured for an individual over 20 days. For the first 5 days, time cues are present in the form of a regular light–dark cycle. For these 5 days, the period of the sleep–wake cycle is 24 h. For days 6–20, no time cues are present and the sleep–wake cycle starts to free-run with a period of 25.5 h, with the result that after 15 days the individual is asleep in the middle of the day.

There can be little doubt that the presence of time cues, particularly a light–dark cycle, is necessary to prevent the circadian system from free-running, but the light–dark cycle has another role, to signal the passing of the seasons. Depending on the latitude at which you live, the length of the day and conversely, the length of the night, varies over the season.

The longer is the night, the longer is the time for which melatonin is secreted. In animals that show distinct seasonal behavior, there are cells that measure the duration for which melatonin is present (Bartness and Goldman, 1989). These cells also regulate seasonal changes in behavior. It is interesting to note that rates of conception in humans exhibit seasonal variations (Roenneberg and Aschoff, 1990a,b), these variations being much larger before the industrial revolution (Bronson, 1995). This suggests that electric lighting can have an impact on the seasonal adjustment of the circadian system. Wehr *et al.* (1995) claim that most individuals who live in modern urban environments at temperate latitudes show no seasonal variation in duration of melatonin secretion. The implication of this is that the consistent use of electric light in the evening, after the sun has set, suppresses melatonin secretion and therefore removes any seasonal variation (Wehr, 2001). Exactly how much exposure to electric light after dark is sufficient to eliminate the effect of the naturally occurring seasonal variation in daylength is an interesting question that has yet to be answered.

3.4   Models of the circadian system

The circadian system can be modeled at two different levels, conceptually and mathematically. Conceptually, the role of the circadian system in all creatures is to produce an internal replication of the external night and day, but one that is stable despite minor variations in the external conditions, such as might be caused by differences in meteorological conditions. A refinement of this concept is that the circadian system should not just be responsive to the actual external conditions but should anticipate them. In other words, it should tell the animal not that dawn or dusk has arrived but rather that dawn or dusk is coming. This has some evolutionary advantage. Twilight, either at dusk or dawn, is the time when predators are most successful in catching prey so, if you are potential prey, twilight is a time best avoided.

There are several mathematical models of the effect of light exposure on the operation of the circadian system. The purpose of these models is to produce accurate predictions and testable hypotheses. One of the earliest models in this field was of the rodent circadian system. The model was used to predict activity and sleep periods, and was based on the mutual phase relationship of two circadian oscillators, one entrained for sunset and one entrained for morning (Pittendrigh and Daan, 1976). Some evidence suggests a similar arrangement in humans (Wehr, 1996; Cagnacci *et al.*, 1997a). Other models applicable to humans have been developed by treating the link between lighting exposure and the phase of the circadian rhythm as a control theory problem. The result has been a number of models based on the van der Pol oscillator (Kronauer, 1990; Kronauer *et al.*, 1999; Forger *et al.*, 1999). The early models were based on data for exposure to single bright light pulses. While these were capable of predicting the outcome of such exposure, they could not accurately deal with the effects of low light level exposure (approximately 100–200 lx) or of very bright (approximately 10,000 lx), short duration (approximately 5 min) exposures separated by lengthy periods of darkness. The most recent models have been refined to accurately predict phase shifts to photopic stimuli of any temporal pattern and any illuminance in the photopic range (Kronauer *et al.*, 1999). Yet other models have been constructed based on the concept that exposure to light not only changes the phase and amplitude of the circadian system but also the period (Beersma *et al.*, 1999). Currently, it is not clear which of these approaches is correct. For that to be established some critical experiments testing the predictions of the models are required, but what is clear is that the response of the human circadian system to a wide variety of patterns of light exposure can now be predicted.

3.5   Effects of light exposure on the circadian system

Figure 3.4 demonstrates that the primary effect of the light–dark cycle is to entrain the circadian system to a 24-h cycle. But what happens if the light–dark cycle is disrupted, say by light exposure occurring during what is normally a period of darkness? The answer is a shift in the phase of the circadian rhythm. The direction of the phase shift depends on the timing of the light exposure. Figure 3.5 shows two cycles of what is called a phase response curve (PRC), based on a light exposure pattern of a 16-h waking period and an 8-h period of bed rest in darkness. The waking period was divided into a total of 11 h spent in dim interior lighting (10–15 lx) and a 5-h period of bright light (7,000–13,000 lx). The timing of the 5-h exposure to bright light occurred at different times within the 16 waking hours (Jewett *et al.*, 1997). What this phase response curve shows is that exposure to bright light during the afternoon has very little if any effect on the phase of the circadian cycle in the next 24 h. However, bright light given early in the night tends to delay the circadian cycle but bright light given late in the night tends to advance the phase of the circadian cycle. The critical time at which the effect of a pulse of bright light changes from a phase delay to a phase advance is around the minimum of the core body temperature. For healthy young people, whose circadian system is entrained by a regular light–dark cycle, this minimum occurs about 1–2 h before awakening. Given the rapid transition from phase delay to phase advance around the core body temperature minimum, what happens when the light exposure straddles the core body temperature minimum? The answer is that there is a reduction in amplitude of the circadian rhythm but no shift in phase, although the circadian system is then more sensitive to subsequent light pulses (Jewett *et al.*, 1991). While the above understanding has been derived from strictly controlled light exposures, in everyday life, people are exposed to light at many different times of the night and day. Fortunately,

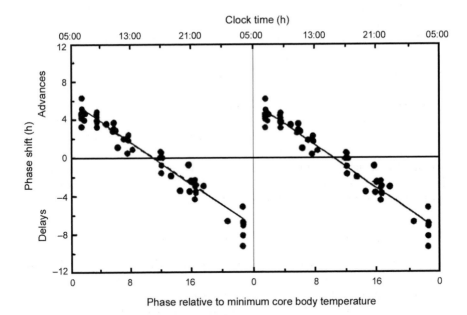

Figure 3.5 Two cycles of a phase response curve for humans. The horizontal axes
show the actual clock time and the time relative to the minimum in core
body temperature. The vertical axis shows the phase shift in hours,
measured as the change in time of occurrence of minimum core body
temperature following exposure to bright light. Exposure to bright light
can advance or delay the phase of the circadian rhythm depending on the
timing of the exposure relative to the timing of the minimum core body
temperature (after Jewett *et al.*, 1997).

the models of the effect of light exposure on the phase and amplitude of the
circadian system discussed above have the potential to be able to predict the
effect of complex patterns of light exposure.

The phase-shifting effect of light exposure occurs many hours after exposure. A more immediate effect of light exposure at night is the suppression
of melatonin synthesis and the consequent increase in alertness measured by
a change in the nature of electroencephalograph patterns, by an increasing
core body temperature and by reported feelings of alertness (Badia *et al.*,
1991; Cajochen *et al.*, 2000). Figure 3.6 shows the effect of exposure to
bright (5,000 lx) and dim (50 lx) light for alternate 90-min periods on core
body temperature, starting at midnight. The overall trend of core body temperature to a minimum around 5 a.m. is obvious as is the modulation of
that trend by the alternate exposure to bright and dim light. Exposure to
bright light tends to increase the core body temperature while dim light
tends to reduce it.

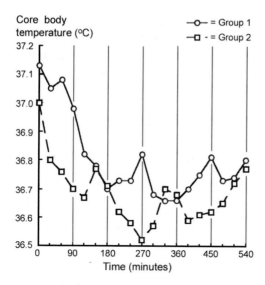

Figure 3.6 Modulation of the core body temperature by exposure to dim and bright light for alternating 90-min periods starting at midnight. Group 1 started with bright light. Group 2 started with dim light (after Badia *et al.*, 1991).

3.6   The amount of light

The importance of the timing of light exposure has been discussed above. Now it is time to consider how much light is required to influence the circadian system. However, before doing this, it is necessary to consider the use of conventional photometric quantities to describe the stimulus to the circadian system. Given that the spectral sensitivity of the photoreceptor used to transmit the external light stimulus to the SCN is not the same as the CIE Standard Photopic Observer (see Section 3.2.1). It is strictly inappropriate to be discussing light at all, light being defined as incident electromagnetic radiation weighted by the combined spectral sensitivity of the long- and medium-wavelength cones in the fovea (cf. Figures 1.2 and 3.2). However, most of the literature on the topic uses illuminance, measured in lm/m^2 (lux), as a metric for the stimulation provided to the circadian system. Therefore, this discussion will use the same terms, bearing in mind that where different light sources have been used the actual stimulus may be different from that implied by the illuminance.

Zeitzer *et al.* (2000) have established a dose–response relationship for light exposure from "cool white" fluorescent lamps, using the phase shifting and suppression of melatonin as markers of the state of the circadian system. They measured the melatonin concentration following exposure to 6.5 h of light at a fixed illuminance centered 3.5 h before the subject's minimum

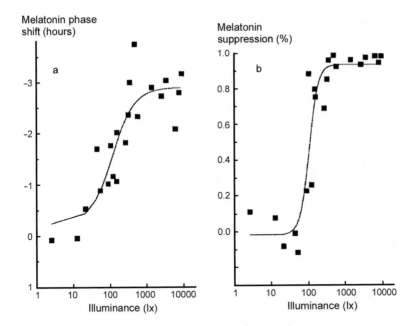

Figure 3.7 The effect of illuminance at the eye on: (a) circadian phase shift of melatonin concentration; and (b) percentage melatonin suppression, for 6.5 h
of light exposure centered 3.5 h before the core body temperature minimum (after Zeitzer *et al.*, 2000).

core body temperature. The illuminances at the eye during the light exposure ranged from 3 to 9,100 lx. Figure 3.7 shows the phase shift in melatonin concentration plotted against illuminance at the eye. The phase shift
is delayed, as would be expected from the timing of the light exposure but,
more interestingly, the phase shift saturates, i.e. reaches 90 percent of the
asymptotic maximum, at an illuminance of 550 lx while the half-saturation
response is produced by an illuminance of about 100 lx.

Figure 3.7 also shows the percentage reduction in melatonin concentration during 6.5-h exposure to different illuminances at the eye. For melatonin suppression, saturation of suppression occurs at about 200 lx and the
half saturation occurs around 100 lx. Similar results have been found by
Brainard *et al.* (1988). Yet others have shown that, given a long exposure
time, illuminances of the same order as those found in conventional lighting installations can delay the onset time for melatonin (Wehr *et al.*, 1995),
have an acute effect on alertness (Cajochen *et al.*, 2000) and phase shift the
sleep–wake cycle (Boivin and James, 2002).

Taken together, these results imply that the dose–response relationship
for the effect of light on the circadian system follows a compressive, non-
linear function. Further, they imply that, given a long enough exposure

time, illuminances that occur in everyday electric lighting installations can be enough to entrain the human circadian system and may be the main source of entrainment for populations in northern climates where daylight is limited for long periods and to many who live in urban areas, with limited exposure to daylight.

While these implications are important for understanding the impact of lighting installations on the circadian system, the illuminances listed should not be treated as definitive because there remain a number of unanswered questions. For example, if exposure to everyday lighting installations is enough to ensure entrainment of the circadian system, why does the duration of melatonin secretion increase from summer to winter for one in three women and one in eight men? Conversely, why does a seasonal change in melatonin secretion duration not occur for two out of three women and seven out of eight men (Wehr, 1997)? Plausible answers lie in the possibility that modern lifestyles ensure that people are shielded from seasonal variations in natural light (Hebert *et al.*, 1998), because they spend most of their time indoors and the amount of light provided by electric lighting installations is enough to determine the duration of melatonin secretion. Another factor to be considered is the difficulty in measuring the relevant light exposure in realistic situations. The quantity that matters for entraining the circadian system is the amount of light that reaches the retina. This is evident from the work of Brainard *et al.* (1997), who have shown that melatonin suppression is greater for dilated pupils than naturally changing pupils and greater for two eyes exposed than for one eye exposed, for the same illuminance on the cornea. Further, Dawson and Campbell (1990) have pointed out that in most lit spaces, the illuminance reaching the retina can vary dramatically depending on the light distribution, the reflectances of the surfaces forming the space, and the direction of gaze. Thus, it is very difficult to know what the actual retinal illuminance is in realistic lighting situations, both indoors and outdoors, even at one moment in time. Given the variability introduced by the ability to look in different directions at different times, the possible errors of measurement become very large indeed. Further, the illuminance at the eye is at best an approximation to the retinal illuminance, ignoring as it does the effects of pupil size and the transmittance of the ocular media. The illuminance on a horizontal plane, which is sometimes used as a measure of light exposure, is even further from the retinal illuminance. Also, it is important to remember that the spectral sensitivity of the photoreceptor that provides a signal to the SCN in humans is different from the CIE Standard Photopic Observer, so illuminance as conventionally measured is not the correct metric to quantify the effective stimulus to the circadian system. Finally, there is the question of individual sensitivity. The maximum concentration of melatonin produced in darkness can vary widely between individuals (Waldhauser and Dietzel, 1985).

Given all this uncertainty, it is not possible to state precisely how much light is necessary to influence the circadian system. However, what is clear is that it is the irradiation of the retina that matters and that whether that irradiance originates from natural light or electric light is immaterial. It is also clear that current lighting practice often provides enough stimulation to the circadian system to modify the timing of the circadian system.

3.7 The consequences of trying to work in circadian night

Humans are diurnal mammals that are active during the day and sleep at night. While the widespread availability of the electric power and efficient light sources at modest cost have eliminated our dependence on natural lighting for vision and may have modified the phase of the circadian system for most people, it has not eliminated the most obvious characteristic of the circadian system, i.e. alternate periods of activity and sleep. People will experience difficulty in performing many sorts of task if they are asked to do it at a time when their circadian system is telling them to sleep, i.e. in its night phase. Figure 3.8 shows the percentage of errors made reading gas meters at different times. The increase in errors at night and around 2 p.m., called the post-lunch dip, are obvious (Minors and Waterhouse, 1981). Similar deteriorations in task performance at night have been shown for the frequency of falling asleep while driving; for the speed of joining threads in textile production, for train drivers missing warning signals, and for the frequency of minor accidents in hospital (Folkard and Monk, 1979). The variation in performance over the 24 h is clear as is the fact that performance during the night is usually worse than during the day.

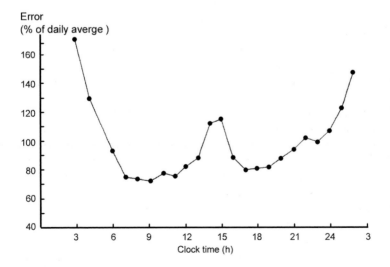

Figure 3.8 Variation in errors made reading gas meters at different times over 24 h (after Minors and Waterhouse, 1981).

At first, the finding that performance of many industrial tasks is worse during circadian night than during the day might be thought inconsistent with the fact that illuminances at a level typically used in interiors at night can entrain the circadian system. There are two reasons why this is not enough to guarantee good night-shift performance. The first is that the extent and speed of phase shifting is dependent on the pattern of how much light exposure occurs and when it occurs, over the whole 24-h period. Exposure to daylight, as might occur on the journey to and from work and which gives a much higher retinal illumination than most interior lighting, will usually determine the phase of the circadian system and may stop the phase shift necessary to reset the circadian rhythm to what is needed for night shift work. The second is that even with carefully controlled exposure to daylight, a 180° phase shift takes a number of days. A rapidly rotating shift system that involves only two or three successive nights of work does not allow enough time for adaptation to occur. Nonetheless, given enough continuous nights on night shift, adaptation does occur. Figure 3.9 shows the phase difference between the maximum core body temperature and the ideal time for the maximum to occur for 21 continuous nights of work in a laboratory (Monk *et al.*, 1978). It is clear that adaptation begins from the first night but it is only after about 15 nights that adaptation approaches completion. Other results of this type (Akerstedt, 1985) support this time scale of adaptation, but core body temperature, as a marker of circadian phase has its limits. Specifically, core body temperature is affected by both

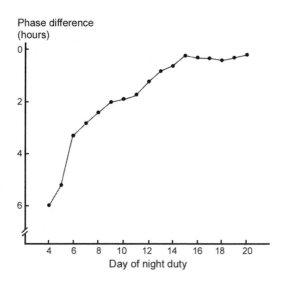

Figure 3.9 The adjustment of circadian phase over 21 night shifts. The vertical axis is the difference between the actual phase of the circadian rhythm and the ideal phase for the shift being worked (after Monk *et al.*, 1978).

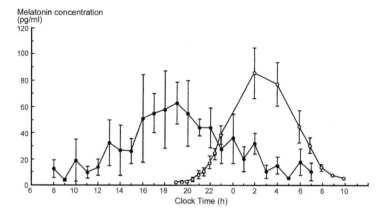

Figure 3.10 Average melatonin concentration plotted against clock time for a group of day-active people (○) and a group of night-active people (●), after five successive night shifts (after Sack *et al.*, 1992a).

sleep and level of activity. Sleep always produces a drop in core body temperature while activity tends to produce an increase. An alternative marker, melatonin concentration, is less susceptible to the effects of sleep. Sack *et al.* (1992a) measured the melatonin concentrations of a group of subjects over the 24 h immediately following the end of 3–5 days of continuous night shifts. Figure 3.10 shows the average melatonin concentration profile of the group of night-shift workers relative to that of a control group of day-workers. Again, there is evidence of partial adaptation occurring, the maximum melatonin concentration occurring about 7 p.m. for the night-workers rather than the 2 a.m. for the day-workers.

This pattern of slow adaptation suggests the possibility of using light exposure to systematically and more rapidly phase shift the circadian system to the required condition. Czeisler *et al.* (1990) have demonstrated that a pattern of exposure to light can be developed that will produce a marked phase shift to an adapted state in four, 24-h days, even when the subjects were exposed to daylight on the journey home. This was achieved by exposure to an illuminance in the range 7,000–12,000 lx during the night shift from 12:15 a.m. to 7:45 a.m. Czeisler *et al.* (1990) also showed that as a result of this adaptation, the subjects had a greater feeling of alertness and achieved better performance on mental arithmetic during the night shift than a control group who showed no adaptation.

Eastman *et al.* (1994) performed a similar study on night-shift work, but in this case, the light exposure and the wearing of dark welder's goggles were combined in all possible ways. This meant that some subjects were exposed to 5,000 lx illumination at night but were free to travel home without wearing the goggles. Others were exposed to less than 500 lx but wore the dark welder's goggles during the journey home. Yet others were exposed to 5,000 lx at night and wore the welder's goggles on the journey home, while yet others were exposed to 500 lx but did not wear the goggles travelling home. All the subjects slept in darkened bedrooms receiving less than 500 lx in daytime. Receiving 5,000 lx at night and wearing the goggles during the day gave the greatest phase shift, either factor alone gave some shift while neither rarely produced any phase shift. This result emphasizes two points. The first is that illuminances typical of conventional interior lighting can produce phase shifts, (i.e. 500 lx during the night), provided exposure to daylight is limited. The second is that to guarantee a phase shift it is necessary to control light exposure throughout the 24-h period.

Both the above studies were conducted in laboratories to determine the practicality of using controlled light exposure to accelerate adaptation to a sudden shift in work time. One application where controlled light exposure has been of practical value is in spaceflight. A NASA space shuttle crew is required to start preparations for launch about 2 a.m. After launch, the crew is split into two teams, both of which work 12-h shifts. Crews on early shuttle flights complained of fragmented and disturbed sleep. In 1990, the conference room in the crew quarantine quarters was fitted with a luminous ceiling capable of producing about 10,000 lx. The crew entered the quarantine quarters 1 week before launch and went through a light exposure pattern of bright light and darkness designed to adapt them to their anticipated work schedule. The returning crew reported much better sleep patterns in flight. Melatonin samples indicated that adaptation had occurred (Czeisler *et al.*, 1991). Reports of the application of light exposure to more mundane industrial activities are rare but one that is available suggests that providing bright light exposure to speed up adaptation has only limited value (Bjorvatn *et al.*, 1999). The fact is that although it has been shown to be possible to rapidly adapt to and from night shift using patterns of light exposure based on the phase response curve, and light–work–sleep schedules for rapid adaptation to night-shift work have been published (Eastman, 1990), controlled light exposure as a means to adapt to night shift work has rarely been used in practice. Possibly reasons for this lack of impact on lighting practice are the difficulty of ensuring compliance with the light exposure pattern throughout the 24 h in a conventional industrial context, the design problem of providing the high illuminance at the eye needed to phase shift rapidly without causing visual discomfort, and the lack of demand for something better from night-shift workers, probably due to their low expectations.

Another situation in which there is a need to rapidly shift circadian rhythm is after rapid travel across several time zones. The outcomes of such travels are similar to those of night-shift workers and include difficulty in the sleeping at a time consistent with the destination, gastrointestinal illness, and decrements in alertness and performance. Collectively, these symptoms are known as jet-lag and occur because of the misalignment between the endogenous circadian clock and the exogenous light–dark cycle. Jet-lag is strictly only associated with travel in an east–west direction. Travel in a north–south direction may cause some similar symptoms due to fatigue and sleep loss but it does not involve any phase misalignment between the endogenous and exogenous components of the circadian system. The time for circadian rhythms to re-synchronize depends on the number of time zones crossed and the direction of travel. For flights across 5–11 time zones, re-entrainment occurs at a mean rate of about 90 min/day for westward travel and about 55 min/day for eastward travel (Klein and Wegmann, 1974; Aschoff *et al.*, 1975). This difference between eastward and westward flight occurs regardless of whether the flights are outward or return, and whether they are day or night flights. The explanation of this difference is that re-entrainment after westward flight occurs by gradual phase delays, i.e. by temporarily having periods longer than 24 h, while re-entrainment after eastward flight occurs by gradual phase advances, i.e. by temporarily taking periods shorter than 24 h. As the free-running endogenous period is greater than 24 h, phase advances represent a bigger change than phase delays, so entrainment will be faster for westward than eastward flights. Of course, these rates of re-entrainment are averages and cover wide individual differences. Re-entrainment rate is proportional to the difference in phase between the endogenous clock and the light–dark cycle, so re-entrainment is greatest immediately after the flight and progressively decreases. Further, different circadian rhythms, re-entrain at different rates, and the old take longer to re-entrain than the young (Boulos *et al.*, 1995).

Given the proven ability of light exposure to shift the phase of the circadian rhythms, it is to be expected that exposure to light at the right time should be able to speed up the process of re-entrainment. Measurements of re-entrainment rates made before the role of light was understood offer some support for this view. Klein and Wegmann (1974) found that re-entrainment after travel across six time zones happened 50 percent faster for subjects allowed outdoor activities every other day than for subjects confined to their hotel rooms. Rapid adjustment has also been shown by military units following eastward airlifts, presumably because of their outdoor activity upon arrival (Graeber *et al.*, 1981). More recent field studies have been less successful, although there is some evidence that exposure to bright light at the destination tends to lead to a consolidation of sleep in one sustained period (Czeisler and Allen, 1987; Cole and Kripke, 1989; Sasaki *et al.*, 1989).

The most plausible reason for the lack of success of these field studies is the lack of control over the subject's light exposure over the whole 24-h period. There can be little doubt exposure to light at the right times can accelerate re-entrainment but whether this is a practical proposition for most travelers seems unlikely. After all, most travelers do not make such a flight simply to test their circadian system. Holidaymakers do not wish to limit their activities outside to times that would be most effective in accelerating re-entrainment, and businessmen often do not have the leisure to do so. Probably the best that can be expected is to modulate light exposure when outdoors by dark sunglasses. Software has been developed for scheduling exposure to light in order to ensure rapid re-entrainment of the circadian system following transmeridianal travel (Houpt et al., 1996). Of course, such an effort is only worthwhile if the traveler is likely to stay at their destination for a number of days. For short duration stays of 1–2 days, it is probably better to try to maintain entrainment on the return destination.

Even if it were possible to accelerate re-entrainment at the beginning and end of a periods of night-shift work or after a long flight across time zones, that would still leave a number of days when work had to be done during the circadian night. It is interesting to consider what the effect of this situation would be on task performance. The first thing to say is that the effect of trying to work in the circadian night can affect all types of task, not just visual tasks. This is because the circadian system affects the "platform" from which we operate and consequently affects all parts of the brain and body. Tilley et al. (1982) studied the sleep patterns and performance of shift workers operating a weekly, alternating, three-shift system. Workers doing night shift, who had to sleep during the day, had shorter duration sleep of degraded quality. As for performance, both simple reaction time and four-choice reaction time were both longer during night shift relative to day and afternoon shift and tended to show a deterioration over the number of days on shift, probably because of the accumulation of sleep debt caused by the inferior sleep duration and sleep quality during the day. Cajochen et al. (1999) studied the sleep and performance patterns of people kept awake for 32 h, i.e. a period covering a conventional day, starting at 8 a.m. and extending to 4 p.m. the next day. Figure 3.11 shows the patterns in core body temperature and plasma melatonin, both well-established circadian rhythms; eye blink rate, slow eye movements and stage 1 EEG patterns, and ratings on the Karolinska sleepiness scale, all of which are related to sleep; and performance on a reaction time task, a mental arithmetic task, and a short-term memory task. It is important to note that during the 32 h of wakefulness, the subjects were exposed to a constant illuminance of 15 lx, i.e. there was no bright light treatment in this study. Figure 3.11 shows the expected pattern of decreased core body temperature and increased plasma melatonin at night. The sleep measures all show an increased propensity to sleep during the night with some recovery the next day. The performance measures all show a decrement in performance over the night with some recovery the following day although not enough to recover to the level of performance achieved at the beginning of the trial. Of particular interest are the results on the reaction time task. What is evident in these data is the increase in range in reaction times at night. The 10 percent fastest reaction times at night are similar to what they are during the day but the 10 percent slowest are 20 times slower. This spread in reaction times is consistent with one of the most commonly observed effects of continuous work

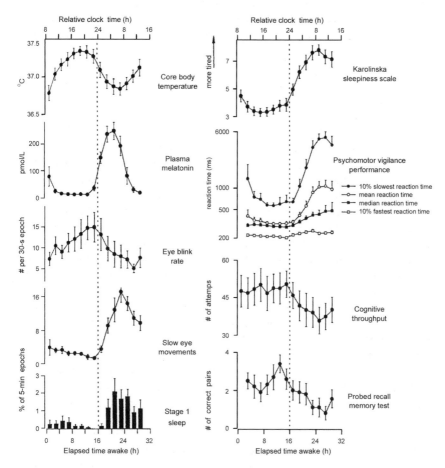

Figure 3.11 The time courses of core body temperature and melatonin concentration over 32 h. Time courses of measures of sleepiness, i.e. mean eye blinks per 30 s, slow eye movements, stage 1 sleep and sleepiness ratings, over 32 h. Time courses of task performance for highly visible tasks requiring vigilance, mental arithmetic (cognitive throughput), and short-term memory, over 32 h. The dotted vertical line represents the subjects habitual bedtime. The error bars are standard errors of the means (after Cajochen *et al.*, 1999).

without sleep, the presence of periods of no response or lapses (Wilkinson, 1969). These periods are correlated with periods of lower arousal and even microsleeps, measured by EEG signals (Cajochen *et al.*, 1999). There are a number of task characteristics which determine the likelihood of lapses occurring. Tasks which are of long duration (i.e. more than 30 min), monotonous, and externally paced seem to be more likely to show lapses during sleep deprivation. Conversely, tasks which are considered of short duration, interesting or rewarding and which are self-paced are less likely to show lapses, although the self-paced task may be done more slowly to maintain the same level of accuracy (Froberg, 1985). It is important to realize that the change in the number of lapses is relative. All tasks show some decline with increasing sleep deprivation but the decline is less for the short duration, interesting, rewarding, self-paced tasks. One aspect of task structure of particular interest is the extent to which short term memory is required. Tasks requiring the use of short-term memory seem particularly sensitive to sleep deprivation, an observation consistent with the finding of Cajochen *et al.* (1999) that the frontal areas of the brain, which are associated with short-term memory, are more susceptible to sleep loss than occipital areas.

Even though the effects on performance of working during the circadian night are likely to be limited when some sleep is possible during the day, there are clear advantages in accelerating re-entrainment. In principle, this can be done by correctly timing the exposure to light over the whole 24 h. However, this requires a degree of compliance that many people are unwilling to give and even if they do, it still takes a period of several days before re-entrainment approaches completion. During this time, performance of sensitive tasks may be reduced. But there may be another possibility. The results shown in Figure 3.11 suggest that the onset of sleepiness and the deterioration of the performance measures are closely related to the increase in melatonin concentration. It is well established that exposure to light can rapidly suppress melatonin. The interesting possibility is that exposure to bright light at night could lead to an improvement in performance of some tasks. There is some evidence to support this possibility. French *et al.* (1990) measured the performance of people on a battery of cognitive tests while working between 6 p.m. and 6 a.m. and exposed to either 3,000 or 100 lx illumination. They showed that oral temperatures were elevated by exposure to bright light between 9:30 p.m. and 5:30 a.m. and that performance on 6 of the 10 cognitive tasks were improved by exposure to bright light, the biggest effect being on serial subtraction and addition. As core body temperature and melatonin secretion are correlated (see Figure 3.11), it can be hypothesized that these changes in performance are due to the suppression of melatonin by the bright light. A similar pattern of results were obtained by Badia *et al.* (1991), with subjects working at night being exposed to alternating 90-min periods of bright (5,000–10,000 lx) and dim (50 lx) light. Of the six cognitive tasks measured, three showed significantly increased levels of performance during the bright light periods, although all tasks showed a deterioration over the night. This fluctuation in performance with light exposure was paralleled by a fluctuation in core body temperature. Boyce *et al.* (1997) showed similar pattern for people working three successive nights under the same lighting installation. Out of seven tasks performed, two showed statistically significant improvements when the work was carried out under 2,800 lx on the work surface from midnight to 8 a.m. or from midnight to 2:30 a.m. with a steadily declining illuminance until 8 a.m. Work over the same time periods at a fixed 200 lx illuminance or with an increasing illuminance that only reached 2,800 lx at 5:30 a.m. produced lower levels of performance.

Unfortunately, none of these studies measured melatonin concentrations over the working period. However, they all measured core body temperature and all three show the expected drop in core body temperature to a minimum around 4 a.m. followed by an increase at later hours, under the low illuminance condition. Further, the negative correlation between core body temperature and melatonin concentration is well established so the link between the suppression of melatonin by bright light at night and task performance is at least plausible if not demonstrated. The other common feature of these results is that some tasks are more sensitive to the effect of sleepiness than others. This may explain why others have failed to show any effect of bright light exposure on performance of a limited range of tests, despite evidence that the phase of the core body temperature had moved further for the subjects exposed to bright light (Campbell, 1995).

To summarize this discussion, the effects of light exposure at night can be divided into those on the immediate physiology and on the more remote performance. That bright light can be used to shift the phase of circadian rhythms and to suppress melatonin rapidly is undoubtedly correct. Such exposure can be used to correct the sudden misalignment between the circadian clock and the light–dark cycle caused either by starting or finishing night-shift work or by rapid travel across time zones. It can also be used to increase alertness at night without necessarily shifting the phase of the circadian clock simply by suppressing melatonin. What the consequences are for task performance during the circadian night is much less clear, some tasks being more sensitive than others to being done at night. Whether suppression of melatonin and/or the shift in circadian phase produced by bright light exposure are sufficient to overcome the negative effects of prolonged work at night depends on the structure of the task and the context in which it is being performed. This implies that there are no easy answers and no guarantees when it comes to predicting the effect of exposure to light at night and on task performance.

3.8 Caveats

Our understanding of the human circadian system is developing rapidly but there remains much still be determined. Until it is fully understood it would be as well not to rush into applications based on manipulating the circadian system with light. There are three arguments for caution. The first is the possibility of long-term side-effects produced by manipulation of the circadian system. The circadian system operates at a very basic level of human physiology and interacts with many other components of that physiology. The nature of these interactions needs to be understood. One example of the concerns that bright light exposure at night can raise is the hypothesis that suppressing melatonin will be associated with an increased likelihood of breast cancer due to an increased level of estrogen (Stevens *et al.*, 1997).

The second argument for being careful about manipulating the circadian system is that the most effective method for doing so is unknown. The amount of light required and the way in which light exposure over 24 h is integrated are still uncertain. Further, there is evidence that while the pattern of light exposure is a potent influence on the human circadian system, it is not the only factor that can influence the entrainment of the circadian system. Until the relative impact of different stimuli and how they interact is known, it will be difficult to determine the optimum pattern of exposure to the various entraining agents.

The third argument is that while it is clearly possible to alter the phase and amplitude of circadian rhythms and to increase alertness at night by exposure to light at the right time, the consequences of these changes for task performance are much less certain. People have an amazing capacity to summon up resources to perform tasks in adverse circumstances, for a limited time, given sufficient motivation. Until the links between the changes in physiology following exposure to light and the effects on task performance in realistic conditions are better appreciated, developing lighting systems to take advantage of the effects on the circadian system may lead to disappointment.

3.9   Summary

Circadian rhythms are a basic part of life and can be found in virtually all plants and animals, including humans. The role of circadian systems is to establish an internal replication of external night and day. This internal representation is not just a passive response to external conditions but rather is predictive of external conditions to come. The human circadian system involves three components: an internal (endogenous) oscillator, located in the suprachiasmatic nuclei; a number of external (exogenous) oscillators that can reset (entrain) the internal oscillator, and a messenger hormone, melatonin, that carries the internal "time" information to all parts of the body through the bloodstream. In the absence of light, and other cues, the internal oscillator continues to operate but with a period longer than 24 h. External stimuli are necessary to entrain the internal oscillator to a 24-h period and to adjust for the seasons.

The light–dark cycle is one of the most potent of these external stimuli used for entrainment. Also, by varying the amount of light exposure and when it is presented, it is possible to shift the phase of the circadian system clock, either forward or backward, as required. In addition, it is also possible to have an immediate alerting effect at night by exposure to light. The amount of light required to cause phase shifts or an immediate alerting effect is within the range of current lighting practice. While this is all true, it is unfortunately also true that the spectral sensitivity of the retinal photoreceptor that provides a signal to the internal oscillator is not the same as the CIE Standard Photopic Observer, used in the most widely used definition of light. This implies that a new "Standard Circadian Observer" is needed before the effects of different spectral power distributions can be evaluated.

The potential value of being able to manipulate the human circadian system is that it would be possible to more rapidly adjust to the need to work at times when one would normally be sleeping, e.g. when starting or finishing night-shift work or after long transmeridianal flights. Evidence to date suggests that while such beneficial effects can be demonstrated in the laboratory they are difficult to achieve in real-world contexts, probably because of the need to control light exposure throughout the 24 h.

Finally, it is important to appreciate that our understanding of the human circadian system is developing rapidly. This is just as well because the human circadian system operates at a very basic level of physiology and influences many other parts of that physiology. Until the possible interactions are understood and any side-effects are identified, it would be as well to be cautions about attempting to use light to manipulate such a fundamental part of our physiology.

Part II
Generalities

4 Lighting and work

4.1 Why lighting matters

Millions of people spend a significant part of their lives working. Electric lighting is provided at their workplaces to ensure that they can see to do their work quickly, accurately, and easily. Thus, the provision of lighting has an economic impact. Further, in many countries, the cost of providing the lighting at a work place is miniscule relative to other costs. For example, in the USA, the annual cost of lighting a $10\,m^2$ office is typically less than 1 percent of the cost of paying someone to occupy that office. This means that only a small change in task performance is required to economically justify a large change in lighting practice. This chapter describes what is known about the relationship between lighting and work.

4.2 An overview

To understand the relationship between lighting and work, it is first necessary to identify the routes by which lighting can affect human performance. There are three such routes: through the visual system, through the circadian system, and through the perceptual system. Figure 4.1 shows a conceptual framework for considering the factors that influence progress down each route and the interactions between them.

The effect of lighting on vision is the most obvious impact of light on humans. With light we can see, without light we cannot. The visual system is an image-processing system. The optics of the eye form an image of the outside world on the retina. At the retina, some image processing occurs. Different aspects of the retinal image are processed through two different channels up to the visual cortex of the brain. The magnocellular channel processes information rapidly but with little detail or color information, while the parvocellular channel provides detail of brightness, color and texture but at a slower rate. In addition, the visual system is organized spatially into two parts, the fovea of the retina, where fine detail is available, and the periphery, which is basically a detection system indicating where in the visual field the fovea should be directed. When there is a lot of light available,

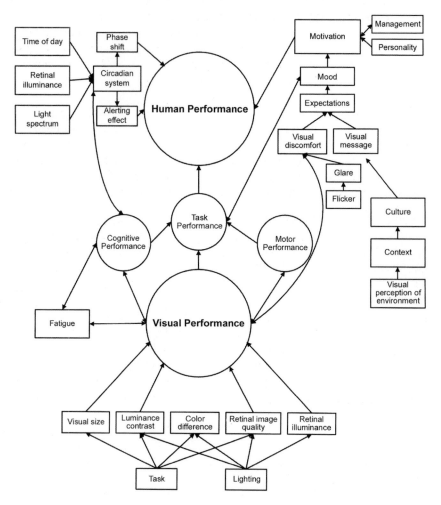

Figure 4.1 A conceptual framework setting out the three routes whereby lighting conditions can influence human performance. The arrows in the diagram indicate the direction of effect.

e.g. in daytime, the whole of the retina is active. When there is very little light, e.g. outside on a moonless night, the fovea is blind and only the peripheral retina operates. A more detailed discussion of the visual system is given in Chapter 2.

Any stimulus to the visual system can be described by five parameters, its visual size, luminance contrast, color difference, retinal image quality, and retinal illumination. These parameters are important in determining the extent to which the visual system can detect and identify the stimulus.

Visual size. There are several different ways to express the size of a stimulus presented to the visual system but all of them are angular measures. The visual size of a stimulus for detection is usually given by the solid angle the stimulus subtends at the eye. The solid angle is given by the quotient of the areal extent of the object and the square of the distance from which it is viewed. The larger is the solid angle, the easier the stimulus is to detect.

The visual size for resolution is usually given as the angle the critical dimension of the stimulus subtends at the eye. What the critical dimension is depends on the stimulus. For two points, the critical dimension is the distance between them. For two parallel lines it is the separation between the two lines. For a Landolt ring, it is the side of the square forming the gap in the ring. The larger is the visual size of detail in a stimulus, the easier it is to resolve that detail.

For complex stimuli, the measure used to express the dimensions is the spatial frequency distribution. Spatial frequency is the reciprocal of the angular subtense of a critical detail, in cycles per degree. Complex stimuli have many spatial frequencies and hence a spatial frequency distribution. The match between the luminance contrast at each spatial frequency of the stimulus and the contrast sensitivity function of the visual system determines if the stimulus will be seen and what detail will be resolved (see Section 2.4.3). Lighting can do little to change the visual size of two-dimensional objects but shadows can be used to enhance the effective visual size of some three-dimensional objects (see Section 8.5).

Luminance contrast. The luminance contrast of a stimulus expresses its luminance relative to its immediate background. The higher is the luminance contrast, the easier it is to detect the stimulus. There are several different forms of luminance contrast (see Section 2.4.1) so it is always necessary to know which definition is being used. Lighting can change the luminance contrast of a stimulus by producing disability glare in the eye or veiling reflections from the stimulus.

Color difference. Luminance quantifies the amount of light emitted from a stimulus and ignores the combination of wavelengths making up that light. It is the wavelengths emitted from the stimulus that influence its color. It is possible to have a stimulus with zero luminance contrast which can still be detected because it differs from its background in color. There is no widely accepted measure of color difference although various suggestions have been made (Tansley and Boynton, 1978) and a number of metrics could be constructed from the location of the object and the immediate background on the color planes of either of the CIE color spaces (see Chapter 1). Color difference only becomes important for detection of an object when luminance contrast is low, although color can be an important factor for enhancing visual search (see Section 8.5). Lighting can alter the color difference between the object and its background when light sources with different spectral contents are used.

Retinal image quality. As with all image processing systems, the visual system works best when it is presented with a sharp image. The sharpness of the stimulus can be quantified by the spatial frequency distribution of the stimulus, a sharp image will have high spatial frequency components present; a blurred image will not.

The sharpness of the retinal image is determined by the stimulus itself, the extent to which the medium through which light from the stimulus is transmitted scatters light, and the ability of the visual system to focus the image on the retina. Lighting can do little to alter any of these factors, although it has been shown that light sources which are rich in the short wavelengths produce smaller pupil sizes for the same luminance than light sources that are deficient in the short wavelengths (Berman *et al.*, 1992). A smaller pupil size produces a better quality retinal image because it implies a greater depth of field and less spherical and chromatic aberrations.

Retinal illuminance. The illuminance on the retina determines the state of adaptation of the visual system and therefore alters the capabilities of the visual system (see Chapter 2). The retinal illuminance produced by a surface luminance is determined by the equation

$$E_r = e_t \tau \ (\cos \theta / k^2)$$

where E_r is the retinal illuminance (lx), τ the ocular transmittance, θ the angular displacement of the surface from the line of sight (degrees), k a constant equal to 15, and e_t = amount of light entering the eye (trolands).

$$e_t = L \cdot \rho$$

where L is the surface luminance (cd/m^2) and ρ the pupil area (mm^2).

The amount of light entering the eye, e_t, measured in trolands, is often referred to as retinal illumination but it does not take the transmittance of the optic media into account and therefore does not truly represent the luminous flux density on the retina. The amount of light entering the eye is mainly determined by the luminances in the field of view. For interiors, these luminances are determined by the reflectances of the surfaces in the field of view and the illuminances on them. For exteriors, the relevant luminances are those of reflecting surfaces, such as the ground, and of self-luminous sources, such as the sky. For both interiors and exteriors, the pupil area is mainly determined by the scotopic retinal illuminance (Berman *et al.*, 1992).

What these five parameters imply is that it is the interaction between the object to be seen, the background against which it is seen and the lighting of both object and background that determine the stimulus the object presents to the visual system and the operating state of the visual system. It is the stimulus and the operating state of the visual system that determine the level of visual performance achieved. This visual performance then contributes to task performance. It is important to point out that visual performance and task performance are not necessarily the same. Task performance is the performance of the complete task. Visual performance is the performance of the visual component of the task. Task performance is what is needed in order to measure productivity and to establish cost/benefit ratios comparing the costs of providing a lighting installation against the benefits of improved task performance. Visual performance is the only thing that changing the lighting conditions can affect directly.

Most apparently visual tasks have three components; visual, cognitive, and motor. The visual component refers to the process of extracting information relevant to the performance of the task using the sense of sight. The cognitive component is the process by which sensory stimuli are interpreted and the appropriate action determined. The motor component is the process by which the stimuli are manipulated to extract information and/or the actions decided upon are carried out. Of course, these three components interact to produce a complex pattern between stimulus and response. Further, every task is unique in its balance between visual, cognitive and motor components and hence in the effect lighting conditions have on task performance. It is this uniqueness which makes it impossible to generalize from the effect of lighting on the performance of one task to the effect of lighting on the performance of another. The effect of lighting on the performance of a specific task depends on the structure of the task and specifically the place of the visual component relative to the cognitive and motor components. Tasks in which the visual component is large will be more sensitive to changes in lighting conditions than tasks where the visual component is small.

Another route whereby lighting conditions can affect work is through the circadian system (see Chapter 3). The most obvious evidence for the existence of a circadian system in humans is the occurrence of the sleep–wake cycle, but this is only the tip of the iceberg. Beneath the surface lie the variations in many different hormonal rhythms over a 24-h period. The organ that controls these cycles in humans is the suprachiasmatic nuclei (SCN). The SCN is linked directly to the retina. When signals are transmitted from the retina to the SCN, no attempt is made to preserve their original location. Rather, the parts of the retina supplying the SCN act like a simple photocell. This means that the aspects of lighting that influence the state of the SCN are the light spectrum and illumination reaching the retina, which in turn depend on the light spectrum of the light source used, the light distribution, the spectral reflectances of the surfaces in the space, the spectral transmittance of the optic media and where the observer is looking.

Our knowledge of how lighting conditions might affect human performance through the circadian system has grown rapidly in recent years. There are two distinct effects; a shifting effect in which the phase of the circadian rhythm can be advanced or delayed by exposure to bright light at specific times (Dijk *et al.*, 1995); and an acute effect related to the suppression of the hormone melatonin at night (Campbell *et al.*, 1995). Both these effects can be expected to enhance human performance in the right circumstances. Attempts have been made to use the phase-shift to more quickly adapt people to night-shift work. These attempts have met with mixed success, the problem being that to get the required shift it is necessary to control exposure to light over 24 h and not just during working hours (Eastman *et al.*, 1995). As for the acute effect, there is clear evidence that exposure to bright light increases alertness at night (Badia *et al.*, 1991) and that this can enhance the performance of complex cognitive tasks (Boyce *et al.*, 1997).

The third route whereby lighting conditions can affect work is through the perceptual system. The perceptual system takes over once the retinal image has been processed by the visual system. The simplest output of the perceptual system is a sense of visual discomfort, which may change the observer's mood and motivation, particularly if the work is prolonged. Lighting conditions in which achieving a high level of visual performance is difficult will be considered uncomfortable as will conditions in which the lighting leads to distraction from the task, as can occur when glare and flicker are present. But perception is much more sophisticated than just producing a feeling of visual discomfort. In a sense, every lighting installation sends a "message" about the people who designed it, who bought it, who work under it, who maintain it, and about the place it is located. Observers interpret the "message" according to the context in which it occurs and their own culture and expectations. The importance of this "message" is sometimes enough to override conditions that might be expected to cause discomfort, as shown by the fact that lighting conditions that would be considered extremely uncomfortable in an office are positively desired in a dance club. According to what the "message" is, the observer's mood and motivation can be changed. Every lighting designer appreciates the importance of "message" but it is only in the context of retailing and entertainment that the "message" a lighting installation sends is given the importance its potential to influence behavior deserves.

Unfortunately, the effect of lighting on human performance mediated through the perceptual system, has rarely been studied. One attempt to demonstrate the effects of lighting quality on the performance of office tasks produced a confusing array of small effects (Veitch and Newsham, 1998a) which may or may not have been associated with differences in visibility. In another study where task visibility was deliberately controlled, different light distributions had no effect on sustained task performance, despite the light distributions being considered very different in quality by lighting experts (Eklund *et al.*, 2000). There can be no doubt that lighting can be used to draw attention to objects (LaGuisa and Perney, 1974) and to modify an observer's mood (Baron *et al.*, 1992; McCloughan *et al.*, 1999) but the extent to which such changes might affect task performance measured as output is subject to so many intervening variables that it has been difficult to demonstrate a robust effect over a wide range of contexts. However, if the definition of human performance were to be widened to include desired behavior then there is considerable opportunity to demonstrate the value of lighting. This opinion is supported by a recent study that established a correlation between the presence of skylights and the value of sales in a supermarket, the presence of skylights leads to higher sales (Heschong *et al.*, 2002). It is unclear whether this effect is due to the presence of daylight, *per se*, or because the illuminances are much higher than in the electrically lit part of the supermarket so the visibility of the merchandise is better or whether the "message" delivered by the higher ceiling and the variation in daylight is important. However, it does suggest that studying the effects of lighting with a wider definition of human performance might be fruitful to the understanding of the economic impact of lighting.

While each of these routes has been discussed separately, it is important to appreciate that they can also interact. For example, someone who is asked to work while sleep-deprived will be fatigued. This will affect task performance through both its cognitive and visual components. Conversely, someone who is performing a task that is visually difficult for a long time will experience fatigue, even if they are not sleep deprived. Another example would be a situation where the lighting provides poor task visibility, so that visual performance is poor. If the worker is aware of the poor level of performance and it fails to meet his or her expectations, then the worker's mood may be altered. There are multiple interactions of this type that can occur. To further complicate the picture, it is necessary to appreciate that while visual performance for a given task is determined by lighting conditions alone, a worker's motivation can be influenced by many physical and social factors, lighting conditions being just one of them (CIBSE, 1999a). As for the circadian system, this can be influenced by such factors as the timing of exercise and social cues as well as the amount and timing of light exposure. It is this complex pattern of interacting effects that has made the study of the relationship between lighting and work so prolonged and difficult.

4.3   Methods of studying light and work

The studies that have been undertaken of the relationship between light and work can be conveniently classified into two broad groups; real-task studies and abstract studies. The differences between these two groups are essentially those of face validity and generality. The real-task studies involve taking a specific task that someone actually does and measuring the performance of the task under different lighting conditions, usually conditions that can be easily changed, such as the illuminance on the task, the spectral power distribution of the light source and the light distribution of the luminaires. These measurements are made either using a real task in the field or a simulated version of the task in a laboratory. Such studies have high face validity for the specific task, particularly when done in the field, but little generality. They lack generality because the results obtained strictly apply only to the task and the lighting installation used. Because other combinations of task and lighting system differ in the stimuli they present to the visual system and have different combinations of visual, cognitive, and motor components, it is not possible to generalize results obtained on one task to any other. It could be argued that by undertaking enough real-task studies, an overall pattern could be detected that would allow generalization to occur but this seems a vain hope given the number of factors involved. Nonetheless, real-task studies have served to demonstrate that changing lighting conditions can alter task performance and, for the task studied, this gives a quantitative basis for deciding on appropriate lighting recommendations.

The abstract studies are characterized by the use of a task which is visually very simple and which no one has ever done for a living. The underlying aim of such studies is to achieve an understanding of how the performance of a visually simple task, with minimal cognitive and motor components, is influenced by lighting conditions. The outcome of such studies is usually a mathematical model that allows performance of the task to be predicted for lighting conditions not studied. By examining the effect of each of the meaningful stimuli a task presents to the visual system, it should be possible to predict the effects of lighting conditions on any task. Thus, the abstract studies follow the classical route to understanding through analysis followed by synthesis.

While these two approaches are clearly very different, they are interdependent. Without the results from the abstract studies, the results of the real-task studies are difficult to understand, a fact that makes generalization from them a matter of chance. On the other hand, if the abstract studies are to produce anything of value, they have to predict the results of real-task studies.

4.3.1 Field studies of light and work

The first thing to say about field studies of light and work is *caveat emptor*. Many claims have been made about the effect of lighting on worker productivity that are little more than assertions, without any of the details necessary to evaluate the claims made. However, there are a few field studies that deserve attention. Among the earliest were studies of such difficult visual tasks as silk weaving (Elton, 1920), linen weaving (Weston, 1922), and typesetting by hand (Weston and Taylor, 1926), the last being a task which has now almost disappeared. Figure 4.2 shows the mean output of 20 linen weavers for the months of October, November, and December, and the illuminance delivered by daylight inside the weaving shed, plotted against time of day. It can be seen that there is a decline in output with decreasing illuminance and that decline is greater for lower illuminances. This study confirms what is common experience; that lighting conditions can become inadequate for a task to be seen clearly, so performance declines. The illuminance at which this happens will depend on the nature of the task.

At about the same time as these early studies were being done, another series of studies were starting which have entered into folklore – the Hawthorne experiments (Snow, 1927; Roethlisberger and Dickson, 1939).

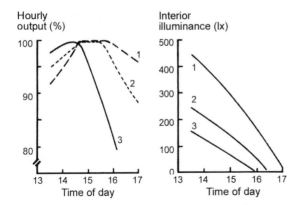

Figure 4.2 Percentage of hourly output of linen weavers plotted against the illuminance provided by daylight inside the weaving shed at different times in the afternoon during: (1) October, (2) November, and (3) December. Output is normalized to the best hourly output in each period (after Weston, 1922).

Initially, these studies were concerned with the effects of lighting on productivity, although, for reasons that will become obvious, they later became more focused on the effects of payment systems, type of supervision, rest times, and total hours of work. The Western Electric Company, based in Hawthorne, Chicago, manufactured telephone apparatus. At the start of the studies, the company conducted three experiments on the effect of lighting on the output of a group of women who inspected parts, assembled relays, or wound coils. In the first experiment, the illuminances on the tasks were varied in a series of steps, both up and down. Output in all three departments changed but showed no clear relationship to the illuminance. The second experiment used only the coil-winding department. The workers were split into two groups of the same level of experience. The control group were exposed to a relatively stable illuminance of 170–300 lx, while the test group were exposed to illuminances in the range 260–750 lx. As the illuminance of the test group was changed, the work output of the both groups increased to a similar extent. The first and second experiments involved both electric lighting and daylight, the presence of daylight explaining some of the variability in illuminance. In the third experiment, daylight was eliminated. The same control group and test group that were used in the second experiment were used in the third experiment but this time the control group worked under a constant illuminance of 110 lx, while the test group experienced illuminances starting at 110 lx and decreasing in steps of 11 lx. After the initial illuminance was set, both groups showed a slow but steady improvement in output. However, when the illuminance experienced by the test group reached 33 lx, the members of the test group protested that they could hardly see what they were doing and their output dropped.

From these studies, the experimenters concluded that lighting was only one, and apparently a minor factor among many that affect worker output, and that there was a whole area of what they called human relations waiting to be explored. Sometimes this conclusion is distorted into a claim that these results imply that lighting has no effect on productivity and that the extent to which workers have control over their own activities, the relationships between workers and management, and the consequences for the worker of their output are the only important factors. While there can be no disagreement with the statement that motivation and relationships are important, the assertion that lighting conditions can never affect work is nonsense. There is a continuum of performance associated with lighting conditions ranging from no light to plenty of light. In the absence of light, we can see nothing, no matter how large or high a luminance contrast the task has. Even when there is some light, the task details necessary to do the work may be below threshold. It is only when there is sufficient light to see the necessary details that performance becomes possible. As the amount of light on the task increases, it becomes possible to see more and more detail, so performance should increase until it becomes limited by some factor other than the visibility of the necessary details. There can be no doubt that lighting affects work, the problem is to identify the range of lighting conditions that allows an improvement in work output to occur. The operative word here is allow. Lighting cannot produce work output. Only the worker can do that. What lighting can do is to make details easier to see and colors easier to discriminate without producing discomfort or distraction. The worker can then use this increased visibility to produce output if he/she is so motivated or is not limited by some other non-visual factor. In the case of the Hawthorne experiment, the cause of the steady increase in output for both groups in the second experiment is generally ascribed to a change in motivation brought about by the greater self-determination of the workers (Landsberger, 1958; Urwick and Brech, 1965), although alternative explanations based on the frequent feedback of output, the consequences output had for pay, and the nature of the supervision have also been proposed (Parsons, 1974; Diaper, 1990). Regardless of which, if any, of these explanations is correct, what has not been explained is why lighting conditions had no consistent affect. A plausible explanation for this lack of effect is that the visual component of the task was less then it appeared. What the workers were doing was winding coils on wooden spools. This was clearly a visual task but how difficult it is impossible to tell. The workers were well practiced at the task so the visual component may have been small. The impact of such practice can be seen when observing an experienced keyboard operator. Logically, it would seem that to press the appropriate key when typing would require the use of the visual system. However, an experienced typist rarely looks at the keyboard and can work well in near darkness.

A similar situation may have applied to the coil winders. The results of the third experiment suggest that it was not until the illuminance fell to 33 lx that the visibility of the task began to limit the performance of the task.

While the initial Hawthorne studies failed to show any effects of lighting conditions on work output, there are other studies in the literature that do. Stenzel (1962) measured the output from a leather factory over a 4-year period, in the middle of which he introduced a change in lighting installation. The work involved punching out fault-free outer leathers from skins for handbags, purses, etc., using iron shapes and mallets. From 1957 to 1959, the lighting was provided by daylight, supplemented by local fluorescent lighting giving an illuminance of 350 lx. From 1959 to 1961, when the investigation stopped, the daylight was virtually eliminated and a uniform 1,000 lx was provided by general fluorescent lighting. The average monthly performance for the 12 people who were present throughout the 4 years is shown in Figure 4.3. There is a statistically significant improvement in performance with the higher illuminance giving the better performance.

Although this study was well controlled, it does illustrate two problems common in field studies. First, only two lighting conditions were used and those were widely different. Thus, the results do little more than demonstrate that changes in lighting can influence the performance of the task studied. Certainly, with only two illuminances it is not possible to identify the optimum lighting conditions. Second, there is considerable uncertainty about the most important aspect of lighting for the change in performance because the change in lighting installation changed several different aspects of the lighting simultaneously. Specifically, the new lighting installation changed the illuminance on the task, the light spectrum and the light distribution. In these circumstances, to ascribe the changes in output to the difference in illuminance alone may be misleading.

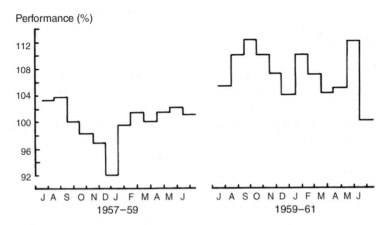

Figure 4.3 Mean monthly performance for cutting leather shapes for the years 1957–59 and 1959–61. Performance is normalized for the average performance over the years 1957–59 (after Stenzel, 1962).

Multiple changes made simultaneously are a problem that bedevils many field studies. Often changes in lighting conditions are made at the same time as changes in decor, furnishings, equipment, working arrangements, and people. If this is the situation, then ascribing any changes in output to lighting alone is also misleading. The basic problem with field studies is the degree of experimental control required (Hartnett and Murrell, 1973). Ideally, the experimenter needs to be able to control the characteristics of the lighting installation, how it is used, the type of work done, the methods of payment and the people involved. It is only rarely that this degree of control is available but, when it is, believable field studies are possible. For example, Buchanan *et al.* (1991) measured the effect of increasing the illuminance in the pharmacists' working area on the dispensing error rate in a high-volume outpatient pharmacy. Based on an examination of 10,888 prescriptions, they found that increasing the illuminance from 485 to 1,570 lx lead to a statistically significant reduction in error rate from 3.9 percent of prescriptions filled to 2.6 percent.

It can be concluded that worthwhile field studies are possible but not probable. All field studies should be carefully reviewed to establish the extent to which the conclusions are justified. If insufficient information is given about the conditions studied, the work done, the way the data was collected, and the nature of the statistical analysis, then it is safer to consider the conclusions as assertions rather than statements of fact, no matter how much they conform to preconceptions.

4.3.2   *Simulated work*

The lack of control possible in many field situations initially turned many experimenters interested in the relationship between lighting and work back to the laboratory, where tasks simulating real life tasks could be done under controlled conditions. Of course, this rather undermines the face-validity of the task because it is now being done in a different context, so people may behave in a different way. Nonetheless, as long as the aim of the study is to determine the effect of different lighting conditions on the performance of the task through changes in task visibility, the gain in sensitivity from having good experimental control is usually worth the decrease in face validity.

There are a number of simulated task studies in the literature. Lion *et al.* (1968) examined the effect of using incandescent and fluorescent lighting on the inspection of plastic disks or buttons on a conveyor belt. The subject had to remove all the disks that had a broken rather than a complete loop marked on them and all the buttons with off-center holes. The two forms of lighting provided the same illuminance on the conveyor belt (320 lx). Significantly better inspection performance was found when sorting the disks with broken loops under the extended fluorescent light source than under the incandescent point light source but no significant effects of lighting were found for the button hole sorting. The probable reason for this difference between the two tasks is their visual difficulty, and hence their relative sensitivity to lighting conditions. These results also serve to demonstrate the difficulty of generalizing from a specific task. Both these tasks could be regarded as sorting tasks but the fact that the two tasks show different effects of lighting conditions mean that the results cannot be applied to sorting tasks in general.

Other simulated work studies have been done by Stenzel and Sommer (1969), who examined sorting screws of different sizes and crocheting stoles; Smith (1976), who studied threading a needle; Bennett *et al.* (1977), who also studied needle probing as well as micrometer reading, map reading, pencil note reading, drafting, vernier caliper measurement, cutting, and thread counting over a range of illuminances from 10 to 5,000 lx; and McGuiness and Boyce (1984) who examined kitchen work. Most of these tasks showed an improvement in performance with higher illuminances, although the amount of improvement varied for different tasks. This should not be surprising because these studies are essentially about task performance rather than visual performance so in addition to the differences in the stimuli the tasks present to the visual system, the tasks also differ in their cognitive and motor components. This is most evident in two studies by Smith and Rea (1978, 1982). In the Smith and Rea (1978) study, a small number of subjects proofread texts for misspelled words. Measurements of the time taken to proofread a passage and the percentage of errors found were take at four different illuminances ranging from 10 to 4,885 lx. Figure 4.4a shows the effect of increasing illuminance, which is to decrease the time taken and an increase the percentage of errors found (hits). In the Smith and Rea (1982) study, the same apparatus was used as was the same range of illuminances, but this time the subjects were asked to read a text and then answer questions about their comprehension of the text. There was little change in either speed or level of comprehension with increasing illuminance (Figure 4.4b). Reading for comprehension has a much larger cognitive component than proofreading.

Simulated work studies are undertaken where the need is to study the effect of lighting on a specific task. They certainly allow for more precise experimental control than is usually possible in the field but they are inevitably limited in that the results are applicable to the specific task and cannot be generalized to other tasks. To use a metaphor; on the road to a general understanding of the relationship between light and work, simulated work tasks are a dead end. There is no reason to go there unless you have business there.

4.3.3 *Analytical methods*

The first attempt to produce a general model of the effects of lighting conditions on work was made by Beutell (1934). The basis of his method was first to define a standard task. The effect of lighting on this standard task

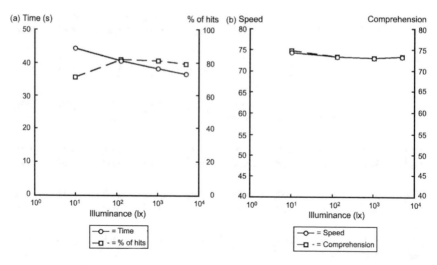

Figure 4.4 Performance on two types of reading task. For both tasks, the printing
was of good quality on white paper: (a) time taken to proofread a passage and percentage of hits, i.e. errors detected, plotted against illuminance; (b) speed and level of comprehension plotted against illuminance
(after Smith and Rea, 1978, 1982).

could then be thoroughly investigated and the illuminance for any desired
level of performance identified. Then the illuminance for any other task
could be obtained by introducing a series of multiplying factors that
allowed for differences between the task of interest and the standard task.
The multiplying factors would be related to the visual size and luminance
contrast of the critical detail of the task, any relative movement between the
observer and the task and the degree of emphasis to be given to the task in
its setting.

Beutell's suggestion was exploited by Weston (1935, 1945), who developed it into a widely used method of investigating the effects of lighting
on work. What Weston did was to devise a very simple task in which the
critical detail was easy to identify and measure. This task is usually known
as the Landolt ring chart, being based on the Landolt ring used in the testing of visual acuity. Figure 4.5 shows an example of a Landolt ring chart. It
consists of a series of Landolt rings, with the gap in the ring being orientated
in one of the four cardinal directions of the compass. The critical detail of
the Landolt ring is the gap. The size of the critical detail is the angular size
of the gap and the critical contrast is the luminance contrast of the ring
against its background. A practical advantage of the Landolt ring chart as
a standard task is that it can be reproduced in large numbers, in many
different media, and the critical size and contrast can easily be changed.

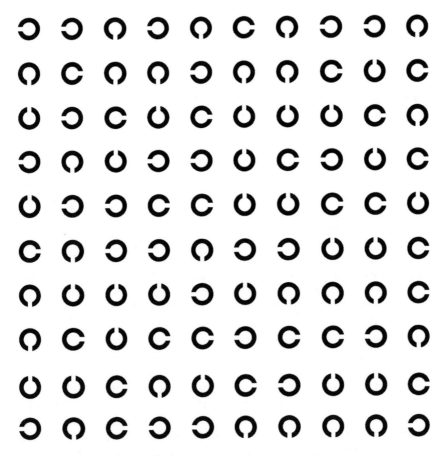

Figure 4.5 A Landolt ring chart.

When doing the Landolt ring chart task, subjects are asked to read through the chart and mark in some way, all the rings that have a gap orientated in a specified direction. The time taken to do this and the number of errors made under different lighting conditions are measured. These measures are then combined to form measures of speed and accuracy of work. Speed is given by the number of rings correctly marked divided by the total time taken, the total time taken being reduced by subtracting the time taken to mark the same number of rings with a gap in the specified direction when they were marked with red ink. The subtraction of the time taken to mark the Landolt rings identified with red ink is an attempt to minimize the contribution of the cognitive and motor components of the task and hence to obtain a measure of visual performance rather than task performance. The rationale behind this is that by marking the rings that need to be identified with red ink, the visual component is minimized so then the time taken is primarily determined by the cognitive and motor components of the task. Accuracy is given by the number of rings correctly marked divided by the total number of rings that could have been marked. Speed and accuracy are then multiplied to form what is called the performance score.

Figure 4.6 shows the results obtained by Weston in his second study (Weston, 1945). There are a number of conclusions that can be drawn from these results. First, the effect of increasing illuminance follows a law of diminishing returns, i.e. that equal increments in illuminance lead to smaller and smaller changes in performance until saturation occurs. Second, the point where saturation occurs is different for different sizes and luminance contrasts of critical detail, saturation occurring at lower illuminances for large size, high contrast tasks than for small size, low contrast tasks. Third, larger improvements in visual performance can be achieved by changing the task, i.e. changing either the size or contrast of the critical detail, than by increasing the illuminance, at least over any illuminance range of practical interest. Fourth, it is not possible to make a visually difficult task, i.e. small size and low luminance contrast, reach the same level of performance as a visually easy task simply by increasing the illuminance. Although these conclusions have been derived from Weston (1945), they have been confirmed many times since, using a number of different visual tasks (Khek and Krivohlavy, 1967; Boyce, 1973; Smith and Rea, 1978, 1982, 1987; Rea, 1981).

This general understanding about the relationship between lighting conditions and work demonstrates the value of an analytical approach based on the concept of critical detail. This observation is reinforced by Boyce (1974). In this study, subjects worked at a ring chart in two forms, one complex and one simple (see Figure 4.7). The visual size and luminance contrast of the gap in both simple and complex rings are the same but the complexity of the task, in terms of the number of alternative locations of the critical detail, is much greater for the complex ring than for the simple ring. The question of interest is how does a change in illuminance affect the performance on the simple and complex rings. The results showed, as expected, that the complex ring chart took significantly longer than the simple ring chart to search but the change in performance with illuminances was similar for both simple and complex rings. In fact, the time taken to do the complex ring task at each illuminance was a constant multiple of the time taken for the simple ring task as the same illuminance. This result suggests that the visual difficulty of the task was adequately described by the size and contrast of the critical detail, i.e. the size and contrast of the gap being sought in both forms of ring. The complex rings were not more visual difficult, because the gap size and contrast were identical to the simple ring, but they took longer to do because there were more possible locations of the gap. The effect of illuminance was determined by the critical detail.

The analytical approach adopted by Weston has served to demonstrate the general form of the relationship between lighting conditions, task characteristics, and visual performance. It has also suggested how the visual difficulty

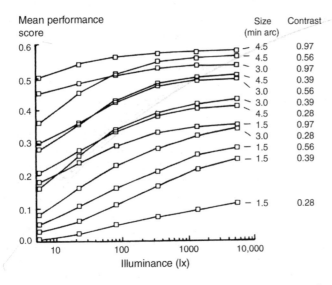

Figure 4.6 Mean performance scores for Landolt ring charts of different critical size and contrast, plotted against illuminance (after Weston, 1945).

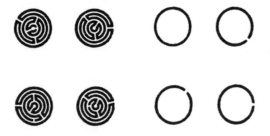

Figure 4.7 Complex and simple Landolt rings.

of a task might be quantified and hence, how the effect of lighting on the performance of the visual component of the task might be determined. However, as used by Weston, the results do have some limitations. Rea (1987) reviewed Weston's studies of 1935 and 1945 and observed that the trends in performance scores with illuminance for Landolt rings of the same visual size and similar luminance contrast were not consistent between the two studies. Rea (1987) also objected to the performance score metric. Specifically, he objected to the fact that the number of correct rejections, i.e. the number of Landolt rings examined and correctly rejected as not having a gap in the specified direction, was ignored. Without considering the number of correct rejections, the measures of speed and accuracy must be imprecise. There is also a more general objection to the performance score metric.

While both speed and accuracy are important aspects of task performance, it is better to treat them as separate but related measures of performance rather than to multiply them together, as is done in the Weston's performance score metric. The ideal approach would be to consider the effect of lighting on speed at a constant level of accuracy or accuracy at a constant level of speed. Unfortunately, the multiplication of speed and accuracy to obtain a single-number measure of task performance is common (Muck and Bodmann, 1961; Waters and Loe, 1973; Smith and Rea, 1978, 1979), but it is nonetheless arbitrary and serves to hide the effect of the experimental conditions on two rather different aspects of task performance. For all these reasons, Weston's results should be considered as indicative of general trends but should not be used as a basis for quantitative models of visual performance.

4.3.4 The visibility approach

While the analytical approach might be considered remote from the real world because it is based on one standard task, it is, at least, based on a measure of suprathreshold performance. The visibility approach is even more ambitious in that it attempts to develop a system for predicting the visual performance of real tasks at suprathreshold levels from threshold performance measurements. The concept behind the visibility approach is that the ease of seeing a task, such as a printed page, can be quantified by the separations of the stimuli the task presents to the visual system from their threshold values. The further the task's characteristics are removed from the threshold surface, the greater is the visibility of the task. It is then assumed that the visibility of the task is consistently related to task performance. The visibility approach was developed over many years by Blackwell (1959), (see also CIE 1972, 1981), based on his extensive measurements of threshold contrast at different luminances (Blackwell, 1946; Blackwell and Blackwell, 1980). The metric of task visibility was visibility level, this being defined as

Visibility level = equivalent contrast/threshold contrast

Threshold contrast was defined as the luminance contrast of the visibility reference task, namely detecting the presence of a 4 min arc diameter luminous disk presented against a uniform luminance field for 0.2 s. Equivalent contrast was the luminance contrast of the visibility reference task that is matched in visibility to the task of interest. Blackwell developed an instrument, called a visibility meter, to measure equivalent contrast, as well as multiplying factors to correct for departures in the viewing conditions of the task of interest from those used in the visibility reference task. He then made visibility level measurements for the stimuli used in other visual performance studies. Initially, these measurements lead to the belief that there was a universal relationship between visibility level and visual performance. However, as more data on more tasks were collected it became obvious that this was not true. It was concluded that the missing factors were the extent of search and scan of the visual field and the need to gather information off-axis. This lead to the development of another standard task, called the visual perform-ance reference task. This consisted of five, 4 min arc Landolt rings, one in a central position and the other four at the four cardinal points of the compass, equidistant from the center. By altering the luminance of the Landolt rings or the luminance of the background, the visibility level of the rings could be changed. By altering the presentation time or the separation of the central and peripheral rings, the difficulty of the task could be altered. Using the understanding gained from the visual performance reference task, a model was developed to predict the effect of lighting conditions on visual perform-ance for a wide range of tasks (CIE, 1981). The model consisted of two sets of transfer functions operating in series. The first was concerned with how the lighting conditions affected the visibility level. The second was concerned with how the visibility level influenced visual performance. The model had three components linked to visibility level. These three components were related to extracting information from the details of the task, the stability of eye fixation and the precision of eye movements. The model could be made to fit independently obtained sets of experimental results but only by varying the weighting of the four components. As more data sets were examined the number of correction factors that had to be introduced to make the model fit increased to such an extent that the model lost all credibility.

Although the visibility approach is rarely mentioned today, it should be appreciated that there is nothing inherently wrong with the concept of vis-ibility level. The extent to which the stimuli a task presents to the visual system are above threshold is a useful way to quantify how well those details can be seen. Indeed it has been used for this purpose in other fields of lighting, such as road lighting (Lipinski and Shelby, 1993). The problem with the concept of visibility level is the tendency for people to assume that equal visibility levels correspond to equal levels of task performance. This is not the case. Different tasks show different relationships between task performance and visibility level depending on the nature of the task (Clear and Berman, 1990; Bailey *et al.*, 1993). The error in the visibility approach developed by Blackwell was to assume that something as complex as the suprathreshold performance of tasks with different visual and non-visual components, occurring on- and off-axis, could be predicted from something as simple as an on-axis threshold measurement. In a sense, the visibility approach was simply "a bridge too far."

4.3.5 *The relative visual performance model*

After the collapse of the visibility approach as a method for predicting the effect of lighting conditions on visual performance, the time was right for a simpler but more rigorous approach. This need was met by the develop-ment of the relative visual performance (RVP) model. The principle adopted in the development of the RVP model was to establish a quantitative link between the physical characteristics of the task and the performance of the visual component of the task. The visibility level of the task was considered to be an unnecessary intervening variable. The visual characteristics of the task would be defined by quantities that could be directly measured.

The origin of the RVP model lies in a study of the effect of luminance contrast on performance of the numerical verification task (Rea, 1981). In this task, two printed pages, the reference page and the response page, each containing a column of 20, five-digit numbers are used. The five-digit numbers on the reference page were random numbers. The corresponding numbers on the response page were the same except that some of the five-digit numbers had one digit different (Figure 4.8). The mean frequency of such discrepancies were three per page. Data were collected for a constant illuminance on the printed pages of 278 lx and for a wide range of luminance contrasts for the reference page. Luminance contrast was varied by changing the reflectance of the ink used to print the numbers; by making the ink either specular or matte by changing the geometry between the luminaire providing the illuminance, the numerical verification task and the observer; and by varying the percentage of vertical polarization in the light incident on the numerical verification task. Luminance contrast was defined as

$$C = (L_t - L_b)/L_b$$

where C is the luminance contrast, L_b the luminance of the background (cd/m^2), and L_t the luminance of the detail (cd/m^2).

The response page was printed in matte ink on matte paper in a high luminance contrast. The visual size of the numbers was held constant by using the same font and point size for the printing and by controlling the location of the subject's head with a chin rest. Data were collected on the time taken to compare the two columns of figures, the number of discrepancies missed (misses) and the number of identical five-digit numbers marked as discrepancies (false positives). Figure 4.9 shows the change in the mean time taken, misses, and false positives for decreasing luminance contrast. It is evident that for a wide range of luminance contrasts, there is very little change in any of the performance measures. However, as luminance contrast drops below about 0.4, the time taken starts to increase, an increase that accelerates as luminance contrast decreases further. A similar pattern is can be seen for the misses and false positive data. These data illustrate that both the speed and accuracy of performance deteriorate with reduced visibility in a non-linear manner. They also demonstrate that luminance contrast is a major determinant of task performance, no matter how that luminance contrast is achieved.

The first complete version of the RVP model (Rea, 1986) was derived from data collected from people doing the numerical verification task using

58313	58313
51424	51424
26538	26538
10508	10508
35148	35148
53427	53427
99147	99147
54483	54483
39154	39155
39417	39417
52807	52807
55394	55394
32393	32393
83118	83118
31510	31510
53009	53409
01632	01632
29394	29394
49619	49619
54101	54101

Figure 4.8 The numerical verification task.

the same experimental materials, experimental room and procedures as in the earlier study (Rea, 1981). Data were collected for a range of illuminances from 50 to 700 lx (giving a range of background luminances from 12 to 169 cd/m^2) and a range of luminance contrasts from 0.092 to 0.894, luminance contrast being defined as in the 1981 study. The data on time taken to compare a set of 20, five-digit numbers, the number of misses and false positives were very similar to those obtained in the earlier study (Rea, 1981) and by others using the numerical verification task (Slater *et al.*, 1983). In developing the RVP model, Rea decided to use only the time taken data. This decision was taken for a number of reasons, the

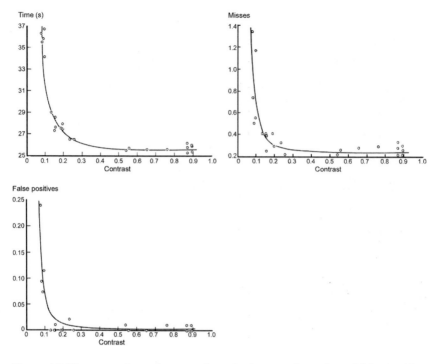

Figure 4.9 The mean time taken, number of misses, and number of false positives
for the numerical verification task, plotted against luminance contrast
(after Rea, 1981).

most important being that the number of misses and false positives were
small and subject to random fluctuations, factors that made misses and false
positives less reliable measures of performance than time taken, and because
the variations of all three measures with luminance contrast were very
similar. Figure 4.10 shows the change in the reciprocal of time taken plotted against luminance contrast at each of the background luminances.
Figure 4.10 shows the compressive function with luminance contrast that
would be expected from the results in Figure 4.9. There are two effects of
increasing luminance that should be noted. The first is that, at the same
luminance contrast, performance is better at higher luminances. The second
is that performance tends to saturate at lower luminance contrasts for higher
luminances.

The time taken to compare the reference and response pages used as the
basis of Figure 4.10 is a measure of task performance, in that it includes
both visual and non-visual components. To produce a time measure that
could reasonably be called a measure of visual performance, Rea subtracted
two elements of time from the total time taken. The first was an estimate of

Work speed (1/s)

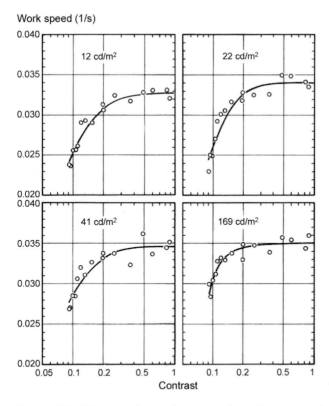

Figure 4.10 Mean work speed, expressed as the reciprocal of time taken to perform the numerical verification task, plotted against luminance contrast, for four background luminances (after Rea, 1986).

the time taken to make a mark against any response page number that contained a discrepancy. The second was the time taken to read the numbers in the response list. The remaining time was then the time taken to read the numbers in the reference list. It is the reciprocal of this time that was taken as a measure of visual performance from which the RVP model was developed. Figure 4.11 shows the form of the RVP model plotted against background luminance of the range 12–169 cd/m^2 and luminance contrast over a range 0.08–1.0. It should be noted that the vertical axis is a relative measure (RVP) calculated from the reciprocal of the time taken to read the reference page, normalized to a value of 1.0 at a background luminance of 169 cd/m^2 and a luminance contrast of 1.0. Memorably, the shape of this model has been described as the plateau and escarpment of visual performance (Boyce and Rea, 1987), the point being that over a wide range of task and lighting variables the change in relative visual performance is slight but at some point it will start to deteriorate rapidly. The evolutionary advantage

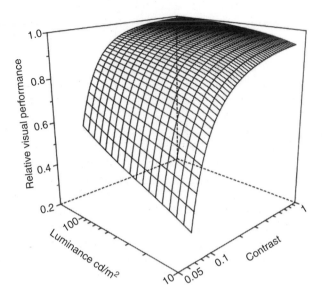

Figure 4.11 The RVP model of visual performance, based on the time taken to read the reference page of the numerical verification task (after Rea, 1986).

of having a visual system which can cope with a wide range of visual conditions with little change in speed of response is obvious.

The second form of the RVP model (Rea and Ouellette, 1988) was developed using a very different approach. Whereas in the first approach, a task with some similarity to tasks actually done by someone, the numerical verification task, was used and the performance measure later adjusted to make it a measure of visual performance, in the second approach the task used was one that could be taken as a direct measure of the speed of visual performance. Specifically, the performance measure was simply the reaction time to the onset of a square stimulus. Detecting the presence of something as against nothing is about the simplest visual task possible. There is very little cognitive component, and as the only motor component is to release a button when the stimulus is presented, the motor component is slight as well. Reaction time measurements were taken for stimuli with a wide range of luminance contrasts, both positive and negative; angular sizes; and adaptation luminances. Luminance contrast was defined as before. Visual size was defined as the solid angle subtended at the eye by the stimulus. An equation of the same form as that used to fit the reciprocal of time taken to do the numerical verification task was applied to the reciprocal of reaction time and fitted the data well. To convert the reaction times into a relative measure, the differences in reaction time between each stimulus condition and the shortest reaction time (obtained for the largest size, highest contrast, and highest adaptation luminance) were calculated. This measure shows the expected increase in reaction time following reductions in visual size, luminance contrast or amount of light entering the eye. Figure 4.12 shows the curves fitted to these differences in reaction time, plotted against luminance contrast, for different amounts of light entering the eye.

At this point there were two alternative forms of the RVP model, one based on the time taken to read the reference page of the numerical verification task (Rea, 1986) and the other based on the difference in reaction time for detecting the presence of a target (Rea and Ouellette, 1988). Fortunately for simplicity, Rea and Ouellette (1991) were able to develop a method for converting the reaction time difference into units of RVP. They did this by identifying a common set of stimulus conditions and then developing a linear transformation between the two measures. What this means is that the differences in reaction times can be expressed in terms of RVP. Figure 4.13 shows the RVP values derived from the reaction time measurements plotted against luminance contrast and retinal illumination, for four different visual sizes of the detection target. The average visual size of the digits used in the numerical verification task is 4.8 µsteradians, so this part

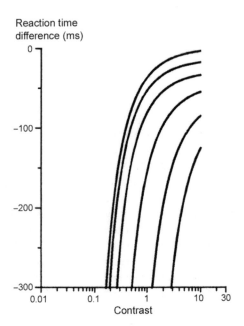

Figure 4.12 Fitted curves showing the difference in reaction time to the onset of a target subtending 2 µsteradians at the eye, plotted against luminance contrast. Each curve is for a different retinal illumination. From left to right, the retinal illuminations are 801, 160, 31, 6.3, 1.6, and 0.63 trolands (after Rea and Ouellette, 1988).

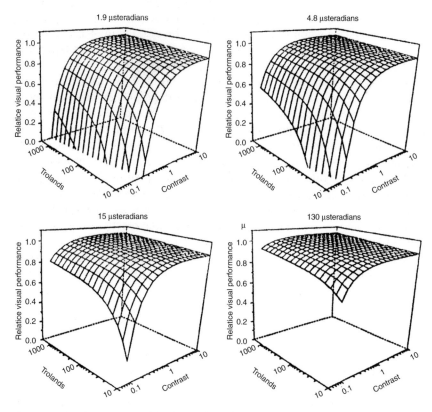

Figure 4.13 The RVP model of visual performance based on reaction time data. Each graph shows the RVP value plotted against luminance contrast and retinal illumination, for a fixed target size (in μsteradians) (after Rea and Ouellette, 1991).

of Figure 4.13 is comparable with Figure 4.11. The similarity between the two figures is obvious.

This has been a complicated discussion, but it is worthwhile because it is important to understand the development of the RVP model. The RVP model represents the most complete method currently available for predicting the effect of both task and lighting variables on visual performance. It has been very carefully developed, using rigorous methodology and careful reporting. What needs to be considered now is how well the RVP model predicts the results of independently collected data.

There are three answers to this question. The first comes from Rea (1987). In this paper, Rea examined a number of experiments in which the performance of visual task had been measured and considered their suitability for acting as a validation of the RVP model based on the numerical verification task. For a variety of reasons, including insufficient documentation, confounding of variables, inappropriate performance measures, and the use of tasks requiring visual search and with significant non-visual components, he concluded that only one of the previous studies was suitable as a check of the accuracy of prediction. This study was done by McNelis (1973). In this study, people were asked to name two briefly presented small letters separated by 10°. The letters were presented at different contrasts and at different background luminances. Predictions of RVP for the luminance contrast and illuminances used showed good agreement with the normalized accuracy scores for naming the letter viewed first, but the agreement was less successful for the lowest contrast used (0.125), possibly because of errors in the measurement of the luminance contrast.

The second answer comes from the study of Bailey *et al.* (1993). In this study, the speed of reading unrelated words was measured, when the words were presented in sizes ranging from 2 to 20 point print, at three different luminance contrasts (0.29, 0.78, and 0.985) and over a range of background luminances from 11 to 5,480 cd/m^2. The reading speed was calculated from the time taken to read each row of words and the number of words in each row. The time taken to read each row was measured from a recording of the eye movements made during the reading. Figure 4.14 shows the measured reading speeds plotted against letter size. The plateau and escarpment shape of visual performance is again evident. Bailey *et al.* (1993) then applied the formula used by Rea to fit his reaction time data (Rea and Ouellette, 1988) to the stimulus conditions in their experiment to predict the reaction times for each combination of background luminance, letter size, and luminance contrast. Then reading speed was calculated as a linear function of reaction time, with two free variables. The resulting fit of the predicted reading speed to the measured reading speeds was good for letter sizes of 5 point and above, for all luminance contrasts and background luminances. However, as letter sizes decreased below 5 point, the predictions became increasingly in error. This is not unreasonable, for two reasons. First, the data on which the RVP model is based did not cover stimulus sizes equivalent to 5-point print and smaller, so such smaller letter sizes were outside the range of the model. Second, it is likely that as letter size decreases, the ability to read the word becomes limited by the ability to resolve the detail of the letter, regardless of luminance contrast. The fact that the basic formula used in the RVP model can be applied to independently collected data on a reading task and give accurate predictions of reading speed over a range of print sizes of practical interest is encouraging as evidence that the underlying concept is correct. However, it is not conclusive proof of the validity of the RVP model because of the two free parameters which were adjusted to maximize the fit.

The third answer comes from the work of Eklund *et al.* (2001). They measured the effect of different lighting and print conditions on the sustained performance of a repetitive, self-paced, data-entry task. Twenty-four subjects worked for almost 4 h at a data-entry task in one of three

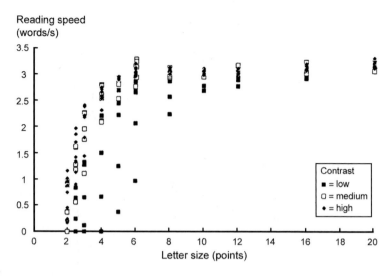

Figure 4.14 Mean reading speed measured in words/s plotted against letter size in points, for three different luminance contrasts (after Bailey *et al.*, 1993).

identical, private, windowless offices. All three offices were lit by similar fluorescent, parabolic lighting systems. The installations were fitted with dimming systems to allow the illuminance on the work to be systematically changed to four levels (29, 103, 308, and 1,035 lx). During the 4 h of work, the subject entered sets of five, 10-symbol, alphanumeric codes, the sets being printed in a cyclical series of print sizes (6, 8, 12, and 16 point) and luminance contrasts (0.10, 0.22, 0.47, and 0.93). In total, task performance measurements were taken for 60 combinations of illuminance, print size, and luminance contrast. The task performance measurements taken were the times taken to enter a block of 50 alphanumeric symbols correctly. Any errors made in data entry were detected by the software running the experiment and had to be corrected before proceeding, thereby increasing the time taken. The result of this procedure is to fix the accuracy of performance at 100 percent. Figure 4.15 shows the mean work speed, derived from the reciprocals of the work times, plotted against illuminance, for each luminance contrast and print size. These data were used to test the precision of the RVP model based on the reaction time data given in Rea and Ouellette (1991). The mean inked areas of the letters and numbers printed in the different point sizes were measured as were the luminance contrasts and the background luminances. These values were then inserted into the RVP model to predict the relative visual performance. The predicted RVP values were then normalized at the value found for the largest size (16 point), highest contrast (0.93), and highest illuminance (1,035 lx). The

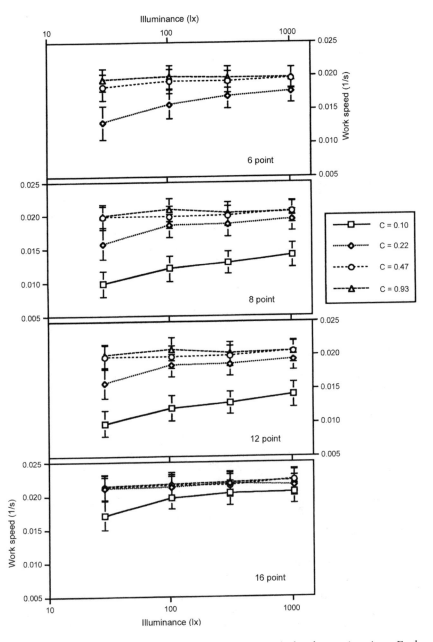

Figure 4.15 Mean work speeds for the data-entry task for four print sizes. Each graph shows the mean work speed plotted against illuminance (lx) for four luminance contrasts. The error bars indicate 95 percent confidence intervals. Note that the plot for the 6-point print contains data for only three luminance contrasts. Six-point print of luminance contrast 0.10 could not be read by a significant number of subjects at any of the illuminances used (after Eklund *et al.*, 2000).

measured mean work speeds on the data entry task were also normalized for same conditions. Figure 4.16 shows the normalized predicted performance plotted against the normalized measured performance. It is clear that the RVP model fits the measured data well, although not perfectly.

These three comparisons serve to demonstrate the robustness and validity of the RVP model. The Eklund *et al.* (2001) study also removes some of the doubts about its utility. Some of these doubts have arisen from the conditions under which the reaction time data used to form the RVP model were collected. For example, the reaction time data were collected monocularly, using an artificial pupil, and with the subject's head at a fixed distance from the stimulus. Further, the only task was to detect the presence of a square target of a fixed size which does not require resolution of detail. In the Eklund *et al.* (2001) experiment, the subject used natural pupils and could move closer or further from the data-entry material as they wished and they had to read alphanumeric characters. Other doubts have arisen because different letters and numbers of the same nominal print size vary in inked area. In the Eklund *et al.* (2001) experiment, the data-entry material was a random collection of letters and numbers of the same nominal size, and hence a collection of letters and numbers that varied in inked area. The fact that the RVP model is consistent with the mean work speed measured over a 4-h work period in real-world conditions suggests that these doubts are unjustified. Over time, the variations in the stimuli presented to the

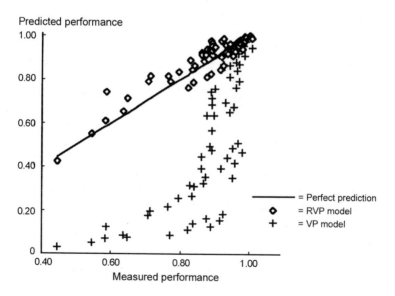

Figure 4.16 Normalized RVP values predicted by the RVP and VP models for the data-entry task plotted against the normalized measured mean work speeds (after Eklund *et al.*, 2001).

visual system caused by different viewing distances and different inked areas average out and hence can be ignored.

4.3.6 *Some limitations*

From the information presented above, it can be concluded that the RVP model represents the closing of the circle, from an abstract model of visual performance to the performance of a realistic task done under realistic conditions. This, in turn, suggests that the RVP model could be used to predict the change in visual performance for other tasks consequent upon changing either the lighting or the task. The existence of this possibility should not be taken to mean that the long study of the relationship between light and work is at an end. A more accurate understanding would be that the RVP model represents a concept that can be applied to a limited range of tasks. The tasks for which it is most suited are those that are dominated by the visual component, that do not require the use of peripheral vision to any extent, that present stimuli to the visual system that can be completely characterized by their visual size, luminance contrast, and background luminance only and that have values for these variables that fall within the ranges used to develop the RVP model.

The limitation of a small non-visual component is derived from the fact that RVP measures visual performance not task performance. These two types of performance are only likely to coincide when the non-visual components are relatively small or when parallel processing of information can occur, as seems likely to have happened in the data-entry task of Eklund *et al.* (2001). If the non-visual component is large and parallel processing is not possible, then RVP predictions will overestimate the consequences of a change in visual conditions on task performance. What is needed to extend the use of the RVP model to tasks with significant non-visual components is the development of a task analysis procedure that could be used to determine the relative impacts of the visual and non-visual components on the performance of any task.

The limitation to tasks using only foveal vision occurs because the tasks used to derive the RVP model and the tasks used to validate it, all used tasks in which the subject knew where to look to gain the necessary information. This is not true in all tasks. Specifically, tasks requiring visual search to find the required information, use peripheral vision. How different peripheral vision was discussed in Chapter 2 and is shown in Figure 4.17. This shows the mean percentage correct identification of three-letter words presented for 100 ms in the fovea and 1° off-axis (Timmers, 1978). Even a slight deviation from foveal viewing causes a reduction in performance. More data are needed before a model of visual performance covering visual search can be constructed.

The limitation to tasks with stimuli that can be completely characterized by the visual size, luminance contrast, and background luminance is necessary because these are the factors built into the RVP model. Other aspects

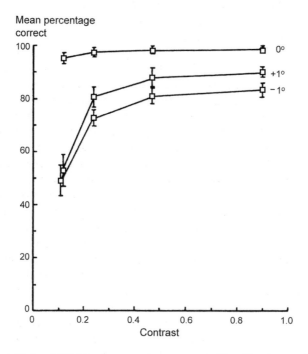

Figure 4.17 Mean percentage correct identifications of three-letter words presented for 100 ms in the fovea and at 1° eccentricity, at different levels of luminance contrast. The error bars are the standard error of the mean (after Timmers, 1978).

of visual stimuli that can be important are color difference and retinal image quality. When applied to print, Smith and Rea (1978) and Colombo *et al.* (1987) have both shown that the effect of illuminance was much greater for fragmented print than for entire print. As for color difference, Eklund (1999) has shown that the ability to read an exit sign at a distance is maintained even at zero luminance contrast, provided there is a color difference between the letters and the background.

Finally, it is worth noting that other models of visual performance have been constructed (CIE, 2002a). One, called the VP model, is based on the data of Weston (1945). Adrian and Gibbons (1994) took these data and performed an extensive curve-fitting exercise to produce a quantitative model to predict the speed and accuracy of performance on the Landolt ring task from the measured visual size, luminance contrast, and background luminance (CIE, 2002a). Unlike the RVP model, the VP model provides a poor fit to the independently-collected task performance data of Eklund *et al.* (2001) (see Figure 4.16).

There are also other models of task performance. Eklund *et al.* (2001) themselves performed a sophisticated curve-fitting procedure on their data to produce what they called the data-entry task performance model. There was a good fit between the predicted mean work times and the measured mean work times for the data-entry task. Clear and Berman (1990) had a more general aim in mind when they put forward a model of task performance in which performance is determined by the addition of two components, one visual that can be related to a visibility level type metric and one non-visual component that is independent of visibility level. They show that a model of this form can be made to fit the Rea (1986) numerical verification task performance data and can handle both speed and accuracy measures of task performance. A similar approach was used by Bailey *et al.* (1993) to fit their reading speed data but in this case the visibility level metric was based on visual size rather than luminance contrast. It is important to appreciate that these models are not models in the same sense as the RVP model. The difference is that the RVP model is a model of visual performance and all the measurements needed to make a RVP prediction can be obtained from physical measurements of visual size, luminance contrast and background luminance of a task. The task performance models are either applicable only to the specific task or require additional information before they can be used. Further, the necessary information cannot be obtained except by measurement of performance of the task, which raises the question as to why is it necessary to make a prediction of the effect of changes in the visual conditions on task performance, when the performance has to be measured directly before the prediction can be made. Really, the value of these models of task performance is the concepts they introduce. It may be that at some time in the future, there will be enough data collected for it to be possible to classify tasks according to their relative importance of their visual and non-visual components and to use some of the task performance models. Until that glorious day arrives, the RVP model marks the quantitative frontier of the search for an understanding of the relationship between light and work.

4.4 Prolonged work

So far this examination of the relationship between light and work has concentrated solely on the effect of lighting conditions operating through the visual system. The effects of lighting on task visibility and hence visual performance will be evident regardless of task duration, but when work is prolonged there is the possibility that other phenomena will occur that will interact with the lighting conditions. These phenomena are eyestrain, fatigue, and mood changes. The nature and causes of eyestrain are discussed in Section 13.3.1. Fatigue and mood changes are considered here.

4.4.1 Fatigue

The definition of fatigue has been the subject of much argument (Bartley and Chute, 1947; Floyd and Welford, 1953; Cameron, 1974; Hockey, 1983;

Craig and Cooper, 1992; Brown, 1994). Partly this is because of the difference between physiological (muscular) fatigue and psychological (cognitive) fatigue. Physiological fatigue is easily measurable, has a clear relationship to energy expenditure, is largely specific to the muscles used and can invariably be demonstrated. According to Grandjean (1969), it occurs in two forms; static, in which the muscles remain in a state of increased tension in order to sustain a particular posture; and dynamic, in which the muscles tense and relax rhythmically. Both types of muscular fatigue can impair perceptual-motor tasks and the static muscular fatigue, in particular, may distract the worker from the cognitive demands of a task. In comparison, psychological fatigue is not easily measured, does not have a clear relationship to energy expenditure, is a general rather than specific response to stress and cannot easily be demonstrated within a conventional working day (Craig and Cooper, 1992). Regardless of how fatigue is defined, prolonged work at mainly cognitive tasks produces a number of different symptoms, and it is these symptoms and their potential for changing task performance that are of interest here.

Early research on sustained cognitive work was dominated by the analogy to muscular fatigue, the assumption being that prolonged mental work would lead to a decrement in performance in the same way that prolonged work with the same muscle always led to a decrement in the strength of muscle contractions. Unfortunately, this did not occur. Poffenberger (1928) shows the results of 5.5 h continuous work on four different cognitive tasks; digit addition, sentence completion, composition judgment, and an intelligence test. The first 2 h showed no change in any of the tasks but after that the addition task declined, sentence completion and composition judgment were unchanged, and performance on the intelligence test improved. This raises the important point that the impact of prolonged work on a cognitive task depends on the nature of the task. The task type which seems most likely to produce a performance decrement is one in which it is necessary to stay alert in readiness to respond to rare events which may occur at unknown times, e.g. a sentry at a military post (Mackworth, 1948). The boredom of such activity is obvious. This suggests that any task which is repetitive, unvarying and uninteresting is likely to produce a performance decrement, if prolonged sufficiently, and particularly if the task is externally paced.

The perceived failure of the simple muscular fatigue analogy and the demands of war changed the emphasis of fatigue studies to the question of how skill deteriorated after prolonged work. Drew (1940) and Bartlett (1953) reported studies of the performance of air crew making long-duration flights in a flight simulator using only instruments. The picture that emerged was of a disintegration of skilled performance and a deterioration in mood. As the flight time increased, alertness decreased and four changes occurred in air crew behavior. They were a deterioration in coordination of responses leading to roughness in handling the controls; a tendency to require larger than normal changes in stimuli before responding; a narrowing of the field of attention, shown by a reduction in the number of instruments checked; and an increased irritability leading to more aggressive responses to other people and the aircraft. Interestingly, a similar picture of break-down in coordination has been obtained for a physical activity. Bates *et al.* (1977) filmed the movements of runners at the start and at the end of a relay race. The coordination between the movements which make for smooth running was much worse at the end of the race.

Since this pioneering work, the same pattern of skill deterioration has been shown for a variety of tasks and environments. Dureman and Boden (1972) showed a similar disintegration of skill for 4 h driving in a simula-tor. Bursill (1958) and Hockey (1970) showed narrowing of attention in hot and humid and noise conditions, respectively. It would be interesting to explore what happens to skilled work requiring coordination, as opposed to simple repetitive work, done in poor lighting conditions.

The variability in response timing implicit in worsening coordination has also been shown in prolonged work of a repetitive nature. Bills (1931) meas-ured the fluctuations of response time to a task requiring fast responses, e.g. color naming. He found occasional much longer response times which he called blocks. Bertelson and Joffe (1963) measured reaction time to a choice task of 30-min duration, which provided no respite in that as soon as a response was made another response was called for. The distribution of reaction times changed over the 30-min period with the number of long reaction times markedly increasing. Of more relevance to the effect of light-ing is a study by Khek and Krivohlavy (1967). In this study, the subject had to report the orientation of the gaps in a series of Landolt rings presented as fast as the subject could handle, for a period of 30 min. Two different lighting conditions were examined, one with and the other without a glare source at 45° to the line of sight. The average speed of working did not change over the 30 min and was no different for the two lighting conditions. However, the variability in speed of working increased dramatically for the glare source conditions but showed no change over time without the glare source. It is important to note that this implied variability in response times does not necessarily lead to a performance decrement. Warren and Clark (1936) report that over 65 h of continuous performance, the frequency of long response times steadily increased but the average response time changed little. This suggests that the long response times allow accumulated fatigue to dissipate and the longer times are made up by shorter response times later. This in turn implies that the increased variability of response is only likely to lead to a worsening of performance in tasks which require a fast response and are externally paced.

This discussion of psychological fatigue serves to emphasize two facts. The first is that there are a variety of changes in the nature of performance which are possible when work is prolonged. The second is that the way in which performance may change depends on the nature of the task.

Tasks which are varied but complex seem likely to deteriorate by a worsening of coordination and a narrowing of attention, i.e. by a deterioration in the quality of work; and tasks which are simple, repetitive and uninteresting seem likely to deteriorate by a performance decrement, i.e. by a simple decline in output.

While this understanding is useful, it ignores the potential people have for changing their work strategies to maintain performance in the presence of increased feelings of tiredness. Schonpflug (1983) has reported a study in which information had to be gathered and held in mind to aid in later decisions or the earlier information could be referenced and consulted while making the decisions, at a cost of longer time taken. When the subjects felt fatigued, either because of working under time pressure or because of distracting noise, they placed less reliance on memory so that they either took longer to complete the task by looking up the information or made more errors by relying on incomplete or poorly remembered information. Welford *et al.* (1950) have shown a similar memory deficit amongst air crews after completing a long flight.

More generally, the concept behind the choices people make when feeling fatigued is that people who are fatigued would rather not make further effort. What this implies is that given a choice between a number of actions which differ in the probability of success and the effort required, as work is prolonged, people will tend to choose actions which require less effort even if they entail greater risk of errors (Shingledecker and Holding, 1974). Of course, this is implicit in the flying simulator studies in that as flight duration increased pilots gradually choose to ignore many of the instruments, including the fuel gauge. Brown *et al.* (1970) have shown an increase in more risk-taking behavior for prolonged driving. Over four, 3-h sessions of driving there was little difference in speed but the number of risky overtaking maneuvers increased by 50 percent between the first and fourth session. Even simple repetitive mental arithmetic has shown the same trade-off between decreased effort and increased risk with increasing time spent in 95 dBA noise (Holding *et al.*, 1983). These results serve to emphasize that the likely effects of prolonged work will depend on the nature of the work and the freedom workers have to modify their strategies.

That lighting conditions inadequate for the task can cause fatigue is suggested by the work of Simonson and Brozek (1948). They had people copy small letters subtending 10 min arc at the eye presented in an aperture, for 0.56 s, for 2 h, at a fixed illuminance. Six different illuminances were used, 20, 50, 150, 500, 1,000, and 3,000 lx. Performance was sampled 5 min after starting work, after 60 min, and after 110 min of work. After completing the work, subjects answered a questionnaire concerned with their feelings of eyestrain and tiredness. At 20 lx, the performance decrement from the beginning to the end of the work period was a maximum, as was the variability in performance and the subjective feelings of eyestrain and tiredness. By comparison, at 1,000 lx, the decrement in performance and the variability of performance were minimal and the feelings of eyestrain and discomfort were least. An illuminance of 20 lx is almost certainly inadequate for this task, so the performance decrement, increased variability and increased feelings of eyestrain and fatigue should be expected, as should the improvement when the illuminance was increased to 1,000 lx. It would be interesting to examine the effects of aspects of lighting other than illuminance on the incidence of fatigue.

4.4.2   Mood changes

All the above discussion has been concerned with the effects of prolonged work on output. However, the flying studies also showed an increase in irritability after prolonged performance. This in itself may affect some aspects of task performance through mood changes known as positive and negative affect. Positive affect, defined as pleasant feelings produced by commonplace events or circumstances, has been shown to influence cognition and social behavior. Specifically, positive affect has been shown to increase efficiency in making some types of decisions, and to promote innovation and creative problem solving. It also changes the choices people make and the judgments they deliver. For example, it has been shown to alter peoples preferences for resolving conflict by collaboration rather than avoidance and also to change their opinions of the tasks they perform (Isen and Baron, 1991). Negative affect has the opposite effects. Eklund *et al.* (2000, 2001) have shown that prolonged work on a monotonous data-entry task produced a reduction in positive affect but little change in negative affect.

The factors which determine positive affect are both small and wide. Small, because the stimuli which have been shown to generate positive affect are low-level stimuli, ranging from receiving a small but unexpected gift from a manufacturer's representative to being given positive feedback about task performance. Wide, because positive affect can be influenced by the physical environment, the organizational structure, and the organizational culture. Lighting is clearly part of the physical environment and lighting conditions such as the illuminance and the correlated color temperature of the lighting have been shown to change mood (McCloughan *et al.*, 1999) and to change behavior in a way consistent with positive affect (Baron *et al.*, 1992). This suggests that lighting can produce mood changes. Lighting conditions which make what needs to be seen difficult to see will likely generate fatigue and negative affect. Conversely, lighting which ensures good visibility of the task and is perceived to be much more attractive than is usual might produce positive affect.

4.4.3   Conclusions

Overall, there can be little doubt that the combination of inappropriate lighting for a task and the need for prolonged performance of the task is likely to produce fatigue and mood changes. These, in turn, are likely to be associated with changes in task performance. What is difficult to predict is what that change in task performance will be. The difficulty arises because people will, where the task allows, manipulate the way they do the task, vary their performance over time, and change the level of errors they are willing to accept. Thus, the consequences of prolonged work in inappropriate lighting conditions depend greatly on the specific nature of the task and the consequences of errors. Fortunately, there is no reason why fatigue and mood changes should occur, at least as long as task duration is limited to a normal working day. By ensuring that the lighting provided allows a high level of visual performance of the task without visual discomfort, the likelihood of the adverse effects of prolonged work occurring will be minimized. A high level of visual performance can be guaranteed by making sure the stimuli presented to the visual system by the task, i.e. visual size, luminance contrast, and color difference, are well above threshold; by making the retinal image quality good by ensuring the observer has the correct optical refraction; and by lighting the task to the level required to produce the necessary retinal illumination. Visual discomfort can be avoided by following the advice given in Chapter 5.

4.5   Summary

Most apparently visual tasks have three components; visual, cognitive, and motor. Every task is unique in its balance between these components and hence in the effect lighting conditions have on task performance. It is this uniqueness which makes it impossible to generalize from the effect of lighting on the performance of one task to the effect of lighting on the performance of another.

The effect of lighting on the performance of a specific task depends on the structure of the task and specifically, the place of the visual component relative to the cognitive and motor components. Tasks in which the visual component is large will be more sensitive to changes in lighting conditions than tasks where the visual component is small.

Lighting conditions can affect task performance through three systems, the visual system, the circadian system, and the perceptual system. The impact of lighting conditions on the visual system and hence on visual performance is determined by the size, luminance contrast, and color difference of the task and the amount, spectrum, and distribution of the lighting. The impact of lighting on the circadian system is determined by the amount and spectrum of the light and the timing and duration of the exposure to the light. The impact of lighting through the perceptual system is determined by the "message" it sends.

The majority of studies of the effect of lighting conditions on work have concentrated on the impact through the visual system. Early work focused on field studies. While these studies have a high face-validity and frequently demonstrate the expected increase in task performance with increasing illuminance, their conclusions are limited to the specific tasks studied. Simulated work in the laboratory has also been done, with better experimental control than the field studies, but with little increase in understanding.

An analytical approach using a standard task measured over a wide range of conditions has served to demonstrate, qualitatively, the effects of increasing illuminance on visual performance. They are that increasing illuminance follows a law of diminishing returns, i.e. that equal increments in illuminance lead to smaller and smaller changes in visual performance until saturation occurs; that the point where saturation occurs is different for different sizes and contrasts of critical detail; that larger improvements in visual performance can be achieved by changing the task than by increasing the illuminance, at least over any illuminance range of practical interest; and, that it is not possible to make a visually difficult task reach the same level of performance as a visually easy task simply by increasing the illuminance over any reasonable range. The overall concept is that the shape of visual performance can be considered as a plateau and an escarpment, the point being that over a wide range of task and lighting variables the change in visual performance is slight but at some point one or more of these variables will decline too far and then performance will start to deteriorate rapidly.

While this understanding is useful, it is not enough to make quantitative predictions of the effect of lighting conditions on visual performance for all tasks although it is possible for some. Specifically, the RVP model of visual performance has been shown to make accurate predictions for tasks that are dominated by the visual component, that do not require the use of peripheral vision to any extent, that present stimuli to the visual system that can be completely characterized by their visual size, luminance contrast, and background luminance only, and that have values for these variables that fall within the ranges used to develop the RVP model, e.g. reading and doing data-entry work. More studies are required before quantitative predictions can be made for tasks outside these boundaries.

While the effect of lighting conditions operating through the visual system will be apparent immediately, there is also the possibility of other phenomena occurring if work is prolonged. These phenomena are eyestrain, fatigue, and mood changes, all of which may change task performance. Unfortunately, it is difficult to predict what the change in task performance will be when such conditions occur because people will, where the task allows, manipulate the way they do the task, vary their performance over time, and change the level of errors they are willing to accept. Fortunately, there is no reason why eyestrain, fatigue, and mood changes should occur, at least as long as task duration is limited to a normal working day. By ensuring that the lighting provided allows a high level of visual performance of the task, combined with visual comfort, the likelihood of the adverse effects of prolonged work occurring will be minimized.

5 Lighting and visual discomfort

5.1 Introduction

In addition to ensuring people can see what needs to be seen, most lighting installations are designed to ensure visual comfort. But what is visual comfort? One view is that visual comfort is simply the absence of visual discomfort. This is logical but not particularly helpful. While it is undoubtedly true that some lighting conditions can cause discomfort, is it also true that there is a positive sense of comfort to be manipulated after all sources of discomfort have been eliminated? Zhang *et al.* (1996) and Helander and Zhang (1997) examined the question of perceptions of comfort and discomfort for seating. They found that perceptions of comfort and discomfort were independent of each other, rather than a continuum. Specifically, the perception of discomfort experienced when sitting was characterized by feelings of pain, soreness, and numbness that changed over time and could be related to the physiological stresses the seating produced. Perceptions of comfort were related to feelings of well-being and esthetics that changed little over time, and could be linked to perceptions of luxury and plushness. Applying this framework to lighting suggests that the most of the recommendations made by authoritative bodies about desirable lighting conditions are concerned with eliminating visual discomfort while how lighting designers make a living is to provide visual comfort. This chapter is devoted to the topic of light and visual discomfort.

5.2 The characteristics of visual discomfort

Visual discomfort has a number of distinctive features. First, visual discomfort is characterized by large individual differences, so much so that some of the early studies of discomfort glare used panels of subjects chosen for the reliability of their responses rather than their representative nature. Part of this individual variability is due to the fact that asking people to identify when a condition becomes uncomfortable involves both a discrimination and a criterion, i.e. the individual has to be able to tell when a condition occurs and then decide if this is uncomfortable or not. For lighting, the discrimination part of this process is likely to be determined by the characteristics of the visual system, which inevitably includes some individual variability, but the criterion part adds another element of variability. The problem of including the criterion is that what lighting is considered uncomfortable, or more importantly, acceptable, is based on past experience and hence on expectations and on attitudes. People have expectations about all sorts of things in life, from relatively simple pieces of hardware such as cars and computers to more sophisticated concerns such as health care and personal relationships, and these expectations are likely to change over time. There is no reason why lighting should be exempt from such shifting expectations. The problem comes about because different people, in the same or different cultures, have different experiences and hence different expectations. The effect of expectations was evident in an attempt to get lighting designers worldwide to rate the quality of lighting in a series of offices furnished with the cubicle systems common in North America. Agreement could only be produced by separating the reactions of designers working in North America, who had experienced similar installations to those presented, from the non-North American designers, who had no such experience (Veitch and Newsham, 1996).

Second, visual discomfort is dependent on context. Lighting conditions that are considered uncomfortable in one application may not be considered uncomfortable in another. For example, flicker in an office is undesirable but in a dance club it is used to generate excitement.

Third, the determinants of visual discomfort cover the whole visual field. This separates visual discomfort from visual performance. The aspects of lighting relevant to visual performance are generally restricted to the immediate task area. The aspects of lighting affecting visual discomfort can occur anywhere within the lit space.

5.3   General causes of visual discomfort

Visual discomfort can be identified by many different measures, ranging from the frequent occurrence of health symptoms that can be linked to exposure to the lighting to such vagaries as whining about the lighting. Among the symptoms that may be taken as markers of visual discomfort are red, sore, itchy, and watering eyes; headaches; and aches and pains associated with poor posture. Of course, there are many other possible causes of these symptoms. Thus, the occurrence of such symptoms should not be taken to mean that the lighting is at fault until the alternatives have been considered. A similar situation prevails for vague whining about the lighting. Photometric measures relevant to the specific aspect of the lighting complained of should be made before accepting the complaints. If the photometric measures are consistent with the complaints then the complaints are probably justified. Systematic procedures for evaluating the lighting of offices and parking lots and using both occupants' opinions and photometric quantities have been developed (Eklund and Boyce, 1996; Boyce and Eklund, 1998).

In its most general sense, lighting is designed to enable the visual system to extract information from the visual environment. Therefore, this examination of the causes of visual discomfort will start by considering the aspects of the visual environment that are likely to influence the ability to extract information. They are:

Visual task difficulty: Any visual task which has visual stimuli close to threshold contains information which is difficult to extract. This in itself leads to headaches and fatigue, but if the problem is of small visual size, there is the possibility of an additional effect. The usual reaction to a small visual size is to bring the task closer. As the task is brought closer, the accommodation mechanism of the eye has to exert pressure on the lens to increase the optical power of the eye and so keep the retinal image sharp. This adjustment can lead to muscle fatigue and hence symptoms of visual discomfort.

Under- and over-stimulation: Discomfort occurs either when there is no or little information to be extracted or when there is an excessive amount of repetitive information. Examples of no-information anywhere in the visual field occur rarely in real life, which is just as well because prolonged exposure to a very uniform field of luminance created by wearing translucent goggles produces severe disturbances of visual perception that can cause anxiety and panic (Corso, 1967). Less extreme conditions, such as might occur in an all-white room lit indirectly, can still be uncomfortable, as anyone who has looked into an integrating sphere for any length of time will know. As for overstimulation, the important point is not only the total amount of visual information, but also the presence of large areas of the same spatial frequency. Wilkins (1995) has associated the presence of large areas of specific spatial frequencies in printed text with the occurrence of headaches, migraines, and reading difficulties.

Distraction: The human visual system has a large peripheral field which detects the presence of objects that are then examined using the small, high-resolution fovea (see Chapter 2). For this system to work, objects in the peripheral field which are bright, moving, or flickering have to be easily detected. If, upon examination, these bright, moving, or flickering objects prove to be of little interest, they become sources of distraction because their attention gathering power is not diminished after one examination. Ignoring objects which automatically attract attention is stressful and can lead to symptoms of visual discomfort.

Perceptual confusion: The visual environment consists of a pattern of luminances, developed from the differences in reflectance of the surfaces in the field of view and the distribution of illuminance on those surfaces. Perceptual confusion can occur when there is a pattern of luminances present which is solely related to the illuminance distribution and conflicts with the pattern of luminance associated with the reflectances of the surfaces.

5.4   Specific causes of visual discomfort

There are many different aspects of lighting that can cause visual discomfort. Insufficient light for the performance of a task has already been considered (see Chapter 4) and will not be discussed further. Rather, attention will be devoted to uniformity, glare, veiling reflections, shadows, and flicker.

5.4.1   Uniformity

While exposure to a completely uniform visual field is undesirable, it is also possible to have too much non-uniformity. For this reason, recommendations of lighting practice usually include something about a minimum illuminance uniformity. Such recommendations are implemented through manufacturers specifying a maximum spacing to mounting height ratio for a regular array of their luminaires; luminaires with different luminous intensity distributions having different maximum spacing/mounting height ratios.

Saunders (1969) measured the acceptability of different illuminance ratios in a windowless room by having subjects occupy two desks successively and then asking them how reasonable it was to have such a difference in illuminance. Figure 5.1 shows the outcome. It can be seen that as the uniformity ratio, measured as lower/higher illuminance, dropped below about 0.7 there was a marked increase in the percentage of people considering the lighting pattern unreasonable. This finding has since been confirmed (Slater *et al.*, 1993), also in a windowless room. However, a moments consideration will suggest that this criterion probably only applies where the lighting installation is perceived by the occupants to be intended to produce a uniform distribution of illuminance. In rooms with large windows, the illuminance on a desk close to the window will be much greater than on a desk well back from the window so the illuminance uniformity ratio will be much less than 0.7, but few complaints are heard. Similarly, studies in offices where the luminaires can be individually switched or dimmed have shown that wide variations in the illuminance on desks can be tolerated, without complaint (Boyce, 1980). What this implies is that where non-uniformity of illuminance is expected or is exchanged for some other benefit, such as a view out of a window or individual control of the lighting, illuminance uniformity requirements can be relaxed. This in turn suggests that illuminance uniformity limitations are more a design requirement adopted to ensure that no one has insufficient illuminance for their work rather than an intrinsic requirement of the visual system.

So far, this discussion of illuminance uniformity has been large scale, i.e. variations between desks and across spaces. However, there is another level at which uniformity needs to be considered – that of the individual work surface. Two potential sources of discomfort for workplace lighting are distraction and perceptual confusion. Distraction can occur where there are areas of high illuminance adjacent to the work area. Studies of peoples'

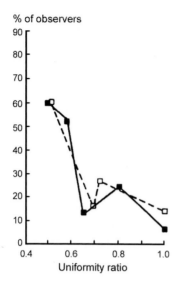

% of observers

Figure 5.1 The percentage of observers considering the uniformity of illuminance between two desks unreasonable, plotted against uniformity
ratio. Uniformity ratio is the lower illuminance/higher illuminance.
For the continuous line the variation in uniformity was achieved by
dimming one luminaire. For the broken line, different spacing of
luminaires was used to produce the change in uniformity ratio (after
Saunders, 1969).

reactions to several different forms of local lighting for desks have shown
that the most preferred form is one that provides a uniform illuminance
over an area of about $1\,m^2$ where the work is to be done and lower illuminances outside that area (Boyce, 1979a). Having high illuminances immediately outside the work area was distracting and irritating. As for
perceptual confusion, this can occur when the illuminance pattern has
a sharp edge so that it could be mistaken for a change in reflectance.

From the above discussion it should be clear that most lighting design has
to do with the distribution of illuminance. However, what the visual system
sees is the pattern of luminance. Fortunately for simplicity, common practice makes this less of a problem for workplace lighting than might be sup-
posed because the common practice is to make the working surface of
uniform reflectance, the variations in reflectance being introduced by the
materials placed on the work surface. Figure 5.2 shows the percentage of
people finding different levels of illuminance uniformity across a task
acceptable for four different tasks using different materials on a desk and
for the desk itself without a task (Slater and Boyce, 1990). Clearly the different conditions lead to slightly different results but the trends are consistent, decreasing illuminance uniformity increases the percentage who find

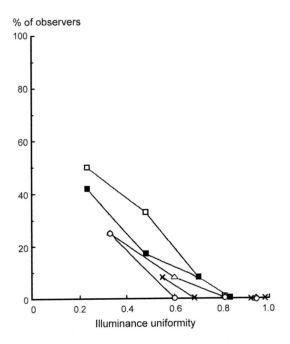

% of observers

Illuminance uniformity

Figure 5.2 The percentage of observers rating the evenness of lighting on the task as unacceptable plotted against illuminance uniformity across the task, for three tasks that required the use of different parts of the desk and for an empty desk (■). Uniformity is expressed as the ratio of the minimum to the maximum illuminance on the task (after Slater and Boyce, 1990).

the conditions unacceptable. Also, it is clear that a minimum illuminance uniformity ratio of about 0.7 will be acceptable for most people.

Given that a uniform illuminance is provided over the work surface, there is still the possibility that discomfort could occur because of a poor choice of desk surface reflectance relative to the reflectance of the task materials. Touw (1951) examined this question by having subjects copy figures onto white paper while sitting at six different gray desks, each with a different reflectance. The results showed that the preferred luminance ratio (desk/paper) was 0.4 although it decreased slightly with higher illuminances. This ratio should be treated as approximate at best. Other studies have produced different preferred surround/task luminance ratios, ranging from 0.1 to 1.0 depending on the specific situation (Rea *et al.*, 1990a). The median luminance ratio for these studies is 0.4. Given that white paper has a reflectance of about 0.75, this implies a desired desk reflectance of about 0.3. If a reflectance of this order is not possible with an existing desk, then the old-fashioned blotter provides a convenient way to provide the desired reflectance for the immediate surround to the work.

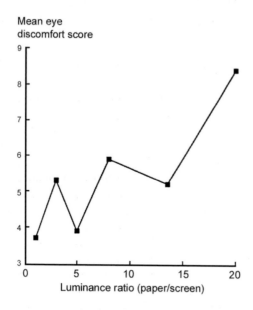

Figure 5.3 The mean eye discomfort score given by women working for several hours between documents and VDTs, plotted against the ratio of the luminances of the paper and the VDT screen (after Wibom and Carlsson, 1987).

One other aspect of uniformity that is important is when it is necessary to move the point of fixation from a low luminance surface to a high luminance surface and back again, repeatedly, such as looking between a computer monitor and a piece of paper. Wibom and Carlsson (1987) examined this question with a field survey of almost 400 workers undertaking such work. Figure 5.3 shows the reported mean eye-discomfort score (a combination of eight reported symptoms of eye discomfort weighted according to the frequency and intensity of symptoms) for 281 women working at such a task for more than 5 h a day, plotted against the luminance ratio of the paper source document and the screen. It can be seen that a luminance ratio of greater than about 15:1 produces a marked increase in eye discomfort.

Clearly, the uniformity criteria that appear in lighting recommendations should be treated as guidance rather than matters of life or death. Complete uniformity of luminance is bad for vision, but the visual system is very tolerant of variations in luminance in the visual scene, indeed it is such variations that make seeing possible. But different degrees of uniformity are desirable in different locations. The most uniform illumination is required in the immediate working area. If working areas are scattered throughout a space, as in a multi-person office, and may move around over time, e.g. when the office is rearranged, then a high level of illuminance uniformity is necessary over the whole working plane. If a working area is stable in location, then away from the working area a greater degree of illuminance non-uniformity will be tolerated and may even be valued for the interest it gives to a space (see Chapter 6).

Given that uniform illumination is desirable for the working area, it is important to note that illuminance uniformity, expressed as a ratio of minimum to maximum illuminance, is not a complete metric. It is also necessary to consider the locations of the minimum and maximum illuminances, the rate of change of illuminance between them and the reflectances of the surfaces illuminated, including the task. This is because it is also necessary to establish a hierarchy of luminances, where the actual work has the highest luminance, its immediate surround is the same or less in luminance and the background is the same or even lower in luminance. It is also necessary to eliminate sudden changes in illuminances, thereby ensuring that any sudden changes in luminance are associated with a change in reflectance.

All the above has been associated with the lighting of work, particularly office work, where the purpose of the lighting is to extract information from a two-dimensional surface. For three-dimensional material, such as often occurs in industry, non-uniform illumination patterns are required to reveal the form and texture of the object. This approach is taken to extremes in the theatre and is widely used in retailing to produce dramatic displays and change people's reactions to the lit object (Mangum, 1998). Thus, as with all the other factors that can cause discomfort, whether a non-uniform illuminance distribution actually causes discomfort will depend on the context in which it occurs.

5.4.2   Glare

The discomfort that may be experienced when non-uniform illumination is used in the wrong context tends to build up over time. Glare is a much more extreme form of non-uniformity and is usually apparent immediately. When faced with a very high luminance in the visual field the usual behavior is to blink and look away, or to shield the eyes from the source of high luminance. This behavior can be taken as an indication that glare is present.

Vos (1999) has suggested eight different forms of glare. Of these eight, four occur rarely. One is flash blindness, which is a temporary state of complete bleaching of retinal photopigment caused by the sudden onset of an extremely bright light sources, e.g. a nuclear explosion. Another is paralyzing glare, so named for the phenomenon in which a person suddenly illuminated by a searchlight at night will tend to "freeze" briefly. Another is exposure to light bright enough to cause retinal damage (see Section 13.2). The last is distracting glare, produced by bright, flashing lights in the peripheral visual field, e.g. lights on emergency vehicles at night. These are all special situations remote from conventional lighting, so they will not be discussed further.

The other four forms of glare are more commonly experienced. The first occurs when a large part of the visual field is bright. This is called dazzle or saturation glare. This is painful and the behavioral response is to shield the eyes in some way, by wearing low-transmittance glasses or shields that restrict the view of the outside world by means of slits. Such devices are used in such diverse settings as the beaches of California and the icy wastes of the Arctic. Despite the differences in culture and lifestyle of the people who inhabit such places, the need to shield the eyes from large areas of very high luminance is common. Saturation glare occurs rarely indoors. It is much more common outdoors. It is probably a spasm of the iris sphincter that causes the sensation of pain (Lebensohn, 1951).

Saturation glare occurs when the large part of the visual field is at a high luminance for a long time. Another form of glare commonly experienced is adaptation glare. This occurs when the visual system is exposed to a sudden, large increase in luminance of the whole visual field, e.g. on exiting a cinema during daytime. The perception of glare is due to the visual system being oversensitive because it is adapted to the darkness of the cinema while being suddenly exposed to the brightness of daylight. Adaptation glare is temporary in that the processes of visual adaptation will soon adjust the visual sensitivity to the new conditions. It can be avoided altogether by designing lighting to give a transition zone of intermediate luminance between the dark area and the bright area, the transition zone being large enough to allow the visual system time to adapt to the prevailing conditions.

The other two forms of glare commonly experienced are essentially a matter of the range of luminances simultaneously present in the visual field. They are disability glare and discomfort glare.

5.4.2.1   Disability glare

Disability glare, as its name implies, disables the visual system to some extent. This disabling is caused by light scattered in the eye (Vos, 1984). The scattered light forms a luminous veil over the retinal image of adjacent parts of the scene, thereby reducing the luminance contrasts of the image of those parts on the retina. The amount of disability glare can be measured by comparing the visibility of an object seen in the presence of the glare source with the visibility of the same object seen through a uniform luminous veil. When the visibilities are the same, the luminance of the veil is a measure of the amount of disability glare produced by the glare source, and is called the equivalent veiling luminance. Numerous studies have lead to several different empirical methods to predict the equivalent veiling luminance (Holladay, 1926; Stiles, 1930; Stiles and Crawford, 1937). Based on this work, an equation was developed to predict the equivalent veiling luminance from directly measurable variables. It is

$$L_v = 10 \Sigma E_n \Theta_n^{-2}$$

where L_v is the equivalent veiling luminance (cd/m^2), E_n the illuminance at the eye from the nth glare source (lx), and Θ_n the angle between the line of sight and the nth glare source (degrees).

The effect of the equivalent veiling luminance on the luminance contrast of the object can be estimated by adding it to the luminance of both the object and the immediate background (see Section 2.4.1 for luminance contrast formulae).

The formula above is adequate for glare sources between about 1° and 30° from the line of sight and for young people. Recently, a series of modifications of the formula have been suggested to extend the range over which accurate predictions can be made from 0.1° to 100°, for age ranges up to 80 years and for eye iris color (Vos, 1999; CIE, 2002b). Age is important because as the eye ages the amount of scatter in the eye increases (see Section 12.2). Iris color only becomes important at angles greater than about 30°, but above this limit, people with blue eyes experience more disability glare than people with brown eyes and non-Caucasian eyes.

Disability glare can be associated with point sources and large area sources. The disability glare formulae can be applied directly to point sources but for large area sources, the area has to be broken into small elements and the overall effect integrated (Adrian and Eberbach, 1969; Adrian, 1976). Disability glare from point sources is experienced most frequently on the roads at night when facing an oncoming vehicle, although it can also occur during daytime when the sun is close to the required line of sight. As for disability glare from an extended source, this can occur outside when approaching a road tunnel during daytime. Then the sky above the tunnel entrance can act as a glare source. It can also occur indoors when a bright sky is visible through a window.

5.4.2.2 Discomfort glare

Disability glare is well understood. It has an effect on visual capabilities that can be measured with conventional psychophysical procedures and a plausible mechanism, light scatter in the eye. Discomfort glare is not well understood. It is said to be occurring when people complain about visual discomfort in the presence of bright light sources, luminaires, or windows. There is no known cause for discomfort glare, although suggestions have been made ranging from fluctuations in pupil size (Fry and King, 1975) to distraction (Lynes, 1977). The separation between disability and discomfort glare should not be taken to mean that disability glare does not cause visual discomfort, nor that discomfort glare does not diminish the capabilities of the visual system. Whenever disability glare makes what needs to be seen more difficult to see, complaints of discomfort are likely to occur. As for the disabling effect of what is conventionally called discomfort glare, the failure to find any effects of visual capabilities is probably more a matter of measurement sensitivity than anything else. In essence, these two forms of glare, disability glare and discomfort glare, are simply two different outcomes of the same stimulus pattern, namely a wide variation of luminance across the visual field. When considering the likelihood of glare occurring for a given lighting situation it is wise to consider both disability and discomfort glare.

Discomfort glare has been studied for more than 50 years (Fischer, 1991). The outcome of this work has been a plethora of systems for predicting the degree of discomfort produced in different lighting situations. Most of these systems use a formula to calculate the glare sensation produced by an array of luminaires. The formulae are all different but, for a single glare source, they all have the following form:

$$\text{Glare sensation} = (L_s^a \cdot \omega_s^b)/(L_b^c \cdot p^d)$$

where L_s is the luminance of the glare source (cd/m^2), ω_s the solid angle subtended at the eye by the glare source (steradians), L_b the luminance of the background (cd/m^2), and p the deviation of the glare source from the line of sight.

Each component of the formula has a different exponent and these differ between the different formulae. The form of the formula indicates the effect of the different components. Increasing the luminance of the glare source, increasing the solid angle of the glare source, decreasing the luminance of the background, and decreasing the deviation of the glare source from the line of sight will all increase the glare sensation. Changes of each component in the opposite direction will decrease the glare sensation. In practice, the components interact. For example, consider the effect of lowering the mounting height of a luminaire in a room, keeping everything else the same. Assuming the direction of view is horizontal along the length of the room, then lowering the mounting height clearly reduces the deviation from the line of sight but it is also likely to increase the luminance of the background and, depending on the shape and luminous intensity distribution of the luminaire, it may also change the angular size and the luminance of the luminaire. To what extent these changes amplify or compensate can only be determined by applying the full equation.

Until recently, there were three major discomfort glare prediction systems in use. The first to be developed was the visual comfort probability (VCP) system used in North America. VCP is based on the work of Guth (1963). For a single glare source, the formula for glare sensation is

$$\text{Glare sensation} = M = (0.50 L_s \cdot Q)/(P \cdot F^{0.44})$$

where L_s is the luminance of the glare source (cd/m^2); $Q = (20.4 W_s + 1.52 W_s^{0.2} - 0.075)$, where W_s is the solid angle subtended at the eye by the glare source (steradians); P an index of the position of the glare source with respect to the line of sight (Guth position index); and F the average luminance of the field of view, including the glare source (cd/m^2).

The glare sensation produced by a number of glare sources is summed to form the discomfort glare rating (DGR) using the formula

$$\text{DGR} = (\Sigma M_n)^a$$

where DGR is the discomfort glare rating and $a = n^{-0.0914}$, where n is the number of glare sources.

The DGR values are then converted to the VCP, which is simply the percentage of people who would be expected to find the conditions represented by the DGR acceptable. Luminaire manufacturers in North America use the VCP system to produce tabular estimates of the level of discomfort glare produced by a regular array of their luminaires for a range of standard interiors. These tables provide all the precision necessary for estimating the level of discomfort glare likely to occur in interiors

By consensus, discomfort glare is believed not to be a problem in lighting installations if all three of the following conditions are satisfied:

- The VCP is 70 or more.
- The ratio of the maximum luminaire luminance (luminance of the brightest 6.5 cm²) to the average luminaire luminance does not exceed 5:1 at 45°, 55°, 65°, 75°, and 85° from nadir, for crosswise and lengthwise viewing.
- Maximum luminances of the luminaire crosswise and lengthwise do not exceed the values given in Table 5.1.

The VCP system is based on empirical relations derived from a variety of experiments. It has been concluded that differences of 5 percent or less are not significant. In other words, it is only if two lighting systems differ in VCP by more than 5 percent that there is a basis for judging that there is a difference in discomfort glare between them.

Table 5.1 Maximum luminance of a luminaire, seen at various angles from nadir, crosswise and lengthwise, if discomfort glare is to be avoided (after IESNA, 2000a)

Angle from nadir (°)	Maximum luminance (cd/m²)
45	7,710
55	5,500
65	3,860
75	2,570
85	1,695

The next system developed was the British Glare Index system (CIBSE, 1985). This system was based on the work of Petherbridge and Hopkinson (1950). For a single glare source the formula is

$$\text{Glare sensation} = G = (0.9L_s^{1.6} \cdot \omega^{0.8})/(L_b \cdot p^{1.6})$$

where L_s is the luminance of the glare sources (cd/m^2), ω the solid angle subtended at the eye by the glare source (steradians), L_b the luminance of the background defined as the uniform luminance of the background that produces the same illuminance on a vertical plane at the eye as the visual field under consideration, excluding the glare source (cd/m^2), and p the Guth position index.

For multiple glare sources, the glare sensations for the individual luminaires are summed using the formula

$$\text{Glare index} = 10\log_{10}(0.5\Sigma G)$$

Glare index typically ranges in value from 10 to 30, a value of 10 indicating imperceptible glare and a value of 30 indicating very uncomfortable glare. Like the VCP system, the glare index system is used by luminaire manufacturers to produce tabular estimates of the level of discomfort glare produced by a regular array of their luminaires for a range of standard interiors. These tables, too, provide all the precision necessary for estimating the level of discomfort glare likely to occur in interiors. Recommendations of glare index suitable for different applications have been given in past lighting standards (CIBSE, 1994). As for accuracy, studies have shown that a difference of 1 unit in glare index is a just detectable difference and that a difference of 3 units will always be seen (Collins, 1962).

The third system arose from dissatisfaction with the summation methods used in the VCP and glare index systems to predict the effects of many luminaires seen together (Arndt *et al.*, 1959). The outcome of this dissatisfaction was a series of experiments using one-third scale models of offices (Bodmann *et al.*, 1966). The lighting of the models was provided by different fluorescent luminaires, with different luminous intensity distributions, in different numbers, in different sizes of rooms and oriented either across or along the direction of view. Observers were asked to view each model and assess the degree of discomfort, using a seven-point rating scale labeled 0 = no glare, 1 = glare between non-existent and noticeable, 2 = glare noticeable, 3 = glare between noticeable and disagreeable, 4 = glare disagreeable, 5 = glare between disagreeable and intolerable, and 6 = glare intolerable. Only four factors were found to affect the glare rating markedly. These were the luminance of the luminaire, the room length and mounting height of the luminaire, the adaptation luminance, and the type of luminaire, specifically whether or not it had luminous sides. The results of these observations for a regular array of identical ceiling-mounted luminaires providing a fixed horizontal illuminance of 1,000 lx were presented as a set of curves of luminaire luminance plotted as a function of emission angle, i.e. the angle between the normal to the central luminaire of the array and the line from that luminaire to the observer. Each curve corresponds to a fixed glare rating (see Figure 5.4). A correction factor was applied to the glare rating to take the effect of different illuminances into account. The correction factor was

$$\Delta G = 1.16 \log_{10}(E/1,000)$$

where ΔG is the quantity to be added to or subtracted from the glare rating achieved at 1,000 lx, and E the horizontal illuminance (lx).

The reliability of these results were later confirmed by data collected in full-size installations and from real-life lighting installations. These curves form the basis for a simple glare prediction system. By plotting the luminance of a luminaire at different emission angles on the curves shown in Figure 5.4, the worst glare sensation can be identified from the highest value iso-rating curve that is not intersected by the plot of the luminaire luminance at different emission angles.

Fischer (1972) used these results to derive an approximate system, for luminaires with and without luminous sides, arranged in a regular array with their main axis parallel to the walls, viewed by an observer standing in the middle of the rear wall and looking in a horizontal direction parallel to the walls. The room surfaces were assumed to have reflectances of: ceiling = 0.7–0.8, walls = 0.4–0.6, and floor = 0.1–0.2. Five maximum levels of glare rating were selected to give five different levels of quality. This method was called the European glare limiting method. As originally developed, it was based on ceiling-mounted luminaires using fluorescent lamps. Later experiments extended it to luminous ceilings (Sollner 1972) and to luminaires using high-pressure sodium lamps (Sollner, 1974).

The data collected during the development of the European glare limiting system place it on the firmest footing of the three systems discussed. Complete lighting installations, using real luminaires, were evaluated by a meaningful number of people. Nonetheless, the European glare limiting system does have a practical disadvantage. The mean glare rating can only be changed by changing the luminaire luminance distribution. The VCP and glare index systems also allow this possibility but, in addition, they allow for the possibility of changing the level of discomfort glare by changing room reflectances and can deal with an irregular array of luminaires. Thus, the VCP and glare index methods reflect the influence of the complete luminous environment on discomfort glare, while the European glare limiting system only allows for differences in luminaires.

Despite these differences, the three methods considered do show some agreement. Manabe (1976) calculated the VCP, glare index, and mean glare rating values for a wide range of installations, lit using different

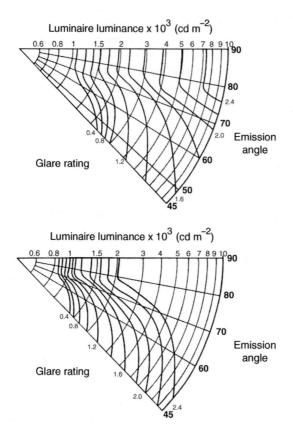

Figure 5.4 Luminaire luminance at different emission angles for different glare
ratings. For the upper diagram, the luminaire is seen end-on; for the
lower diagram, the luminaire is seen side-on. The emission angle is
the angle between the downward normal to the luminaire and the
line joining the luminaire and the observer (after Fischer, 1972).

luminaire types, in Tokyo. He found the following correlation coefficients –
VCP versus glare index $= -0.67$, VCP versus mean glare rating $= -0.31$,
and glare index versus mean glare rating $= 0.32$. Higher correlation coefficients were found when only one type of luminaire was used, but that rather
misses the point which is that any discomfort glare prediction system that is
of value has to be able to handle different luminaire types.

This state of affairs was obviously unsatisfactory and lead to a number
of initiatives by the CIE. In 1983, the CIE adopted a modified glare index
system based on the work of Einhorn (CIE, 1983b) but this proved to be
rather impractical in use. However, it did resolve one problem with the original glare index system. This was that the exponent of the solid angle subtended at the eye by the luminaire was not unity but rather 0.8. This means

that if a lighting installation of large luminaires was treated as consisting of twice the number of luminaires of half the size, the calculated glare index would be different. To avoid this "subdivision problem" the exponent of the solid angle has to be unity.

In 1987, a new discomfort glare formula was proposed by Sorensen (1987). This formula took the form

$$UGR = 8 \log_{10} (0.25/L_b) \Sigma (L_s^2 \cdot \omega/p^2)$$

where UGR is the unified glare rating, L_b the background luminance (cd/m^2), excluding the contribution of the glare sources (this is numerically equal to the indirect illuminance on the plane of the observer's eye, divided by π), L_s the luminance of the glare source (cd/m^2), ω the solid angle subtended at the observer's eye by the glare source (steradians), and p the Guth position index of the glare source.

The Sorensen formula forms the basis of what is called the unified glare rating (UGR) system. It is indeed a unified system, because the underlying formula combines aspects of the formulae from which the VCP and glare index system were derived and incorporates Einhorn's requirement that the exponent of the solid angle should be unity. The constants were chosen so that the UGR formula would give good agreement with the values that would have been obtained by the original glare index system for the same installation. The CIE has adopted the UGR formula (CIE, 1995b) and from it has developed a series of methods for presenting the results in forms familiar to users in different countries. There are now tabular UGR presentations for use by those who are familiar with the glare index system. There are UGR curves for those who in the past used the European glare limiting system. There are even proposals for converting the UGR values to VCP values for North America (Sorensen, 1991), a step that would be sensible given the high correlation between UGR and VCP ($r = 0.82$) obtained from calculations of the discomfort glare experienced in a room lit by regular arrays of 30 different types of fluorescent luminaires, including both prismatic and parabolic types (Mistrick and Choi, 1999).

In a sense, the UGR system represents a triumph of mathematical manipulation over human perception. The adoption worldwide of the UGR system would undoubtedly make life easier for those seeking to globalize the lighting business but whether it would represent an improvement in the accuracy with which the perception of discomfort glare can be predicted is open to question. This is because the UGR system has been developed from the original discomfort glare formulae with the twin aims of mathematical consistency and practical applicability in mind. Unfortunately, the accuracy with which these original formulae predict subjective responses is poor. Figure 5.5 shows data for a number of different Japanese lighting installations. The vertical axis on each graph is the mean subjective rating of glare, while the horizontal axis represents the calculated VCP, glare index, or

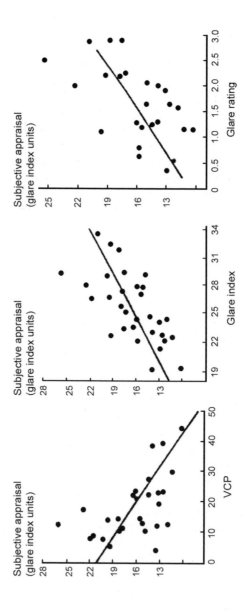

Figure 5.5 Mean subjective ratings of glare plotted against the calculated VCP, glare index, and mean glare rating for a number of different lighting installations in Tokyo. The subjective rating of glare is expressed on a glare index scale: 13 = perceptible glare, 16 = just unacceptable glare, 19 = unacceptable glare, 22 = just uncomfortable glare, 25 = uncomfortable glare, and 28 = just intolerable glare (after Manabe, 1976).

mean glare rating. The correlation coefficient between the mean subjective ratings and the calculated vales are, for VCP = −0.64, for glare index = 0.64, and for mean glare ratings = 0.52 (Manabe, 1976). What this means is that VCP and glare index explain about 41 percent of the variation in mean subjective ratings and the mean glare rating explains 27 percent. As for UGR, Akashi *et al.* (1996) examined the relationship between mean subjective ratings of 56 non-experts in lighting and the calculated UGR values for an office, using one type of bright-sided luminaire. The correlation coefficient between the UGR values and the mean subjective ratings was 0.89, although there was a bias such that the mean subjective ratings consistently indicated less glare than predicted by the UGR values. Given the origins of the UGR system, this high correlation is surprising. It would be interesting to know how much it would be altered by introducing different luminaire types and different room types into the data. Until this is done, a provisional conclusion is that UGR is the best system currently available for predicting peoples' perceptions of discomfort glare in interiors. Indeed, it may represent the best that is achievable, given that discomfort glare studies have consistently shown very large individual differences (Stone and Harker, 1973).

The existence of the UGR system should not be taken to mean that we know all we need to know to predict and hence control discomfort glare from lighting installations. Unfortunately, there remain a number of questions of application that need answering. For real luminaires, there can be problems in determining the luminance and solid angle of the glare source. For example, parabolic luminaires, which are one of the most widely used types of fluorescent luminaires, do not have a uniform luminance with sharply defined edges. Rather, they are non-uniform in luminance. This makes it difficult to determine the glare source area and hence the solid angle. As for glare source luminance the conventional method of calculation is to divide the luminaire luminous intensity in a given direction by the projected area of the luminaire in the same direction. If the area of the glare source is ambiguous, then the glare source luminance must be ambiguous also. Even if the glare source area is defined, it must be doubted if the resulting average luminance is the true measure of the glare source luminance for non-uniform luminance luminaires (Waters *et al.*, 1995). CIE (2002b) offers some advice on how to deal with such luminaires.

Another factor to be considered is the effect of the luminance of the immediate surround to the glare source. In his original formulation of the glare sensation, Hopkinson (1963) included a fifth term for the immediate surround luminance. He found that an immediate surround, intermediate in luminance between the glare source luminance and the background luminance, produced a marked reduction in glare sensation. None of the glare prediction systems consider the immediate surround although luminaire designers are certainly aware of the potential of the immediate surround to reduce glare sensation.

There are also questions about glare source size. Einhorn (1991) points out that all the glare control systems in use predict intolerable glare for very small sources, such as occur in chandeliers, yet these are tolerated very well. Likewise, there is problem with large sources because then the area of the glare source is large enough to interact with the adaptation luminance. CIE (2002b) recommends a simple classification of size whereby small sources have areas less than $0.005 \, \text{m}^2$; and large sources have areas greater than $1.5 \, \text{m}^2$. Paul and Einhorn (1999) have shown that discomfort glare for small sources more than 5° off-axis is determined by the luminous intensity in the direction of the eye, not by the luminance of the source; and the UGR formula can be made applicable to such small glare sources by replacing the summed term $(L_s^2 \cdot \omega)$ in the formula with $(200 \, I^2/R^2)$ where I is the luminous intensity in the direction of the eye (cd) and R the distance from the source to the eye (m). For large sources, CIE (2002b) recommends a transition formula for use with luminaires having areas just greater than $1.5 \, \text{m}^2$ up to really large luminaires. The formula is

$$GGR = UGR + (1.18 - (0.18/CC)) \, 8\log(2.55(1 + (E_d/220))/(1 + (E_d/E_i)))$$

where, GGR is the large room glare rating, UGR the unified glare rating, CC the ceiling coverage equal to A_0/A_1, where A_0 is the projected area of the glare source towards the nadir (m^2), A_1 the area lit by one glare source (m^2) = the room area/number of glare sources, E_d the direct illuminance at the eye from the glare source (lx), E_i the indirect illuminance at the eye (lx).

The same GGR and UGR values represent the same level of discomfort glare. Where a luminous ceiling or uniform indirect lighting is used, CIE (2002b) proposes a limit on the average illuminance provided. Specifically, if a UGR value of 13 is desired then the average illuminance provided should not exceed 300 lx, for UGR = 16, the average illuminance should not exceed 600 lx and for UGR = 19, the average illuminance should not exceed 1,000 lx.

Finally, there is a question about the limits on deviation from the line of sight. The Guth position index is usually limited to deviations less than 53°, it being assumed that there is no sensation of glare at greater angles, but it has been shown that given sufficient luminance, discomfort glare can be experienced from a luminaire at much larger deviations, i.e. overhead (Ngai and Boyce, 2000; Boyce *et al.*, 2003a).

All the above are questions about applying the glare prediction systems to real luminaires in real lighting installations. These are relatively simple to solve compared to the major problem with the study of discomfort glare. This is that there is no established physiological or perceptual mechanism for discomfort glare. Until a plausible mechanism is identified, the study of discomfort glare is likely to remain a matter of playing catch-up, as new technology and new design trends throw up unexpected discomfort glare problems. Perhaps, inspiration can be gained from a related area. In the last

20 years, Wilkins (1995) has developed an understanding of temporal variations in luminance that lead to visual discomfort and headaches. Is it unreasonable to hope that some effort given to examining the spatial distribution of luminances that lead to visual discomfort would be similarly fruitful?

5.4.3 Veiling reflections

Veiling reflections are luminous reflections from specular or semi-matte surfaces that physically change the contrast of the visual task and therefore change the stimulus presented to the visual system (Figure 5.6). Veiling reflections and disability glare are similar in that both change the luminance contrast of the retinal image but differ in that veiling reflections change the luminance contrast of the task itself while disability glare changes the luminance contrast of the retinal image of the task.

The two factors which determine the nature and magnitude of veiling reflections are the specularity of the surface being viewed and the geometry between the observer, the surface, and any sources of high luminance. If the surface is a perfectly diffuse reflector, no veiling reflections can occur because then the distribution of light reflected from the surface is independent of the direction from which the light is incident. If the surface has a specular reflection component, veiling reflections can occur. Veiling reflections occur at positions where the geometry between the observer, the surface and any sources of high luminance is such that the angle of incidence equals the angle of reflection and the reflection is in the direction from the surface to the eye of the observer.

The magnitude of veiling reflections can be quantified by the contrast rendering factor (CRF). The CRF of a surface at a specific location and viewed from a particular direction is the ratio of the luminance contrast of the object under the lighting of interest to the luminance contrast of the object under completely diffuse lighting. Completely diffuse lighting produces only weak veiling reflections so a CRF value close to unity is desirable if veiling reflections are to be avoided. Unfortunately for simplicity, where sources of high luminance are surrounded by areas of low luminance, e.g. most installations of recessed parabolic luminaries, the magnitude of veiling reflections, and hence the CRF can vary dramatically within the installation (Boyce, 1978). There is no such thing as a single value of CRF for a lighting installation. There is always a distribution of values.

The effect of veiling reflections on the luminance contrast of a specific target may be quantified by adding the luminance of the veiling reflection to the appropriate components in one of the luminance contrast formulae discussed in Chapter 2. What the appropriate components are depends on the reflection properties of the material being viewed. For glossy ink writing on matte paper, the luminance of the veiling reflections should only be added to the luminance of the ink. For a glossy magazine page or a visual

Figure 5.6 A glossy book, without and with veiling reflections.

display terminal (VDT) screen, where there is a specularly reflecting transparent coating over the whole surface, veiling reflections occur over the whole surface. In this case the luminance of the veiling reflections should be added to all terms in the luminance contrast formula.

The usual effect of adding the luminance of the veiling reflections to the luminance contrast formulae is to reduce the luminance contrast of the target. However, this is not always the case. For example, given a very specular black ink printed on matte white paper, it is possible for veiling reflections to increase the luminance of the ink so much that it becomes higher than the luminance of the paper, i.e. the polarity of the print is reversed and luminance contrast starts to increase again. This will lead to an increase in the CRF, despite the fact that the strength of the veiling reflections is increasing. This possibility means that CRF values should be evaluated with care wherever the material consists of both specular and matte reflecting elements.

The extent to which changes in luminance contrast change visual performance can be estimated using the relative visual performance model discussed in Chapter 4 but the extent to which it causes discomfort is different. Bjorset and Frederiksen (1979) measured the acceptability of the contrast reduction produced by veiling reflections for a wide range of hand-written and printed materials. Figure 5.7 shows the percentage reduction in contrast that 90 percent of their subjects found acceptable, plotted against the contrast of the material when no veiling reflections occurred. It is apparent that 90 percent of the subjects accepted a contrast reduction of about 25 percent regardless of whether the materials had high or low luminance contrast in the absence of veiling reflections. This result suggests two conclusions. The first is that discomfort can occur whenever lighting interferes

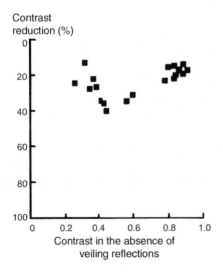

Figure 5.7 Percentage reduction in luminance contrast that 90 percent of observers found acceptable, plotted against the luminance contrast of the material when no veiling reflections occurred (after Bjorset and Frederiksen, 1979).

with the visibility of the task even if the reduction in visual performance is slight. The second is that weak veiling reflections are not critical for comfort, a conclusion supported by Reitmaier (1979), at least for pencil, pen, and printed text on matte and moderately glossy paper. However, for very glossy papers, people were very sensitive to the presence of even slight veiling reflections, something that De Boer (1977) also found.

Although veiling reflections are usually considered a negative outcome of lighting that can cause discomfort, they can be used positively, but when they are, they are conventionally called highlights. Physically, veiling reflections and highlights are the same thing. Display lighting of specularly reflecting objects is all about producing highlights to reveal the specular nature of the surface.

5.4.4   Shadows

Shadows are cast when light coming from a particular direction is intercepted by an opaque object. If the object is big enough, the effect is to reduce the illuminance over a large area. This is typically the problem in industrial lighting where large pieces of machinery cast shadows in adjacent areas. The effect of these shadows can be overcome either by increasing the proportion of inter-reflected light by using high reflectance surfaces or by providing local lighting in the shadowed area. If the object is smaller, the shadow can be cast over a meaningful area which in turn can cause perceptual confusion, particularly if the shadow moves. An example of this is the shadow of a hand cast on a blueprint. This problem can also be reduced by increasing the inter-reflected light in the space or by providing local lighting which can be adjusted in position.

Although shadows can cause visual discomfort, it should be noted that they are also an essential element in revealing the form of three-dimensional objects. Techniques of display lighting are based around the idea of creating highlights and shadows to change the perceived form of the object being displayed. Many lighting designers insist that the distribution of shadows is as important as the distribution of light in achieving an attractive and meaningful visual environment (Lam, 1977; Tregenza and Loe, 1998).

The number and nature of shadows produced by a lighting installation depends on the size and number of light sources and the extent to which light is inter-reflected around the space. The strongest shadow is produced from a single point source in a black room. Weak shadows are produced when the light sources are large in area and the degree of inter-reflection is high.

5.4.5   Flicker

Virtually all electric light sources that operate from an alternating current supply produce regular fluctuations in the amount and spectrum of light emitted. When these fluctuations become visible they are called flicker. A lighting installation which produces flicker will be almost universally disliked, unless it is being used for entertainment and then only when exposure is short. For some people, exposure to flicker can be a health hazard (see Section 13.3.2).

The main factors that determine whether a fluctuation in light output will be visible are the frequency and percentage modulation of the fluctuation at the eye, the proportion of the visual field over which the fluctuation occurs and the adaptation luminance. The higher is the adaptation luminance and the larger is the area, the more likely it is that a given frequency and percentage modulation fluctuation will be seen. Temporal modulation transfer functions (see Section 2.4.4) can be used to predict whether a given frequency and percentage modulation of fluctuation occurring over a large area will be visible at a given adaptation luminance. It is important to appreciate that there are wide individual differences in sensitivity to flicker (Hopkinson and Collins, 1970). This, together with the fact that electrical signals associated with flicker can be detected in the retina, even when there is no visible flicker (Berman *et al.*, 1991), implies that a clear safety margin is necessary to avoid discomfort from flicker.

The probability that a lighting installation will be seen to produce flicker can be minimized by ensuring a stable supply voltage and by the use of high-frequency electronic control gear for discharge lamps. Incandescent light sources do not require control gear but they are particularly sensitive to fluctuations in supply voltage. Where the local electricity network has equipment attached to it that can impose sudden large loads, e.g. the motors of a steel rolling mill, local fluctuations in supply voltage are likely and, in consequence, so are fluctuations in light output of incandescent light sources. These can be minimized by using a voltage regulator between the electricity supply and the light source. Discharge lamps are less sensitive to supply voltage fluctuations than incandescent lamps because the electricity supply is filtered through the control gear. It is the output of this control gear that determines whether a lighting installation using discharge lamps will produce flicker. Older-style electromagnetic control gear typically produces an output at the same frequency as the electricity supply frequency, i.e. at 50 or 60 Hz, depending on the country. Modern electronic control gear typically produces an output around 25–50 kHz. Given the time constants of the light producing processes in most discharge lamps, this increase in supply frequency not only produces a higher frequency in light output it also produces a smaller percentage modulation in light output. The use of such high-frequency control gear on fluorescent lamps has been associated with a reduction in the prevalence of headaches and eyestrain (Wilkins *et al.*, 1989).

Another approach used to reduce flicker is to combine light from lamps powered from different phases of the electricity supply on the working plane. This results in an increased frequency and a reduced percentage modulation and hence a decrease in the probability of flicker being seen when looking at the working plane. Obviously, it does nothing for the probability of flicker being seen when looking directly at an individual light source.

Although flicker occurring over a large area is almost always disturbing, localized flicker does have its uses. Localized flicker is a potent means of attracting attention because peripheral vision is sensitive to changes in the retinal illumination pattern, either in space or time. Localized flicker is widely used to attract attention to essential and non-essential information, such as the presence of emergency vehicles and the location of coupon dispensers in supermarkets.

5.5   Discomfort, performance, and behavior

While lighting conditions which make it difficult to achieve good visual performance will almost always be considered uncomfortable, lighting conditions which allow a high level of visual performance may also be considered uncomfortable. Figure 5.8 shows the mean detection speed of a group of 20–30-year-olds trying to find a specific two-digit number from 100 such numbers printed in black ink on gray paper and laid out at random on a table. Figure 5.8 also shows the percentage of the same group considering the lighting "good" (Muck and Bodmann, 1961). As might be expected, increasing the illuminance on the table increases mean detection speed and the percentage considering the lighting "good." However, as the illuminance exceeds 2,000 lx, the percentage considering the lighting "good" declines even though the mean detection speed continues to increase.

These results have three interesting implications. The first is that lighting conditions that are considered uncomfortable do not necessarily lead to a decrease in task performance. The reason for this lack of effect is probably motivation. In laboratory experiments, people are usually highly motivated to succeed and will try to ignore any discomfort. It must be doubted if the same degree of motivation occurs frequently outside the laboratory, in which case lighting that causes discomfort may lead to a decline in task performance. The important word here is "may." There can be no doubt that motivation can affect task performance and little doubt that lighting conditions can affect motivation, but so do many other factors. These other influences ensure that while lighting may sometimes affect performance by changing motivation, it is unlikely to have any effect that is consistent across individuals or even across time with the same individual, until extreme conditions are reached.

The second implication is that to achieve a lighting installation that will be liked by the people who work under it, it is necessary to provide lighting which allows easy visual performance and avoids discomfort. This may not be as simple as it sounds because the visual performance component of task performance is determined solely by the capabilities of the visual

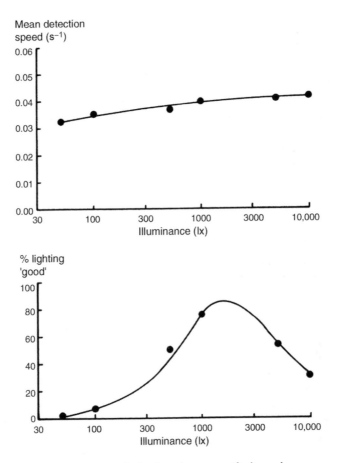

Figure 5.8 Mean detection speeds for locating a specified number amongst others at different illuminances and the percentage of subjects who consider the lighting "good" at each illuminance (after Muck and Bodmann, 1961).

system but visual comfort is linked to peoples' expectations. Any lighting installation that does not meet expectations may be considered uncomfortable even though visual performance is adequate; and expectations can change over time.

The third is that as lighting conditions vary, perceptions of visual discomfort will be more sensitive to changes than task performance. This, in turn, implies that judgements of visual discomfort are the most effective way to determine the quality of a lighting installation, a conclusion that has also been reached by Roufs and Boschman (1991) for visual displays.

Finally, it is interesting to note that visual discomfort can have behavioral consequences when discomfort is severe. People are not stupid. If there is an obvious source of visual discomfort in the environment, people will act to remove it or diminish its effect. It is not uncommon to find pieces of paper taped to the windows of offices. The usual reason for this is to prevent the light from the early morning or afternoon sun falling directly on a fixed computer monitor. Similarly, people will adjust vertical blinds during the day to prevent direct sunlight falling on their desks while preserving a view out (Maniccia *et al.*, 1999). Also, given that veiling reflections can be diminished dramatically by changing the geometry between the source of high luminance, material and observer, it is common for people to tilt the material or move their heads to reduce veiling reflections (Rea *et al.*, 1985a). The occurrence of behavioral modifications to the environment are a clear indication of visual discomfort occurring in their absence.

5.6   Visual discomfort and lighting quality

Lighting designers frequently claim that what they deliver is lighting quality and it seems obvious that avoiding visual discomfort is an inherent part of lighting quality, but is that all there is to it? This question is difficult to answer because there is no clear definition of lighting quality. Like pornography, we know it when we see it but what it is differs from one individual and culture to another. A number of different approaches to defining lighting quality have been suggested; from single-number photometric indices calibrated by subjective responses (Bean and Bell, 1992); to the results of a holistic design process based on lighting patterns (Loe and Rowlands, 1996); to the lighting conditions which have desirable impacts on task performance, health and behavior (Veitch and Newsham, 1998b); to lighting which enhances the ability to discriminate detail, color, form, texture, and surface finishes without discomfort (Boyce and Cuttle, 1998). The definition that seems most generally applicable is that lighting quality is given by the extent to which the installation meets the objectives and the constraints set by the client and the designer. Depending on the context, the objectives can include facilitating desirable outcomes, such as enhancing the performance of relevant tasks, creating specific impressions, and generating a desired pattern of behavior, as well as ensuring visual comfort. The constraints are usually the maximum allowed financial and power budgets, a maximum time for completion of the work and, sometimes, restrictions on the design approach to be used.

To many people, such a definition must be a disappointment. It is both mundane and obvious. It is not expressed in terms of photometric measures, but rather in terms of the impact lighting has on more distant outcomes. There are three arguments in favor of such an outcome-based definition of lighting quality rather than any of the alternatives based directly on lighting variables. The first is that lighting is usually designed and installed as a means to an end, not as an end in itself, so the extent to which the end is achieved becomes the measure of success. The retailer does not care about lighting, *per se*, but only about lighting as a tool for increasing sales. The second is that what is desirable lighting depends very much on the context. Almost all the aspects of lighting that are considered undesirable in one context are attractive in another. The third is that there are many physical and psychological processes that can influence the perception of lighting quality (Veitch, 2001a,b). It is this inherent variability that makes a single, universally applicable recipe for good quality lighting based on photometric quantities an unreal expectation.

So where does visual discomfort fit into lighting quality? A simple concept that offers a place for visual discomfort is that lighting installations can be divided into three classes of quality: the good, the bad, and the indifferent. Bad-quality lighting is lighting which does not allow you to see what you need to see, quickly and easily and/or causes visual discomfort. Indifferent quality lighting is lighting which does allow you to see what you need to see quickly and easily and does not cause visual discomfort but does nothing to lift the spirit. Good-quality lighting is lighting that allows you to see what you need to see quickly and easily and does not cause visual discomfort but does raise the human spirit. On this scale, much of what constitutes current standard lighting practice falls into the class of indifferent lighting.

This in turn raises another question, what is it that causes lighting to be classified as good, bad, or indifferent quality? The outcome-based definition of bad quality lighting implies that such lighting occurs when it is inappropriate for what the visual system is being asked to do. For example, if a particular task with specific visual size and contrast characteristics has to be performed, then lighting which makes the signal-to-noise ratio between the task and its background low, or the signal-to-noise ratio between irrelevant stimuli and their backgrounds high, will be considered bad lighting; the former making the visibility of the task poor and the latter causing distraction. Among the phenomena which can contribute to such effects are insufficient light, too much light, excessive non-uniformity, veiling reflections, shadows, flicker, and disability and discomfort glare, i.e. all the phenomena which we currently think of as being responsible for visual discomfort. Eliminating these phenomena, remembering always that in some contexts they may be desirable, will generally lead to indifferent lighting. This would not be a mean achievement. Indeed, it may be the best that can be expected from the use of guidelines and quantitative lighting criteria. It may be that once bad lighting is avoided, the difference between indifferent lighting and good lighting is a matter of context, fashion, and opportunity. Context is important because what would be considered attractive lighting for an office seems unlikely to be so attractive in an intimate restaurant. Fashion is important because we often crave the new to provide interest and variety. There is no reason to suppose that lighting should be any different in this respect than most other aspects of life. As for opportunity, that is partly a matter of technology and partly a matter of being in the right place at the right time.

And what is the right place? An eminent lighting designer, J.M. Waldram, once said "If there is nothing worth looking at, there is nothing worth lighting" so the right place is presumably, a place which contains something worth looking at. Also, given that to be really good the lighting has to be matched in some way to the particular environment, each lighting solution would be specific and not generally applicable. This combination of fashion and specificity suggests that the conditions necessary for good lighting quality are liable to change over time and space and hence will not be amenable to determination by scientific methods. At the moment, good quality lighting most frequently occurs at the conjunction of a talented architect and a creative lighting designer, neither of whom is given to slavishly following numerical lighting criteria.

But this is looking at current lighting practice through rose-tinted spectacles. Table 5.2 shows the percentage of office workers agreeing with statements about the lighting of their office (Eklund and Boyce, 1996). The data comes from 1,259 people working in 13 different offices in the northeastern region of the US. The offices had lighting installations of different types and ages. The questionnaire used had been shown to be reliable and valid. From Table 5.2, it is clear that there is still a long way to go before visual discomfort at work is eliminated. Only 69 percent of the office workers agreed that the lighting of their offices was comfortable – yet it is possible to achieve much higher values. Studies in individual offices using the same questionnaire have produced percentage agreements in the range 90–95 percent comfortable. Table 5.2 also suggests what the causes of discomfort are. The basic problems appear to be light distribution and veiling reflections. The highest percentage agreements, apart from the overall assessment of comfort are for the statements "The light is poorly distributed here" (25 percent) and "Reflections from the light fixtures hinder my work" (19 percent). Relative to these, discomfort caused by flicker is rare.

Table 5.2 The average percentage of office workers agreeing with a statement about the lighting of their offices. The data was collected from 1,259 people working in 13 different offices (after Eklund and Boyce, 1996)

Statement	Average agreeing (%)
Overall, the lighting is comfortable	69
The lighting is uncomfortably bright for the task I perform	16
The lighting is uncomfortably dim for the tasks I perform	14
The lighting is poorly distributed here	25
The lighting causes deep shadows	15
Reflections from the lighting hinder my work	19
The light fixtures are too bright	14
My skin is an unnatural tone under the lighting	9
The lights flicker throughout the day	4

5.7 Summary

Where lighting is intended to facilitate the extraction of information from the visual environment, there are four situations in which visual discomfort is likely to occur:

- visual task difficulty, in which the lighting makes the required information difficult to extract;
- under- or over-stimulation, in which the visual environment is such that it presents too little or too much information;
- distraction, in which the observer's attention is drawn to objects that do not contain the information being sought;
- perceptual confusion, in which the pattern of illuminance can be confused with the pattern of reflectance in the visual environment.

The occurrence of visual discomfort is made manifest by the occurrence of red, itchy eyes, headaches, and aches and pains associated with poor posture. The aspects of lighting that can cause visual discomfort are too little light, too much light, too much variation in illuminance between and across working surfaces, disability glare, discomfort glare, veiling reflections, shadows, and flicker. Whether any or all of these aspects of lighting do cause visual discomfort depends on the context in which the lighting is installed. Virtually all of them can be used positively in the right context. For example, veiling reflections are undesirable when they mask what you want to see but when they are used to reveal the nature of the specular surface of a silver plate, and are called highlights, they are just what is wanted.

There is a lot of guidance published on how to avoid visual discomfort in common working interiors. Some of this advice is qualitative, e.g. for avoiding shadows and minimizing flicker, while other advice is quantitative, e.g. limiting values for discomfort glare and luminance ratios. The existence of quantitative recommendations should not be taken to mean that exact compliance is necessary. The quantitative guidance for providing visual comfort should be treated as approximate at best.

Eliminating visual discomfort is not a recipe for good quality lighting. Rather, it is a recipe for eliminating bad quality lighting and replacing it with indifferent lighting. This would be no small achievement. Surveys of poccupants' opinions about current office lighting practice indicate that only about 70 percent find the lighting comfortable. There is clearly a long way to go before the blessings of even indifferent quality lighting are available to all.

6 Lighting and the perception of spaces and objects

6.1 Introduction

The route from a luminous environment to the perception of that environment by a human being is long and convoluted. As formulated by Cuttle (2003):

> The *luminous environment* is transformed into the *retinal image* which is the stimulus for the *visual process* that provides information to enable the *perceptual process* to recognize objects and surfaces which form the visual basis for the *perceived environment*.

From this formulation, it should be apparent that the photometric and colorimetric quantities described in Chapter 1 and that are used to quantify the luminous environment are effectively measures of the stimulus to the visual system. The response of the visual system, in terms of the perception, is related to the stimulus received but not to the stimulus alone. This is for three reasons. First, perception depends on the state of adaptation of the visual system. Different states of adaptation have different capabilities. For example, when the visual system is in the mesopic state its ability to discriminate color is reduced from what it is in the photopic state and when it is in the scotopic state it has no ability to discriminate color at all. Also, as the state of adaptation changes, so does the sensitivity of the visual system to light. Consider a vehicle headlight seen by day and by night. The headlight has the same luminance under both conditions, but it will be seen as much brighter by night than by day because of the increased sensitivity to light produced by adaptation of the visual system to the generally lower luminances at night.

Second, the stimulus for perception in the real world is rarely a single item, seen in isolation. Rather it is a complex structure in which objects are seen against different backgrounds. Variations in the background can alter the perception of objects seen against that background. Figure 6.1 is a demonstration of this. It shows a gray annulus, the background of which is either black or white. The reflectance of the gray annulus is the same

Figure 6.1 A demonstration of the interaction between elements of a scene affecting perception. The two parts of the square annulus have the same reflectance but appear to be of different lightnesses because of the influence of the background. The effect is enhanced by drawing a line across the annulus at the black/white boundary of the background.

throughout, but the gray seen against the black background is perceived as lighter than the gray seen against the white background. This effect is enhanced if a dividing edge is placed along the black/white border, thereby demonstrating that perception is also influenced by the way the luminous environment is organized into patterns. It is these interactive aspects of perception, where the perception of each element in the luminous environment is dependent on other elements, that makes the link between the luminous environment and the perception of that environment so labile.

Third, perception is guided by our present knowledge and past experience of the luminous environment which determine the assumptions we make about objects and the ways they are usually lit. The well-known moon illusion is an example of the influence of such assumptions. When the moon is seen close to the horizon, it appears to be much larger than when it is seen high in the night sky. Measurement of the angle subtended at the eye by the diameter of the moon will show that the moon is actually the same size in both positions. Perception of size is changed because when the moon is close to the horizon, it is assumed that it is much closer than it actually is.

There can be no denying that perception of the luminous environment is a complex process and that a knowledge of the luminous environment alone is not enough to make an accurate prediction of how that environment will be perceived. Nonetheless, the luminous environment is the starting point of perception and lighting can be used to change the luminous environment. This chapter is devoted to describing the impact of lighting conditions on both simple and higher-order perceptions.

6.2   Simple perceptions

Given the number of factors that may intervene in the relationship between the stimulus provided by a luminous environment and the resulting perception, it is necessary to first consider the extent to which different people have the same perception in the same luminous environment. The first thing to say is that there is stability of perception. The occurrence of the perceptual constancies (see Section 2.6.1) and the existence of visual illusions that are seen by everybody is evidence that perception can be stable, even if it is distorted. The question then becomes under what conditions does stable perception occur. There is no quantitative answer to this question but there is a statement of principle that can be made. This is that the greatest stability of perception will be obtained when there is least opportunity for factors such as knowledge and past experience to intervene and where the perception is closely linked to operation of the visual system. Thus, it is relatively easy to show a stable relationship between the luminance of a small disk in a dark field and the brightness of the disk, and the change in brightness of the disk when the spectrum of the light is changed. Least stability of perception will occur when there are many opportunities for knowledge and past experience to intervene or when the operating characteristics of the visual system have only a minor impact on the perception. Thus, the impact of light distribution on the perception of the spaciousness of an interior is much less stable than the perception of brightness, because spaciousness is associated with factors other than the luminous environment, such as the amount and arrangement of furniture in the space. Simple perceptions, such as the brightness and colorfulness of a stimulus, are more likely to be stable than higher-order perceptions such as the spaciousness and attractiveness.

6.2.1   Lightness

As discussed in Chapter 2, surfaces that are seen by having light reflected from them have a perception of lightness, related to their reflectance. Sources that emit light have a perception of brightness, related to their luminance. Lynes (1971) discusses the conditions in which lightness constancy is maintained, i.e. over which the perception of the lightness of a surface will remain stable. He concludes that diffuse, uniform lighting, which is what is provided in most commercial and industrial interiors, ensures lightness constancy. Nonetheless, it should be appreciated that lightness constancy is not perfect. Over a wide range of illuminances, the lightness of a surface changes (see Section 2.6.1).

6.2.2   Brightness

6.2.2.1   Luminance and brightness

The study of the basic relationship between luminance and brightness has a long history, starting with Fechner (1860) and coming to fruition in the work of Stevens (1961). Over a number of years, Stevens was able to show that there was a consistent relationship between stimulus intensity and a number of different perceptions, such as loudness, smell, heaviness and taste, as well as brightness. The relationship between luminance and brightness was studied using a uniform, self-luminous target whose brightness was to be judged, surrounded by a uniform luminance field. The perception of brightness was quantified using a direct magnitude estimation procedure in which the subject was told to award a number to the first luminance presented. If the number given was 10, then the subject was instructed that if the next stimulus was seen as twice as bright as the first, then it should be given the value 20 but if it was one-tenth as bright as the first, it should be given the value of 1. In this way, a ratio scale of brightness perception was constructed. Another method used was cross-modality matching, in which subjects adjusted the magnitude on one perceptual dimension until it matched the perceived intensity of the stimulus from another dimension, e.g. they adjusted the force exerted on a handgrip until it appeared to match the brightness of the stimulus. Both measurement methods showed that the link between brightness and luminance is not linear but rather follows a power law of the form:

$$B = kL^n$$

where B is the brightness magnitude, k a constant, L the luminance of the stimulus (cd/m^2), and n an exponent.

Another consistent finding was that there are large differences between individuals in the value of the exponent, although whether this is due to some inherent difference in the sensitivity of the individuals' visual systems or to differences in the way the measurement method was used by different individuals is unclear. It was also found that the exponent of the power law was influenced by the size and luminance of the surround field, relative to the luminance of the target whose brightness was being judged, as well as the state of adaptation of the observer's visual system and the color of the stimulus and/or surround. Marsden (1969) reports that for small targets subtending less than 2°, in dim surrounds, the exponents found by different experimenters range from 0.23 to 0.31.

Bodmann and La Toison (1994) report a general model of brightness/luminance that covers a range of background luminances. Specifically, they use the model of Haubner (1977) which is:

$$B = cL_t^n - B_0 \quad \text{and} \quad B_0 = c(s_0 + s_1 L_u^n)$$

where B is the brightness magnitude, set to $B = 100$ at $L_t = L_u = 300 \, \text{cd/m}^2$; L_t the luminance of a test field (cd/m^2); L_u the luminance of a uniform background (cd/m^2); n an exponent $= 0.31$; and c, s_0, s_1 are constants. For a $2°$ test field, $c = 22.969$, $s_0 = 0.07186$, and $s_1 = 0.24481$.

When L_t is much greater than L_u, this equation reduces to Stevens power law. As L_u increases, the brightness of the test patch decreases, ultimately reaching zero, i.e. blackness.

While such studies in a simple context have generated a basic understanding of the relationship between brightness and luminance, there must always remain some doubt about their relevance to a complex luminance patterns typical of those found in lighted interiors. Fortunately, Marsden (1970) examined the perception of brightness for a real-size room furnished with both opaque reflecting objects and translucent self-luminous objects. The range of luminances available from surfaces within the room was $3–4,000 \, \text{cd/m}^2$, a range typical of surface luminances in an interior. His studies have shown that the perceived brightness of any single surface in the room increases with luminance according to a power law with an exponent of 0.35 but that the brightness of a number of surfaces seen simultaneously follows a power law with an exponent of 0.6. These relationships can be used to estimate the change in brightness of surfaces in an interior following a change in luminance by assuming that the brightest surface in the room has a brightness given by

$$B_{max} = L_{max}^{0.35}$$

Then, other surfaces in the room will have a brightness given by:

$$B = aL^{0.6}, \quad \text{where } a = B_{max}/L_{max}^{0.6} = L_{max}^{-0.25}$$

This simple system gives estimates of the average perception of brightness for a group of observers but underestimates the brightness of highly saturated colored surfaces and overestimates the brightness of translucent surfaces. It is important to appreciate that these relationships are approximate at best, and should be used for guidance only. There are large individual differences in the perception of brightness.

6.2.2.2   Luminous intensity distribution and brightness

It is implicit in the finding that the constant in the expression linking the luminance of a surface in a room to its perceived brightness is determined by the maximum luminance of any surface in the room that the overall perception of brightness in a room is influenced by the distribution of light in the room. This has implications for the perception of room brightness produced by different types of lighting installation. Direct lighting is lighting that directs all the light emitted from the luminaire downward to the working plane. Indirect lighting is its opposite, all the light leaving the luminaire being directed upwards so that light only reaches the working plane after reflectance at one of the room surfaces, usually the ceiling. It is widely believed that indirect lighting is perceived as less bright than direct lighting when both deliver the same illuminance to the working plane, because the luminance of the luminaires used for direct lighting is typically greater. Whether this is true or not will depend on the light distribution from the direct lighting luminaire. If the luminous intensity distribution is narrow so that very little light reaches the walls and the luminaire has a low luminance from most viewing directions, then direct lighting may produce a perception of lower brightness than indirect lighting.

Loe *et al.* (1994) examined the effect of light distribution on the perception of room brightness. Using 18 different lighting installations in a conference room, they found that the perception of brightness was determined not only by the range of luminances present in the space but also by the location of those luminances, some locations being more important than others. Specifically, it appeared that the luminances within a 40° horizontal band about the line of sight were most important in determining the perception of brightness. For someone looking down a room, a 40° band emphasizes the luminance of the walls relative to the luminances of the ceiling and floor. However, these results are inconsistent with the results of Miller *et al.* (1995) who conducted a similar study using a direct lighting, with and without wall-washing, and indirect lighting, for a range of illuminances on the working plane. Attempts to relate the brightness of the room to the luminance of different surfaces were most successful when both wall and ceiling luminances were considered. Figure 6.2 shows the relationship between the perception of brightness and a metric called volumetric brightness, i.e. the average luminance of ceiling and walls. The weakness of this study, as admitted by the authors, was that all the subjects used were experts in lighting design, and hence might be expected to have prejudices about the different lighting types. It would have been interesting to know how a group of people naive in lighting would have responded to the same installations. For the moment, it will have to be sufficient to note that both studies justify the practice of using wall-washing to enhance the perception of brightness in a room where direct lighting puts little light on the walls.

Despite this uncertainty about the relative importance of different room surfaces, there can be no doubt that the distribution of light in a room can alter the perception of brightness of the room. Shepherd *et al.* (1989, 1992) confirmed this in a series of studies devoted to the perception of what might be called the opposite of brightness, gloom. Using a word list from which

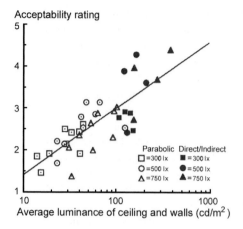

Figure 6.2 Mean acceptability ratings (1 = completely unacceptable, 3 = barely acceptable, 5 = completely acceptable) for different lighting installations plotted against the average luminance of ceiling and walls produced by the installations. The installations used recessed parabolic luminaires or suspended direct/indirect luminaires, to produce 300, 500, or 750 lx on the working plane, in the same office setting (after Miller *et al.*, 1995).

subjects had to choose the words that applied to the space, they showed that the perception of gloom was associated with several different situations. Specifically, Shepherd *et al.* (1992) concluded that a perception of gloom could be produced by:

• low surround luminances, irrespective of task illuminances;
• conditions in which small details in the periphery are obscured;
• high task illuminances with low luminances on the peripheral surfaces;
• adaptation luminances in the mesopic range.

Again, these results imply that direct lighting that leaves the walls poorly illuminated will result in a perception of gloom and using wall-washing to increase the illuminance on the walls will increase the perception of brightness. Interestingly, the same conditions produced the same perception of gloom for students of lighting and architecture and for members of the general public, which suggests that the concern about the use of experienced subjects by Miller *et al.* (1995) may be unwarranted.

The mixed use of luminance and illuminance in these conclusions raises another interesting question. Is it possible to make a room with low-reflectance surfaces, such as mahogany paneled walls, look bright? The answer to this question is positive, but only when the amount and distribution of light can be understood. As explained in Section 2.6.1, provided the amount and distribution of light in the room can be understood, the perceptual system can split the pattern of luminances received at the retina into a pattern of reflectances and a pattern of illuminances. Then, provided brightness in this situation can be taken to relate to the perception of the amount of light in the room, which is what Ishida and Ogiuchi (2002) have shown, a room with low-reflectance surfaces can indeed be perceived as bright.

6.2.2.3 *Luminaire luminance and brightness*

Another aspect of lighting practice that can affect the perception of brightness is the design of luminaires. The question of interest here is whether high luminance elements in a luminaire can alter the perception of brightness in a space or do they simply make the lighting installation uncomfortable by causing discomfort glare (see Section 5.4.2.2).

Bernecker and Mier (1985) examined the effect of changing the luminance of an element of fixed size in an indirect lighting luminaire. They found that increasing the luminance of the element led to a perception of greater room brightness. Akashi *et al.* (1995) followed up this work but included the area subtended at the eye by the luminous element and the background luminance as variables. To measure brightness, a subject looked at two, one-fifth scale model rooms, furnished with a desk and a valence luminaire. The only difference between the two rooms was that in one room, the test room, the valence luminaire had a slot forming a luminous element cut in it. The illuminance on the desk in the test room, and hence the background luminance, was set to a fixed level. The subject then adjusted the amount of light in the other room until the overall brightness of the two rooms looked the same. The ratio of the illuminances on the desks in the two rooms when they appeared to be of equal brightness was taken as a measure of the impact of the luminous element in the luminaire. If the ratio was greater than unity, it meant that the effect of the luminous element was to increase the perception of brightness. The mean illuminance ratios for the small number of subjects used, for three different background luminances, five different luminous slot sizes, and five different luminous slot luminances ranged from 0.8 to 1.3. These values imply that the presence of a luminous element in a luminaire can both enhance and diminish the perception of room brightness, depending on the conditions. Figure 6.3 shows the combination of the luminance of the luminous element and its size, measured as solid angle subtended at the eye, and the associated illuminance ratio for equal brightness, for a background luminance of $50 \, cd/m^2$. The shape containing the illuminance ratio indicates whether that combination of luminance and size was described as glaring, sparkling, or merely bright. From Figure 6.3 it can be seen that reducing the size of the luminous element and increasing its luminance will, within limits, enhance the perception of brightness, primarily when the luminous slot is described as sparkling. However, increasing the luminance too much or making the luminous element too large will lead to

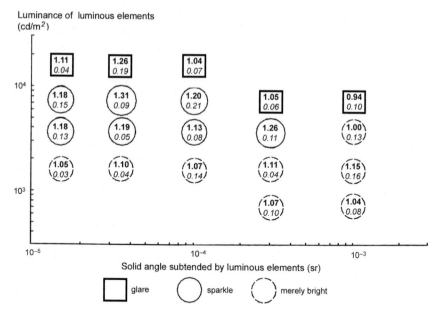

Figure 6.3 The impression given by different combinations of luminance and solid
angle of the luminous element in a luminaire. The impression is classified
as either glaring, sparkling, or merely bright. The upper number within
each symbol is the mean illuminance ratio for the perception of equal
brightness for two spaces that were identical except for the presence of
the luminous element in the luminaire in one of them. The lower number
is the associated standard deviation. An illuminance ratio greater than
unity implies that the presence of the luminous element enhances the
perception of brightness (after Akashi *et al.*, 1995).

a perception of glare which makes the rest of the room look less bright,
while having too little luminance relative to the background produces little
effect. It should be clear that using a luminous element in a luminaire to
enhance room brightness is something that needs to be handled with care.

6.2.2.4 *Light spectrum and brightness*

There are four reasons why light spectrum can be expected to have an affect
on the perception of brightness, over and above any effect of luminance. To
appreciate this it is simply necessary to remember that luminance is based on
the spectral sensitivity of the fovea, containing medium- and long-wavelength
cones, in photopic conditions, and measured using flicker photometry, a
method that emphasizes the magnocellular channel (see Section 2.2.7).
Therefore, increasing the size of the area being examined so the peripheral
retina is included will introduce the effect of short-wavelength cones and rod

photoreceptors and will reduce the impact of the macula that covers the central part of the retina. Decreasing the adaptation luminance so that the visual system moves from the photopic state to the mesopic and ultimately to the scotopic will change the active photoreceptors. Viewing a steadily presented scene will allow both the magnocellular and parvocellular channels to operate which means that both the achromatic and color channels of the visual system will be active. All these effects will be applicable to an abstract object consisting of a disk of light and more complicated scenes such as the rooms and streets that make up the visual environment for most people. The one remaining reason why the light spectrum might affect brightness over and above luminance will not be applicable to a disk of light but it will be important for real scenes. This is the effect of light spectrum on the colors of objects and surfaces that form the scene. Light spectra that make the colors of surfaces constituting a scene more saturated can be expected to influence the perception of the brightness of the scene.

The fact that the spectral content of the light reaching the eye influences the perception of brightness has been known for many years and goes under the name of the Helmholtz–Kolrausch effect (Wyszecki and Stiles, 1982). This effect is simply that when two fields of different color but the same luminance are placed side-by-side, the one with the greater color saturation will appear brighter. The Helmholtz–Kolrausch effect is different for different colors. Specifically, the effect is much weaker for a perception of yellow than for a perception of red, green, or blue (Padgham and Saunders, 1975).

This pattern is also evident in the work of Ware and Cowan (1983). These authors took data from 29 different studies involving heterochromatic brightness matching for small fields (less than 2°), and derived an empirical formula to calculate conversion factors for determining the relative brightnesses of colors. For colors where the y chromaticity coordinate is greater than 0.02, the conversion factor can be calculated from the equation

$$C = 0.256 - 0.184y - 2.527xy + 4.656x^3y + 4.657xy^4$$

where C is the conversion factor and x, y are CIE 1931 chromaticity coordinates of the stimulus.

To rank order a number of light sources for brightness, at the same luminance, or even at different luminances, all that is required is to obtain the sums of

$$\log(L) + C$$

where L is the luminance (cd/m^2), and C the conversion factor.

For the same luminance, the light source with the highest value of the conversion factor will appear brightest. Figure 6.4 shows the iso-conversion factor contours based on the Ware and Cowan equation. The iso-conversion

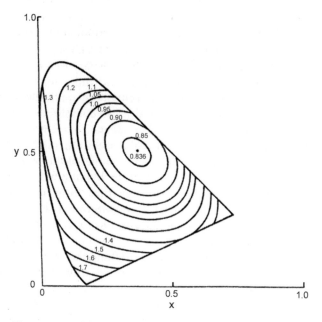

Figure 6.4 Iso-conversion factor contours plotted on the CIE 1931 (x,y) chromaticity diagram (after Ware and Cowan, 1983).

factor contours of Ware and Cowan, and those of other authors with a more theoretical approach (Nakano *et al.*, 1999), have an interesting implication for lighting practice. The conversion factors increase with greater saturation of the color apart from toward the yellow part of the CIE 1931 chromaticity diagram. This implies that, for the same luminance, nominally white light sources with a higher correlated color temperature (CCT) will be seen as brighter than those with lower CCTs, a conclusion also reached by Harrington (1954). The only doubt about this conclusion is that the data on which it is based are for small disks of light, and so, presumably, are due to the action of the chromatic channels, but not to photoreceptors other than the medium- and long-wavelength cones. The interesting point is how well this conclusion will stand-up when applied to a real room which forms a large visual field and so stimulates all the photoreceptors in the retina, and which may contain numerous colored surfaces.

One study that addresses this question is that of Boyce *et al.* (2002). This study used two offices identically furnished and lit to the same illuminance by fluorescent lamps with different CCTs and CRIs, housed in luminaires with similar light distributions. Specifically, one light source had a CCT of 6,500 K and a CRI of 98 and the other had a CCT of 3,500 K and a CRI of 82, i.e. the two light sources had very different light spectra. For the same distribution of illuminances in the offices, the light source with the

CCT of 6,500 K and a CRI of 98 produced a perception of greater brightness than the other. Another study in the same area was that of Boyce and Cuttle (1990), who had 15 observers carry out a color discrimination task in a small office and then give an assessment of the lighting. The office was lit by fluorescent lamps with virtually the same CRI value (CRI = 82–85) but very different CCTs (2,700–6,300 K), producing four different illuminances on the desk (30, 90, 225, and 600 lx) from the same luminaire. The major factor in determining the impression given by the lighting was the illuminance. Increasing the illuminance made the lighting of the office appear more pleasant, more comfortable, more warm, more uniform, less hazy, less oppressive, less dim, and less hostile. In this study, the different CCTs had virtually no effect on the observer's impression of the lighting of the room. The divergence between the results of these two studies serves to emphasize the limitations of some of the widely used single-number metrics of light source color properties. The light sources used in the Boyce and Cuttle (1990) study differed less in lamp spectra than those used in the Boyce *et al.* (2002) study, and the former showed no effect of light source while the latter did. For perception it is the difference in light spectra that matters and hence the differences in the mixture of wavelengths that reach the eye, either directly, or indirectly after reflection from the room surfaces.

One study that had large differences in light spectrum and that showed a clear effect of light spectrum on brightness, even in an achromatic interior, was that of Berman *et al.* (1990). In this study, subjects viewed a spectrally neutral interior lit by pairs of light sources with different light spectra but the same chromaticity, each for a 5-s period. The results showed that subjects consistently perceived a difference in brightness for two light sources that was in opposition to the photopic luminances. These results imply that, in large visual fields, brightness perceptions are different for light sources with different light spectra, even when those light spectra lead to the same chromaticity and hence the same CCT.

One remaining question is: What is the relative effect of light spectrum directly on brightness and indirectly through its effect on the saturation of surface colors in a space? This question can be answered by considering the magnitude of the effect in achromatic interiors and interiors filled with surface colors. Berman *et al.* (1990) had observers looking at a uniform spectrally neutral interior and found large differences in brightness for different light sources. This implies that the effect of light spectrum on brightness is present even in the absence of surface colors. But does the presence of surface colors enhance the effect? Boyce and Cuttle (1990) measured the effect of introducing a range of colors, in the form of fruit and flowers, on the perception of an achromatic room and a room decorated in one color. It was found that introducing fruit and flowers into the rooms increased the perception of brightness at the same illuminance. This implies that both the direct and indirect effects of light spectrum are important for the perception of brightness. Obviously, there is still much to learn about how and why different light spectra affect the perception of brightness of interiors but that they do is no longer a matter of argument.

Also not a matter of argument is the fact that light spectrum has an effect on the perception of brightness in mesopic conditions such as are produced by many forms of exterior lighting. In mesopic conditions, both rod and cone photoreceptors are active. It is well known that rod and cone photoreceptors interact in highly non-linear ways to affect brightness perception (Wyszecki and Stiles, 1982) and that rods and short-wavelength cones are not involved in luminance. Using a diorama of a scene involving a grassy knoll and a number of brightly colored model vehicles, Rea (1996) had observers compare high-pressure sodium and metal halide lighting viewed alternately. Specifically, the observers were asked to adjust the illuminance provided by the high-pressure sodium lamp until the scene looked equally as bright as the scene illuminated by the metal halide lamp. The mean luminance ratios for the two lamp types for equal brightness at a luminance of $0.1 \, cd/m^2$ for the metal halide lit scene was 1.4 while for a luminance of $0.01 \, cd/m^2$ it was 2.1. It can be seen a higher luminance is required for high-pressure sodium lighting to be seen as equally as bright as metal halide lighting and that the luminance ratio increases as the adaptation luminance decreases further into the mesopic state. How much the advantage of the metal halide would decline if the scene was essentially achromatic remains to be determined.

6.2.3   *Visual clarity?*

Another perception that is often claimed to be associated with light spectra is that of clarity. The first paper in this field was that of Aston and Bellchambers (1969). These authors had people stand in front of two cabinets identically furnished with colorful objects and surfaces. The cabinets were lit by two different fluorescent lamps in identical luminaires so the light distribution was the same. The illuminance on the working plane of one cabinet was set to a fixed level and the subject was asked to adjust the illuminance of the other "... so that the overall clarity of the scene is the same in both cabinets." Further, "overall clarity" was defined as "... the satisfaction gained by you personally, discounting as far as possible any obvious differences in color and brightness." It was found that lamps with high CRIs were consistently set to a lower illuminance than lamps with low CRIs. In a later study, Bellchambers and Godby (1972) extended this research to full-size rooms and found very similar results indicating that, on average a 24 percent higher illuminance was required for a lamp with moderate color rendering properties (CRI = 55–65) to be seen as giving equal visual clarity to a lamp with very good color rendering properties (CRI = 92), for a reference illuminance of 500 lx.

Over the years since the publication of this work a number of researchers have followed the same path [see Fotios (2001) for a review]. One problem in comparing the resulting data has been that different studies have asked the observers to make the adjustment using apparently different criteria. Boyce (1977) asked observers to adjust the illuminances in two model rooms seen side by side for equal satisfaction with visual appearance. Fotios and Levermore (1997) asked the observers to match for visual equality. Despite the vagueness of these terms, there can be little doubt that the illuminance ratios set for each criterion do provide a measure of the relative effectiveness of different light spectra in providing some holistic impression, although whether that impression is based on the perception of brightness, clarity, or something else is impossible to say.

The question of interest now is what can be used to predict the illuminance ratio for the same perception, for the same interior, given a choice between two different light spectra providing the same luminance. There are two answers to this question, one empirical and one physiological. The empirical answer is based on existing color matching functions (Fotios and Levermore, 1998a). Figure 6.5 shows the fit between the measured mean illuminance ratios for 15 different pairs of nominally white lamps plotted against the predicted illuminance ratio based on the ratio of gamut areas for each lamp pair. The data used is taken from Boyce (1977) and Fotios and Levermore (1997). The equation for predicting the illuminance ratio from gamut area ratio is

$$E_1/E_2 = 1.0 - 0.61 \log_{10}(G_1/G_2)$$

where E is the illuminance for a light source (lx) and G the gamut area of the light source.

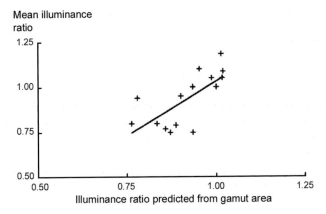

Figure 6.5 Mean illuminance ratios for equal visual clarity, plotted against the illuminance ratios predicted from gamut areas, for 15 different pairs of light sources (after Fotios and Levermore, 1998a).

The gamut area of a light source is the area on the CIE 1976 UCS diagram formed by a line joining the positions of the eight CIE standard test colors, lit by the light source, on the diagram (see Section 1.6.3.4). This equation enables the illuminance ratio for any given pair of nominally white light sources to be calculated from a knowledge of the spectral power distribution of the light sources using conventional colorimetry.

The physiological answer can be given by modeling the level of activity in the achromatic and chromatic channels of the visual system given a specific light source. Fotios and Levermore (1998b) show a very good fit to the measured illuminance ratios for the Fotios and Levermore (1997) data using activity in the chromatic channels. The problem with this answer is that the weighting function applied to the spectral power distribution of the light source is determined by the goodness of fit to the data. Until the accuracy of prediction made by this model is tested on independent data, the value of this model must remain open to question. Further, there is argument over the photoreceptors involved in different perceptions. Berman *et al.* (1990) argue that it is rod activity that explains the effect of light spectrum on brightness, even in photopic conditions, and that the brightness perception of two light sources with different light spectra, at a fixed luminance, can be approximately predicted from the square root of the lamp scotopic/photopic ratio. Worthey (1985) argues that it is the stimulus to the red–green channel of the color vision system that determines the differences in the perception of clarity for different light spectra providing the same luminance because red–green contrasts contribute to the sharpness of borders. Fotios and Levermore (1998c) argue that the illuminance ratios for visual equality can be best explained by the activity of the short-wavelength cones. Until this conflict about photoreceptors and perceptions is resolved, it will be difficult to know how to construct suitable predictive models based on the physiology of the visual system. More generally, it is not until the basis on which subjects are matching two scenes lit to the same luminance by different light spectra is made specific that the rather nebulous field of visual clarity will itself be clarified.

6.2.4   Color appearance

The most obvious factor in determining the effect of lighting conditions on the color appearance of a surface or a scene is the choice of light source. The choice of light source can have two effects of the color appearance of a space. The first is to shift the overall color appearance of the space. The second is to change the appearance of colors in the space relative to each other. Provided there is time for chromatic adaptation to occur, the overall shift in color appearance has only a limited effect on the perception of the space. For example, Boyce and Cuttle (1990) showed that after more than 20 min, the perception of brightness and colorfulness in an achromatic room was the same for good color rendering fluorescent lamps with CCTs ranging from 2,700 to 6,300 K. The effect on the relative color appearance of different colors in the space is a much more important effect, regardless of the time available for color adaptation. Light sources that increase the saturation of surface colors, and hence that have large gamut areas, will increase the colourfulness of an interior.

There are other aspects than the choice of light source that matter for color appearance. For example, there is the Bezold–Brucke effect. This effect is simply that changing luminance can change the hue of a color. Increasing luminance makes reds yellower and violets bluer. Figure 6.6 shows the direction and magnitude of the Bezold–Brucke effect. The vertical axis shows the monochromatic wavelength of a field of luminance $7\,\text{cd/m}^2$ that matches the color of the same-size field at $120\,\text{cd/m}^2$. The horizontal axis is the monochromatic wavelength of the higher luminance field (Boynton and Gordon, 1965). These luminances cover the range usually

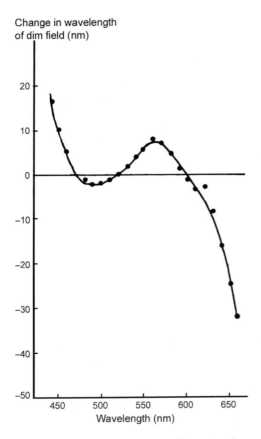

Figure 6.6 The Bezold–Brucke effect. The change in wavelength needed for one half of a bipartite field at $7\,\text{cd/m}^2$ to match the other half at $120\,\text{cd/m}^2$ in hue (after Boynton and Gordon, 1965).

found in interior lighting. One point of particular interest in Figure 6.6 is the fact that there are three wavelengths at which there is no shift in hue, approximately 475, 507, and 570 nm.

Another factor that influences color appearance is field size. The importance of field size is evident from the fact that CIE has found it necessary to introduce two standard observers for colorimetry. The CIE (1931) 2° Standard Observer is used for color fields subtending less than 4° at the eye and the CIE (1964) 10° Standard Observer is used for fields of color subtending more than 4° (Wyszecki and Stiles, 1982). Figure 1.2 shows the spectral sensitivities of the CIE 2 and 10° Standard Observers. It is clear that increasing the field size from 2 to 10° makes the visual system more sensitive to the part of the visible spectrum below 550 nm.

There is another aspect of field size and color appearance that will be evident to anyone looking at a large field, subtending say 10°, filled uniformly with light of a single color. The appearance of the field will not be uniform. Rather, there will be a spot of diameter about 4°, with an ill-defined boundary, which differs in color from the rest of the field. This spot is called the Maxwell spot and it will move around as the point of fixation moves. The existence of the Maxwell spot is mainly due to the existence of the macular, a yellow pigment that covers the fovea and the area immediately around it and acts a filter to the light reaching the retina.

These few examples should suffice to show that, like most other perceptions, the effect of lighting on color appearance is not a simple matter, and certainly not a simple matter when the possibility of an interaction between light source color and the colors of the decor is considered (Mizokami *et al.*, 2000). However, this complexity has not stopped attempts to provide a model of how the stimulus to the visual system affects the appearance of colors, seen separately or in combination with others. The most extensively developed of these has been produced by Hunt (1982, 1987, 1991). In 1982, Hunt proposed a set of cone spectral sensitivity functions that enabled predictions to be made of hue, lightness, and chroma of colors seen under a medium photopic level of illumination. In 1987, the model was extended to provide, in addition, predictions of brightness and colorfulness at any illuminance, no matter whether the visual system was operating in the photopic, mesopic, or scotopic state. In 1991, the model was revised to make it easier to use and to cover situations in which a color is seen amongst other colors and in isolation. The final model provides correlates of hue, lightness, and chroma, brightness and colorfulness, saturation, relative yellowness–blueness, relative redness–greenness and whiteness–blackness. The model has been shown to be consistent with physiological data in the form of changes in cone responses in monkeys as luminance is increased. It has also been shown to fit perceptions of brightness, hue, lightness, and colorfulness, the effect of applying color filters to part of a projected slide and the Bezold–Brucke effect for humans. The development of this and other models (Nayatani *et al.*, 1994) lead the CIE to produce an interim model of color appearance (CIE, 1998) and development is still proceeding. This interest in the topic of color appearance is a good thing because it may lead to the development a metric of light source color properties based on what it is that people care about, i.e. how the light source will make colors appear. Attempts have already been made to use Hunt's model to develop a better color rendering metric (Pointer, 1986), but with little acceptance to date. Fortunately, the increase in the number of nominally white light sources being produced by the lighting industry that cannot be meaningfully separated by the CIE general CRI is likely to keep up the pressure to develop a comprehensive and reliable model of color appearance.

6.3 Higher-order perceptions

6.3.1 The correlation method

Most of the studies of the effect of luminous conditions on the simple perceptions of lightness, brightness, and color appearance have been undertaken using what has been called the correlation method. The correlation method is simple. It involves establishing a relationship between some subjective judgment of the lighting and a physical measure of the lighting, e.g. between brightness and luminance. While the correlation method has been widely used it has also been extensively criticized. These criticisms are related to accuracy of simulation, plasticity of ratings and relevance and independence of the conclusions.

The criticism of accuracy of simulation arises from the fact that many of the experiments attempting to determine the effect of lighting conditions on perception have used abstract conditions, i.e. uniform fields of luminance. This inevitably leaves some doubt about the validity of the relationships found for real interiors, and particularly interiors that have a definite context.

The criticism of the plasticity of ratings arises from the observation that subjects tend to match the center of a rating scale with the center of the range of conditions experienced (Poulton, 1977), from the idiosyncratic responses produced on rating scales when there is ambiguity in the instructions given to the observers (Rea, 1982) and from the variability in the use of rating scales introduced when there is no link between the rating scale and an obvious feature of the luminous environment (Tiller and Rea, 1992). These criticisms can be overcome by showing each subject only one condition, by being very careful to avoid ambiguity in instructions, and by selecting the words used to label the ends of a semantic differential rating scale so that the scale relates to an obvious feature of the luminous environment. The basic problem with using semantic differential rating scales is that people are willing to apply them even when they make no sense. This places a heavy load on the experimenter when selecting rating scales. Further, if rating scales are to be restricted only to those that have a link to an obvious feature of the luminous environment, then there is little prospect of ever understanding the impact of lighting on what might be called higher-order perceptions, such as complexity, formality, and interest. Tiller and Rea (1992) suggest that one way out of this dilemma is to use semantic differential rating scales, not as an end in themselves, but rather as a means of developing hypotheses about higher-order perceptions that could then be tested more stringently, perhaps using behavior as a dependent variable. For example, if one lighting condition was shown to be considered more formal than another by using rating scales it could be hypothesized that people would behave more formally in it. Such a hypothesis could easily be tested. Despite the ingenuity of this two-step approach, there have been few attempts to implement it.

The criticisms of relevance and independence remain, even when the rating scales are carefully used. The point is that obtaining a correlation between a physical measure and a subjective judgment is no guarantee that the psychological attribute represented by the judgment is relevant to people's everyday experience and, even if it is, when a number of such relationships are established there is no reason to suppose that they represent independent dimensions.

All these criticisms might suggest that there is little to be gained by applying the correlation method. This suggestion would be unfair. The correlation method has been used to identify lighting conditions that are considered uncomfortable by large numbers of people. Further, experience in the field suggests that the criteria developed to express this understanding are robust. What this implies is that the correlation method is useful for identifying visually uncomfortable lighting conditions. What the criticisms do suggest is that the correlation method is not suitable for exploring the impact of lighting on the higher-order perceptions. To explore higher-order perceptions, it is necessary to put people first, rather than lighting equipment, and to place them in a specific context, not in an abstract setting.

6.3.2 *Multi-dimensional methods*

One seminal study in which a real interior was used was that of Flynn *et al.* (1973, 1979). In this study, 96 observers experienced a conference room lit in six different ways. The observers' responses were collected on 34 rating scales. The instruction to the observers was to evaluate the room. Factor analysis was then applied to the data. Factor analysis is a statistical technique by which an enormous quantity of data can be ordered so that the minimum number of underlying independent dimensions on which the observers are basing their responses can be revealed. Factor analysis also gives a measure of the extent to which each individual rating scale is related to the independent dimensions. Figure 6.7 shows the six lighting installations used by Flynn *et al.* to light the conference room, the room being furnished with the same conference table and chairs for all six lighting installations. Figure 6.8 shows the five independent dimensions on which

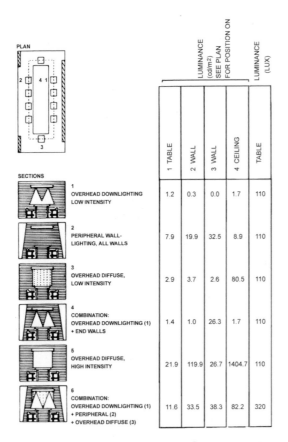

Figure 6.7 A plan of the conference room, six schematic illustrations of the six lighting installations used and some photometric measurements of the lighting conditions used (after Flynn *et al.*, 1973).

the impressions of the room under the six lighting installations are based. Also given are the individual rating scales that were most strongly related to each dimension. Based on the rating scales that are most strongly related to each dimension, the five dimensions were identified as evaluative, perceptual clarity, spatial complexity, spaciousness, and formality. The numbers located along each scale indicate the mean rating on that scale for each lighting installation. A simple examination of Figure 6.8 shows that only three of these dimensions have much separation between the lighting conditions, these being the evaluative, perceptual clarity, and spaciousness dimensions.

A closer examination of the different rating scales in Figure 6.8 reveals the richness of the information available. For example, the pleasant–unpleasant scale, which is the scale most strongly related to the evaluative dimension, shows that the installations that are most pleasant (installations 4 and 6)

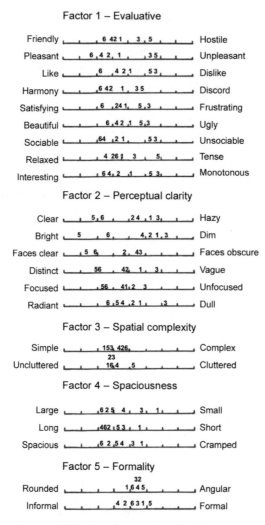

Figure 6.8 The five factors identified from the rating of the conference room lit by the six lighting installations shown in Figure 6.7. The rating scales most strongly related to each factor are shown under each factor title. The mean rating for each installation on each scale is given by the appropriate installation number, i.e. installation 6 has the mean rating closest to the pleasant end of the pleasant–unpleasant scale (after Flynn *et al.*, 1973).

have a common feature, as do the two installations that are most unpleasant (installations 3 and 5). Installations 4 and 6 provide lighting on the working surface, i.e. the table, and on the room surfaces. Installations 3 and 5 provide diffuse lighting on the table only. This supports the view that for good quality lighting it is necessary to light both the task and the room, something that has long been argued by practitioners of lighting design.

The second dimension, perceptual clarity, is apparently related to the illuminance on the table. As might be expected, the higher illuminance installations are considered to give greater perceptual clarity. Further, the separation on the clear–hazy rating scale, which is the one most strongly linked to the dimension of perceptual clarity, is much less between 320 lx (installation 6) and 1,100 lx (installation 5) than between 110 lx (installations 1–4) and 320 lx (installation 6). This is just what would be expected from what is known about the non-linear response of visual performance to illuminance and about peoples' preference for illuminance in an office setting, measured using the correlation method (see Section 7.2).

The third dimension, spaciousness, is more difficult to understand. The separation between the different lighting installations is smaller and the number of scales related to the dimension is fewer, although the placing of the different lighting installations is consistent across scales. What makes it difficult to understand is that the lighting installations that provide a perception of greater spaciousness have several different features. The installations that provide light on the table alone, particularly at a low illuminance, make the room appear to be small and cramped. The installations that light the walls alone, or both the walls and the table give an impression of a large and spacious room.

Such information can be used to answer questions of interest to the designer. For example, suppose the designer has a choice of either lighting the conference table to a high illuminance or having a lower illuminance on the table but some light on the walls. From the data discussed above it is apparent that the former approach will give the impression of a room which is clear but rather unpleasant, whilst the latter will be assessed as less clear but more pleasant and spacious. Such conclusions do not tell the designer what to do but they do supply some information on which to base a decision.

It can be concluded that the rating scale/factor analysis produces much useful information. It also overcomes some of the criticisms of the correlation procedure. It ensures that the dimensions identified are independent and by using rating scales related to the impression of the room rather than just the lighting, ensures that the results reflect the importance of the lighting variables to the particular context. For example, for the range of conditions used in the conference room, it is clear that changing the lighting has a much bigger effect in the perceptions of perceptual clarity than on the perception of spaciousness.

However, there is still one big limitation of this approach. It still rests on the use of a finite number of rating scales selected by the experimenter. If there are no scales about a particular aspect of the visual environment then there is no way a dimension related to that aspect can be found. Flynn *et al.* also examined another multi-dimensional approach that eliminates the need to select rating scales by eliminating their use. Specifically, all they asked the observers to do was to rate how different two installations were from each other, for all possible pairs of installations. Flynn *et al.* used this method to examine the same lighting installations in the same conference room used for the rating scale/factor analysis study, but with a different set of 46 observers. By applying a multi-dimensional scaling (MDS) analysis, the number of independent dimensions that are needed to describe the differences between the installations can be identified. These *n* dimensions are then used to form an *n*-dimensional space in which the individual installations can be located. Flynn *et al.* found that for the six installations in the conference room, a three-dimensional space was the minimum necessary to explain the data. Figure 6.9 shows the three dimensions and the locations of the individual installations. From the positions of the installations relative to each axis, the three dimensions were named peripheral/overhead, uniform/non-uniform, and bright/dim. As the dimensions are identified from the installation characteristics there is no possibility of directly evaluating the meaning of the dimensions. The best that can be done is to tell which installations are considered more or less different from each other and the independent dimensions on which they differ. This may seem a rather meager return for such a large investment but it is only when the ratings scale/factor analysis and the difference/MDS approaches are combined that their full value becomes apparent. From the rating scale/factor analysis approach, the three dimensions that most clearly separate the

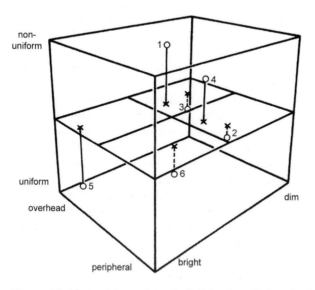

Figure 6.9 The positions of the six lighting installations in the conference room shown in Figure 6.7 in the three-dimensional space derived from the MDS analysis (after Flynn *et al.*, 1973).

installations are evaluative, perceived clarity and spaciousness. Choosing a rating scale strongly related to each of these dimensions allows stepwise multiple regression equations to be calculated for the mean ratings on each scale and on the three MDS dimensions. The evaluative dimension gave a correlation of 0.83 on the overhead/peripheral MDS dimension; the perceptual clarity dimension gave a 0.99 correlation on the bright/dim MDS dimension; and the spaciousness dimensions gave a correlation of 0.69 with the uniform/non-uniform MDS dimension. These correlations demonstrate that there is good but not perfect agreement between the dimensions obtained by the two approaches and the location of the various installations on those dimensions. This agreement allows the MDS dimension which the observers use to distinguish between the installations to be related to the observers' rated impressions of the conference room. For example, it is clear that the extent to which the room is considered pleasant is mainly related to the existence of peripheral lighting, although as the introduction of the uniform/non-uniform MDS dimension increases the correlation with rated pleasantness to 0.92, having a non-uniform distribution of light is also of value. This conclusion could have been reached from examination of the factor analysis alone but this result was obtained by asking the observer specific questions. It is good to have the conclusions confirmed by a procedure that allows the observer to use whatever he or she wishes to use to make judgments.

It can be concluded that both the rating scale/factor analysis and the difference/MDS analysis procedures provide rich information about the perceptions of lit spaces, particularly if used together. However, before adopting these procedures and applying them wholesale, it would be as well to consider the stability of the dimensions identified and the position of different installation types on them. Flynn *et al.* (1975) examined this question in two stages. First, they applied the rating scale/factor analysis method to five different lighting installations in three different sized rooms, but all with the same context, i.e. conference rooms. The same factor structure was obtained for all three rooms and the locations of individual lighting installations on the dimensions was similar. They also examined the dimensions found for different lighting installations for rooms with a different context. In this study, they used an auditorium. Based on the ratings obtained, it was clear that there was some consistency in the impression given by different lighting installation types. High illuminances were related to greater clarity; peripheral lighting and low illuminances produced a pleasant impression and the use of high illuminances and peripheral lighting produced an impression of spaciousness. Such results suggest stability of impression and support Flynn's concept that lighting provides a number of cues that people use to interpret a space and that these cues are at least partly independent of the room that is being experienced (Flynn, 1977). But all these results come from one source. Before accepting such an important conclusion it would be as well to examine what others have found.

An obvious comparison is with the results of Hawkes *et al.* (1979). The subject of the evaluation was a small rectangular office lit in 18 different ways (Table 6.1). The illuminance on the desks was always 500 lx but the distribution in the rest of the office varied greatly. Factor analysis of the data collected, omitting the evaluative scales, revealed two independent dimensions. One was the brightness dimension, so called because it had such rating scales as bright/dim, strong/weak, and clear/hazy associated with it, although it also had the cramped/spaciousness scale related to it. The other dimension was simply named interest, because the rating scales strongly related to this dimension were simple/complex, mysterious/obvious, uninteresting/interesting, and commonplace/special. The positions of the different installations on the plane formed by these two dimensions are shown in Figure 6.10. From this figure it can be seen that the brightness dimension is clearly related to the amount of light in the room whilst the interest dimension appears to be primarily related to light distribution. Also shown in Figure 6.10 are iso-preference contours indicating those areas on the plane that are preferred to different extents. The top right-hand corner contains the installations that are most preferred. Such installations are both bright and interesting. It is important to appreciate that these two dimensions are independent. Making an uninteresting situation interesting

Table 6.1 Lighting installations used by Hawkes *et al.* (1979)

1 Regular array of ceiling recessed fluorescent luminaires with opal diffusers
2 Incandescent downlights in a regular array plus fluorescent wall-washing of the two end walls
3 Regular array of ceiling recessed fluorescent luminaires with prismatic panels
4 Fluorescent wall-washing of the two side walls
5 Fluorescent desk lights at either side of each desk
6 Incandescent spot lights at the end of the room and on the desks
7 Incandescent downlights in a regular array
8 Incandescent spot lighting of the side walls plus fluorescent wall-washing of the right-hand wall
9 Regular array of ceiling recessed fluorescent luminaires with prismatic panels
10 Fluorescent desk lights at either side of each desk plus fluorescent wall-washing of the left-hand wall
11 Regular array of incandescent downlights plus incandescent spot lighting of the two side walls.
12 Regular array of ceiling recessed fluorescent luminaires with specular louvres plus fluorescent wall-washing of the right-hand wall
13 Fluorescent wall-washing of the right-hand wall
14 Regular array of ceiling recessed fluorescent luminaires with specular louvres plus incandescent spot lighting of the two side walls
15 Incandescent spot lighting of all walls and desks
16 Regular array of ceiling recessed fluorescent luminaires with specular louvres
17 Fluorescent wall-washing of all four walls
18 Fluorescent desk lights at either side of each desk plus incandescent spot lighting of the two side walls

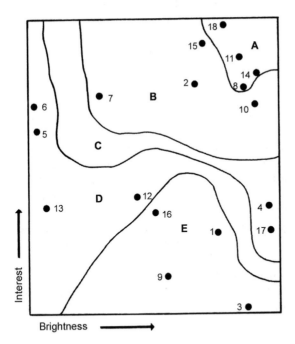

Figure 6.10 A map showing the location of the types of office lighting listed in Table 6.1 on two dimensions, interest and brightness, identified by factor analysis. Superimposed on this map are iso-preference contours based on preference ratings of the same lighting installations. These contours define areas for equal preference from area A (most preferred) to area E (least preferred) (after Hawkes *et al.*, 1979).

does not make is brighter. Making a dim situation brighter does not make it more interesting. The value of the data shown in Figure 6.10 is that the most preferred installations have certain common features, as do the least preferred installations. All the most preferred installations contain some element of variety in lighting produced by spot lighting or wall lighting. All the least preferred installations consist only of recessed fluorescent lighting providing uniform lighting. While this is useful for design purposes, there is one caveat that should be noted. All the installations provided a similar illuminance on the desks, i.e. on the work area. It is doubtful if the installations producing variety of lighting would have been preferred if the variety had lead to a much lower illuminance on the work surface. What this caveat implies is that the results in Figure 6.10 apply to the lighting of the space, given that the lighting of the task is adequate.

One criticism of the studies of Flynn *et al.* and Hawkes *et al.* is that the number of subjects was too small for a robust factor analysis, given the number of scales used. Veitch and Newsham (1998a) overcame this problem by having 292 observers rate the appearance of an open-plan office lit by one of nine different lighting installations. The evaluation was made with the subject sitting in a cubicle in the office after having worked on a computer and on paper for about 5 h. Factor analysis of the ratings revealed three dimensions named brightness, visual attraction, and complexity, although in total they only explained 46 percent of the variance, probably because the differences between the lighting installations were smaller than in the earlier studies. Examination of how the different lighting installations related to each of these dimensions showed that the direct lighting system produced a greater impression of brightness than indirect lighting and that lighting installations using parabolic luminaires were considered less complex and less bright than installations using lensed luminaires, and that task-ambient lighting installations were considered less bright than direct lighting installations.

The interesting point about these three sets of studies, separated by almost 30 years and all using functional interiors where work is to be done, is the similarities and differences in their factor structures. All show that brightness is one dimension on which people evaluate a visual environment. Another appears to be something to do with variety, meaning non-uniformity in light distribution away from the work surface. After that there is no agreement. Flynn *et al.* (1973) have a dimension of spaciousness. Veitch and Newsham (1998a) have a dimension of complexity and Hawkes *et al.* (1979) do not have a third dimension. The fact that brightness is a consistent dimension on which the visual environment is evaluated should not be too surprising, given that brightness is related to the amount of light in the space and how well we can see is determined by the amount of light available. The importance of brightness is also consistent with an information-processing model of environmental appraisal (Kaplan, 1987). This model rests on the idea that the information available in a scene is central to our appraisal of it, and brightness is a marker for how much information is revealed. As for variety, there is no doubt that humans need some variety, but not too much. Monotonous environmental conditions, if taken to extremes, lead to cognitive breakdown (Corso, 1967). On the other hand, too much variation in environmental conditions can lead to confusion and distress.

Another thing to say about variety is that, unlike brightness, variety can be introduced through aspects of the visual environment other than lighting. For example, variety can be introduced into a room by changing the decor, by changing the furnishings and by changing the people in it. Using a non-uniform light distribution is but one way of introducing variety and may not be the most potent. This applies even more strongly to higher level perceptions, such as whether the interior is friendly or formal or patriotic. This does not mean that lighting does not matter, rather that the more the perception of a space is determined by people's expectations and hence by their past experience, the less likely it is that simple lighting variables, such as light distribution, are the only factors that matter.

What this suggests is that any attempt to try and link specific lighting conditions to higher-level perceptions and to apply them across many different contexts is doomed to fail. There are too many other factors beyond the photometric conditions that influence higher-order perceptions, and those factors will vary with different contexts for different cultures. It may still be possible to make a link between lighting conditions and higher-order perceptions for specific contexts in specific cultures but how valuable such a finding would be would depend on the how specific the conclusion was. There is little merit in determining what type of lighting makes a meeting room look formal if that perception changes dramatically when the furniture is changed.

6.4   The perception of objects

This rather dismal conclusion about the impact of lighting on higher-order perceptions in functional spaces should not be taken to mean that lighting cannot make a dramatic difference to how a non-functional space or an object is perceived. Rather it should be taken to mean that the perception will depend on the nature of the space and the object, and the context in which they are viewed, as well as the lighting. This is evident in the work of Mangum (1998). He examined the perception of three museum objects under six different lighting installations, all subject to the constraint that no illuminance on the object could exceed 50 lx. The perception of the object was identified by the use of a balanced (positive and negative) word list from which observers had to select all the words that applied to their perception of the object. When viewing a doll, clothed in materials varying in texture, color and reflection properties, under diffuse illumination, the words most frequently chosen to describe the observers' perceptions were unattractive, unpleasant, obscured, veiled, bland, boring, mundane, and ordinary. When the same object was lit with directional lighting, using key-light, side-light, and back-light, the words most frequently used were interesting, attractive, eye-catching, clear, pleasant, revealing, dramatic, and spectacular. Similar results were found for a plain vase, although other forms of directional lighting produced more dramatic perceptions for this object. Clearly, changing the lighting can change the higher-order perceptions of an object, which is just as well for the practitioners of display lighting. Whether it does or not will depend on the nature of the object and to what extent the form, texture, color, and reflection properties of the object are modified by the change in lighting. Of course, objects can vary enormously in their form, texture, color, and reflection properties. This is not often the case for functional interiors. For most functional interiors, the surfaces are diffusely reflecting, with limited variations of texture, color, and form, and the distribution of light is much more limited than in display lighting. What this means is that the higher-order perceptions in functional interiors are most likely to be influenced by the lighting when the architect has provided a space that has an interesting form, containing surfaces that vary in texture and color, and the lighting designer has used his or her skills to complement and enhance the architecture, thereby producing something of visual interest.

6.5　Summary

The link between the stimuli provided to the visual system by a luminous environment and the perception of that environment is often tenuous. This is because the perception of the luminous environment depends on the state of adaptation of the visual system, the background against which objects in the environment are seen and the observers' past experience and knowledge. Nonetheless, the luminous environment is the starting point of perception and lighting can be used to change the luminous environment. Therefore, lighting can change the perception of spaces and the objects in them.

How stable the perception of a space will be will depend on how closely the perception is linked to the operation of the visual system and how much opportunity there is for non-lighting factors and for past experience and knowledge to intervene. Simple perceptions, such as lightness, brightness, and color appearance show robust links to the luminous environment. Higher-order perceptions, such as formality, spaciousness, and complexity show more tenuous links.

The simple perception of lightness is linked to reflectance, but given a very wide range of luminance, even the perception of lightness will change. Specifically, as the luminance decreases, the lightness of a surface decreases. The simple perception of brightness is linked to luminance by a power law, but it is also influenced by the light distribution in a room, the luminance of the luminaire, and the spectrum of the light. The results in this area show that increasing the luminance of the room surfaces will increase the perception of brightness, as will choosing a higher CCT light source or one with a larger gamut area. As for luminaire luminance, depending on the luminance and area of any bright patches on the luminaire, the brightness of the room can be enhanced or diminished. Balancing the luminance and area so that the bright patch on the luminaire is perceived as sparkling will also enhance the brightness of the room. Increasing the luminance further so that the luminaire becomes glaring will diminish the brightness of the room. The simple perception of color appearance is linked to the spectrum of the light source and the luminance. How strong an effect the choice of light source has depends on whether the space is essentially achromatic or one containing many colored surfaces. The effect of light source will be much greater for the latter conditions than the former because chromatic adaptation can offset some of the difference due to different light spectra in an achromatic room but cannot offset the effect of the light spectrum on the saturation of colors in the room.

The effect of the luminous environment on higher-order perceptions is much less certain. This is because higher-order perceptions, such as formality, are influenced by the whole of the environment, not just the luminous environment, as well as the context of the space and the culture of the observer. The only consistent evidence suggests that almost all functional spaces are evaluated on the dimensions of brightness and visual interest, the former being related to the amount of light in the space; the latter being enhanced by a non-uniform distribution of light away from the work area.

This should not be taken to mean that lighting always has such limited effects on the perception of non-functional spaces and objects. There is clear evidence that by changing the lighting, the perception of objects can be changed from drab and boring to eye-catching and dramatic. Everyday experience suggests that the same is true of non-functional spaces, such as restaurants, shops, and theaters. The problem with functional spaces is that the possible lighting effects and the materials they have to work with are often limited by the need to provide good visibility for work over a large portion of the space. In such spaces, lighting is most likely to have an effect on higher-order perceptions when the architect has generated an attractive space and the lighting designer has produced lighting that provides suffi-cient brightness in the task area and enhances the architecture to provide some visual interest elsewhere.

Part III
Specifics

7 Lighting for offices

7.1 Introduction

Over the last 50 years, the office has become the place where more and more people spend their working lives. The technology available to agriculture and industry in developed countries is such that large numbers of people are no longer required to produce food or manufactures, and if they are, then globalization of business means that production is moved to less developed countries where wages are lower. In 1995, in the US, the largest economy in the world, 41 percent of people employed were office workers (Carnevale and Rose, 1998).

Further, over the last 20 years the nature of office work has changed dramatically, not because the purpose of office work has changed but rather because the means to do the work have changed. The purpose of office work remains the collection, recording, and distribution of information, together with the making of decisions based on that information and the direction of effort to carry out the decisions made. What has changed over the last 20 years has been the immense growth in the ability to collect, record, and distribute information rapidly, over vast distances, electronically. This process began with the introduction of the personal computer, gained strength with the development of local networks and reach its full flowering with the growth of e-mail and the World-Wide Web.

There can be little doubt that the consequences of these changes will be of great interest to the social historian, but what do they have to do with lighting? The answer is a great deal. In the paper-based office, the primary surface to be viewed is horizontal and increasing the amount of light in the office makes any information on that surface more visible. In the computer-based office, the primary surface to be viewed is vertical and increasing the amount of light in the office makes the information displayed on the self-luminous screen less visible. Therefore, the widespread introduction of computer-based technology into the office has made a fundamental change in the requirements for lighting an office. Early-on in the introduction of computer-based technology, it appeared that a completely different approach to office lighting would be needed because an entirely screen-based office was technically possible and the visibility of the displays produced by early screen technology was very sensitive to the lighting conditions. A walk around any office today will reveal very few that are completely screen-based as well as showing that modern screen technology is much less sensitive to lighting conditions. This means that any lighting installation designed for an office today has to be satisfactory for materials that are self-luminous, i.e. computer screens, and seen by reflected light, i.e. paper, and for lines of sight that can be both across the office and down at the desk.

This chapter is devoted to the lighting of offices and is organized around the choices the lighting designer has to make, namely the illuminance to be provided, the light source to be used, the type of luminaire and its layout and the nature of any lighting controls. This chapter is only concerned with the features of a display that make it more visible and/or more comfortable to use in as far as they interact with the lighting of the office. Anyone who is interested in the features of the display that affect visibility and comfort independent of the lighting is referred to Roufs (1991).

7.2 Illuminance

The most widely used criterion of good lighting in an office is the average illuminance on a horizontal plane at desk height, a surface conventionally called the working plane. Every national and international set of recommendations for lighting contains such illuminance recommendations (Japanese Standard Association, 1992; CEN, 1996; IESNA, 2000a; CIBSE, 2002a). The history of illuminance recommendations shows considerable variation, in different countries and at different times (Mills and Borg, 1999). Table 7.1 summarizes the recommended illuminances for different activities in offices from 19 different countries. The differences between countries in the form and magnitude of the recommendations is obvious. As for the changes over time, the trend in illuminance recommendations for virtually all countries is an increase in recommended illuminances for office work from the 1930s until the early 1970s, followed by either a stabilization or a decline. This trend reveals the illuminance recommendations are not simply a matter of logical deduction. Rather, making illuminance recommendations is a human activity that is influenced by both practical and political considerations, in addition to the state of knowledge of how illuminance affects task performance and visual comfort (Boyce, 1996). The practical considerations are matters of technology. For example, it can plausibly be argued that one of the driving forces that led to the general increase in recommended illuminances from the 1930s to the 1970s was the introduction and development of the fluorescent lamp and the high-intensity discharge lamps, light sources with dramatically higher luminous efficacies than the incandescent lamp they replaced. The political considerations are financial and consequential. The financial consideration is the cost of providing a given illuminance relative to the benefits obtained. Increases in this

Table 7.1 Recommended maintained horizontal illuminances (lx) for different activities in offices in different countries (after Mills and Borg, 1999)

Country and year	General	VDT tasks	Reading tasks	Drafting
Australia, 1990	160	160	320	600
Austria, 1984	500	500	—	750
Belgium, 1992	300–750	500	500–1,000	1,000
Brazil, 1990	750–1,000	—	200–500	3,000
Canada, 1993	200–300–500	300	200–300–500	1,000–1,500–2,000
China, 1993	100–150–200	—	75–100–150	200–300–500
Czech Republic	200–500	300–500	500	750
Denmark	50–100	200–500	500	1,000
Finland, 1986	150–300	150–300	500–1,000	1,000–2,000
France, 1993	425	250–425	425	850
Germany, 1990	500	500	—	750
Japan, 1989	300–750	300–750	—	750–1,000
Mexico, proposed	200	—	900	1,100
Holland, 1991	100–200	500	400	1,600
Russia, 1995	300	200	300	500
Sweden, 1993	100	300–500	500	1,500
Switzerland, 1997	500	300–500	500	1,000
UK, 1994	500	300–500	300	750
USA, 1993	200–300–500	300	200–300–500	1,000–1,500–2,000

cost tend to lead to declines in recommended illuminances. The consequential considerations refer to such consequences of burning fuel to generate electricity for lighting as increased global warming and increased dependence on foreign fuel suppliers. These financial and consequential considerations can lead to sudden changes in lighting practice as the perceived costs of lighting weigh much more heavily than the benefits in the mind of the user.

The observation that illuminance recommendations, and presumably therefore the illuminances installed, have varied over time make it interesting to consider if there is any stability in the illuminances preferred for offices. Saunders (1969) had a windowless room equipped as an office and lit by conventional ceiling-mounted luminaires. Subjects came into the room, were given time to look around and adapt to the conditions and then asked to sit at a desk and read from a book. After reading a few paragraphs, the subjects were asked to give their opinions on the suitability of the lighting for reading and the general quality of the lighting. Figure 7.1 shows mean ratings of the quality of the lighting plotted against the illuminance on the desk. It can be seen that lighting that provided illuminances less than about 200 lx was considered of poor quality but increased illuminances produced opinions of increased quality following a law of diminishing returns.

A similar result, but obtained using data collected in a field survey and from a much larger number of subjects, was reported by Van Ierland (1967). Figure 7.2 shows the percentage of about 2,000 office workers in different offices in The Netherlands rating the illuminance on their desk "low," "right," or "high." Again, it is evident that the percentage considering the

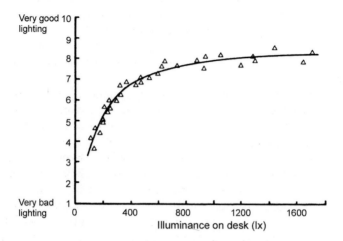

Figure 7.1 Mean ratings of the quality of lighting in a windowless office at different desk illuminances (after Saunders, 1969).

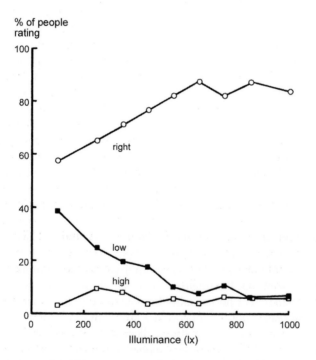

Figure 7.2 The percentage of people rating the illuminance on their desk, low or rather low, high or rather high, or just right (after Van Ierland, 1967).

lighting "right" increases up to about 500 lx after which it shows little change. Also evident is the fact that as the illuminance decreases below about 500 lx, the percentage considering the illuminance "low" increases. Another feature of Figure 7.2 is that no single illuminance is considered "right" by all the subjects. It appears that the best that can be achieved is about 80 percent.

While these two studies support each other, it is undeniable they were both done many years ago, before the introduction of the computer into the office. Given the reduction in visibility of the self-luminous screen when the ambient illuminance is increased, it might be expected that preferred illuminances for offices would now be somewhat lower. Newsham and Veitch (2001) examined this question using a novel technique. Their experiment took place in a windowless office equipped with a lighting installation that allowed the illuminances to be adjusted by dimming and switching. Two people occupied the office at the same time, but only one had control of the lighting system. This person was asked to adjust the lighting to what he/she preferred at the start of the day. After that the lighting was unchanged throughout the day. During the day, both subjects performed paper- and computer-based office tasks. At the end of the day, the subject who had not had the opportunity to adjust the lighting was given a chance to set the lighting to his/her preference. Figure 7.3 shows the linear regression of the change in illuminance on the desk made at the end of the day by 47 subjects, plotted against the illuminance they had experienced throughout the day. The regression line crosses $y = 0$ at 392 lx, i.e. this is the illuminance experienced during the day after which no change was desired. For higher illuminances, the change at the end of the day is to reduce the illuminance and for lower illuminances the change is to increase the illuminance. The null point of 392 lx is somewhat lower than the earlier studies and can be attributed to the fact that many of the subjects adjusted the lighting to reduce the luminance of reflected images of the overhead luminaires on the computer screen. When Newsham and Veitch (2001) eliminated the changes in illuminance made to reduce the reflected images in the computer screen, the null illuminance increased to 458 lx. This illuminance is close to the value recommended for offices in many countries (500 lx) but it is also lower than the asymptotic values shown in Figures 7.1 and 7.2. This is probably because the lower illuminance represents a compromise between the effect of a high illuminance making reflecting material more visible and making self-luminous material less visible.

Another evident feature of Figure 7.3 is that the spread of data points is large, indicating a wide variation in individual preference for illuminance. These individual differences matter because Newsham and Veitch (2001) also found that the deviation between the subjects' illuminance preferences and the illuminance they experienced during the day was a statistically significant predictor of the subject's mood and satisfaction with the lighting. Overall, it can be concluded that there is some stability in preferred

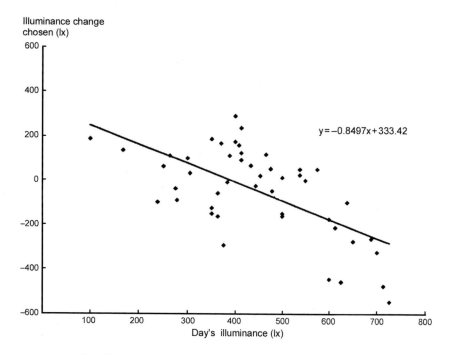

Figure 7.3 The illuminance change chosen at the end of a day's work plotted against the illuminance on the desk experienced during the day (after Newsham and Veitch, 2000).

illuminance in offices for groups of people but that there are also large differences in what individuals prefer. This suggests an opportunity for improving satisfaction with lighting by the use of individual control (see Section 7.5.2).

But illuminance alone is not enough to ensure good office lighting. Figure 7.4 shows the percentage of people in each of 20 different deep, open-plan offices who considered the lighting comfortable, plotted against the mean illuminance (Kraemer *et al.*, 1977). The variation in the percentage considering the lighting comfortable and the lack of any clear relationship with illuminance is obvious. Given the time when this study was done, before the widespread introduction of computers into offices, the results from Figures 7.1 and 7.2 can be used to explain the lack of effect of illuminance. It is simply that all the illuminances in the offices are above 400 lx and the vast majority lie between 600 and 1,000 lx. The results shown in Figures 7.1 and 7.2 indicate no dissatisfaction with illuminance over this range. Nonetheless, Figure 7.4 shows considerable variation in the percentage of people considering the office lighting comfortable. This variation suggests that it is not so much what you do in providing an illuminance but

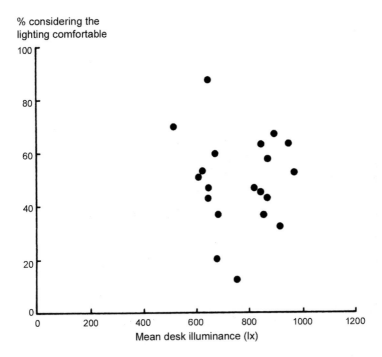

Figure 7.4 The percentage of people in each of 20 deep, open-plan offices who considered the lighting comfortable, plotted against mean desk illuminances in the offices (after Kraemer *et al.*, 1977).

the way that you do it, that determines comfortable lighting conditions. Some of the aspects of lighting conditions that can cause visual discomfort, such as illuminance non-uniformity, glare, and veiling reflections, are discussed in Chapter 5. All these aspects of lighting require the designer's attention if comfortable lighting is to be achieved.

Despite the fact that illuminance alone is not a secure basis for designing comfortable office lighting, it remains the primary design criterion. Attempts have been made to get the people responsible for many lighting installations to consider other factors than illuminance by providing simple advice in codes of practice (IESNA, 2000a; CIBSE, 2002a) but with little obvious effect so far. Illuminance remains the primary design criteria because it is easy to predict and to measure and it can be clearly linked to the costs of the lighting installation. It is also, undeniably, an important factor in determining the acceptability of a lighting installation. As discussed in Chapter 6, the one dimension on which people consistently evaluate a lighting installation is brightness, and the illuminance provided is related to the perception of amount of light in the space and hence to brightness.

7.3   Light sources for office lighting

By choosing the light source, the lighting designer determines the spectrum of the light used to illuminate the office and hence influences the appearance of the office and the people in it. The first major decision to be made on light source is the balance between daylighting and electric lighting. The second major decision is the type of electric light source to be used.

7.3.1   Daylight

Daylight always deserves consideration. It deserves consideration because people desire it. The fact that daylight is desired can be shown by evidence from four sources. From research, comes the fact that almost any study which asks office workers about which light source they would like to illuminate their work area reveals a strong preference for daylight (Markus, 1967; Cuttle, 1983). From behavior comes the observation that those of higher status in an organization are commonly given offices closer to windows or with more windows (Cuttle, 1983). From advertising comes the frequent claim that one electric light source or another is "like daylight"; the rationale being that a light source "like daylight" is what people desire. Finally, from finance, comes the fact that the rent charged for daylit office spaces is more than for non-daylit office spaces. There is little doubt that daylight is desired by people in offices, and in other spaces. But why is daylight so desirable? The usual responses to this question range from surprise that it should be asked at all, through comments about daylight being natural, to quasi-religious statements about the Sun. Such answers are not helpful. Rather, it is necessary to consider daylight from the points of view of physics, physiology, and psychology.

Physically, daylight is simply electromagnetic radiation in the wavelength range that is absorbed by the photoreceptors of the human visual system. In this respect it is the same as all other light sources. The actual wavelengths present in daylight will vary over the day, with meteorological conditions, with latitude, and with season. It is this variability in spectral content and the fact that, at all times, it is a continuous spectrum with power in all parts of the visible wavelength range, that separates daylight from the fluorescent lamp used in nearly all offices. Nonetheless, there are electric light sources available, such as the xenon lamp, the sulfur lamp, and some filtered incandescent lamps, that have a spectral content similar to that which occurs with daylight on some occasions. Thus, there is no unique physical characteristic of daylight which separates it from all other light sources.

Physiologically, the response of the human visual system to the light spectrum is determined by the spectral sensitivity of the three cone photoreceptors and the rod photoreceptor (see Section 2.2.4). All these photoreceptors have a broad spectral response, i.e. the response of each photoreceptor type covers a wide range of wavelengths. This implies that the visual system should be capable of functioning equally well using light consisting of many different wavelength combinations. This belief is supported by measurements of visual acuity, contrast sensitivity, and other threshold visual functions made under different light sources (Boff and Lincoln, 1988). The only threshold performance which is strongly influenced by the spectral content of the illuminant is color discrimination; a fact which is not unexpected given that changes in the spectral content of the illuminant will change the stimulus presented to the visual system. Likewise, measurements of suprathreshold visual task performance have shown little difference between light sources, provided the different light sources produce the same illuminance and the tasks did not require fine color discrimination (Smith and Rea, 1979).

The conclusion to be drawn from these observations is that there is no physical or physiological reason for daylight to be so desirable. This leaves some psychological factor as a possible source of the desire for daylight. For daylight to be attractive psychologically, it has to fulfill some basic human needs. One such is the ability to see well. Daylight is characterized by the high illuminance it is capable of providing and the excellence of its spectrum for color discrimination and color rendering. These two properties give daylight the potential to produce good vision. Whether it does or not depends on how the daylight is delivered. Daylight delivered through windows can produce a bright interior in which people look their best and can see details and colors well. Equally, daylight delivered through windows can produce uncomfortable solar glare and very high luminance reflections on display screens, both of which hinder vision.

Another human need is for variety. Monotonous environmental conditions, if taken to extremes, lead to cognitive breakdown (Corso, 1967). In the context of lighting, Hawkes *et al.* (1979) measured peoples' responses to the same office lit by electric lighting, in 18 different ways (see Section 6.3.2). Their results showed that peoples' preference for the lighting was based on two independent dimensions; brightness and interest. The brightness dimension was related to the illuminance on the working area while the interest dimension was related to the variation of light patterns in the space. The most preferred office lighting was that which provided both brightness and interest. Daylight can be used to provide lighting which is both bright and interesting, in the sense that it can provide a lot of light and that light has a distribution which shows meaningful variation over space and over time.

Given that variety is what is being sought, it is important to note that the usual means by which daylight is provided, namely a window, provides variety through a view-out in addition to variety in the lighting of a space. This raises the question of whether the commonly expressed preference for daylight is really a preference for a window with a view-out. I believe this question poses an artificial dichotomy. It seems much more likely that both aspects are important. Certainly, when asked about important aspects of windows, both the provision of daylight and the provision of a view-out are frequently considered important. Further, it is necessary to recognize that a view-out can sometimes have a psychological cost, in terms of loss of privacy. Heerwagen (1990) argues that what people seek from windows is visual access without visual exposure. To meet this aim, a careful balance between view-out and view-in is required.

Given that daylight appears to be strongly desired by most people, it is reasonable to ask what happens when people are asked to work without daylight, when it is available outside. These conditions, which occur in windowless rooms and underground spaces, have produced a consistent pattern of response. When the windowless space is small and the occupant has little possibility to leave the space, the lack of variety is noticeable and the lack of windows is disliked (Ruys, 1970). However, in large spaces, such as school classrooms (Larson, 1973) and factories (Pritchard, 1964), the lack of windows has a much more variable impact. This may be because in a large space, where there are many other activities going on and there is a lot of interaction between people, there is often plenty of stimulation in the environment. In a small office, it may be that the view out of the window is the only source of environmental stimulation. Some support for this view comes from the work of Heerwagen and Orians (1986). They observed that people occupying small private offices without windows used twice the number of visual materials to decorate their offices as did people whose offices had windows, and that those materials were dominated by views of nature. Overall, the work on peoples' reactions to windowless buildings supports the idea that a view-out is valued, but it does not necessarily support the primacy of view-out over the provision of daylight. In windowless offices, surrogate daylight is provided by the electric lighting, it is a surrogate view that is missing.

A supplementary question to that concerning windowless buildings is, if daylight is desired so strongly, how willing are people to give it up when it is available? Observation of almost any multi-story office building will reveal that people are willing to give up daylight when it also causes visual and thermal discomfort. Measurements of the use of window blinds in multi-story office buildings have shown two trends. The first is that window blinds are increasingly likely to be pulled down as the Sun begins to shine on the window and so cause solar glare and radiant heating. This might be expected. Less expected is the second trend, that many of these blinds are kept down even after the Sun has ceased to shine directly on the window; and that, in some cases, the blinds are left in the down position for days, months, or years (Rea, 1984). This latter trend suggests that the desire for daylight, or at least direct sunlight, is limited when it causes discomfort.

Yet another supplementary question is, if daylight, *per se*, is strongly desired, it might be expected that when there is plenty of daylight available in a space, people will switch off or dim the electric lighting to minimize its impact on the daylighting. However, prolonged measurements in a set of deep, private offices with large windows have shown that, given the opportunity, people tend to increase the amount of electric light as the amount of daylight in the office increases (Begemann *et al.*, 1994, 1995). This behavior is consistent with a desire to balance the luminance of the window and the surfaces near it with the luminances of the surfaces deep in the room. It also implies that the overall pattern of light distribution in the room is more important than the purity of daylight. This lack of concern with the absolute purity of daylight is again evident in the widespread use of tinted glazing in buildings and the wearing of tinted sunglasses by people.

So why is daylight desired? The observations above suggest that the reason why daylight is so popular as a means of lighting is that, if carefully designed, daylighting delivered through windows provides a comprehensive package which can meet the requirements of good lighting by revealing both the task and the space clearly and the requirement of environmental stimulation by variation of lighting conditions in the space and a view-out through the window. By comparison, electric lighting installations, if well designed, can provide good visibility of the task but often do little for the space and rarely provide any variation over time or space. Even when electric lighting installations are designed to produce some variation over time and space, the variation is usually simple, repetitive, and arbitrary compared to the complexity of daylight variation and the meanings that such variations carry. But the desire for daylight is not unlimited. When daylight through windows is inappropriate, either because it causes visual or thermal discomfort or because of excessive environmental stimulation, or because of loss of privacy, it is common to reduce daylight and eliminate the view-out by covering the windows in some way.

What all this suggests is that the desirability of daylight is not so much a matter of its inherent superiority but of the limitations of electric lighting systems. This conclusion is consistent with the results reported by Cuttle (1983). From a series of questionnaire surveys of office workers in England and New Zealand, he concluded that the preference for daylight could be attributed to the belief that working by daylight results in less stress and discomfort than working by electric light. The belief was not so much that daylight was beneficial but rather that working by electric lighting was deleterious to health, particularly in the long term. Of course, this may have been due to the specific forms of electric lighting used in the offices surveyed, which may have been visually uncomfortable, but it is interesting to realize that similar beliefs about the negative effects of fluorescent lamps on people were found in a survey of 2,950 members of the public attending a New York State Fair in 1991 (Beckstead and Boyce, 1992). A bias against some widely used forms of electric lighting can be expected to produce a desire for daylight, some form of lighting being essential and daylight being the most obvious alternative.

Given that daylight is desired, could the provision of daylight in a workplace, as opposed to electric light, have a positive effect of task performance?

Existing knowledge offers little support for the idea that the provision of daylight, *per se*, can guarantee an improvement in productivity (Cuttle, 1983; Norris and Tillett, 1997). However, there is a conceptual framework which can be used to identify if and how the provision of daylight might enhance productivity (see Section 4.2). Lighting conditions can have an impact on task performance through three routes. The first is by changing the stimulus the task presents to the visual system or by changing the operating state of the visual system. How changes in the amount of light, the spectral content of light and the distribution of light alter either the stimuli a specific task presents to the visual system or the operating state of the visual system are well understood. Further, the effects of these changes on the performance of the visual component of a task can often be predicted through a model of visual performance (Rea and Ouellette, 1991). For this route, daylight is just another form of lighting, which, in a given situation, will produce a specific amount of light with a known spectral content and, depending on how it is delivered, with a known light distribution. Therefore, depending on the form of daylight, visual performance can be enhanced or degraded.

The second route is through mood and motivation. It has been established that when people are in a good mood, they tend to be more positive about the work, more cooperative, and more creative; attributes which are considered desirable in many working situations (Isen and Baron, 1991). A large number of environmental and psychological factors can influence a person's mood, such as an unexpected gift or compliment, an attractive smell, and even lighting, but one common feature of the factors which promote a good mood is that they are unexpected. Daylight delivered through windows, because of its variability, has some potential for delivering the unexpected, both directly in the space and through the view-out. A fixed electric lighting installation has much less potential for generating a good mood because what it does is entirely predictable. Even a variable electric lighting system is inferior to daylight variation because the variations carry no implications for what is happening outside. While daylighting and electric lighting may differ in their potential to create a good mood, it should be appreciated that both daylight and electric lighting installations can generate a bad mood, particularly when they induce eyestrain and headaches. A bad mood is associated with negative, obstructive behavior; a situation which is not good for productivity.

Unfortunately for simplicity, there are many factors which can determine mood, and lighting is only one of them. Therefore, even though daylight has the potential to generate a good mood, its effects are likely to be swamped by all the other aspects of working life which also influence mood. This may explain the variability which occurs in the attempts that have been made to measure the effects of daylight availability on task performance. Hedge (1994) measured the performance of a clerical task on a computer in a room lit by different electric lighting systems, with and without windows.

He found a small but statistically significant improvement in task performance when windows were present. Whether this occurred because the presence of windows improved the stimuli the tasks presented to the visual system or changed the operating state of the visual system or because the subjects were in a better mood or if all three occurred is not clear. Stone and Irvine (1991) also measured the performance of a managerial task in a room which had a window or was windowless. In this case, no statistically significant effects of having a window were found for task performance or mood. It is clear that the provision of daylight through windows is not a sure recipe for an improvement in task performance.

Daylight also has photobiological effects (see Chapter 3). Reduced exposure to daylight for a long time is known to be associated with a form of clinical depression, called seasonally affective disorder (SAD) (see Section 13.4.2). This usually occurs in the winter but not in the summer and increases in prevalence at higher latitudes (Terman, 1989). The symptoms associated with SAD are feelings of sadness and irritability, as well as lethargy, oversleeping, and overeating. These symptoms tend to lead to difficulties at work and in personal relationships. These symptoms have been successfully treated by exposure to electric light, either as a massive daily light dose (Terman *et al.*, 1989) or as a simulation of the Sun rising at dawn (Avery *et al.*, 1994). The former might be expected, it is simply providing the missing dose of light. The latter is not expected but it has been suggested that exposure to daylight in the early morning may be a very effective way of reducing the concentration of the hormone melatonin, that is associated with sleep, and hence kick-starting the working day. How effective daylight is compared with a cup of strong coffee remains to be determined. Clearly, we have some way to go before we fully understand this impact of light exposure on people. However, there is no doubt that the symptoms of SAD are real, are associated with reduced exposure to light, and are disruptive to everyday life. There is also the possibility that many more people suffer from sub-clinical SAD in which the symptoms of SAD are present but in a mild form (Kasper *et al.*, 1989b). If this is so, then the impact of lack of daylight on productivity may be underestimated, particularly where daylight exposure before and after work is limited for several months of the year.

This analysis of why daylight is desired and how it might affect productivity carries a number of implications, both conceptual and practical. In conceptual terms, one implication is that to understand the impact of daylight on peoples' feelings and behavior, it is necessary to stop treating daylight as an end in itself and to consider its provision as a means to an end. This suggests paying attention to the more profound psychological aspects of life, such as the desire for knowledge, social isolation, privacy, and stimulation, rather than the physical characteristics of the form of daylight provision.

In practical terms, the implications are that the designers of daylight-delivery systems need to control solar glare and radiant heating from sunlight, provide a view-out, perhaps limit the view-in, and pay attention to the lighting of the space as well as the lighting of the task. Designers of electric lighting systems need to remember to work with daylight as much as possible and recognize that both the lighting of the task and the lighting of the space matter.

7.3.2   Electric light sources

The great disadvantage of daylight as a light source is that it fails every day, for a long period. Therefore, every office is designed with an electric lighting system that is used after dark and, frequently, throughout the day as well. The most commonly used electric light source in offices is the fluorescent lamp. Other electric light sources, such as the incandescent lamp or the metal halide discharge lamp, could be used but rarely are. The reason why the incandescent lamp is rarely used is that it has a much lower luminous efficacy and a much shorter life than the fluorescent lamp, both factors that increase the costs-in-use of a lighting installation (see Chapter 1). The metal halide lamp is rarely used because the low ceiling heights typical of offices, usually less than 3 m, impose limits on the maximum possible luminance of luminaires, restrictions that are difficult to meet with a metal halide lamp without using a low wattage version that has a low luminous efficacy or using a higher wattage version in an inefficient luminaire.

Fortunately for the lighting designer, fluorescent lamps come in many different sizes and shapes, and with many different light spectra. It is this last attribute that is most obvious to the office worker. The effect of light spectrum on office work has to be considered on three levels. The first is the effect on the performance of chromatic tasks. The second is the effect on the performance of achromatic tasks. The third is the preference for different light spectra.

7.3.2.1   Light spectra and chromatic tasks

There are three main roles for color in office tasks. The first is to increase the visibility of the task by providing a color difference between the task and its immediate background, in addition to any luminance contrast, e.g. printing in different colors. The second is to increase the conspicuity of the task by marking it with a different color from others around it, e.g. highlighting a paragraph on a document. The third occurs where the color itself is meaningful, e.g. the color of a warning sign or a filing label. For the first and second roles, the color itself is immaterial. The only thing that matters is how different it is from the color of the immediate background and its surroundings. For the third role, the accurate naming of the color is important The light spectrum can influence the effectiveness of colored stimuli for all these roles because light spectrum is one of the factors that determine color appearance. Therefore, light spectrum can increase or decrease the color difference between the task and its background and change the appearance of colors.

Eklund (1999) examined the ability of people to read an exit sign when the letters and the background either were or were not different in color and could be varied in luminance contrast. Figure 7.5 shows the accuracy with which the sign was read at different luminance contrast values, both positive and negative. When the letters and the background had the same color so that the letters could be distinguished from the background only by luminance contrast, the accuracy of reading starts to decline to chance level (50 percent) as luminance contrast falls below about 0.25, but when there was a clear color difference there was no decline in the accuracy of reading the sign, even when the luminance contrast was zero. This result implies that light spectrum will only matter for the performance of achromatic tasks when the luminance contrast is low. In these conditions, a light spectrum that enhances the color difference between the task and the background will also enhance the task performance.

Williams (1966) studied the search times for finding a specified item from a display of 100 items, that could vary in size, shape, color, and the information they contained. In fact, the information within each item was a two-digit number. The inspector was asked to locate a particular number, the number being specified either alone or together with various combinations of the size, color, and shape of the item it was in. For example, the inspector could be asked "Find the number 45 which is in a large blue square,"

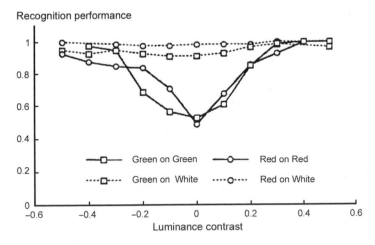

Figure 7.5 Recognition performance, measured as the proportion of times the orientation of the word EXIT was correctly recognized, plotted against negative and positive luminance contrasts. The legend is read as "letter colour on background colour." The light sources used are LEDs. The green LED has a peak emission at 530 nm. The red LED has a peak emission at 660 nm. The white LED has a spectral emission similar to that shown in Figure 1.17 (after Eklund, 1999).

Table 7.2 Mean time to find a target for different
target specifications (after Williams, 1966)

Target specification	Mean time (s)
Number only	22.8
Number and shape	20.7
Number and size	16.4
Number, size, and shape	15.8
Number and color	7.6
Number, color, and shape	7.1
Number, color, size, and shape	6.4
Number, color, and size	6.1

or simply "Find the number 45." The mean search times for the different target specifications are given in Table 7.2.

It can be seen that the longest search times occurred when the number alone was specified and the shortest search times occurred when the color and size of the item in which the number was located were also specified. This is not simply a case of the more factors that are specified, the fewer the items that have to be inspected. The point is that some parts of the specification are more important than others. Table 7.2 shows that whenever the color of the item in which the desired number lay was specified, short search times were achieved. Specifying the color reduces the items where the number might be to 20 percent of the total but so does specifying the shape and yet the latter has much less effect on search time. This apparent effectiveness of color as an aid to visual search is consistent with one of the claimed evolutionary advantage of color vision, namely to rapidly detect targets against dappled backgrounds (Mollon, 1989). However, the finding that including color in the specification of the search item had a major impact on the speed of visual search should not be taken to mean that color coding is a guaranteed way to enhance visual search performance. If the differences between the colors used by Williams (1966) had been slight, it is doubtful if specifying the color would have been anywhere near as effective. A more general way to identify what dimension is important is to consider the signal-to-noise ratio between the target items and non-target items, on each dimension. The higher the signal-to-noise ratio on a given dimension, the greater is the importance of including that dimension in the specification of the search item. It is worth noting that the signal-to-noise ratio of colors can be enhanced by the use of a light source with a high CIE general CRI (> 80) because the colors are then more widely separated in color space.

Turning now to tasks where the color of the target is itself meaningful, the choice of light source depends on the degree of color discrimination required. Where only course discrimination is required, e.g. telling red from blue or green, then only light sources with low CIE general CRIs, such as the high-pressure sodium and mercury vapor lamps, will cause confusion about colors (Collins *et al.*, 1986). Where fine color discriminations are required, e.g. when examining color printing, great care is required in the choice of light source. Recommendations of the light source to be used for fine color discrimination, in a number of different industries, have been made (IESNA, 2000a). If no specific recommendation is available, then a general rule is that the higher is the CIE general CRI the better will be the ability to discriminate colors. The extent to which different light sources will make it possible to discriminate colors can be estimated by using the MacAdam ellipses (see Figure 2.25). Each MacAdam ellipse sets the boundary at which a given percentage of people are able to determine that two colors, one with chromaticity coordinates at the center of the ellipse and the other with chromaticity coordinates on the ellipse, are just noticeably different (MacAdam, 1942; Wyszecki and Stiles, 1982). MacAdam's ellipses were determined in conditions that offer the maximum sensitivity to color differences: side-by-side comparison, unlimited observation time, foveal viewing, and photopic operation of the visual system. Changing any of these factors and adding distracting or confusing stimuli, can be expected to increase the difference in color needed to reach discrimination threshold (Narendran *et al.*, 2000). There can be little doubt that the spectral content of the light used becomes of increasing importance as the degree of color discrimination required becomes finer and the outcome of that discrimination is more important for the successful completion of the task.

7.3.2.2   *Light spectra and achromatic tasks*

At first, it might seem odd to consider what effect light spectrum would have on the performance of an achromatic task, particularly as previous studies have failed to show any effect of light spectrum (Rowlands *et al.*, 1971). Specifically, Smith and Rea (1979) systematically investigated the effects of both light level and light spectrum on suprathreshold task performance. Illuminances ranged from about 7 to 2,000 lx; cool white fluorescent, metal halide, and high-pressure sodium lamps were used as illuminants. Moreover, they studied, in combination with the two lighting variables, the impact of the achromatic luminance contrast and the "quality" of the visual task as well as the subjects' age. Subjects were in two age groups, younger than 30 years and between 50 and 60 years. Task materials were number lists printed in high and low luminance contrast (0.8 and 0.3) and were both typed (8-point type; 12 characters per inch) and handwritten (numbers were approximately the same size and spacing). All subjects were required to find mistakes in these task materials under the different light conditions. Light level, subject age, task contrast, and task "quality" all had statistically significant effects, both in terms of task performance and subjective ratings of difficulty. Light spectrum showed no statistically significant effect for either measure.

More recently, Berman *et al.* (1993) showed that light spectrum can influence performance for small, briefly flashed, low luminance contrast, achromatic tasks, specifically the accuracy with which the orientation of a Landolt ring can be identified. Figure 7.6 shows the proportion of correctly reported orientations for Landolt rings with a gap size subtending approximately 2 min arc at the subject's eye and presented at four levels of luminance contrast and four levels of background luminance, for two different light sources. One light source was greenish-blue in color and had a scotopic/photopic ratio of 4.31. The other light source was a mixture of red and pink fluorescent lamps, the combined effect having a scotopic/photopic ratio of 0.24 (see Table 1.4 for scotopic/photopic ratios for some commonly used light sources). The marked non-white color appearance of the light sources used ensures that they are unlikely ever to be adopted for real-life applications. However, they did produce differences in task performance. The greenish-blue light source consistently has better task performance than the red–pink light source, for different luminance contrasts and background luminances, until the performance reaches the maximum possible. A similar pattern of results has been found for elderly subjects exposed to the same light sources (Berman *et al.*, 1994).

The proposed explanation of these findings rests on the role of pupil size. Specifically, pupil size in a large visual field is determined predominantly by the response of the rod photoreceptors, even in photopic conditions; the greater the response from the rods, the smaller the pupil area (Berman *et al.*, 1992). For the light sources described above, the pupil area under the greenish-blue light source was 40 percent smaller than under the red–pink light source. A smaller pupil area has three effects; it reduces the retinal illumination, it increases the depth of field and it reduces distortion of the retinal image by spherical and chromatic aberrations. The first of these effects, the reduction in retinal illuminance, can be expected to degrade visual performance. The other two, increasing the depth of field and reducing aberrations, can be expected to improve the quality of the retinal image and hence to improve visual performance. All these effects are small, and the trade-offs they produce will depend on the inherent quality of the individual's optical system. An individual who is perfectly refracted will gain little from increasing the depth of field, so might be expected to experience deterioration of visual performance under a light source that produces smaller pupil sizes. However, most people do not have perfect refraction. For these people, the results suggest that light sources that promote smaller pupil sizes can increase visual performance somewhat for achromatic resolution tasks, i.e. where the task conditions place it close to threshold; e.g. small size, low luminance contrast, and limited exposure time. In the experiment described above, the gap in the Landolt rings subtended approximately 2 min arc at the subject's eye, the highest luminance contrast used was 0.4 and the Landolt ring was only displayed for 200 ms. The question now of interest is whether smaller pupil sizes are of benefit for suprathreshold

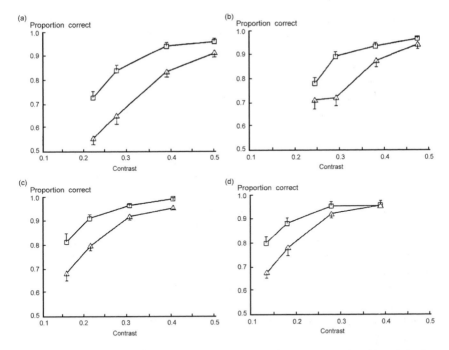

Figure 7.6 Means and associated standard errors of the proportion of Landolt ring orientations, presented for 200 ms on a spectrally neutral background, that were correctly identified, plotted against luminance contrast, for four different background luminances: $a = 11.9$, $b = 27.7$, $c = 47.0$, $d = 73.4$ cd/m^2. In all four diagrams, the upper curve ($\square$) is for the scotopically enriched greenish-blue illuminant (surround field scotopic luminances = 228 cd/m^2) and the lower curve ($\triangle$) is for the scotopically diminished red–pink illuminant (surround field scotopic luminance = 13 cd/m^2). Both illuminants produced a surround field photopic luminance of 53 cd/m^2 (after Berman *et al.*, 1993).

tasks. As noted above, the effect of light spectrum on task performance was not apparent in the Smith and Rea (1979) study which used a suprathreshold task. Moreover, Rea *et al.* (1990a) manipulated pupil size over a large range by changing the reflectance and size of the area surrounding the numerical verification task (see Section 4.3.5). They found no statistically significant effect of surround size and reflectance and, thus, pupil size, on the performance of the task, a task that was large in size, continuously viewed and presented at two levels of luminance contrast, one low and one high (0.15 and 0.86). Finally, Boyce *et al.* (2003b) tested the hypothesis that light sources that produce smaller pupil sizes ensure better visual task

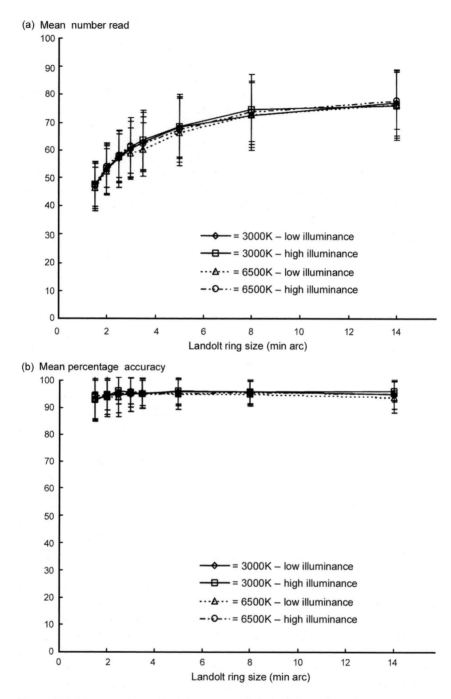

Figure 7.7 Means and standard deviations of the number of Landolt rings examined in 20 s and the proportion of Landolt rings of the specified orientation correctly identified, plotted against Landolt ring gap size, for two
lamp types with CCTs of 3,000 and 6,500 K, at 344 and 500 lx (after
Boyce *et al.*, 2003b).

performance at the same photopic illuminance, for both near threshold and suprathreshold conditions, when the task is done under realistic conditions. Subjects performed a Landolt ring task for eight different gap sizes (1.5–14 min arc) at a high fixed luminance contrast (0.80), two different illuminances (344 and 500 lx), and two lamp spectra covering the range of fluorescent light sources used in offices (CCTs of 3,000 and 6,500 K). The speed and accuracy of performance of the task was determined by the gap size, and to a much lesser extent, by the illuminance. Lamp spectrum had no statistically significant effect on the performance of the task (Figure 7.7).

To summarize, there can be little doubt that light spectrum operating through pupil size can influence the performance of small, briefly flashed, low-contrast achromatic tasks which are close to threshold, i.e. where everything about the stimulus matters, but it has little effect on the performance of suprathreshold tasks done under realistic conditions.

7.3.2.3 *Preferred light spectrum*

The appearance of a uniformly lit office will be greatly influenced by its color scheme. Surface finishes of different colors can produce large changes in the appearance of the office that the choice of fluorescent light source can modify only slightly. However, the properties of the light source cannot be entirely ignored.

Given that the light source most commonly used in offices is the fluorescent lamp, the range of possible CIE general CRIs for office lighting is 65–98 and the range of CCTs is between 3,000 and 6,500 K. The question of interest is whether these fluorescent lamp types are equally acceptable.

Cockram *et al.* (1970) studied peoples' preferences for different fluorescent light sources for lighting an office, at night and when lit in combination with daylight. They used eight similar offices in the same building in the UK, lit by four different fluorescent lamp types in the same type of luminaire: a ceiling-mounted, louvered troffer. Because the same number of lamps was used in each case, installations with the different lamp types produced different illuminances. Forty office workers rank-ordered four of the rooms, each with a different light source, for preference, at night and during the day. The characteristics of the four light sources and the overall order of preference are shown in Table 7.3. These results suggest that the only clear preference, or rather lack of preference, is for the color matching fluorescent lamp, despite its high CRI. It is least preferred both day and night. There are two possible reasons for this. The first is that the color matching lamp produced the lowest illuminance of the four light sources and illuminance is known to be important in the evaluation of lighting. The second is its high CCT (6,500 K), which almost certainly conflicts with the office workers' previous experience in the UK. Comments such as "the room looks like a refrigerator" indicate that when it comes to expressing a preference there is a large cultural and contextual element in the decision.

Table 7.3 Rank-ordering of preferred fluorescent lamps in real offices assessed by office workers (after Cockram *et al.*, 1970)

Lamp name	CCT (K)	CRI	In daytime, with clear glazing	In daytime with tinted glazing	At night, with tinted glazing
Color matching	6,500	>85	4	4	4
Deluxe daylight	4,000	>85	2	3	3
Daylight	4,300	<70	1	2	2
White	3,500	<70	3	1	1

Interestingly, a recent study in the US has examined the preference for two very different fluorescent light sources, providing the same illuminances from the same luminaire (Boyce *et al.*, 2002). Of the two light sources, one had a CRI of 98 and a CCT of 6,500 K, while the other had a CRI of 82 and a CCT of 3,500 K. It was found that at illuminances representative of those used in offices (350 and 500 lx) the lighting provided by the lamp with the higher CRI and CCT was seen as brighter than the other, as would be expected (see Chapter 6). However, the lamp with the CRI of 82 and a CCT of 3,500 K was liked more, at both illuminances. A number of subjects commented that the lamp with the CRI of 98 and CCT of 6,500 K made the office look like a hospital, which was considered inappropriate. What this suggests is that given a high enough CRI (above 65) and an illuminance appropriate for offices, i.e. generally satisfactory lighting conditions, it is expectations of appropriateness based on previous experience that determines the acceptability of different fluorescent light sources. This implies that the acceptability of different light sources will be different in different cultures. All the above preferences for light spectrum have both been obtained in Western cultures at illuminances that were typical of office lighting at the time the studies were made. It would be interesting to know the preference for light source spectrum under other conditions. For example, it would be interesting to know how acceptable a fluorescent lamp with a CCT of 3,500 K would be in Japan, where lamps with a CCT of 5,000 K are commonly used in offices.

Despite the variability in the available data, national lighting standards are replete with recommendations about the choice of lamp CCT at different illuminances. Specifically, the advice is not to use lamps with high CCT at low illuminances, usually below 300–500 lx. These recommendations are based on the work of Kruithof (1941). Figure 7.8 shows a schematic derived from his results. Combinations of illuminance and CCT that lie in the lower shaded area are perceived as cold and dim; those that lie in the upper shaded area are perceived as overly colorful and unnatural. It is only

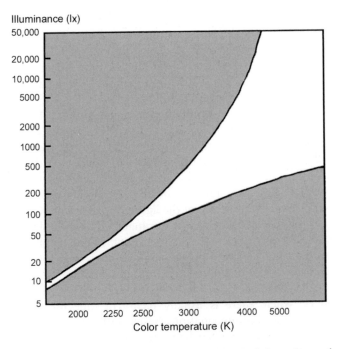

Illuminance (lx)

Color temperature (K)

Figure 7.8 The Kruithof curve: the white area defines the preferred combinations of the color temperature of a light source and illuminance. Color temperature/illuminance combinations in the lower shaded area are claimed to produce cold, drab environments, while those in the upper shaded area are believed to produce overly colorful and unnatural environments (from IESNA, 2000a).

combinations of illuminance and CCT that lie in the unshaded area that are considered to be pleasing. Unfortunately, the work on which this summary is based has not been extensively reported. What is apparent from the details that have been given is that different light sources; incandescent lamps, fluorescent lamps, and daylight, were used to produce the different CCTs so it is likely that the light distribution and color rendering of the light varied with CCT. As both distribution of light and lamp color properties are known to affect people's perceptions of an interior, the validity of Kruithof's boundary conditions is open to question. Bodmann *et al.* (1963) examined peoples' assessments of a conference room lit by fluorescent lamps of different color temperatures to provide uniform illuminances over a wide range (from <700 to >3,000 lx). A trend in the impressions similar to what would be expected from Kruithof's results was found, although it only became evident at the highest illuminances.

A direct test of Kruithof's boundary conditions was made by Boyce and Cuttle (1990). They had 15 observers carry out a color discrimination task in a small office and then give an assessment of the lighting. The office was lit by fluorescent lamps with virtually the same CRI value (CRI = 82–85) but very different CCTs (2,700–6,500 K), producing four different illuminances on the desk (30, 90, 225, and 600 lx) from the same luminaire. From Figure 7.8 it can be seen that, over this illuminance range, the different lamp types should produce a significant change in perception, if Kruithof's boundary conditions are correct. This did not prove to be the case. The major factor in determining the impression given by the lighting was the illuminance. Increasing the illuminance made the lighting of the office appear more pleasant, more comfortable, more warm, more uniform, less hazy, less oppressive, less dim, and less hostile. The CCTs had virtually no effect on the observer's impression of the lighting of the room. One plausible explanation for this lack of effect is chromatic adaptation. Each subject experienced each lighting condition for about 20 min before giving their evaluation. During this time, the subject's visual system would have adapted to the chromaticity of the light source, an adaptation that would have diminished the difference between the light sources. What this suggests is that for offices, where there is usually plenty of time for chromatic adaptation to occur, the recommendations based on Kruithof's results are unnecessarily restrictive. As for what happens when their is little time for adaptation, Davis and Ginthner (1990) examined peoples' perceptions of a conference room, lit to three different illuminances (250, 550, and 1,250 lx) by fluorescent lamps of similar CRI (89 and 90) but different CCT (2,750 and 5,000 K). In this experiment the subjects had only 1 min to adapt to the chromaticity of the light source before giving their evaluation. Nevertheless, the subjective ratings of preference were influenced only by the illuminance and not by the CCT. Han (2002) used a model office to examine the effect of illuminance, CCT, and color tone of decor on the perception of the lighting of an office. As would be expected, she found that illuminance was the major factor determining the perception of brightness and the acceptability of the lighting for an office, but she also found that the use of light sources with a higher CCT produced a perception of greater brightness at the same illuminance (see Section 6.2.2). The acceptability of the lighting for an office was also influenced by the CCT but in this case it was the CCT that was most commonly used in office lighting practice that was considered most appropriate (4,100 K rather than 3,000 or 6,500 K).

Basically, these results indicate that the CCT of fluorescent lamps is a minor factor in determining satisfaction with the lighting of an office. The illuminance provided is much more important. As for the light source CRI, there is no doubt that choosing a light source with a CRI above 80 will tend to produce more saturated surface colors and a perception of greater brightness and visual clarity (see Section 6.3). There is equally no doubt that choosing a light source with a CRI below about 60 will be unsatisfactory because of the non-white color appearance of the lighting and the unattractive rendering of skin tones. People can use skin tone to make fine discriminations between fluorescent lamp types (Rea *et al.*, 1990b). Given that all that is required is to avoid light sources with a CRI below 60, then there is a wide range of acceptable fluorescent light sources which can all be preferred in different conditions (see Table 7.3).

Although the discussion above has been in terms of the color properties of the light source, it is important to appreciate that the impact of these properties may be modified by the color content of the office decor. This is because the spectral content of the light reaching the eye is a combination of the light received direct from the luminaires and the light received after reflection at the surfaces in the interior. Mizokami *et al.* (2000) showed that when the color of walls, floor, and furniture were orange, subjects perceived a room to be illuminated by incandescent lamps though it was actually illuminated by fluorescent lamps of a higher CCT. The color content of the office may also influence satisfaction with the lighting. Boyce and Cuttle (1990) showed that introducing natural color, in the form of fruit and flowers, into what was essentially an achromatic space, enhanced the positive impressions created by the lighting, particularly at high illuminances.

Taken together, these findings support the view that provided the light source has a CRI above about 60 and its chromaticity coordinates are close to the Planckian locus so that it can be called a nominally white light source, almost any fluorescent lamp type will be acceptable for office lighting. Given that a number of different fluorescent light sources are acceptable, the choice between them is best made on the basis of the application. If the work involves accurate color judgments then a light source with a high CRI is desirable. If accurate color judgments are not part of the work, then the choice should be made after taking into account the effect of the lighting on the office decor and the "message" it sends, as well as the occupants experience and expectations of office lighting. Where care is necessary is where a low illuminance is being proposed because then the perception of greater brightness associated with high CCT light sources may be valuable; or where a very warm or very cool decor is proposed, in which case a light source CCT that offsets the warmth and coolness of the decor would be appreciated (Han, 2002).

One final matter regarding light spectrum that deserves to be mentioned is the issue of what are called full-spectrum lamps. There is no official definition of what full-spectrum means but the *de facto* definition is a fluorescent lamp with spectral emissions in all parts of the visible spectrum and in the near ultraviolet, with a CCT of 5,000 K or more and a CRI of 90 or more. All sorts of extravagant claims have been made for such lamps to match their extravagant price. Most of the claims are based on the idea that they are in some way "like daylight" despite the fact that even the most cursory examination of the spectrum of such lamps will show that their light spectrum is much more like that of a fluorescent lamp than anything produced by sunlight passing through the atmosphere. Basically, a full-spectrum lamp is a fluorescent lamp with a high CRI and a high CCT. All the major lamp manufacturers produce such lamps but do not market them as full-spectrum lamps. Apart from the effect on brightness, color perception, and pupil size that would be expected for lamps with a high CRI and CCT, the claims made for full-spectrum lamps regarding beneficial effects on mood, behavior, and health have been found to be, at best, unproven (McColl and Veitch, 2001; Veitch and McColl, 2001).

7.4   Lighting systems

By choosing a lighting system, the lighting designer determines the distribution of light in the office and hence such aspects of visual comfort as discomfort glare, veiling reflections, and shadows (see Section 5.4). In addition, the distribution of light has an influence on the perception of the office (see Section 6.3). One of the major differences in perception is whether the office is perceived to be primarily lit by daylight or by electric light. As a rule of thumb, any office where the average daylight factor is more than 5 percent will be perceived as daylit and the electric lighting will not be needed during the daytime. Conversely, any space where the average daylight factor is significantly less than 2 percent, will be perceived as electrically lit, even in daytime (CIBSE, 2002a) and daylight will only be noticeable on room surfaces close to the windows. What the average daylight factor is will depend on the architecture of the space and the daylight delivery system used.

7.4.1   Daylight delivery systems

Daylight can be delivered into an office through conventional windows, clerestory windows or skylights, as well as a number of remote distribution systems, such as light shafts, ducts, and pipes (Littlefair, 1990; Littlefair *et al.*, 1994; Shao *et al.*, 1998). By far the most common is the conventional window. Windows have the advantage of providing both daylight to the interior and a view-out. Their disadvantage is that the amount of daylight delivered to the office decreases dramatically as the distance from the window increases, although the view-out is preserved over a large distance. Clerestorys, which are essentially windows mounted high up close to the ceiling, are one way to improve this situation but they suffer from the disadvantage that the view-out is then limited to the sky. Skylights offer the possibility of delivering daylight over the whole of the office but, again, the view-out is limited to the sky. In addition, skylights only effectively deliver daylight to the room immediately beneath them. This is not a problem for a single story building but for multiple stories, the only alternatives are either windows in the outside walls or a light shaft of some sort and these latter are notoriously inefficient.

The important aspects of windows as far as people are concerned are their size, shape, spectral transmittance, and solar shielding. Ne'eman and Hopkinson (1970) examined the minimum acceptable size of window using a model open-plan office through which subjects could look at real views. The subjects were asked to adjust the window width to give the minimum acceptable window size. It was found that they could do this consistently except when the window was uniformly bright and featureless. For more usual views, the minimum acceptable window size was determined by the amount of visual information provided by the view, near views requiring larger window sizes than far views. The judgment of minimum acceptable window size was not much influenced by the amount of daylight or sunlight that was admitted to the model, the interior illuminances or by the position of viewing the window. These findings, together with the fact that subjects could not make consistent adjustments when the window was featureless, suggest it was the view-out that was determining the minimum window size. About 25 percent of the window wall area was the minimum acceptable window size for 50 percent of observers, rising to 32 percent if 85 percent of people were to be satisfied.

Keighly (1973a) also measured the response to window size. Using a model of an open-plan office, 40 observers were shown a range of windows that differed in size (11–65 percent of the window-wall area) and in the number and layout of apertures. Through these windows a number of different views were presented on film. The observers were asked to rate their satisfaction with the windows on a five-category scale. Figure 7.9 shows the mean ratings of satisfaction for single windows of different sizes for

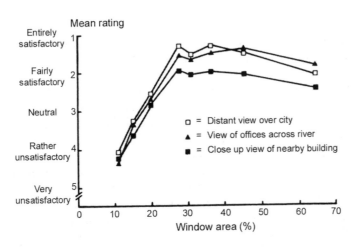

Figure 7.9 Mean ratings of satisfaction with windows for different window areas and views. The window area is expressed as a percentage of the window-wall area (after Keighly, 1973a).

the three views used. From Figure 7.9 it can be seen that glazed areas of 15 percent or less of the window-wall area are considered unsatisfactory but above 30 percent almost compete satisfaction occurs. These results are in reasonable agreement with those of Ne'eman and Hopkinson (1970), although the importance of the type of view is much less, possibly because of the static and limited nature of the views used by Keighly.

Keighly (1973a) also examined the effect of dividing a given glazed area into different numbers of apertures. This produced some very marked differences. People disliked the use of several windows of different sizes, a large number of regularly arranged narrow windows, and wide mullions. The common characteristic of these aspects of window design is that they break up the perception of the view. In general, people preferred windows to be large, regularly arranged, and horizontal.

The preference for horizontally orientated windows is in disagreement with the conclusions of Markus (1967) who found a preference for vertically orientated windows. Keighly (1973b) again used a model open-plan office to examine the preferred window shape. Specifically, a rectangular window covering 20 percent of the window-wall area was adjusted until it was in the preferred position and of the preferred shape. The most frequently preferred condition was a central horizontal aperture, with the elevation being determined by the skyline of the view, until the skyline approached the level of the ceiling. The views used were predominantly orientated horizontally which suggests that the shape of the window is being treated as frame for a view. Ludlow (1976) reached the same conclusion. Using a similar apparatus to Keighly, he asked 20 observers to adjust a window aperture to the preferred shape and size for the view. He concluded that the specific view did have a large effect on the preferred size and shape. Ludlow (1976) also found that the preferred window size was between 50 and 80 percent of the window wall, values that are much higher than considered satisfactory by Keighly's observers. This difference may have been simply the difference between what is satisfactory and what is preferred or it may have been due to differences in the views from the models. Nonetheless, the basis for the usual practice in temperate climates of having window sizes somewhere between 20 and 40 percent of the window-wall area is clear. Below 20 percent of the window-wall area, dissatisfaction with the windows is likely to arise, particularly if they are concentrated into one or two large areas so that many people do not have a view at all. Above 40 percent of window-wall area, satisfaction with the window size will be high but unless care is taken to control the admission of sunlight, thermal, or visual discomfort is likely to increase. It is important to appreciate that these results have all been collected from people who are used to a temperate climate. It is at least plausible that people who feel daylight is rationed, e.g. the inhabitants of countries such as Finland, and those who have an excess of daylight, e.g. the inhabitants of India, would have different preferences. Also, it is worth noting that these results have been obtained in the context of an office.

Preferences for window size may be different in other contexts (Butler and Biner, 1989).

As for spectral transmittance, there are two aspects that need to be considered, the total transmittance of light and the color appearance of the light transmitted. Boyce *et al.* (1995) used a model office to examine the acceptability of different total transmittances for three types of glass with different spectral transmittances, when looking at a real scene under clear and overcast skies, from an interior lit to either 500 or 1,000 lx. The window size was fixed at 42 percent of the window-wall area. Figure 7.10 shows the percentage of subjects finding glazing acceptable for use in a modern office plotted against the percentage transmittance. Two conclusions can be drawn from Figure 7.10. The first is that percentage transmittances above about 50 percent are highly acceptable but that as the percentage transmittance decreases, percentage acceptance also decreases. The second is that the major factor determining acceptability is the visible transmittance but there is some variation associated with the spectral transmittance of the glass, the nature of the sky, and the illuminance of the office. Support for these data comes from a field study by Cooper *et al.* (1973). In the

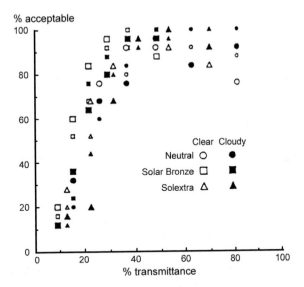

Figure 7.10 Percentage of observers considering the glazing acceptable, plotted against the transmittance of the glazing, for three glass types, under clear and cloudy sky conditions. The filled symbols indicate a cloudy sky; the open symbols, a clear sky. The larger symbols of each type are for the 1,000 lx interior illuminance and the smaller symbols are for the 500 lx interior illuminance (after Boyce *et al.*, 1995).

Boyce *et al.* (1995) study, 50 percent of subjects found the glazing unacceptable in the percentage transmittance range 15–22 percent. Cooper *et al.* (1973) found adverse reactions to glazing in offices occurred with percentage transmittances of 12 and 15 percent.

Before accepting the results from Figure 7.10 as gospel, it is important to appreciate that they have a number of limitations. The first is the short time, only a minute or two, for which each window condition was viewed. The second is that the measurements were all taken during the middle part of the day. The third is that measurements were not taken when the Sun was incident on the window. The fourth is that all the measurements were made in a temperate climate. The short time for viewing means that the levels of acceptability for different transmittances are representative of the subjects' immediate reaction on entering a room. This is important because decisions about whether to switch on electric lighting in a daylit room are frequently based on the minimum illuminance experienced when first entering the room (Hunt, 1979). However, it is undeniable that much longer exposures might change the minimum glazing transmittance that is acceptable, although whether minimum glazing transmittance would increase or decrease depends on whether sensitization or habituation occurred. The fact that the measurements were taken in the middle part of the day is a limitation because it could be argued that low-transmittance glazing has its largest effect around dawn and dusk, because it effectively shortens the day, particularly on dull, overcast days. It seems likely that a glazing transmittance that is just acceptable at midday would not be acceptable around dawn or dusk. The fact that measurements were only taken without the Sun being incident on the window is not much of a limitation because when the Sun is visible through a window, the tendency is to use blinds to mask the Sun (Rubin *et al.*, 1978; Rea, 1984) in which case the glazing transmittance is irrelevant. Finally, the fact that the measurements were taken in temperate climates implies a certain attitude to daylight. In high northern or southern latitudes, where daylight is in short supply for part of the year, the minimum acceptable transmittance will probably be increased, while in latitudes close to the equator where daylight, and particularly sunlight, is plentiful and needs to be controlled, the minimum acceptable transmittance will probably be decreased and a decrease in acceptability of high glazing transmittances may occur. These predictions about the effects of latitude are matters of conjecture. It would be interesting to determine if they are true or false.

There can be no denying that there is a lower limit to the transmittance of glass used in windows if the glazing is to be acceptable. There are also limits on the colors of the glass that are acceptable. Glass is available in several different tints, typically, bronze, green, blue, and gray, the tint often being chosen for architectural reasons rather than anything to do with the effect on the visual conditions inside the building. Cuttle (1979) examined the limits of tinting of glass used for windows. Figure 7.11 shows the dissatisfaction contours for glass to be used in windows for daylighting, based on judgments of people sitting in a room lit by light transmitted through a window filled with a liquid filter that could be continuously varied over a wide range of colors. From Figure 7.11 it is clear that glass types with chromaticities that depart from the central part of the Planckian locus risk being considered unsatisfactory by a significant number of people.

Any daylight delivery system also needs to be considered in terms of its solar shading because solar shading can have an impact on both the visual and thermal environment. The impact on the visual environment is through

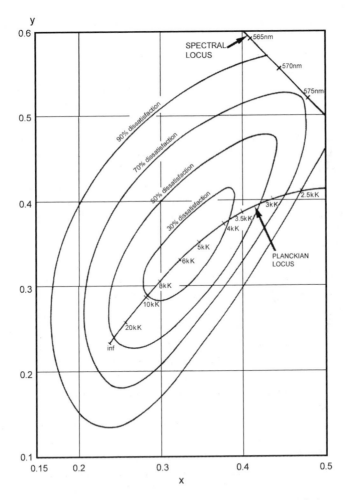

Figure 7.11 Percent dissatisfaction contours for the chromaticity of glazing plotted on the CIE 1931 (x, y) chromaticity diagram (after Cuttle, 1979).

the admission of sunlight to the office. The impact on the thermal environment is through the heat gain and heat loss of the whole building and, locally, on the likelihood of thermal discomfort caused by overheating due to excessive thermal radiation (sunlight) or overcooling, due to radiant heat loss to a cold window or the generation of draughts. Markus (1967) found that 86 percent of the people in the 12-story office he studied preferred sunlight in the office all the year round but there was a tendency for those who sat nearest to the windows to be less enthusiastic about it than those who worked further away. Ne'eman *et al.* (1976) claimed a similar result. Over the four different office buildings they studied, there was a strong emphasis on the desirability of sunlight but there was also a clear underlying complaint about visual and thermal discomfort. The strength of the complaints depended on the facilities for controlling the visual and thermal environment. Where blinds were available and the air-conditioning was adequate, few complaints were heard. Where there was no means to control sunlight falling on the working area or where the air-conditioning system was inadequate, complaints were common.

Clearly, there are multiple factors that determine the acceptability of a window as a daylight delivery system and all of them should be considered when designing windows. But it would be as well not to stop there. Butler and Biner (1989) found that the factor most strongly predicting peoples' preferences for window size was the view-out for temporal information. Other aspects of the view-out were also closely correlated with the preference for window size. This finding emphasizes the need to evaluate any window design in two directions, for its effect on the lighting of the interior and for its impact on the view-out. Both these aspects need to be satisfactory for a window design to be considered satisfactory.

7.4.2 *Electric lighting delivery systems*

Electric lighting delivery systems used in offices consist of luminaires of different types arranged in different ways. By far the most common layout is the regular array designed to provide a uniform illuminance across the whole of the assumed horizontal working plane. Such regular arrays are popular because they allow the owner of the office to arrange the furnishing however desired with the confident expectation that the illuminance will be similar everywhere. Given the frequency with which office layouts are changed in some rapidly expanding industries, the fact that there is no need to rearrange the lighting everytime the furniture layout is changed is valued as one less thing to worry about.

Another constraint on office lighting delivery systems is the maximum lighting power density set by some authorities (see Section 14.4.1). Such limits more or less exclude the use of some possible lighting systems from widespread use because they cannot provide the required illuminances without exceeding the lighting power density limit. Examples of currently excluded systems are cove lighting and luminous ceilings. Even when there is no legal limit on lighting power density, there is a *de facto* limit set by expectations about how much it will cost to run an electric lighting system. Most businessmen will be reluctant to pay much more than usual to operate an overhead, literally and financially.

The luminaires used in regular arrays can be classified into three types; direct, indirect, and direct/indirect. A direct luminaire is one in which all the light is emitted below the horizontal plane where the luminaire is located, usually the ceiling. An indirect luminaire is one in which all the light is emitted above the horizontal plane where the luminaire is located, usually a plane some distance below the ceiling, the luminaire being suspended from the ceiling. A direct/indirect luminaire emits light both above and below the horizontal plane where the luminaire is located. Of course, these simple classifications are not absolute. Luminaires are assigned to one or other class according to where the majority of their light is directed. Some indirect luminaires have a perforated metal lower plate to give the luminaire some brightness, so some light is emitted below the horizontal plane. Provided the amount emitted downwards is less than about 10 percent of the total luminaire light output, the luminaire can be classified as an indirect luminaire. A similar consideration applies to the direct/indirect luminaire. It is not really appropriate to classify a luminaire as a direct/indirect luminaire unless the minimum light output above or below the horizontal plane is more than 10 percent of the total light output. Care should also be taken with advertising claims. One type of ceiling-recessed luminaire is marketed as a recessed indirect luminaire because there is no light emitted directly from the lamp, all the light emitted being first reflected from the interior of the luminaire. Calling such a luminaire a recessed indirect is misleading. The light distribution from a regular array of such luminaires recessed into a ceiling is virtually identical with that produced by a regular array of direct luminaires fitted with a prismatic lens.

Regular arrays of direct, indirect, and direct/indirect luminaires can give very different appearances to an office. The appearance produced by a regular array of direct luminaires depends on the luminous intensity distribution of the luminaire and their spacing. The narrower is the luminous intensity distribution, the closer the luminaires have to be placed to each other and the more vertical is the flow of light. Such an installation will produce a lot of light on the horizontal working plane but very little on any vertical plane, especially the walls. Further, the only light that reaches the ceiling does so by reflection from the other surfaces in the room. There are two risks with such an installation. The first is that the office will appear as gloomy and rather cave-like. The second is that the there is a strong possibility of uncomfortable shadows and veiling reflections occurring. This emphasizes another point; that a luminaire is a complete package, and not just in its mechanical form. It is almost always a mistake to select a luminaire on the basis of one criterion alone. It is much better to consider all the possible impacts of a luminaire on the visibility of the task and the appearance of the office before making a selection.

As for indirect lighting, this effectively uses the ceiling as a large area, low-luminance luminaire. To avoid reflected images of the inside of the luminaire being formed, ceilings used with indirect lighting are almost universally diffuse reflectors with the result that the distribution of light in an office lit by an indirect lighting installation tends to be diffuse. Thus, the walls are almost always well lit, as is the working plane. Further, diffuse lighting means that shadows are few and veiling reflections very slight. Although such conditions are unlikely to be considered uncomfortable, they may not do much to make the office attractive. In terms of the lighting quality classifications laid out in Section 5.6, indirect lighting is an almost infallible means to achieve indifferent lighting.

Direct/indirect lighting is a compromise between direct and indirect lighting, but it is a good compromise in that it avoids the worst features of both. The indirect component softens any shadows and veiling reflections and provides some light on the walls. The direct component provides stronger modeling and offers some relief from the boring uniformity of indirect lighting.

It should not be thought that adopting direct/indirect lighting is the only means to avoid the disadvantages of direct and indirect lighting. The low wall illuminances produced by direct lighting can be overcome by using wall-washing or by placing the luminaires close to the wall, although the latter may still cause a problem if the sharp cut-off of the direct luminaire produces a shadow at the top of the wall. Task lighting can also be used to reduce the shadows and veiling reflections in the work area (Japuntich, 2001). As for indirect lighting, some accent lighting of features in the office or some decorative lighting can go some way to make the space more interesting.

7.4.2.1   Preferences

It is now appropriate to consider if there is clear preference between one type of electric lighting delivery system and another. Harvey *et al.* (1984) had people perform a task for an hour and then give ratings for various aspects of the quality of the direct and indirect lighting systems used. The two indirect lighting systems were clearly preferred over the direct lighting system, using a recessed parabolic luminaire. Leibig and Roll (1983) and Roll (1987) examined people's reactions to direct, indirect, and direct/indirect lighting while reading text on paper and on a computer screen. For paper-based work, there were no statistically significant differences in the assessment of the installations. For the computer-based work, the direct and the direct/indirect lighting installations were considered much better than the indirect lighting installations. Hedge *et al.* (1995) carried out an evaluation of a building where the renovated offices were lit by either indirect or direct lighting. In this case, there was a clear preference for the indirect lighting. Further, it was observed that 1 year after the lighting was installed the occupants had modified much of the direct lighting by disconnecting or removing lamps, whereas the indirect lighting was virtually untouched. Bedocs *et al.* (1982) examined people's reactions to indirect lighting and found wide acceptance for one installation and only moderate acceptance in another.

The only real conclusion that can be drawn from these studies is that any of the three approaches can be preferred depending on how it is done. There is no feature of direct, indirect, or direct/indirect lighting that always excludes it from consideration.

7.4.2.2 *Performance*

Another possibility is that the distribution of light in the room can change task performance. If this could be shown to be true then there would be a real reason for choosing one type of electric lighting delivery system over the others. There are two reasons why the distribution of light in the room might be expected to change task performance. The first is that different distributions of light in the room may make the task more or less visible because of shadows, disability glare, or veiling reflections and hence affect task performance directly. The other is the possibility that different distributions of light in the room will produce different perceptions of the space which, in turn, will influence the mood of the subject and hence affect the motivation to perform the tasks.

Harvey *et al.* (1984) had people perform a task involving checking lists of numbers printed on paper against a similar list displayed on paper against a similar list displayed on a visual display terminal (VDT), under direct and indirect lighting. There was no difference in the performances of the task for the different lighting systems.

Veitch and Newsham (1998a) measured the performance of an array of computer-based tasks, done by a large number of temporary office workers, over a full working day, in a large windowless office furnished with individual cubicles. On each day the office was lit by one of nine different lighting installations classified according to three levels of power density and three levels of perceived lighting quality. The low lighting quality installations consisted of a regular array of direct, prismatic luminaires. The medium-quality installations used a regular array of direct parabolic luminaires. The high-quality installations used various arrangements of indirect or direct/indirect luminaires, sometimes suspended from the ceiling and sometime furniture-mounted. The low lighting power density installations all used an adjustable task light and an under-shelf task light. The medium and high lighting power density lighting installations did not use any form of task lighting. The subjects were able to discriminate between the lighting installations. When the office was lit by direct lighting systems, it was considered as brighter than when it was lit by indirect lighting systems and

lighting conditions that included task lighting and lower ambient illuminances were considered as less bright and less glaring than lighting systems that did not have task lighting. As for task performance, a complex
pattern of interactions was obtained, with some tasks being performed better under one lighting installation and others under another. Most of the
changes in task performance were small (1–3 percent of variance explained).
Interestingly, the strongest effect any visual condition had on the performance of any task was for the change of the VDT screen polarity on a typing
and a proofreading task. Changing the screen type from negative polarity
(bright text on a dark background) to positive polarity (dark text on a bright
background) explained 7–9 percent of the variance in task performance,
probably because the higher background luminance of the positive polarity
screen made the display less subject to interference from reflected images of
the rest of the room. This is a direct effect lighting on task visibility.

One limitation of these two studies is that the different lighting installations produced different illuminances and different light distributions on
the paper tasks and on the VDT which may have changed the visibility of
the tasks. This makes it impossible to determine the extent to which
changes in visibility or changes in the perception of the space are responsible for any changes in task performance. A study by Eklund *et al.* (2000)
avoided this problem. They had temporary office workers work for 8 h
doing a data-entry task in three private, windowless offices, all with the
same decor and furniture. Three different lighting installations were used,
one in each office. All three lighting installations provided a similar illuminance on the task, without veiling reflections or disability glare, so for
the same task they provided similar task visibility. However, the three
lighting installations were very different in light distribution over the room,
ranging for very uniform indirect lighting to very concentrated overhead
lighting. There was no difference in performance of the task under the three
lighting installations. Somewhat more surprisingly, there was no difference
in the workers' perceptions of the three lighting installations, despite the
opinions of lighting experts that the lighting installations clearly differed in
quality. However, changes in point size of the material to be entered, and
hence in task visibility, did produce statistically significant changes in task
performance.

Eklund *et al.* (2001) had people perform the same data-entry task for 4 h,
but this time they always had the same lighting installation with the same
light distribution, but with two different levels of decor. In one condition,
the office was bare and achromatic. In the other, some colorful decor was
added to the room. Again, there was no difference in task performance
between the two decors, although the subjects did perceive the chromatic decor as more colorful, attractive, and interesting. Again, changes in
the task visibility created by changing print size, luminance contrast, or
illuminance on the task did produce statistically significant changes in task
performance.

These studies clearly demonstrate that changes in task visibility can be reliably expected to change the performance of visual tasks but the effects of differences in the perception of the lighting of the space are much less certain. There are several reasons why this might be so. One is that perhaps people naive in lighting are much less sensitive to lighting conditions that do not affect task visibility or visual comfort than lighting experts, in the same way that wine connoisseurs are more sensitive to wine characteristics than are people who do not drink wine frequently. If people are not sensitive to changes in lighting conditions that do not affect task visibility or visual comfort then it is unlikely that changes in such conditions will change their mood and hence their motivation to perform the task. Another is that given a long enough exposure to lighting conditions that do not affect task visibility or visual comfort, people habituate to them and so the conditions become of less and less significance to them and hence less and less likely to change their mood. Another is that the range of lighting conditions may have not been extreme enough, although the three lighting installations used in Eklund *et al.* (2000) were selected to cover the extremes of current lighting practice. Finally, the nature of the tasks may have been too strongly visual. It might be that tasks that involve less visual and more cognitive and judgmental elements would be more sensitive to the changes in mood than are highly visual tasks, the performance of which would be dominated by the task visibility. It might also be important that the data-entry task focused attention on a small area in that all the information needed to do the task was available in that small area. The rest of the room contained no information relevant to the task so the lighting of the rest of the room was irrelevant to the worker. There is little doubt that for tasks in which there is information spread over many different parts of the room, the light distribution over the whole room will matter but that will be because the light distribution then affects the ability to extract information to do the task, i.e. it affects task visibility. Despite all this speculation, there is one conclusion that is clear. This is that lighting distributions that do not affect task visibility or cause visual discomfort are not certain to affect task performance. They may, because light distributions can also affect the mood and motivation of the office worker, or they may alter the "message" sent by the lighting and this "message" may in turn alter behavior (see Section 4.4.2). Unfortunately, there are many other factors besides lighting that can affect mood and "message" and light distribution may be of little importance relative to these other factors. More research is needed.

7.4.2.3 *Lighting and VDTs*

Over the last 20 years, the nature of office work has changed dramatically. From being almost entirely based on paper, it has moved to being heavily dependent on computers, although any count of the number of printers and copiers in use will indicate that paper-based tasks are still an important part of office work. This widening of activity from paper alone to paper plus computer displays has lead to the design of luminaires specifically for use in what originally were called VDT offices but now constitute every office. The reason for these changes lies in the nature of the VDT display. First, the VDT is usually positioned in a vertical or near vertical plane. Second, the display is self-luminous, which means it can be seen in an unlit room. Third, being self-luminous means that increasing the amount of light from the room falling on the display decreases the visibility of the display. Therefore, to enhance the visibility of the display it is necessary to minimize the incidence of light from the room onto the display, which in practice means reducing the illuminance on vertical planes. The result of such an analysis has been a rash of recommendations concerning the light distribution of luminaires suitable for use in offices containing VDTs. These recommendations take two forms. For direct lighting, the recommendations limit the luminance of the luminaire in specified directions, usually above angles of 55° or 65° from the downward vertical. For indirect lighting, the recommendations consist of a maximum average luminance and a maximum peak luminance of the ceiling. For direct/indirect luminaires, both forms of recommendation apply. The objective of both these forms of recommendation is to limit the luminance of the reflected image of the luminaire or ceiling seen in the VDT. The recommendation for direct lighting assumes that the luminaires are mounted in a ceiling and the VDT is in a vertical or almost vertical plane. If either of these two assumptions is incorrect, then limiting the luminance of the luminaire above 55° or 65° is of little value.

Before accepting any recommendations, it is a good idea to consider what lies behind them. The basis of the VDT luminaire recommendations lies in the effect light incident on the display has on the display's visibility. Light incident on a VDT display can have three undesirable effects: reduce the contrast of the display, increase competition for the users' attention between the display and the reflected images, and cause changes in accommodation due to the fact that the display and the reflected images are at different distances from the observer (Boyce, 1991; Rea, 1991). The key to avoiding these problems is to reduce the luminance of the reflected images relative to the luminance of the display. None of the recommendations explicitly state the source of their recommendations but one of the most influential sets of data has been that of Leibig and Roll (1983). They asked people to identify the borderline between distracting and non-distracting reflections in displays generated on screens with different reflection properties. The luminance of the reflections was varied by changing the luminance of the luminaire being reflected. The luminance identified as representing the borderline between disturbing and non-disturbing reflections for each subject was used to calculate the percentage of subjects who would find a given luminance non-disturbing, i.e. who would be satisfied with the display. Figure 7.12 shows the percentage of people satisfied with the display

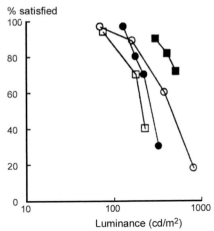

% satisfied

Luminance (cd/m²)

○ = Bright characters on dark screen/specular reflection/direct lighting
□ = Bright characters on dark screen/specular reflection/indirect lighting
● = Dark characters on bright screen/specular reflection/indirect lighting
■ = Dark characters on bright screen/diffuse reflection/direct lighting

Figure 7.12 Percentage of people satisfied with the display plotted against the luminance of the luminaire that is reflected in the screen. Data are given for both positive- and negative-polarity displays on screens with different reflection properties seen under different lighting conditions (after Leibig and Roll, 1983).

for various display polarity/screen reflectance/lighting type combinations plotted against the luminance of the luminaire that is reflected in the display. By comparing the results for the same display; bright characters on a dark specularly reflecting screen seen under two forms of lighting, direct and predominantly indirect lighting, it can be concluded that an acceptable luminaire luminance for one lighting installation will not necessarily be acceptable for another. By comparing different displays; bright characters on a dark screen and dark characters on a bright screen, both seen on screens with specularly reflecting surfaces lit by predominantly indirect lighting, it can be concluded that negative polarity displays (dark characters on a bright background) are less sensitive to reflections than positive polarity display (bright characters on a dark background). By comparing the results for the most sensitive conditions; the specularly reflecting, positive-polarity display seen under predominantly indirect lighting, with the least sensitive condition; the diffusely reflecting, negative-polarity display seen under direct lighting, it can be seen that what constitutes an acceptable luminaire luminance can vary widely, e.g. from 130 to 500 cd/m² for 80 percent of the subjects. From these data, Leibig and Roll recommended a maximum luminaire luminance of 200 cd/m² in any direction that will lead to a reflection from a VDT screen.

One of the unfortunate outcomes of such recommendations is the implication that it is only by changing the lighting that the problems of VDT reflections can be overcome, although it is evident from Figure 7.12 that the reflection characteristics of the VDT are also important factors in determining the degree of disturbance. Lloyd *et al.* (1996) attempted to address this problem by developing a quantitative model to predict observer's ratings of how disturbing the reflections on a VDT were from a set of photometric measurements of both the luminaire and the VDT. The three variables that were used to predict the observer's response were the luminance contrast of the display under the lighting installation, the luminance contrast of the reflected image of any source of high luminance, such as a luminaire or window, and the width of the blur function, i.e. the degree to which the reflection from the display is blurred. For the lighting installation, the parameters that have to be measured are the luminance of the source that is seen by reflection in the VDT, the luminance of the background to that source and the illuminance incident on the display. For the VDT, the parameters to be measured are the minimum and maximum luminances of the display in the absence of any room lighting, the diffuse and specular reflection properties of the display, the proportion of the display at the maximum luminance, and the blur width. These parameters are all that is needed to calculate the three variables in the model. The three variables were combined into an equation that explained 85 percent of the variance in a set of ratings of disturbance for reflections seen in both cathode ray tubes (CRTs) and liquid crystal displays (LCDs). Of course, this might be expected because the model was constructed based on the goodness-of-fit to these data. To really test a model it is necessary to determine how successful it is in predicting the results from an independently gathered set of data. Miller (2000) provided such a test using 10 different direct lighting luminaries and three different VDT screen types. Figure 7.13 shows the mean ratings of the conspicuity of the reflected image in the display plotted against the predicted responses on the Lloyd *et al.* (1996) disturbance scale. These results provided some support for the model, in that it explained 79 percent of the variance in the ratings of conspicuity, but also revealed some limitations. The most important was the fact that the model does not take into account the size of the luminaire and hence the size of the reflected image of the luminaire. Miller *et al.* (2001) showed that the mean ratings of disturbance, for each screen type, could be most closely related to the luminous intensity of the luminaire raised to the power 0.4 (Figure 7.14). Luminaire luminous intensity is the product of the luminance of the luminaire in the given direction and the projected area of the luminaire in the same direction. This finding provides a very practical means of identifying luminaires suitable for use with VDTs because the luminous intensity of the luminaire in any given direction is given in the standard photometric data already supplied by manufacturers. Miller (2000) suggests a maximum luminous intensity of 300 cd for direct lighting, when the quality of the VDTs is unknown.

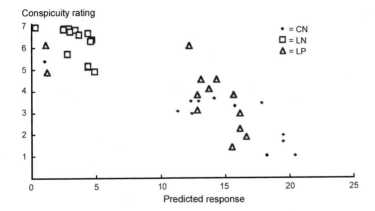

Figure 7.13 Mean ratings of the conspicuity of the reflected image of a luminaire in a VDT screen plotted against the ratings predicted using
the Lloyd *et al.* (1996) model. The conspicuity scale is 1 = very
conspicuous, 4 = moderately conspicuous, 7 = not at all conspicuous. The three different types of display are CN, which is a CRT
display with negative polarity; LN, which is a LCD display with
negative polarity; and LP, which is a LCD display with positive
polarity (after Miller, 2000).

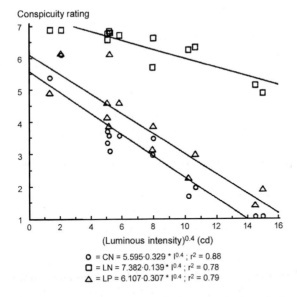

Figure 7.14 Mean ratings of the conspicuity of the reflected image of a
luminaire in a VDT screen plotted against the luminous intensity
of the luminaire in the direction of the screen, raised to the power
0.4. The conspicuity scale is 1 = very conspicuous, 4 = moderately
conspicuous, 7 = not at all conspicuous. The three regression lines
through the points are for three different types of display: CN is
a CRT display with negative polarity; LN is a LCD display with
negative polarity; LP is a LCD display with positive polarity (after
Miller *et al.*, 2001).

This raises another point, namely, the quality of the VDT. Figure 7.13 suggests that the difference in properties of the screens is a major factor in determining the conspicuity of reflections. The best quality of display from this point of view is one with very low diffuse and specular reflectivities and a high background luminance. Given these characteristics, Figure 7.14 suggests that there is really no need to limit the choice of luminaires to those recommended for use with VDTs. This is important because the quality of displays, in terms of maximum background luminance and low diffuse and specular reflectivities, has improved dramatically over the last decade. It may be that improvements in display technology and the replacement of older technology will soon make it unnecessary to modify the lighting to avoid disturbing reflections from VDTs.

7.5   Lighting controls

Lighting controls are installed in offices to reduce energy use and/or to provide a means of adjusting the lighting conditions to ensure individual comfort. The controls can be either discrete or continuous in that the light output of the electric lighting can either be switched or dimmed. The controls can also be automatic or rely on the occupants' actions. How effective and how useful lighting controls are depends on their ease of use and how satisfactory occupants find the outcome. Automatic controls require minimal human intervention, once they have been commissioned successfully, but they may not always produce satisfactory outcomes. Manual controls require human intervention and also may not produce a satisfactory outcome.

7.5.1   *Window lighting controls*

The controls applied to windows are usually some form of blinds; venetian, vertical, roller, etc. One reason blinds are provided is to enable people to eliminate the discomfort experienced when they have a direct view of the Sun or a bright sky. When blinds are not provided, it is not unknown to find pieces of paper taped to the window at positions where a view of the Sun occurs. But the use of blinds is more than just a simple open/shut operation. Maniccia *et al.* (1999), in a study of private offices, found that people adjusted the slant angle of their vertical blinds to occlude the Sun and to allow daylight to enter while preserving a view-out; in other words they adjusted the blinds to optimize the performance of the window. Further, the pattern of blind use was different on the different sides of the building. Figure 7.15 shows the average percentage of time with the blinds in each position for offices on the four sides of the building. Clearly, the blinds are most frequently closed on the south and west sides of the building, less often closed on the east side and rarely closed on the north side. This pattern is consistent with the exposure to the Sun over the working day.

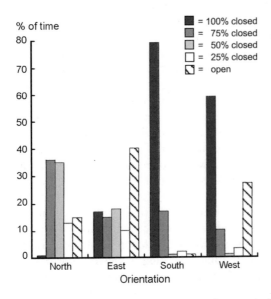

Figure 7.15 The average percentages of time window blinds were in four categories of position, for four different window orientations (after Maniccia *et al.*, 1999).

Overall, blinds should be considered as an essential component of a window system. They provide a necessary means to control the lighting provided by the window. Certainly, people value window blinds just as much as the windows themselves (Maniccia *et al.*, 1999).

7.5.2 *Manual electric lighting controls*

The most common form of manual lighting control is the switch-by-the-door. Ideally, manual switching should be used to ensure that the electric lighting is on only when there is someone present and daylight is not available. Unfortunately, as in so many other human activities, this ideal is rarely achieved. Crisp (1978) reported a series of field observations of the patterns of lighting switching in multi-occupant offices in the UK. He found that the lighting was only switched at the beginning and end of the working day, it was rarely switched in the middle of the day, a pattern also found in California by Rubinstein *et al.* (1999). However, Crisp (1978) also found that in a classroom, switching was most likely to occur at the beginning and end of each period of occupation, something that occurred several times a day. What this suggests is that the probability of switching will be related to the social organization of the space. In the multi-occupied office it is likely that no one was willing to take responsibility to switch off the lighting while there were still people in the office, while in the classroom, the teacher took the responsibility to switch the lighting on and off. This observation can be considered as a difference between "public" and "private" lighting. "Public" lighting is lighting serving several people and switching the lighting will affect all of them. There is a social inhibition against switching the lighting because of the potential for aggravation. "Private" lighting is lighting serving only one person or a group of people where one person is clearly in charge. In this situation, the lighting may be considered the property of the individual in charge. It is likely that switching will occur much more frequently in "private" lighting than in "public" lighting situations.

However, even in a private office, where one person is clearly in charge, there is no guarantee that manual switching of lighting will always occur. Love (1998) found a wide range of switching behavior in private offices that was more a characteristic of the occupant than of the daylighting conditions. For the same daylight conditions, one occupant rarely used the electric lighting, others kept the electric lighting on for most of the working day while yet others switched the lighting according to the amount of daylight available. It is clear that switching behavior is driven by many factors besides the photometric conditions.

One factor that definitely affects switching behavior is the ease of understanding what luminaires are controlled by what switches (Crisp, 1978). Where the layout of the switches bears no obvious relationship to the layout of luminaires, or where the switch panel is not even in the same room as the luminaires it controls, the probability of switching is low. The message of these observations is simple. To ensure that lighting will be switched off when not required, make the switching simple to do. Put the switch panel where anyone operating it can see the luminaires being switched; put a plan of the luminaire layout and the switches adjacent to the switch panel; and wire the luminaires to the switch panel so that all the luminaires controlled from a single switch are receiving similar amounts of daylight.

It might be thought that one way to encourage switching in a multi-occupied space is to convert "public" lighting to "private" lighting by giving each luminaire its own switch. Boyce (1980) observed such an installation, in which each luminaire was fitted with a pull-cord switch. These were little used. This was partly because the pull-cords were not easily accessible to some of the occupants and partly because the luminaires were not often located directly above someone's desk. Again, this is a matter of "public" and "private" lighting. What is considered "private" depends on the location of the luminaire relative to the user's work space. The pull-cord switch is most likely to be used when the luminaire is directly above the desk and hence can be considered "private" lighting.

These observations suggest that pull-cord switches are of limited use for encouraging manual switching of electric lighting and the switch-by-the-door is likely to remain the most common means of providing such action. Given that lighting is most likely to be switched when someone enters an

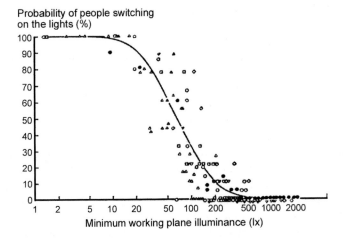

Figure 7.16 The probability of people switching on the lighting on entering a room plotted against the minimum illuminance on the working plane at the time of entering the room (after Hunt, 1979).

office, what is it that determines whether a person will use the switch-by-the-door to turn on the electric lighting? The answer appears to be the amount of daylight in the working area of the office when the person enters. Hunt (1979) carried out a series of field studies of patterns of lighting switching in deep-plan offices and the photometric conditions at the time of switching. He found that the probability of the lighting being switched on when entering a room was closely related to the minimum illuminance on the working plane (Figure 7.16). Specifically, the probit curve fitted through the data took the form:

$$y = -0.0175 + 1.0361/[1 + \exp(4.0835(x - 1.8223))]$$

where $100y$ is the percent switching probability and $x = \log_{10}$ (minimum daylight illuminance in working area in lux).

According to this equation, the probability of switching on is 100 percent when the minimum illuminance is 7 lx, is zero at 658 lx, and is 50 percent at a minimum illuminance of 67 lx. Such information allows for estimates to be made of the number of hours for which electric lighting is likely to be operated in daylight spaces, where manual switching is used (Lynes and Littlefair, 1990). It also suggests an opportunity for a simple energy-saving system in daylight spaces. Such a system involves a simple time switch that turns off the electric lighting at a time when there is a lot of daylight available, say at noon. Given that there is a lot of daylight, it is unlikely that anyone will trouble to switch on the electric lighting until dusk so energy will be saved (Hunt, 1980). Such a system represents a positive use of human inertia.

In recent years, developments in dimming electronic ballasts for fluorescent lamps and in computer networks have opened up the possibility that individuals in both private and in multi-occupied offices could be given control of the illuminance in their working areas. A number of studies have been made of the way in which office workers use such manual dimming controls. The first thing to say is that people tend to use such controls in a rational way. Boyce *et al.* (2000a) conducted a laboratory study of how temporary office workers doing a variety of tasks in small, private, windowless offices used a hand-held dimming control. The lighting system used could provide two ranges of illuminances; 12–1,240 lx, and 7–680 lx. Measurements showed that the workers choose lower illuminances when working at a VDT than when doing the same task on paper. This is rational because for paper-based work, increasing the illuminance increases the visibility of the task materials while for the self-luminous VDT, increasing the illuminance reduces the luminance contrast of the display. The second thing to say is that there is a very wide range of illuminances chosen by different people doing the same tasks. Figure 7.17 shows the range of illuminances chosen by each subject during the day, when they all did the same tasks. Four facts are apparent in Figure 7.17. First, some subjects use the dimming control a lot while some rarely use it. Second, different subjects have very different preferences for illuminance. Third, many of the subjects prefer a lower illuminance than that commonly used in offices (500 lx). Fourth, the illuminances selected depend on what the range of illuminances available are. This last point implies that any hope of saving energy by giving people control of the lighting is dependent on the choice of the maximum illuminance installed. The most effective approach would be to make the maximum illuminance the same as would be installed for a fixed lighting installation. Then, any use of the individual dimming control is guaranteed to reduce the energy consumption.

While this study is interesting, it is limited by the short time for which the subjects worked in the office (1 day). A longer-term field study of different forms of lighting controls, including manual dimming, has been made by Maniccia *et al.* (1999) in a building containing mainly private offices. The most obvious difference in the use of the dimming control was the frequency with which the dimming control was used. In the laboratory study, the subjects used the dimming control about 8–10 times a day. In the field study, the subjects made an average of one dimming adjustment every 3.9 days. Some of this difference can be attributed to the novelty of having the dimming control in the laboratory experiment. Another plausible explanation that would explain why the frequency of use was low in the field experiment is that people use the dimming control to adjust the illuminance to the level they prefer for the sort of tasks they do and once that is established they use the dimming control rarely. Regardless as to which of these explanations is correct, the subjects in both Maniccia *et al.* (1999) and Boyce *et al.* (2000a) valued having the dimming control.

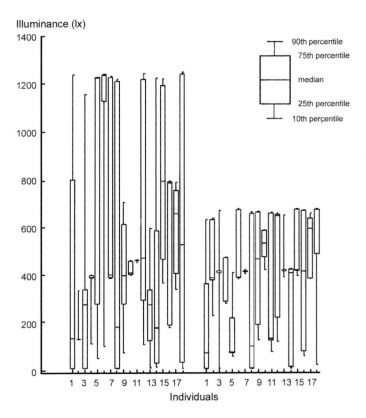

Figure 7.17 The range of illuminances chosen by 18 subjects carrying out a number of different office tasks in a private office for a day and having available a hand-held device for dimming the lighting. Two different offices were used, one with a maximum illuminance on the workstation of 1,240 lx and the other with a maximum illuminance of 680 lx (after Boyce *et al.*, 2000a).

Carter *et al.* (1999) examined the use of dimming controls in 11 offices in the UK. Some of the offices were narrow, providing a lot of daylight. Others were deep-plan in which little daylight reached the middle of the office. Interviews with the occupants revealed a preference for daylight, the control systems being seen mainly as devices for reducing the illuminance from the electric lighting. The illuminances set by using the controls were generally below the levels recommended. As was found in earlier studies, the window blinds were primarily used to control glare and as would be expected from the wide individual differences in preferred illuminances, there were conflicts between individuals over the choice of illuminance for groups of luminaires serving more than one occupant. Besides individual preferences, the factors influencing the choice of dimming level were the amount of daylight available at the location, the activity being undertaken, and the luminances of the surroundings.

7.5.3 *Automatic electric lighting controls*

There are three common forms of automatic lighting controls; time switches, photosensor dimming, and occupancy sensors. Time switches are simple to install and commission. What they do is to switch the electric lighting between states at set times. The states most commonly used are simply on and off, although intermediate levels are sometimes used. Time switching (also known as scheduling) is a valuable means of reducing energy waste in buildings where there is a consistent pattern of use making it possible to predict when lighting will not be needed.

Photosensor dimming (also known as daylight harvesting) involves an automatic control system, consisting of a photosensor, a control algorithm, and a dimming ballast, that dims the electric lighting as the amount of daylight increases. In principle, such a system can ensure significant and sustainable reductions in electrical energy consumption (Lynes and Littlefair, 1990; Rubinstein *et al.*, 1999). However, there are four factors that make it difficult to achieve the possible savings in practice. First are the wide differences in individual preferences for illuminance. In multi-occupied spaces, the individual who likes the most light is likely to be the first to complain about the electric lighting being dimmed. Facility managers dislike occupant complaints and so will usually increase the illuminance at which dimming starts, thereby reducing the potential for energy saving. Second, the inertia in the use of window blinds used to control glare from the Sun and sky means that the daylight available may be less than expected. Third, the commissioning of the control system is not simple (Rubinstein *et al.*, 1997) partly because of the frequently unknown control algorithm of the system (Bierman and Conway, 2000). Fourth, photosensor dimming systems are not cheap. There is an interesting paradox in that the more efficient the electric lighting system becomes and hence the less energy there is to be saved, the more difficult it is to justify the cost of an elaborate control system. What all this means is that photosensor dimming systems need to be applied with care if they are to be successful.

Occupancy sensors (also known as motion detectors) are probably the most widely used automatic electric lighting controls. Using either passive infrared or ultrasonic technology, they detect movement in a space. Passive infrared sensors can only detect movement within the direct line of sight. Ultrasonic sensors can detect movement out of the line of sight. For both technology types, if no movement is detected over a fixed time, the electric lighting is switched off, or sometimes set to a dimmed condition. Conversely, if movement is detected, the lighting is automatically switched on. Occupancy sensors use different technologies, provide different coverage areas, have different sensitivities to movement, and can be arranged to have different combinations of automatic/manual and on/off actions. The inertia about lighting conditions evident in the use of window blinds suggests that the automatic-off/manual-on arrangement is likely to save more energy than any other combination.

All three forms of automatic electric lighting control have the potential to eliminate energy waste by minimizing the electric lighting when there is no one present or when there is an alternative light source available. The problem with them all is that they are automatic and therefore do not take into account the vagaries of human nature and office life. In offices, people expect to be provided with the environmental conditions that they need in order to do their work, and appropriate lighting is one of the most important of those conditions. Any automatic electric lighting control system that fails to provide the expected conditions when called for, or which is distracting because of the nature and frequency of its operation, stands the risk of producing complaints or being sabotaged. Occupancy sensors that are insufficiently sensitive to movement so that they repeatedly switch the electric lighting off when there are people present in the space are not likely to be operating for long. To avoid such a situation, it is necessary to understand the characteristics of the occupancy sensor, to place it in an appropriate location and to commission it so that it has appropriate sensitivity and delay times. Also, as a general principle, all automatic electric lighting control systems should be equipped with a manual override to allow the occupants of the space to take control of the lighting when necessary. It is also a good idea to explain the purpose of the automatic control system. Maniccia *et al.* (1999) found that the reasons why people adjusted the illuminance in their offices were primarily to do with the work being performed, the need to compensate for daylight, and to create an appropriate atmosphere for work. Saving energy was not a consideration. This might be changed by propaganda in favor of energy saving but until it is, and given that the objective of most automatic lighting control systems is to save energy, there is a need to explain why the controls are operating. This is particularly so when operation occurs while people are present and particularly if the operation makes the visual conditions less satisfactory. An application where this is likely is that of peak-lopping. Peak-lopping reduces electric lighting so as to reduce the maximum demand on the electricity supply. In doing so, it will reduce the illuminance in some areas. If this is understood by the occupants to be for the general good and as being something necessary to preserve the electricity supply in the community then it might be accepted but if it is seen as the company degrading working conditions to make more money to pay for the CEO's stock options, it is likely to be resisted. Guidance on the occupancy patterns for which the different types of automatic electric lighting control systems are most appropriate is given in Crisp and Henderson (1982).

7.6 Summary

Office lighting today is more complex than it used to be because of the arrival almost everywhere of the computer. Prior to the arrival of the computer, lighting designers concentrated on lighting the desk because that was where the work was done. Today, any lighting installation designed for an office has to be satisfactory for materials that are self-luminous, i.e. computer screens, and seen by reflected light, i.e. paper, and for lines of sight that can be both across the office and down at the desk. This chapter is devoted to the lighting of offices and is organized around the choices the lighting designer has to make, namely, the illuminance to be provided, the light source to be used, the type of luminaire and its layout and the nature of any lighting controls.

The illuminance on the working plane is a major determinant of satisfaction with the lighting in an office. There is support from a number of sources for the current average illuminance recommendation for offices in North America and Europe of about 500 lx on the task. However, it is important to note that simply providing such an illuminance is not enough to guarantee comfortable office lighting. It is also necessary to consider all the aspects of lighting conditions that cause discomfort and that are discussed in Chapter 5.

The first major decision to be made on the light source to be used for office lighting is the balance between daylighting and electric lighting. Daylight always deserves consideration because people like it, as long as it does not cause them thermal or visual discomfort. People like it because, if carefully designed, daylighting fully reveals both the task and the space and provides environmental stimulation by meaningful variation of lighting conditions. By comparison, electric lighting installations usually provide just good visibility for the task and rarely provide any environmental stimulation. Electric lighting installations are, in a sense, always playing "catch-up" to daylighting.

However, electric lighting is always necessary because daylight fails reliably everyday. Operating costs and ceiling heights combine to ensure that the fluorescent lamp is almost universally used for office lighting. But what spectrum should that fluorescent lamp have? The answer to this question depends on what the tasks are. The most universal answer is a fluorescent lamp with a high CIE general CRI (>80). Such a lamp ensures that tasks that require fine color discrimination, or tasks where the color differences are important, or tasks where the accurate naming of colors is important can all be done easily. Where only achromatic tasks are involved, then there is little reason to chose any specific spectrum other than by individual preference. There is evidence that a light source with a scotopically enriched spectrum can enhance the performance of an achromatic task when the task is close to threshold, due to the spectrum-induced decrease in pupil size, but the effect disappears in suprathreshold conditions. As for preferred light spectra for offices, the results to date indicate that provided the CRI is high enough (>65) and the light source has a nominally white color appearance, lamp color properties are a minor factor in determining the satisfaction felt with the lighting of an office. The illuminance provided is much more important. This conclusion should not be taken to mean that light source color never matters. If the light source color moves too far from what peoples' expectations are, it is likely that complaints will be heard.

The distribution of light has an influence on the perception of the office (see Chapter 6). One of the major differences in perception is whether the office is perceived to be primarily lit by daylight or by electric light. As a rule of thumb, any office where the average daylight factor is more than 5 percent will be perceived as daylit. Conversely, any space where the average daylight factor is significantly less than 2 percent, will be perceived as electrically lit, even in daytime. Daylight can be delivered into an office through conventional windows, clerestory windows, or skylights, as well as a number of remote distribution systems, such as light shafts, ducts, and pipes. By far the most common is the conventional window. The important aspects of windows as far as people are concerned are their size, shape, spectral transmittance, and solar shielding. All these aspects are subject to limits and all should be considered when designing windows.

As for electric lighting systems, the most common arrangement in offices is a regular array of either direct, indirect, or direct/indirect luminaires, with or without some form of local task lighting. Attempts have been made to demonstrate that one or other of these approaches is better than the others but with very mixed results. Basically, the reaction to the type of lighting system used depends on how it is done. From the point of view of the occupant, there is no feature of direct, indirect, or direct/indirect lighting that always excludes it from consideration. There is also little evidence for the proposition that light distributions that do not affect task visibility or cause visual discomfort but do change the appearance of the space affect task performance.

The introduction of large numbers of computers into offices has led to the design of a class of luminaires with restricted luminances above certain angles, known as VDT luminaires. These luminaires are designed to minimize the number and magnitude of reflected images seen in the computer displays standing on desks. The unthinking use of these luminaires has created offices that are seen as gloomy and cave-like. Fortunately, there is less and less need for such special luminaires. Modern display technology is becoming less sensitive to the ambient lighting conditions. This is particularly so when the display has a high luminance background and the screen on which the display is presented has low diffuse and specular reflectances.

As for controls, windows where a direct view of the Sun is possible should always be fitted with blinds of some sort. Blinds are used to optimize the performance of a window and are valued by people sitting close to windows as much as the window itself. Electric lighting controls can be automatic or manual. Automatic controls require minimal human intervention, once they have been commissioned. Manual controls require human intervention. How frequently manual controls are used will depend on whether the lighting being controlled is seen as "public" or "private." "Public" lighting is lighting serving several people and altering the lighting will affect all of them. "Private" lighting is lighting serving only one person or a group of people where one person is clearly in charge. Manual lighting controls will be used much more frequently in "private" lighting than in "public" lighting situations. The other factor determining the probability of use of manual controls is their ease of use. To ensure manual lighting controls will be used, keep them simple and located where the person using the controls can see what effect they have.

There are three common forms of automatic lighting controls: time switches, photosensor dimming, and occupancy sensors. All three forms have the potential to eliminate energy waste by minimizing the use of electric lighting. The problem with them all is that they are automatic and therefore do not take into account the vagaries of human nature and office life. Automatic electric lighting control systems should be like the perfect butler; they should perform without being noticed. The closer any specific system gets to this ideal, the more likely it is to be acceptable to the occupants of the office.

8 Lighting for industry

8.1 Introduction

The study of lighting for industry is the "Cinderella" of lighting research. Compared to the effort that has been put into studying lighting for offices, lighting for industry had been sadly neglected. The reason for this neglect is not the lack of importance of lighting for industry, but rather the difficulty in generalizing any conclusions. The nature of the work done in an office is similar in all offices, so any conclusions reached from research are relevant to all offices. The same cannot be said for industry. Study of the optimum lighting conditions for any given industrial activity is likely to be specific to that activity. This makes it difficult to justify research for industrial activities unless there is a specific problem to be solved.

This chapter will review the aspects of lighting that need to be considered for all industrial activities, although whether they are important for a specific industry will depend on the situation in that industry. Anyone seeking general advice on lighting for industry is referred to the guidance documents published by professional associations (IESNA, 2001; CIBSE, 2002b) and to books published by authors with extensive experience of industrial lighting (Lyons, 1980, 1981). Advice about lighting for specific industries and activities is available, although many of these publications were prepared many years ago so care should be taken that the processes described are still relevant to that industry (IES Industrial Lighting Committee, 1949–90; CIBSE, 1979, 1983). Examples of good, current industrial lighting practice are also published (Philips Lighting, 1999).

8.2 The problems facing lighting in industry

The basic problem facing anyone designing lighting for industry is the wide variability in the amount and nature of visual information required to undertake work in different industries. Some industrial work requires the extraction of a lot of visual information, typically the detection and identification of detail, shape, and surface finish. Other types of industrial work require accurate eye–hand coordination and the judgment of color. Yet other types of industrial work can be done with very little visual information at all. The materials from which visual information has to be extracted can be matte or specular in reflection or some combination of the two, and the information can occur on many different planes, implying many different directions of view. Further, the material from which the information has to be extracted can be stationary or moving.

None of these requirements pose insuperable problems, given that the lighting designer has a clear idea of the visual information needed and the nature and location of the material from which the visual information has to be extracted. However, the physical situation within which the work takes place may set constraints on the lighting equipment that can be used. A common constraint is the extent of obstruction. Many factories have overhead conveyors that obstruct light from high-mounted luminaires. In some factories, the machinery is so large that it obstructs the lighting of the work-stations in and around it. Even in small-scale assembly operations, the ability to see inside a box may be limited by shadows cast by the worker. Other situations where lighting equipment with specific characteristics is required are those where there is an explosive, flammable, or corrosive atmosphere. Another factor to be considered is the extent to which the atmosphere is clean or dirty. The ability of the lighting to deliver the specified lighting conditions will be limited to a short period if the atmosphere is very dirty. There is lighting equipment available to deal with all of these conditions, but the lighting designer has to be aware of the need.

The fact that different levels of necessary visual information occur at different locations in different environmental conditions in different industries implies that the design of industrial lighting is inevitably a matter of tailoring the lighting to the situation. There is no "one size fits all" solution to industrial lighting. Having said that, it is only fair to point out that there is a limit to how closely the lighting can be tailored. This limit is set by the fact that many different tasks are likely to occur on the same industrial site, within the same building, on the same production line and, certainly, within the area lit by one general lighting installation. The usual solution to this problem is to provide general lighting of the whole area appropriate for the average level of tasks; localized lighting where work is concentrated, e.g. on an assembly line; and task lighting where fine detail needs to be seen, e.g. on a lathe in a machine shop, or where obstruction reduces the visibility of the task, e.g. on the workpiece of a hydraulic press, or where there is an obvious hazard, e.g. on the feed to a circular saw. The only place where this general/localized/task lighting approach is impossible is where the scale of the equipment is so large that both the people and the lighting work within the equipment, e.g. a chemical plant. For such applications, lighting equipment is integrated into the plant.

8.3   General lighting

Despite the variability faced by the designer of industrial lighting, the objectives of the lighting are the same everywhere. They are:

- to facilitate quick and accurate work;
- to contribute to the safety of those doing the work;
- to create a comfortable visual environment.

The principles of how to use lighting to enhance visual work have been set out in Chapter 4. From the discussion there, it should be clear that the factors that determine the level of visual performance are the visual size, luminance contrast and color difference of the visual information that has to be extracted and the retinal image quality and the operating state of the visual system. The visual size, luminance contrast and color difference are determined by the task and its interaction with the lighting. The retinal image quality is determined by the characteristics of the worker's optical system and the operating state of the visual system is determined by the adaptation luminance and hence the effect of the illuminance provided by the lighting and the reflectance properties of the surfaces illuminated. The amount of light provided, i.e. the illuminance on the working plane is usually determined by the inherent size of detail of the task and its luminance contrast. The smaller is the size and the lower is the inherent luminance contrast, the higher the illuminance required. This trend is evident from a cursory inspection of any of the illuminance recommendations for industrial lighting (IESNA, 2001; CIBSE, 2002b). Of course, the working plane may not be horizontal and there may be more than a single working plane. For example, in a warehouse some areas will be devoted to unpacking and repacking pallets of goods. In these areas, both horizontal and vertical planes may contain necessary visual information, but in the storage aisles, the primary working plane will be vertical, with the horizontal plane containing little visual information (Figure 8.1).

The locations of the working planes are important in determining the placing and desirable distribution characteristics of the luminaires used in the lighting installation. Carlton (1982) argues that the emphasis given to the illuminance on a horizontal plane by the desire to meet illuminance recommendations at minimum cost has lead to the development and use of industrial luminaires that are apparently energy efficient but in fact are unsuitable for many industrial situations, where the illuminances on vertical planes are important.

There can be no doubt that a knowledge of where the light is needed is essential to the successful design of industrial lighting, but delivering the recommended illuminance in the right place is not enough to guarantee success. The directions from which the light comes are also important, for two reasons. The first is the degree of obstruction. Obstructions cast shadows.

Figure 8.1 In industry it is common for the necessary visual information to be located on several different planes.

The densest shadows are formed when all the light reaching a point comes from one direction. Therefore, to minimize shadows it is desirable to have light incident on a point from many different directions. This ideal can be approached by using a larger number of smaller wattage light sources rather than a smaller number of larger wattage light sources, by using luminaires with a widespread light distribution, and by having high-reflectance surfaces in the space. Figure 8.2 shows a small workshop where shadows have been minimized by this approach. At the very least, a proportion of the light emitted by luminaires should be emitted upward to be reflected from a high-reflectance ceiling or roof. The more obstructed is the space, the more valuable this approach is. It should be noted that the obstruction may not always be obvious. A study in a letter sorting office in Sweden, where the letter sorting was almost entirely automatic with very limited human involvement, hardly seems a good candidate for a determining the benefits of different types of industrial lighting. However, it was found that a lighting installation based around a continuous light pipe was preferred over an array of individual luminaires partly because the widespread light distribution produced by the light pipe made it easier to see inside the sorting machines to carry out maintenance (Figure 8.3) (Boyce and Eklund, 1997).

The other reason why the distribution of light is important is the occurrence of veiling reflections. What these are and how they may change the

Figure 8.2 A small workshop with high-reflectance walls and lit by a regular array
of luminaires with a wide luminous intensity distribution. The result is
a shadow-free environment.

luminance contrast of a task are fully discussed in Section 5.4.3 but basically, veiling reflections are images of high luminance objects, such as a
luminaire or a window, superimposed over the task. Veiling reflections can
decrease task luminance contrast, in which case the visual performance of
the task will deteriorate. Veiling reflections may also be used to reveal the
nature of a surface, in which case they are called highlights, and can be beneficial for the performance of some tasks. Whether veiling reflections are
beneficial or not will depend on the work being done. If they are beneficial
then they are more easily provided by a local lighting installation mounted
close to the task. It is generally unwise to try to provide veiling reflections
from general lighting. The approach used to minimize shadows, i.e. to use
luminaires with a widespread luminous intensity distribution and an interior
with high reflectance surfaces, also works to minimize veiling reflections.

The other factor that the designer of the general lighting needs to consider is the spectrum of the light used. Where color is of no importance,
many different light sources have been used in industry. Some, such as
the high-pressure sodium lamp, have poor color rendering and a distinctly
non-white color appearance. Where color is used to convey meaning, e.g. in
electrical wiring, or where color is an important determinant of value, e.g.
when grading diamonds, or where matching colors is an important element
of quality, e.g. fine color printing, care should be taken with the selection

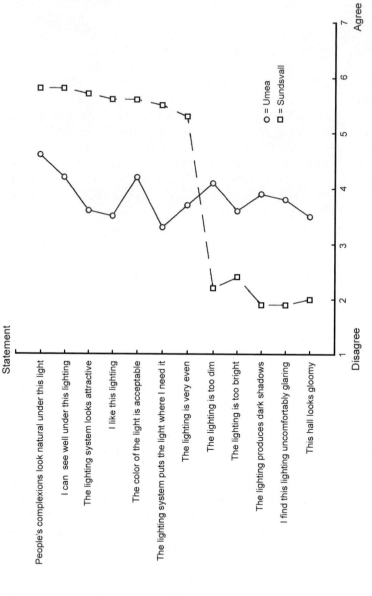

Figure 8.3 Mean ratings of agreement with a series of statements about the general lighting of two postal sorting offices in Sweden, both of which use the same automated sorting machinery. The sorting office in Umea is lit by a regular array of specular louvre luminaires containing fluorescent lamps. The sorting office at Sundsvall is lit from a number of rows of continuous light pipe, with the light provided by sulfur lamps (after Boyce and Eklund, 1997).

of the light source. Collins and Worthey (1985) have demonstrated the consequences of choosing an inadequate spectrum for lighting in a study of lighting for meat and poultry inspection. They found that when working under high-pressure sodium lighting, the inspectors were more likely to pass diseased meat and poultry than when working under fluorescent or incandescent lighting. In general, for the discrimination of colors that are well separated in color space (see Section 1.6.1) a light source with a CIE general color rendering index (CRI) of 60 or higher is desirable. Where more accurate color judgments are required, such as discriminating electrical color codes, a light source with a CIE general CRI of 80 or greater is required (CIBSE, 1994). Where the judgment of color has an major impact on the value and/or quality of a product, then specific lighting recommendations are made and the lighting is provided in either special rooms or booths. For example, for grading cotton, an illuminance in the range 600–800 lx, provided by a light source with a spectral power distribution simulating daylight is recommended (ASTM, 1996a). For judging textile colors, an illuminance greater than 1,000 lx from a light source with a high correlated color temperature (CCT) is recommended (Nickerson, 1948). For the examination of colors of opaque objects, national standards have been developed (ASTM, 1996b).

All the above has been concerned with the objective of facilitating quick and accurate work, and it has been implicit that the only way to do this is to improve the lighting. However, it should always be remembered that it may also be possible to make the task easier by changing the task characteristics. For example, Ruth *et al.* (1979) carried out a thorough analysis of the work processes and lighting conditions in foundries in Sweden. They found that workers doing manual casting worked in very poor visual conditions, because of the low illuminances in the work area, the low contrast between the molten metal and the mold, and the use of low transmittance eye-protectors. They certainly suggest increasing the illuminance as a means of improving working conditions but they also suggest redesigning the mold so as to increase the contrast between the part of the mold into which the molten metal is poured and the rest of the mold. It is always worth considering if the task can be changed to make it visually easier before undertaking a major lighting change.

Another objective of industrial lighting is to contribute to the safety of the workers. Some consideration of the impact of lighting on safety is appropriate in all lighting applications but it is particularly important in industrial situations. This is because of the complex layout of many plants, the hazards associated with some manufacturing processes and the dangers from moving equipment. Minimum illuminances are recommended for safety whenever the space is occupied, ranging from 5 to 54 lx depending on the level of hazard (IESNA, 2000a). But illuminance alone is not enough. Hazardous situations can arise whenever seeing is made difficult by disability glare, strong shadows, and sudden changes in illuminance.

Safety will also be enhanced by marking hazards with the appropriate colors (ANSI, 1998) but this will be of little use if the light sources used do not allow the safety colors to be correctly named. Jerome (1977) examined people's ability to name safety colors at an illuminance of 5 lx, the lowest illuminance recommended for safety lighting. He showed that light sources with low CRIs, such as high-pressure mercury discharge and high-pressure sodium discharge lamps, made it difficult to correctly identify some of the safety colors at such a low illuminance. How much these findings are due to the low illuminance and how much to the light spectra of the different light sources remains to be determined. What is known is that the ability to correctly name colors with poor color rendering lamps improves as the illuminance increases (Saalfield, 1995) and that accurate color naming is possible with light spectra covering a wide range of color rendering properties, given an illuminance of more than 100 lx (Boynton, 1987; Boynton and Purl, 1989).

One aspect of safety that needs to be considered where there is rotating or reciprocating machinery is the possibility of a stroboscopic effect. A stroboscopic effect is evident when oscillations in the illumination of a moving object cause that object to appear to move at a different speed from the speed it is actually moving or even to appear to be stationary. All light sources operating from an alternating current electrical supply produce oscillations in light output. Whether these oscillations are enough to produce a stroboscopic effect will depend on the frequency and amplitude of the oscillation. The closer the fundamental frequency of light oscillation is to the frequency of rotation and the larger the amplitude of light oscillation, the more likely a stroboscopic effect is to occur. The probability of a stroboscopic effect occurring can be reduced by using electronic control gear for discharge lamps, because such control gear changes the frequency of light oscillation and reduces the amplitude of the oscillation; by mixing light from light sources operating from different phases of the electricity supply before it reaches the relevant machinery, because such mixing also alters the frequency of light oscillation and the amplitude of oscillation; and by supplementing the general lighting of machinery with task lighting using a light source with inherently small oscillation in light output, such as an incandescent lamp.

The final objective of industrial lighting is to create a comfortable visual environment. This objective is not always given the attention it deserves. This will be evident to anyone familiar with industrial lighting, the design of which is too frequently dominated by a desire to maximize lighting system efficacy. This desire is often consummated through a lighting installation which uses the smallest number of luminaires, containing the highest wattage light source, positioned at the widest allowed spacing, and directing most of their light downwards to an assumed horizontal working plane. This is a recipe for deep shadows, strong veiling reflections and possibly discomfort glare, as well as inadequate illumination on vertical planes (Carlton, 1982). That these disadvantages are recognized is demonstrated by the preference of industrial workers for luminaires with a significant proportion of upward light over luminaires with no or little upward light (Subisak and Bernecker, 1993). Such upward light will be reflected from the ceiling or roof of an industrial building. Provided the reflectance of the ceiling or roof is high, the diffusely reflected light will weaken both shadows and veiling reflections, diminish any discomfort glare and provide some light on any vertical working surfaces.

The lower priority given to eliminating visual discomfort in industrial lighting is also evident in commonly used lighting recommendations. For example, the criteria for discomfort glare are usually much less stringent for industrial applications than for offices. Further, a wider range of light sources is considered acceptable for industrial use than for offices, including some discharge light sources that have poor color rendering properties but high luminous efficacies. There is no logical reason why this should occur. People who work in industry have the same visual system as those who work in offices. Rather, it is a matter of expectations. Many aspects of the physical environment are less comfortable in industry than in offices and lighting is just one of them. The quality of industrial lighting would undoubtedly be better if all the aspects of visual discomfort discussed in Chapter 5 where considered by those designing industrial lighting.

8.4   Localized and task lighting

Localized and task lighting can take many different forms and serve many different purposes. Probably the most common forms are the fixed luminaire that provides additional illuminance in a localized work area and the adjustable task luminaire that allows the worker to have some control over the lighting of the task. Fixed localized lighting is common where the work area is in shadow. Adjustable task lighting is common where the tasks to be done are much more visually difficult than average. For large-scale manufacture, localized lighting can be moveable, consisting of luminaires mounted on a wheeled frame, so that lighting can be moved into position when work demands it. Fixed localized lighting rarely does more than provide a higher illuminance. Adjustable task lighting also does this, but in addition, allows some modification of the distribution of light falling on the task.

8.5   Visual inspection

One type of local lighting that can take many different forms, each form being designed for a particular function, is lighting for visual inspection. Visual inspection work involves two separate but successive components. The first is the search for and identification of any defects. The second is deciding on what to do about the identified defects. Lighting can only directly affect the first component.

Studies of eye movements made while searching for defects in products have revealed a common pattern of fixation and saccade. The observer

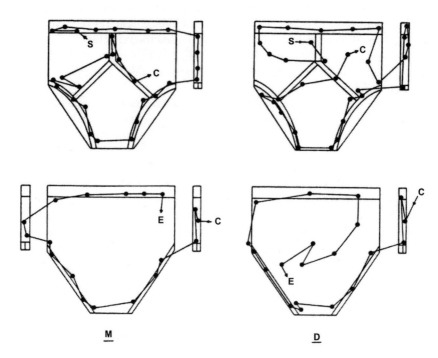

Figure 8.4 The pattern of fixations made by two inspectors examining men's briefs held on a frame. S = start of scan path. C = end of scan of front and one side, rotation of frame, and continuation of scan across back and sides. E = end of scan. Inspector M examines only the seams while inspector D examines the fabric as well (after Megaw and Richardson, 1979).

searches through a series of fixation pauses with rapid saccadic eye movements between them. Figure 8.4 shows such an eye movement pattern made by an inspector examining men's briefs held on a frame. Observations of this sort illustrate that the search pattern made by experienced inspectors is often systematic rather than random, the search pattern being based on the inspector's expectations about where the defects are likely to occur (Megaw and Richardson, 1979). The fixation and saccade pattern of visual search implies that the defect, or something that may be a defect, is likely to be first detected off-axis, i.e. in the peripheral visual field, and subsequently confirmed by bringing the fovea to bear on it through a saccadic eye movement. Therefore, the essential requirements for rapid visual search are off-axis detection of a defect, preceded by a clear definition of what constitutes a defect. Lighting cannot provide a clear definition of a defect for the inspector, although it can sometimes be designed to reveal the visual characteristics that define a defect, once that definition is available. What lighting can always be designed to do is to enhance the probability of off-axis detection of a defect. For a uniform field, where any departure from uniformity

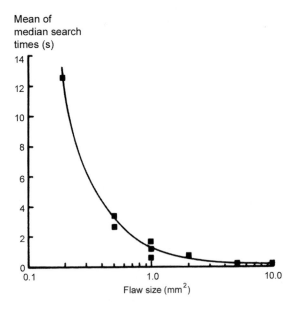

Figure 8.5 Mean of the median search times for detecting a single flaw in a sheet of glass, plotted against the flaw size (after Drury, 1975).

is a defect, the probability of off-axis detection can be related to the visibility of the defect. Figure 8.5 shows search time plotted against defect size for the inspection of a sheet of glass for a single defect (Drury, 1975). It is clear that as the defect size increases, which will make it more visible, the search time decreases.

The concept used to model the effect of lighting conditions on search time is the visual detection lobe, i.e. a surface centered on the fovea that defines the probability of detecting the defect at different deviations from the fovea within a single fixation pause (Lamar, 1960; Bloomfield, 1975a). Figure 8.6 shows some probability data for detecting targets of different sizes and luminance contrasts. From such results, it is possible to calculate a visual detection lobe for each target by assuming radial symmetry about the visual axis. As would be expected, such visual detection lobes have a maximum at the fovea; the probability of detecting the defect decreasing as the defect is located further off-axis. Clearly, different defects will have different visual detection lobes. A large-area, high-contrast hole in some sheet material will have a large visual detection lobe while a small-size, low-contrast hole will have a small lobe. The size of the visual detection lobe matters because, provided the interfixation distance is related to it and the total search area is fixed, the total time taken to cover the search area is inversely proportional to the size of the visual detection lobe. Visual detection lobes can be measured directly by psychophysical procedures or estimated from threshold

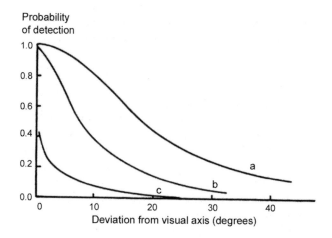

Figure 8.6 The probability of detection of targets of: (a) contrast = 0.058, size = 19 min arc; (b) contrast = 0.08, size = 10 min arc; (c) contrast = 0.044, size = 10 min arc; within a single fixation pause, plotted against deviation from the visual axis. Each curve can be used to form a visual detection lobe for each target by assuming radial symmetry about the visual axis.

performance data available for peripheral vision (Boff and Lincoln, 1988). Howarth and Bloomfield (1969) have suggested a simple equation, based on a random search pattern, which can be used to predict search times. The basic form of the equation is

$$t_m = t_f \times (A/a)$$

where t_m is the mean search time (s), t_f the mean fixation time (s), A the total search area (m²), and a the area around the line of sight within which the target can be detected in a single fixation, i.e. the visual detection lobe for the fixation time (m²).

For the inspection of a given article, the total search area is likely to be fixed and the mean fixation time is likely to be reasonably constant so the mean search time becomes proportional to the reciprocal of the size of the visual detection lobe.

Inditsky *et al.* (1982) have proposed an alternative model of visual search performance using a measure of target visibility based on visibility lobe per glimpse and assuming a random search pattern. Bodmann (1992) has reported that a model based on the visibility lobe concept can be made to fit visual search time data for high luminance contrast targets. It would be interesting to know if such models can accurately predict visual search times for defects of different sizes and contrasts, in realistic conditions. There must be some doubt about their ability to do so because experienced inspectors tend to use stereotyped search patterns, which may or may not be suitable for the distribution for defects but which are unlikely to be random (Megaw and Richardson, 1979). Also, inspectors rarely have to search for one type of defect, usually there are several.

There is also the question of what happens when the area to be searched contains other detail. For searching uniform, empty fields, it is the visibility of the defect off-axis that determines the search time. However, for many inspection tasks, the defect appears not in a uniform, empty field but in a cluttered field, i.e. one in which many different items are present. In this situation, the visibility of the defect alone is not enough to predict the search time. The other factor that must be considered is the conspicuity of the defect, i.e. how easy it is to distinguish the defect from the other items in the search field. High visibility is not enough to guarantee high conspicuity. As an example of this consider searching for a person in a crowd. All the people are equally visible, but if the person being sought is wearing a red hat and the rest of the crown is hatless, then the person being sought is conspicuous as well. For high conspicuity, the defect should differ from the other items in the field on as many dimensions as possible. Figure 8.7 shows the mean search times for two observers searching for rectangular or square targets among an array of square non-targets, plotted against an index of

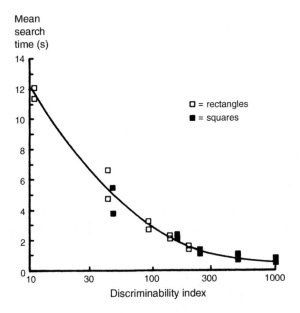

Figure 8.7 Mean search times for two observers searching for rectangular or square targets amongst an array of square non-targets, plotted against an index of discriminability. The discriminability index is given by $(A_1^{0.5} - A_2^{0.5})^2$, where A_1 and A_2 are the areas of the targets and non-targets, respectively (after Bloomfield, 1975b).

discriminability. Discriminability is given by the square of the difference in the square roots of the areas of the targets and non-targets (Bloomfield, 1975b). This result, and others like it, suggest that it should be possible to estimate an effective visual detection lobe where the lobe is determined not only by the target but also by the other items amongst which it is seen, i.e. not only by the visibility of the defect but also its conspicuity. Engel (1971, 1977) has shown how such an effective visual detection lobe can be measured and has demonstrated that it is related to the probability of finding a target within a fixed time.

It is important to appreciate that there are many different dimensions besides size on which the target can differ from the items around it. As discussed in Section 7.3.2.1, Williams (1966) studied search times for finding a specific item in a display of 100 items that could vary in size, shape, color, and the two-digit number contained. The inspector was asked to locate a particular item, where either the number alone was specified, or the number and various combinations of the size, color, and shape of the item on which the number was printed were specified. The results showed that some aspects of the specification were more important than others. Specifically, whenever the color of the item was specified, short search times were achieved, but specifying the shape showed little reduction in search time from what it was when the number alone was specified (see Table 7.2). The explanation for these results is that the differences in color give a much larger effective visual detection lobe than do differences in shape. This explanation is supported by measurements of the eye movement patterns made during the search. Whenever the color was specified, fixations were made predominantly on items of that color. When the shape was specified, there was little change in the eye movement patterns from when the number alone was given. This difference in pattern of eye movements can be explained if the items of a specified color have a large enough visual detection lobe so that fixation on one item with the specified color allows off-axis detection of an adjacent item of the same color.

Given that the efficiency of visual search is determined by the actual or effective visual detection lobe, the role of lighting conditions as regards visual search is to increase the size of the visual detection lobe. Many of the lighting techniques used for visual inspection are aimed at either increasing the visual size or luminance contrast of the defect, either by casting shadows (Figure 8.8) or by using specular reflections (Figure 8.9). Faulkner and Murphy (1973) list 17 different methods of lighting for inspection. Their methods can be classified into three types; those that rely on the distribution of light, as shown in Figures 8.8 and 8.9; those that rely on some special physical property of the light emitted that interacts with the material being inspected, e.g. ultraviolet radiation for detecting the presence of some types of impurities in a product; and those that call for the projection of a regular image onto or through the material being studied. Figure 8.10 shows this last approach. Any distortion of the grid when it is viewed

Figure 8.8 A cut in textured material lit by directional lighting delivered at a glanc-
ing angle to the surface of the material. The cut is visible under the directional lighting because of the high luminance contrast. The high
luminance contrast occurs because of the highlights on the sides of the
cut and the deep shadow in the cut.

through the beaker indicates a defect in the glass. A summary of the
lighting techniques used to reveal defects in or on materials of different
reflection characteristics is given in the IESNA *Lighting Handbook* (IESNA,
2000a).

Probably the most widely applicable aspect of lighting which aids visual
inspection is to increase the illuminance on the search area. Figure 8.11
shows the mean search times for finding a specific two-digit number located
among 100 such numbers randomly arranged on table, plotted against
the illuminance on the table. Increasing the illuminance leads to shorter
search times, particularly for the small-size, low-contrast target (Muck and
Bodmann, 1961).

While illuminance is generally a useful method of reducing search times, it
should not be used without thought. If the effect of increasing illuminance is
to decrease the luminance contrast, or effective visual size of the defects or to
produce confusing visual information in the search area, visual inspection
performance will be worse with higher illuminances. An early example of this
is shown in Figure 8.12, which gives the time taken for the inspection and pack-
ing of cartons of 25 shotgun cartridges during the period immediately before
and after lighting was switched on in the afternoon of a winter working period

Figure 8.9 A specular aluminium surface with a cross scribed into it, lit by directional lighting from above and behind the camera. The scribed cross is easily seen because the scribed marks cut into the surface and thereby alter the reflection characteristics of the surface. The result is a high-luminance reflection towards the camera for the cut and a high-luminance reflection away from the camera for the undamaged surface.

(Wyatt and Langdon, 1932). The sudden reduction in the speed of inspection with the onset of lighting from a single incandescent lamp overhead is obvious. The important point to note is this onset of lighting almost certainly increased the illuminance on the task but this caused a worsening of performance. The inspectors stated that the electric lighting produced an element of reflected glare from the brass caps of the cartridge cases and the cases were less uniformly lit. Thus, in this case, the increased illuminance was provided in such a way that the defects became more difficult to see. These results demonstrate the need to understand the whole impact of a lighting installation on visual search, rather than just one part, the illuminance.

Another example of the need to understand the whole impact of a lighting installation occurs is the inspection of topographic defects in painted automobile body shells (Wiggle *et al.*, 1997). There are many different forms of defect in automobile paint finishes, ranging from unwanted mixing of colors to surface defects, such as runs, sags, and "orange peel." However, one of the most difficult to see is the presence of dirt particles that fall into the paint before it hardens. These are difficult to see in the factory because they are small, they are the same color as the paint because they are enrobed

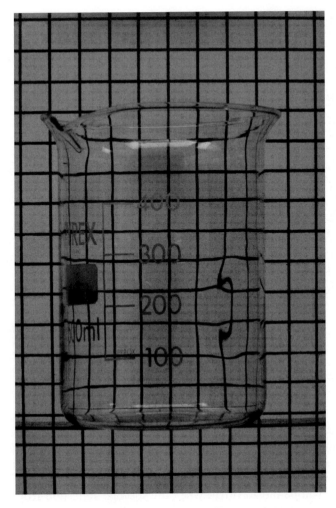

Figure 8.10 A distortion in a transparent glass beaker is revealed by the distortion in the grid seen through the beaker.

in the paint and, for the same reason, they have no luminance contrast. Yet, such dirt defects are seen by purchasers in sunlight and their presence leads to warranty claims against the manufacturer. There is therefore consider-able interest in using lighting to make it easier for the inspectors in the paint shop to detect and rectify such dirt defects. To design appropriate lighting, it is first necessary to understand the physics of the situation. A typical painted automobile shell has multiple layers of paint. For the purposes of visibility, it is only the top two layers that matter. These layers are usually a pigmented top-coat that gives the surface its color, and a clear sealing coat

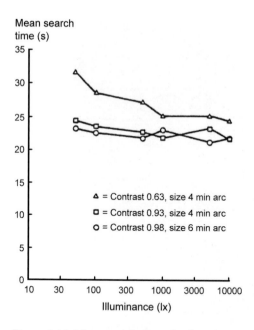

Figure 8.11 Mean search times for locating a specified two-digit number from a random array of 100 such numbers, plotted against illuminance, for numbers of three different size and contrast combinations (after Muck and Bodmann, 1961).

that gives the paint its gloss finish. Light incident on the surface of the paint is partially specularly reflected from the clear-coat layer and partially transmitted through the clear-coat to the pigmented layer, where it is diffusely reflected. This structure is the key to designing appropriate lighting. A speck of dirt in the paint provides a local deflection in the paint surface. This deflection will be most evident when the specular reflection from the clear-coat surface is emphasized and the diffuse reflection from the underlying pigmented surface is diminished. This can be achieved by using a series of discrete, high-luminance points or lines to illuminate the painted surface. The areas between each discrete, high-luminance point or line should be of low luminance. Figure 8.13 shows such an installation. When the inspector looks at the vehicle, the effect is to see the specularly reflected image of the installation, but, in addition, any deflections in the painted surface close to the image of the light source now have a highlight attached to them, a highlight that itself has a high luminance contrast and hence is likely to be more easily detected off-axis. Figure 8.14 shows a close-up of the image of a fluorescent tube reflected from a piece of black-painted automobile. The highlight to one side of the reflection of the lamp is caused by a dirt defect. When the dirt defect is beneath the image of the lamp, then the deflection has a much lower luminance than the image. In both situations,

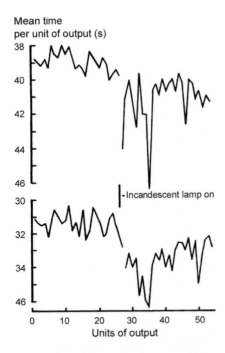

Mean time
per unit of output (s)

-Incandescent lamp on

Units of output

Figure 8.12 Mean time taken by two workers to inspect and pack cartons of 25 shotgun cartridges before and after an incandescent lamp was switched on (after Wyatt and Langdon, 1932).

the dirt-defect has a high luminance contrast and, hence, is much more likely to be detected off-axis. Obviously, this approach will only work within a short distance of the reflected image of the lamp (Lloyd *et al.*, 1999). To overcome this problem, the usual approach is to space fluorescent luminaires at regular intervals so the vehicle is seen as being covered with stripes of light (see Figure 8.13). The movement of the vehicle through the inspection area ensures that the reflected images sweep over the whole body shell, giving the inspector multiple opportunities to detect the dirt defect.

The problem of how to light a paint inspection area for the automobile industry has been considered in some detail, because it demonstrates the sort of process that is necessary to develop a successful solution. Whenever visual inspection lighting is under consideration, it is necessary to have a clear understanding of the physical nature of the defect and how it interacts with light, and the constraints imposed by the conditions in which the inspector works. Simply providing more light, without thinking about the consequences for the visibility and conspicuity of the defect, may make visual inspection more difficult. It is also necessary to consider the consequences of providing lighting to make it easier to detect one type of defect for the ability to detect other types of defect. The striped image shown in

Figure 8.13 A lighting installation designed to make dirt defects in painted automobile body shells easier to detect (courtesy of Holophane, a Division
of the Acuity Lighting Group Inc.).

Figure 8.13 is very good for revealing dirt defects, and other defects that
cause local deflections in the paint surface, but may make it more difficult
to detect fine changes in color and large area defects such as swirl marks.
Finally, it is necessary to appreciate that visual inspection by humans may
not be the only possibility. Automated inspection is possible for many
simple, repetitive inspection tasks and is steadily increasing in sophistication.
The main advantage of automated inspection is that the detection criterion
is clear and automated inspectors do not become bored and inattentive
(Lippincott and Stark, 1982; Batchelor *et al.*, 1985; Reynolds *et al.*, 1993).

To summarize, the general problem with identifying the best lighting conditions for visual search is that they are likely to be specific for each situation. It is clear that lighting that increases the effective visual size or
luminance contrast or color difference of the item being sought, or that
makes the visual system more sensitive to differences in visual size, luminance contrast, or color differences, is likely to improve the performance of
a visual search task (Kokoschka and Bodmann, 1986). However, the
specifics of such lighting depend critically on the area to be searched, what

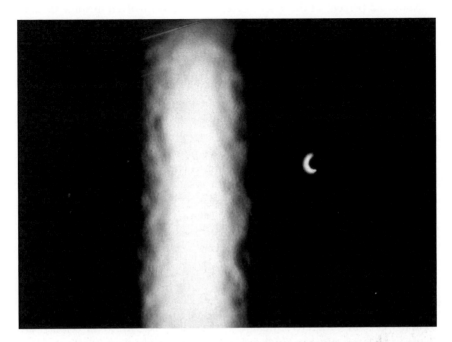

Figure 8.14 A close-up photograph of the image of a fluorescent lamp seen by reflection in a painted automobile surface. The large bright stripe is the image of the lamp. The small bright crescent to one side of the image of the lamp is a local highlight produced by light from the lamp striking a small bump in the surface caused by the dirt in the paint.

else that area contains, and the luminous and color characteristics of the items in the area, including the defect. Finally, it is important to appreciate that the complete visual inspection task involves a lot more than the just the ability to see the defect. Inspection is often done at a set speed, which limits the time available for searching, and once the defect has been detected there comes the decision as to what to do about it, a decision that is influenced by social, organizational, and psychological factors. Lighting has a part to play in visual inspection but it is a limited part. Other factors such as the time allowed to inspect the item and the manner of presentation are also important. Megaw (1979) gives an interesting review of these factors.

8.6   Special situations

There are three features of industrial lighting applications that deserve special consideration from the lighting designer. They are the widespread practice of night-shift work; the increasing number of self-luminous displays used in industry; and the use of exterior storage for valuable material.

About 22 percent of the work force in the US is involved in some form of shift work (Mellor, 1986). The amount and quality of work done during the night shift is usually worse than is done during a normal working day (Monk and Folkard, 1985), presumably because the workers are having to be attentive and alert while their physiology is telling them to go to sleep. Further, prolonged shift work can have adverse consequences for the workers' health (Rutenfranz *et al.*, 1985) and may disturb their social life (Walker, 1985). Lighting cannot solve all these problems but it can, at least in principle, do something about the major complaint of people doing night-shift work, the accumulating feeling of tiredness caused by poor quality sleep during the day. The possibility of speeding up adaptation to working at night by using light to shift the phase of the human circadian system is discussed in Section 3.7.

Another feature of industrial work that is growing rapidly is the use of self-luminous displays as parts of control systems for machines. The impact of lighting conditions on the visibility of such displays deserves careful consideration. Inappropriate lighting can reduce the luminance contrast of all parts of the display and/or produce discomfort by providing high-luminance reflections on the display that are distracting. The lighting conditions necessary to avoid these problems are fully discussed in Section 7.4.2.3 in the context in offices. The principles discussed there also apply to industrial situations, although it is important to remember that the lines of sight to the displays are likely to be much more variable in industrial situations than in offices.

One application that combines both night-shift work and a large number of displays is the control room. The consequences of errors in control rooms can be large, in both societal and financial terms, e.g. the Chernobyl disaster. Therefore, there is every reason to attempt to make the control room a place where the lighting helps rather than hinders the collection of visual information, improves the alertness of workers during the night shift, and minimizes stress and fatigue. What needs to be done to ensure good visibility for sources of information is well understood, although not always implemented. The possibility of using lighting to increase alertness during the night shift by the acute suppression of the hormone melatonin is discussed in Section 3.7. As for minimizing stress and fatigue, Sato *et al.* (1989) describe a study of the perceptions of the visual environment in a control room. The visual environment was varied from a standard windowless control room design by changing the lighting system, the illuminance, the presence of a window, the ceiling height, the colour of the floor, the size of the room, the color of the control panels, and the presence or absence of decorative, such as potted plants. The observers' appraisal of the visual environment was made on two dimensions, spaciousness and friendliness. Table 8.1 shows the effect of changing these various features on the perceptions of spaciousness and friendliness. Clearly, many different aspects of the visual environment, including the type of lighting, can change the perception of spaciousness and friendliness, for better or worse. Too often, lighting designs for control rooms are dominated by need for information

Table 8.1 The impact of various modifications of the visual environment of a control room on the perceptions of spaciousness and friendliness

New feature (standard)	Change	Spaciousness	Friendliness
Lighting system (recessed louvers with 80% reflectance)	Louvers (40%)	−	−
	Louvers (100%)	−	0
	Luminous ceiling	+	−
	Luminous ceiling with recess	0	+
Illuminance (1,000 lx)	2,000 lx	+	+
	Non-uniform	−	0
Window (windowless)	Inside window	+	0
	outside window	+	+
Ceiling height (2.8 m)	3.5 m	+	+
	4.2 m	+	+
Floor color (gloss N8)	Matte N8	−	0
	Matte N6	−	0
	Beige	0	+
Panel color (green)	Ivory	+	+
Decorative items (none)	Potted plants	0	+ +
	Accessory colors	0	+

Source: After Sato *et al.*, 1989.

Notes
+ + = Much improved.
+   = Improved.
0   = Unchanged.
−   = Worsened.

to be easily seen. While this is undoubtedly important; alone, it is not enough. The lighting design for a control room also needs to consider the possible effect of the lighting on the circadian system and the effects of the visual environment on perception and hence on mood and behavior if it is to be successful.

Yet another situation that is common in industry and that deserves special consideration is the use of exterior storage for valuable materials (Lyons, 1980). Sometimes such areas are worked at night, e.g. a container terminal. In this situation, the lighting needed to ensure safe movement of workers is usually sufficient to make security dependent on other aspects than lighting. In other applications, the site is empty at night, e.g. a storage yard for new automobiles. What type of lighting is needed to assist with security in this situation will depend on the type of security system in use, but usually a fence is involved. The factors that should be considered when lighting fenced areas are discussed in Section 11.5.1.

8.7 Summary

The visual requirements for industrial work can vary greatly. Some industrial work requires the extraction of a lot of visual information, typically involving the detection and identification of fine detail and fine differences in color. Other types of industrial work require different forms of visual information, for example, shape and texture rather than detail and color. Yet other types of industrial work can be done with very little visual information at all. The materials from which visual information has to be extracted can be matte or specular in reflection, the information can occur on many different planes, and the material containing the information can be moving or stationary. Further, the nature of the process may impose constraints on the type of lighting that can be used, e.g. where obstruction is extensive and where the atmosphere is hazardous, corrosive, or just plain dirty. This variability means that good industrial lighting is inevitably tailored to the application.

Despite this variability, the objectives of industrial are the same everywhere. They are to facilitate quick and accurate work; to contribute to the safety of those doing the work and to create a comfortable visual environment. The principles of lighting for quick and accurate work are discussed in Chapter 4. Applying these principles to industry requires an understanding of the information that needs to be obtained to do the work, where it is likely to be found and the constraints imposed by the application. Once this information is collected, the necessary amount, distribution, and spectrum of light delivered can be determined.

Minimum illuminances are recommended for safe movement but illuminance alone is not enough. Care needs to be taken to avoid disability glare and strong shadows. Light sources should also be chosen to make the correct naming of safety colors easy. Where rotating or reciprocating machinery is in use, care should be taken to minimize any stroboscopic effect.

As for comfort, the aspects of lighting that can cause discomfort are discussed in Chapter 5. In principle, the same comfort conditions should be applied to lighting installations wherever they are used. Unfortunately, this is sometimes not the case in industry. Many aspects of the physical environment are less comfortable in industry than in offices and lighting is often one of them.

Many industrial lighting installations are designed around a general/localized/task lighting approach, the localized lighting being used where activity is intense, e.g. on an assembly line, and the task lighting being used where tasks are either critical or more difficult than usual. One form of task lighting that requires special care is lighting for visual inspection. Rapid visual inspection calls for off-axis detection of defects. How well this can be done will depend on the visibility of the defect and, if there are other objects in the area to be searched, the conspicuity of the defect. There are many different methods of lighting for visual inspection. All depend on the use of lighting to make the defect more visible and more conspicuous.

9 Escape lighting

9.1 Introduction

Most countries have legal requirements that make it obligatory to provide an adequate means of escape in buildings where work is done and/or to which the public routinely have access (NFPA, 1997). Emergency lighting is an essential part of an adequate means of escape. Emergency lighting can have three roles; escape, shutdown, and continued operation. Escape lighting is lighting that is designed to ensure either the safe and rapid evacuation of a building or the ability to move to a place of refuge. To achieve this, escape lighting is designed to define the escape routes so that the occupants know which way to go, and to illuminate the escape routes so that the occupants can move along them quickly and safely. Escape lighting is not designed to enhance the ability of either the occupants or the rescue services to deal with the emergency, other than to illuminate the positions of alarm points and fire-fighting equipment. Shutdown lighting is emergency lighting designed to enable the people involved in a potentially dangerous process or situation to carry out an appropriate shutdown procedure. CEN (1999) recommends that shutdown lighting should provide an illuminance of not less than 10 percent of the normal lighting and never less than 15 lx. Standby lighting is used in parts of a building or site where, even in an emergency, activities should continue substantially unchanged, such as an operating theatre at a hospital. Standby lighting is usually powered from an emergency generator and should provide an illuminance similar to that provided under normal operating conditions. This chapter is concerned with escape lighting.

9.2 Escape lighting in context

Escape lighting is part of an emergency escape system. A well-researched emergency is the occurrence of fire. Research on the behavior of people in fires (Canter, 1980; Anon, 1998; Proulx, 1999) has revealed a consistent pattern of response. The first stage is recognition, the second, action, and the third, escape. Recognition is usually associated with a high level of ambiguity. Investigations of fires occurring in prisons, nursing homes, hospitals, and hotels have all shown that recognition of the existence of a severe fire is often dangerously delayed. Once the occurrence of a fire is recognized, there are a number of actions open to the people in the building: contacting others, fighting the fire, seeking refuge, or leaving the building. A person tends to choose a course of action depending on the role the individual plays in an organization. For example, Best (1977) reported that in the Beverly Hills Supper Club fire, in which 164 people died, the waitresses showed people out through the smoke but only people who were seated at the tables for which they were normally responsible. Another factor that influences the choice of action is the presence of other members of the social group. In the Summerland Leisure Complex fire, in which 50 people died, there is evidence that parents looking for children were more likely to escape because the search took them away from the fire (Sime and Kimura, 1988). The third stage comes down to a choice between attempting to escape, seeking out a refuge, or staying and protecting oneself in place until rescued. If the decision is to try to escape or to seek a refuge, it is necessary to identify the escape route. Escape lighting is important for defining and revealing the escape route, but this alone is not enough to ensure escape. The maximum volume of traffic that an escape route can handle also needs to be considered. The prediction of evacuation times and how they vary with different features of an escape route has been the subject of much argument and study (Pauls, 1980, 1988; Hadjisophocleous *et al.*, 1997).

To summarize, three types of information are needed for building occupants to escape in an emergency. They are:

- information on the presence of a hazard, including its nature and location;
- information on the recommended course of action; and
- information on how to carry out the recommended course of action.

9.2.1 *Information on the presence of a hazard*

Ideally, information on the presence of a hazard should be both immediate and complete. In some situations, the necessary information is provided by the hazard itself. A domestic fat fire is usually obvious to the person doing the cooking. Where it is not obvious is when the hazard occurs remotely from many of the occupants. A fire beginning on the tenth floor of an office building may not be apparent to the occupants of the twentieth floor for some time. It may be argued that the sounding of an audible fire alarm in the building and/or a sudden power failure should be interpreted as a reason for leaving the building. However, Tong and Canter (1985) and Geyer *et al.* (1988) have shown that only 10–20 percent of people interpret the sounding of an audible fire alarm as a reason for immediately leaving the building. A plausible reason for this lack of response is the ambiguity of the message (Proulx, 2000a). Possible interpretations of the sounding of the fire alarm, without any other signs of fire, are that there is a false alarm, an unscheduled fire drill, a test of the fire alarm system, a small fire which can be easily controlled, or a fire a long way away representing no hazard. Unless there is other evidence of the seriousness of the hazard, such as smoke, most people's response is either to investigate further or to carry on as normal until further information is available.

If this interpretation of the lack of response to a fire alarm is correct, providing more information would probably increase the number of people responding. Geyer *et al.* (1988) examined people's interpretation of a number of different modes of presenting messages concerning an outbreak of fire. The modes examined were three- and two-dimensional graphic displays, a text display, a speech warning, and a conventional fire bell. The three- and two-dimensional displays indicated the location of the fire and the location of exits. The text displays and the audible message gave the location of the fire and the instruction to leave immediately. The subjects took longer to acquire the message from the three- and two-dimensional graphic displays than from the other modes, but there was no difference in the total response time. Thus, the time taken to receive the message and decide on an appropriate action was the same for all modes of presenting the message. What was different was the percentage of subjects interpreting the display as a genuine fire warning and the percentage choosing to leave the building (see Figure 9.1). This study suggests that giving more information is

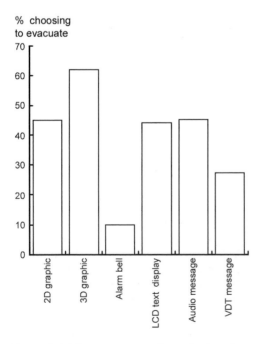

Figure 9.1 The percentage of people choosing to evacuate for different means of presenting information about a fire (after Geyer *et al.*, 1988).

likely to increase the frequency of the desired response. Given the importance of time to the probability of success in escaping from a fire or other
hazard, this is a significant finding.

Given that more extensive information on a fire is desirable, the next
question to consider is who should have that information. An approach
sometimes recommended is to limit the information given to the occupants
of the space in order to avoid panic and to allow an organized evacuation.
Sime (1980) suggests that panic is rather a rare event in fires. Best (1977),
in his report on the Beverly Hills Supper Club fire, says that prior to the
entry of thick smoke into the room the evacuation was orderly. It is also
worth noting that flight behavior, which retrospectively tends to be called
panic, may be forced on people who have received only limited information
on the location and extent of a fire. Sime (1980) argues that panic is used
as a portmanteau explanation of any ineffective behavior in fire. As such, it
is not a useful concept. Rather, behavior in fires can be interpreted in terms
of the individual's knowledge of the options available at different stages of a
fire (Canter *et al.*, 1980). Certainly, there is little support for limiting the
information presented to the occupants of a building in order to avoid panic.

These findings as to how people react to fire alarms leads to the conclusion that the more information presented to building occupants about a fire
and the faster it is presented, the more likely they are to choose a rational
action. Present day communication technology makes it relatively easy to
provide much more information about a hazard than is available through
the sounding of a fire bell.

9.2.2   Information on the recommended course of action

Once the reality of the situation is recognized, the building occupant has to
decide on an appropriate course of action. The decision made is likely to be
determined by a number of different factors, such as

- the occupant's knowledge of the situation;
- the occupant's concerns; and
- the occupant's place in the social organization in the building.

The occupant's knowledge of the hazard includes such information as
where it is, whether it is likely to spread, and how much time is available.
The occupant's concerns may vary from the need to flee from immediate
danger to tackling the hazard, warning other people, summoning help,
securing belongings, and ensuring the safety of close relatives. The occupant's place in the social organization will influence his/her concerns. Canter
et al. (1980) showed that in hospital fires, who investigates and who assists
is determined by the organizational hierarchy. Wood (1980) showed that the
roles of husband and wife tended to be maintained in domestic fires. Both
Bryan (1977) and Wood (1980) collected data on behavior from a large

number of people involved in fires. Their results show a wide diversity of behavior, as might be expected from the list of influences given above.

These observations again suggest that it would be useful to provide more information so that occupants could choose the appropriate behavior. At present, the most widely used approach is to teach only one behavior, evacuation. Unfortunately, the rigor with which this is carried out and the advice given varies from place to place. Large facilities with internal communication systems have used auditory messages to guide people to an appropriate action (Proulx and Koroluk, 1997; Proulx, 1998). Such systems can be automated or manual. Automated systems can provide guidance to building occupants but if they are to be helpful, they need to be sophisticated enough to match the guidance to the specific emergency situation. A manual system allows emergency personnel to talk to building occupants through a public address system. This certainly provides flexibility, but its value depends critically on the information available to the person controlling the communication network. If he/she is uninformed or uncertain what to do, then the danger presented by the hazard can be enhanced rather than diminished. Proulx (2000b) reviews a number of strategies whereby occupants of a large building can be guided to an appropriate response to a fire alarm signal.

9.2.3   Information on carrying out the recommended course of action

Given that the recommended course of action is to evacuate the building or to move to a specified refuge, there are two pieces of information required by the occupant; which way to go and how to move safely over the chosen route. This is where escape lighting has a role to play. The exit sign is the primary means of providing information on which way to go and the escape route lighting is important in allowing people to see well enough so that they can move safely over the chosen route. The loss of life that can occur when insufficient or misleading information is provided by the escape lighting is revealed in Willey (1971), Lathrop (1975), Bell (1979), and Anon (1983). Of course, the presence of an exit sign simply conveys the information that there is a route out of the building. It does not indicate whether or not that route is a safe one to follow. To know that requires that much more information about the location and nature of the hazard be given to the occupant, as discussed above.

9.3   Operating conditions

The conditions in which escape lighting has to operate can be considered on two dimensions, the availability of electrical power and the presence of a turbid medium, such as smoke. There are four possible combinations of these conditions.

- The power–no smoke condition corresponds to the conventional operating state of the building. In this condition, the normal electric lighting is available and visibility is unaffected by smoke, so there is no need for escape lighting to be operating.
- The no power–no smoke condition corresponds to a power failure or shut-down in which the building needs to be evacuated but there are no turbid media present. In this condition, the conventional electric lighting is not available but visibility is unaffected by the absorption and scattering of light. In this condition, the escape lighting should be operating.
- The power–smoke condition occurs when there is a fire but the electrical circuits are still operating. In this condition, the normal electric lighting is available, but visibility is limited by the scattering and absorption of light due to smoke. In this condition, the escape lighting should be operating.
- The no power–smoke condition corresponds to a fire in which the power supply is disabled or shut-down. In this condition, the conventional electric lighting is not available, and visibility is further restricted by the scattering and absorption of light due to smoke. In this condition, the escape lighting should be operating.

This brief consideration of the possible operating conditions shows that for three out of the four possible combinations, escape lighting is required.

9.4 Exit signs

Exit signs are designed and positioned to indicate which way to go to get out of the building. Detailed specifications exist for exit signs in many countries. In the United States, the word "EXIT" is commonly used. The specification determining what constitutes an exit sign covers the height and width of the letters forming the word EXIT, the spacing between the letters and the size, shape, and location of any directional indicators. Details of these quantities, together with the required photometric characteristics, are given in the *Life Safety Code* (NFPA, 1997). In the European Union, which currently has 15 members speaking almost as many languages, a pictogram, consisting of a stickperson running towards a solid rectangle signifying a door and an arrow indicating the direction of movement, is used. Details of the required form and size of the pictogram are given in the European Communities Council Directive 92/58/EEC. The photometric requirements are given in *European Standard 1838* (CEN, 1999).

But simply specifying the necessary physical and photometric characteristics for an exit sign is not enough. Schooley and Reagan (1980a,b) conclude that in order to assess the visibility of an exit sign it is also necessary to specify the maximum distance at which the sign is to be read. The *Life Safety Code* covers this requirement by stating that no point on an escape route should be more than 30 m (100 ft) from an exit sign. Collins (1991) examined the distances at which people could detect and correctly identify the words on an externally illuminated exit sign conforming to the dimensions of the *Life Safety Code* as part of a larger study of the visibility of directional indicators. The signs used were illuminated to 54 lx, the minimum specified for externally illuminated signs in the *Life Safety Code*. All 20 observers used were able to correctly identify the word "EXIT" at distances greater than 30 m (100 ft). These results suggest that the physical and photometric specifications for exit signs given in the *Life Safety Code* (NFPA, 1997) are consistent with the demand that no one should be more than 30 m (100 ft) from an exit sign when on an escape route.

The question that now needs to be considered is how commercially available exit signs perform. Collins *et al.* (1990) evaluated 13 different internally illuminated exit signs using various light sources. All the exit signs met the *Life Safety Code* specifications for physical dimensions and contrast, but they varied greatly in luminance. Even in a dark room, some of the signs viewed from a distance of 19 m (63 ft) were barely visible. The situation became worse when lighting was introduced into the room. When an illuminance of 54 lx fell on the face of these internally illuminated signs, as might occur when a luminaire forming part of the escape route lighting is operating close to the exit sign, the luminance contrast of some exit signs was reduced to below the *Life Safety Code* standard of 0.5. This suggests that should an emergency arise requiring evacuation some currently used exit signs would be of limited value.

This conclusion is supported by the measurements of readability of exit signs made by the Lighting Research Center (LRC, 1994, 1995). In these measurements, observers saw 60 different commercially available, internally illuminated exit signs. The exit signs used five different light sources; compact fluorescent lamps, incandescent lamps, an electroluminescent panel, LEDs, and radioluminescent tubes (see Section 1.7). The exit signs were in four different formats; panel, stencil, matrix, and edge-lit. In a panel exit sign, both the letter and the background are luminous. In a stencil exit sign, the letters are luminous and the background is opaque. In a matrix exit sign, the letters are formed by points of light and the background is opaque. In an edge-lit exit sign, light from an enclosed source is directed through a transparent plate that has the letters etched in or attached to its surface. The sign face that forms the background to the letters appears luminous although how luminous will depend on what lies behind the sign. The 60 exit signs used were presented at the end of a long corridor, one at a time, in a mirror box (Figure 9.2). By placing the exit sign in the mirror box at different positions, four different orientations of the word "EXIT" could be obtained. For each exit sign the observer started to walk towards the exit sign from a distance of 50 m (165 ft). The distance at which the observer could correctly identify the orientation of the exit sign was recorded. Figure 9.3 shows the percentage of observers who could correctly identify the orientation of the exit sign at 30 m (100 ft), plotted against the mean letter luminance, with the corridor lighting on and off. The corridor lighting provided an average illuminance of 340 lx on the floor of the

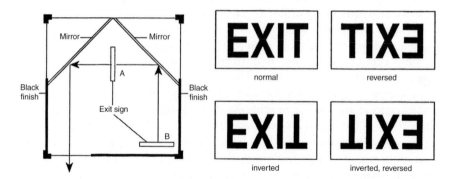

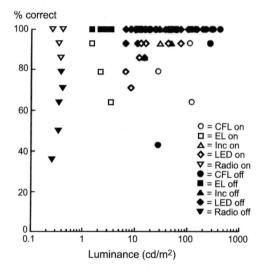

Figure 9.2 The mirror box used in the measurement of readability of exit signs. By placing the exit sign in the mirror box at positions A or B, the right way up or upside down, the word EXIT can be made to appear in the four different orientations shown (after LRC, 1994).

Figure 9.3 The percentage of observers able to detect the correct orientation of an exit sign at 30 m (100 ft) plotted against the average luminance of the legend of the exit sign. Five different light sources were used in the exit signs: CFL = compact fluorescent; EL = electroluminescent; Inc = incandescent; LED = light emitting diodes; Radio = radioluminescent. Measurements were made with the normal corridor lighting on (open symbols) and off (filled symbols) (after LRC, 1994, 1995).

corridor when on. It is clear from Figure 9.3 that a low letter luminance leads to difficulty in reading the exit sign at 30 m (100 ft). The letter luminance can be closely related to nature of the light source used to internally illuminate the exit sign. The radioluminescent and electroluminescent exit signs are difficult to read at 30 m (100 ft) because of their low letter luminance. Radioluminescent and electroluminescent exit signs are allowed under the *Life Safety Code* (NFPA, 1997) by an exception. Exit signs using these technologies are not required to have an equivalent visibility to an exit sign externally illuminated to 54 lx, but simply to have a minimum letter luminance of $0.21 \, \text{cd/m}^2$.

As for exit signs lit by the other light sources, in some form, all can be read at 30 m (100 ft) by all observers. However, at the same letter luminance, some exit signs fail to achieve readability by 100 percent of observers, while others using the same light source, succeed. This suggests that other factors besides mean letter luminance are important to readability. Inspection of the detailed measurements indicated that such aspects as the uniformity of the letter luminance and the luminance contrast of the letters against the background also influence the readability of the sign. There was no consistent difference in readability for different sign formats, i.e. for panel, stencil, matrix, or edge-lit formats.

These measurements serve to demonstrate an important point; that the readability of an exit sign is determined by the stimulus it presents to the visual system, not by the technology or format used to create that stimulus. From a number of studies similar to those described above, a specification for a visually effective exit sign has been developed (EPA, 2001). Table 9.1 sets out this specification, the dimensions being the same as in the *Life Safety Code* but the photometric requirements being more stringent. How the specified photometric conditions are achieved is irrelevant to the readability of the exit sign. If improvements in technology, such as have happened to light-emitting diodes and electroluminescent panels since this work was done, make it easy to meet the photometric requirements, then

Table 9.1 Dimensions and photometric specification for the US Environmental Protection Agency's Energy Star exit sign (from EPA, 2001)

Letter size and letter spacing: The sign shall have the word "EXIT" or other appropriate wording in plain legible letters not less than 6 in. (15.2 cm) high with the principal strokes of letters not less than 3/4 in. (1.9 cm) wide. The word "EXIT" shall have letters of a width not less than 2 in. (5 cm) except the letter "I", and the minimum spacing between letters shall not be less than 3/8 in. (1 cm). Signs larger than the minimum established in this paragraph shall have letter widths, strokes, and spacing in proportion to their height

Luminance contrast: Greater than 0.8

Average letter luminance: Greater than $15 \, \text{cd/m}^2$ measured at 0 and 45° viewing angles

Minimum letter luminance: Greater than $8.6 \, \text{cd/m}^2$ at 0 and 45° viewing angles

Maximum to minimum luminance within letters or background: Less than 20:1, measured at 0 and 45° viewing angles.

---

exit signs using these technologies should be more than adequate. In general, it is only when a specific technology or format consistently limits some critical aspect of the visual stimulus that the technology or design should be considered unsuitable for use in exit signs.

Given that the function of an exit sign is to tell the observer the way to go, then some exit signs also need to have one or two directional indicators. Collins (1991) examined the distance at which various forms of directional indicator could be correctly detected. She established that out of five different forms of directional indicator arrow, all of the same vertical size; the chevron was most often correctly identified at the greatest distance. In further studies, she showed that the factors influencing the distance at which chevron directional indicators could be correctly identified were the area and color of the chevron, the greater the area and the use of red or green rather than gray tended to increase the distance at which the direction indicated by the chevron could be correctly identified. Boyce and Mulder (1995), using the mirror box approach discussed above, confirmed the superiority of the chevron as a directional indicator, within the constraint that it had to fit within conventionally sized exit signs.

While readability is an important function of an exit sign, it is of little value if the sign cannot be found first among all the other information presented to the visual system. This is unlikely to be a problem in the no-power condition because then the exit sign will be one of the few signs still operating. However, when normal electrical power is available all the other signs and luminaires will be operating. Jin *et al.* (1987) examined the factors affecting exit sign conspicuity using a computer image of a scene. As might be expected, the location and the size of the sign were important determinants of conspicuity. What was also important was the presence or absence of other signs of the same color as the exit sign. Conspicuity was enhanced when none of the other signs were the same color as the exit sign. This supports the value of having an exit sign of a unique but well-known colour. Jin *et al.* (1985) also examined the effect on conspicuity of flashing an exit sign twice a second. People were asked to rate the conspicuity of the flashing sign relative to a non-flashing sign in a large shopping mall. The results obtained showed that flashing was effective in increasing conspicuity for signs which where visible but not conspicuous. Flashing made little difference to the conspicuity of large signs which were already conspicuous because of their size relative to other signs in the mall, nor to small signs which were too small to be recognized as exit signs from the viewing position. These data were obtained in a power – no smoke condition. It would be interesting to know how effective flashing would be when applied to exit signs in a no power – smoke condition, where the scattering of light in the smoke might cause confusion rather than provide information (Malven, 1986). The *Life Safety Code* (NFPA, 1997) encourages conspicuity of exit signs by insisting that "no decorations, furnishings, or equipment that impairs visibility of an exit sign shall be permitted, nor shall there be any brightly illuminated sign (for other than exit purposes), display or object in or near the line of vision to the required exit sign of such a character as to detract attention from the exit sign," and by allowing exit signs to flash at about one cycle per second upon activation of the fire alarm system.

This brief discussion of conspicuity suggests that a colored exit sign is more likely to be seen than a colorless exit sign, but what color is best? Different countries used different colors for exit signs. Some states in the United States demand red letters in an exit sign while others insist on green. The European Union pictogram has white elements on a green background. Enthusiasts for one color or another will produce rationalizations such as "red means stop and green means go" or "red means danger and green means safety." There is no evidence to support any of these rationalizations.

Rather, what seems to matter most is familiarity with the color of an exit sign. It is for this reason that the International Standardization Organization (ISO) has produced a list of recommended safety colors (ISO, 1984). The colors in this list are well separated in color space and hence are easily discriminated. The *European Standard EN 1838* (CEN, 1999) requires that exit signs conform with the color specified in the ISO standard.

9.5   Escape route lighting

9.5.1   Ceiling- and wall-mounted luminaires

Escape route lighting is designed to allow people to move over the escape route quickly and safely. Escape route lighting is usually provided either by defining some of the luminaires used in the normal lighting as emergency luminaries and arranging for them to continue to operate when the normal power supply fails; or by providing special luminaires that only operate when the normal power supply fails or a fire is detected. In both approaches, the luminaires are designed and spaced to provide a minimum illuminance along the escape route. Different countries have different criteria for escape route lighting. In the United States, the *Life Safety Code* specifies an initial average illuminance of 10 lx on the floor of corridors, passageways, stairways, stair landings, ramps, and escalators, with a minimum illuminance at any point of 1 lx (NFPA, 1997). In Europe, the horizontal illuminance on the floor along the center of an escape route has to be not less than 1 lx, although some countries are allowed lower illuminances (CEN, 1999). In the United Kingdom, the horizontal illuminance at floor level on the center line of a defined escape route should be not less than 0.2 lx and in Ireland it should not be less than 0.5 lx. Such differences raise the question of what is an appropriate illuminance for an escape route.

Ouellette and Rea (1989) examined a series of measurements of peoples' ability to move over escape routes lit to different illuminances by ceiling-mounted luminaires, with exit signs to indicate the way to go. Independent studies by four authors (Nikitin, 1973; Simmons, 1975; Jaschinski, 1982; Boyce, 1985) showed that a mean illuminance on the floor of the escape route of at least 0.5 lx was sufficient to ensure movement over the escape route without collisions with objects. The results of Simmons (1975), Jaschinski (1982), and Boyce (1985) can also be compared by using speed of movement over the escape route as a common measure. Figure 9.4 shows the relationship between the speed of movement and the mean illuminance on the floor of the escape route, for young and older subjects. Simmons (1975) had an escape route that was effectively a network of corridors with steps and occasional obstructions formed from large cardboard boxes. Jaschinski (1982) formed an escape route through several interconnected small rooms with steps between the rooms. Boyce (1985) used a large open-plan office where subjects had to find their way through the furniture to the door. Given that the three studies were carried out using different subjects moving over very different courses, the agreement shown in Figure 9.4 is remarkable. What is clear from Figure 9.4 is that there is an accelerating decline in movement speed as illuminance is reduced from the highest mean illuminance used which is typical of normal office lighting. At the initial mean illuminance specified by the *Life Safety Code* of 10 lx, there is

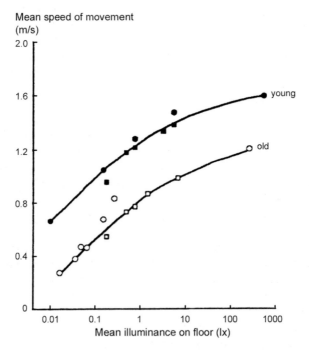

Figure 9.4 Mean speed of movement in cluttered or furnished spaces by young and old people, plotted against mean illuminance on the floor (after Ouellette and Rea, 1989).

about a 10 percent reduction in movement speed for younger people and a 18 percent reduction for elderly people. At the minimum illuminance specified by the European Standard (CEN, 1999) of 1 lx, the reduction in speed is about 25 percent for the young and 32 percent for the elderly. Jaschinski (1982) and Boyce (1985) also obtained subjective assessments of satisfaction with the lighting of the escape route. Jaschinski (1982) found his subjects were satisfied at a mean illuminance of 3 lx. Boyce (1985) found that a mean illuminance of 7 lx was sufficient to ensure a high level of satisfaction.

From these studies, it can be concluded that the initial mean illuminance recommended in the *Life Safety Code* (10 lx) is sufficient to ensure safe and speedy movement over the escape route in a clear atmosphere, provided a 10–20 percent reduction in movement speed from that achievable in normal room lighting is acceptable. While this is useful information it does not indicate the reason for the decline in movement speed. That there should be such a decline is obvious because in complete darkness the only way the occupant can move over the escape route is by touch and this is a slow and uncertain process. Figure 9.4 shows the compressive function expected for any relationship between an aspect of performance dependent on vision and illuminance, but what determines the illuminance at which the decline in movement speed begins to accelerate? One plausible answer to this question is the range of luminances over which neural adaptation can occur (see Section 2.3.1). Neural adaptation to a sudden change in adaptation luminance is very fast, of the order of fractions of a second. For larger changes of adaptation luminance, photochemical adaptation is necessary and this takes of the order of minutes, especially if vision is dependent on the operation of rod photoreceptors. Neural adaptation can cover a range of about two to three log units of luminance, which implies that for normal lighting providing an average illuminance of 400 lx on the floor, neural adaptation can take place down to an average illuminance of 4 to 0.4 lx. At lower illuminances there will be a marked decrease in movement speed because of the delay caused by the time taken by the visual system to adapt to the lower luminances. This explanation implies that in situations where the occupants can be expected to already be adapted to low luminances, such as in cinemas and theaters, lower illuminances would be acceptable for escape route lighting.

Despite the emphasis given to mean illuminance on the escape route, it should not be thought that simply providing the specified mean illuminance is sufficient to ensure adequate escape route lighting. Other factors to be considered are the illuminance uniformity, the possibility of disability glare, the color properties of the light sources used, the time duration for which the escape route lighting shall be provided and, where the escape route lighting is not permanently lit, the time delay between the failure of the normal power supply and the ignition of the emergency lighting. The European Standard (CEN, 1999) specifies that the ratio of maximum to minimum illuminance on the floor on the center line of the escape route should not be more than 40:1. Disability glare has the potential to make it difficult to see exit signs and any obstructions along the escape route. Disability glare is controlled by setting maximum luminous intensities from the escape route lighting luminaires within 60–90° from the downward vertical, at all angles of azimuth, for level escape routes, and at any angles for non-level escape routes (see Table 9.2). To ensure accurate identification of safety colors, the minimum CIE general CRI of any light source used for emergency lighting is 40. To allow adequate time for evacuation of a building, the European Standard recommends the escape route lighting should be above the minimum mean illuminance for at least 1 h. As for time delay, the European Standard recommends that the escape route lighting should reach 50 percent of the mean illuminance within 5 s and the full-required illuminance within 60 s of the failure of normal lighting. The *Life Safety Code* (NFPA, 1997) deals with these factors by requiring that the maximum to minimum illuminance ratio of the escape route lighting should not be more than 40:1; the escape route lighting should be provided for 90 min in the event of a failure of normal lighting, although the mean illuminance is allowed to decrease to 6 lx with a minimum at any point of 0.6 lx, by the end of the 90 min. The *Life Safety Code* neither mentions disability glare, nor the color properties of the light sources to be used. It does consider the delay time between failure of the normal power supply and the onset of the escape route lighting in general terms by saying that there should be no appreciable interruption of illumination during the changeover to another source of electricity. The only indications of what is appreciable is the requirement that where emergency lighting is provided by a generator, a delay of not more than 10 s is permitted.

The effect of different time delays on the ability to move over an escape route was studied by Boyce (1986) in an open-plan office. He found that given a low mean illuminance on the escape route of 0.16 lx, delaying the onset of the emergency lighting until 5 s after the normal lighting was extinguished ensured more rapid, steadier movement with fewer collisions once

Table 9.2 Maximum luminous intensity (cd) from emergency lighting luminaires (from CEN, 1999)

Mounting height, h, above floor (m)	Maximum luminous intensity for escape route lighting and open area (anti-panic) lighting (cd)
$h < 2.5$	500
$2.5 < h < 3.0$	900
$3.0 < h < 3.5$	1,600
$3.5 < h < 4.0$	2,500
$4.0 < h < 4.5$	3,500
$h > 4.5$	5,000

movement started than if the subject moved immediately following an instantaneous change from normal lighting to escape route lighting. However, the total time taken to leave the room was slightly longer than if the subjects moved immediately following the instantaneous changeover. The faster, steadier movement with the 5 s delay occurs because the delay allows some visual adaptation to occur. How much adaptation is needed will depend on the difference between the adaptation luminance provided by the normal lighting and that provided by the escape route lighting. The smaller is this difference the less adaptation is needed, so the less the benefit, in terms of smoother and faster movement, and the greater the penalty, in terms of longer escape times, of having a delay between the failure of the normal lighting and the onset of the escape lighting. While such understanding is interesting academically, given the usual behavior of people in fires is to hesitate before responding to an emergency signal, unless there is obvious cause for alarm, worrying about whether the delay time should be 5 s or less, is not the most important question to anyone concerned with improving fire safety in buildings.

Finally, it is necessary to consider how an occupant might get to an escape route. For small spaces, such as the rooms in a hotel where the corridor outside is the escape route, this is not a problem and no special consideration is required, but for large open areas where there are many different possible directions of movement the European Standard 1838 (CEN, 1999) calls for what it terms anti-panic lighting, in the form of a minimum illuminance at floor level of 0.5 lx, with the same illuminance uniformity, light source color properties, illuminance duration, and maximum time delay as for the escape route, although the luminous intensity limits are relaxed (see Table 9.2).

9.5.2 Path-marking

An alternative approach to the lighting of escape routes is that of path-marking. This approach, which is widely used in commercial aircraft and cruise ships, aims to mark the escape route at frequent intervals, typically separated by a few centimeters, from a low mounting position, and to rely on the light from the path-marking to illuminate the escape route.

One possible technology for path-marking is photoluminescent panels. These panels typically consist of a zinc sulfide phosphor with copper and cobalt activating agents. It is produced as a powder and applied to paint, tape, or plastic sheet. The phosphor is excited by a range of wavelengths determined by its chemistry. When there is no light incident on the panel, the photoluminescent panel continues to emit light, although the luminance of the panel declines over time (Garlick, 1949; Webber *et al.*, 1988). Fortunately, this decline in luminance under conditions of darkness is paralleled by an increase in sensitivity of the visual system. A path-marking system based on photoluminescent panels is attractive because it requires no power supply, only incident illumination prior to the emergency.

Webber *et al.* (1988) studied the ability of people to move along a corridor and down a flight of stairs using either conventional ceiling-mounted escape route lighting or path-marking using photoluminescent markers. They found that the mean movement speed of people under the photoluminescent path-marking was similar to that achieved at an illuminance of 0.2–0.3 lx from conventional escape route lighting on the stairway, but was comparable with only about 0.05–0.10 lx for the corridor (Figure 9.5). Subjective assessments of how difficult it was to see where to go showed that on the staircase the photoluminescent installation was considered less difficult than the conventional escape route lighting but in the corridor it was more difficult. The reason for this changeover in difficulty between the two escape lighting types was the density and placing of the photoluminescent material. This an important point. For any path-marking system to be effective, it has to mark the path completely and unambiguously. Thus, the placement of photoluminescent material calls for considerable care to ensure continuity of guidance. Given this, there is considerable potential for

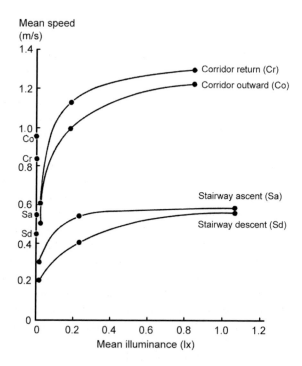

Figure 9.5 Mean speed of movement along a corridor and up and down a flight of stairs, under conventional ceiling-mounted escape route lighting and using photoluminescent path-marking alone, plotted against the mean illuminance on the floor. The mean speeds for the photoluminescent path marking are plotted at zero illuminance (after Webber *et al.*, 1988).

the use of photoluminescent materials in path-marking systems (Proulx *et al.*, 2000), particularly with some of the new photoluminescent materials becoming available, with their greater light output and longer decay times. But this is in the future. At the moment, concerns over the decay of light output over time and the dependence on illumination from normal lighting prior to that lighting being extinguished, have contributed to a reluctance by authorities to accept the use of photoluminescent materials as the sole basis of escape route lighting. While this is understandable, the low cost of the material and the ease with which it can be located without access to a power supply suggest it would be a useful supplement to conventional escape route lighting in many situations.

Other forms of path-marking use powered electroluminescent panels, miniature incandescent lamps and LEDs as light sources. Aizlewood and Webber (1995) measured people's ability to move over an escape route consisting of a corridor and a staircase, for three different path-marking installations and traditional escape route lighting from ceiling-mounted luminaires. The path-marking systems used photoluminescent panels, electroluminescent track, and miniature incandescent track, all mounted close to the skirting board in the corridor and parallel to the pitch line of the stairs on the staircase. Figures 9.6 and 9.7 show the measured mean speeds of movement in the corridor and on the stairs for the different lighting systems plotted against the mean illuminance on the floor or steps. It can be seen that there is little difference in mean speed between the conventional ceiling-mounted escape lighting and the path-marking systems at mean illuminances of 1 lx and more but as the mean illuminance was decreased below about 0.2 lx, the path-marking systems allowed speed to be maintained at a higher level than did the ceiling-mounted escape route lighting. As for the differences between the path-marking systems, the only consistent difference is the slightly lower speeds achieved under the unpowered photoluminescent path marking relative to the powered incandescent and electroluminescent systems.

The subjects also walked the escape route with four obstacles placed on the path, the four obstacles being a life-sized dummy, a wooden stool, a wastepaper bin, and a buff-colored folder. Video recordings of the subjects' movements over the escape route showed that under the powered electroluminescent and incandescent path-marking systems and the traditional ceiling-mounted escape route lighting system, many subjects saw the obstacles from a distance and deliberately avoided them, but under the photoluminescent system, many subjects did not see the obstacles until they were very close to them or actually touched them. For the dummy, stool, and bin, the percentage of subjects detecting these obstacles was 88, 73, 69, and 46 percent for the miniature incandescent path-marking system, the ceiling-mounted escape route lighting, and the electroluminescent and photoluminescent path-marking systems, respectively. From these results, it was concluded that low-mounted path-marking systems could ensure a speed of movement as good as or better than traditional ceiling-mounted escape

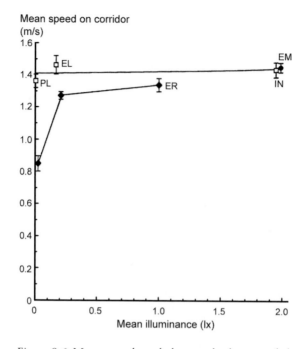

Figure 9.6 Mean speed, and the standard error of the mean, for movement along a corridor lit using electroluminescent (EL), incandescent (IN), or photoluminescent (PL) low-mounted path-marking systems or conventional ceiling-mounted escape route lighting (EM), plotted against the mean illuminance on the floor. Also shown for comparison purposes are the mean speeds over the same corridor lit by ceiling-mounted escape route (ER) luminaires to lower illuminances. These mean speeds were measured for a different group of subjects in an earlier experiment (Webber *et al.*, 1988) (after Aizlewood and Webber, 1995).

lighting. However, the difficulty in detecting the obstacles under the photoluminescent path-marking system suggests that simply marking the path is not enough. The escape route also needs to be illuminated. Aizlewood and Webber (1995) suggest a minimum illuminance on the floor of an escape route of 0.1 lx. This can be provided by a path-marking system if it has sufficient light output. Guidance on both electrically powered and non-electrically powered path-marking systems is available (Webber and Hallman, 1989; Webber and Aizlewood, 1993a; PSPA, 1997; UL, 1997; British Standards Institution, 1998, 1999). One application where powered path-marking systems have been required is in passenger ships (IMO, 1993).

9.6   Special situations

So far, this consideration of escape lighting has ignored a number of situations that may make current escape lighting systems inadequate. These

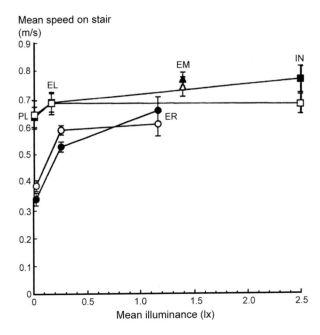

Figure 9.7 Mean speed, and the standard error of the mean, for movement up
and down a flight of stairs lit using electroluminescent (EL), incandescent (IN), or photoluminescent (PL) low-mounted path-marking
systems or conventional ceiling-mounted escape route lighting
(EM). Also shown for comparison purposes are the mean speeds
over the same flight of stairs lit by ceiling-mounted escape route
(ER) luminaires providing lower illuminances. These mean speeds
were measured for a different group of subjects in an earlier experiment (Webber *et al.*, 1988) (after Aizlewood and Webber, 1995).

situations cover both the physical and the physiological. The main physical
situation that is often ignored is the presence of smoke, a situation where
escape is often a necessity. The physiological situations include the presence
of people with defective color vision, very limited visual capabilities, and
limited mobility. Each of these situations will be considered in turn.

9.6.1 Smoke

Watanabe *et al.* (1973) studied the movement of firemen through smoke.
Not surprisingly, movement speed decreased as the density of the smoke
increased until speed became constant at a level equal to that measured in
complete darkness. This result implies that, even in smoke, providing some
information about the way to go and how to get there will facilitate faster
evacuation.

Physically, smoke consists of small particles suspended in air. Light
incident on these particles is both scattered and absorbed. The simplest

approach to quantifying the effect of smoke on light is to ignore the distinction between scatter and absorption and treat their combined effect on light loss as absorption alone. In mathematical terms, this approach is expressed in Lambert's law which states that the luminous intensity, I, of light propagating a distance d, through a uniform medium is given by

$$I = I_0 \, e^{-Ad} \tag{1}$$

where I_0 is the unattenuated luminous intensity (cd) at distance d (m) equal to zero and A is the absorption coefficient. The effect of applying Lambert's law to an exit sign is to reduce the luminance of all parts of the sign by the same proportion, without any blurring.

Although Lambert's law is simple to apply, it does oversimplify the effects of smoke. Scatter is not the same as absorption. Light that is absorbed when it interacts with a particle is eliminated. Light that is scattered when it interacts with a particle is simply moved to another location. Scattered light can be divided into two types: large-angle scatter and small-angle forward scatter. Large-angle scattering can remove light from the field of view, resulting in an overall reduction in luminance. However, in the case of high particle density smoke, multiple, large-angle scattering can cause some of the scattered light to reach the eye resulting in a luminous veil over the entire retinal image. When the object being viewed is the only source of illumination, this luminous veil caused by large-angle scatter is slight because the amount of light reaching the eye is a small fraction of the light that has been scattered equally over all angles. When there are other light sources present, the luminous veil caused by large-angle scattering can be large if the light output of the other sources is large compared with the light output of the object being viewed. This is why the presence of escape route lighting can reduce the visibility of exit signs in smoke.

When the scattering angle is very small, the scattering is described as small-angle forward scatter. Small-angle forward scatter changes the path of the light slightly but the light usually still reaches the eye. Such forward-scattered light contributes little to light loss but rather degrades the retinal image quality by smoothing out the luminance distribution of the retinal image.

Light loss and retinal image quality are both important for exit sign visibility, because light loss reduces average sign luminance and image quality affects the luminance contrast. Schooley and Reagan (1980a) examined the effect of smoke on the distance at which the two types of exit sign were visible. The two exit signs were an internally illuminated sign and a self-luminous sign. Both signs conformed to the then requirements of the *Life Safety Code*. In an unlit room, the distance at which the sign was visible increased as the luminance of the sign increased. This is consistent with the results of Rea et al. (1985b) as well as Collins et al. (1990). Rea et al. (1985b) had subjects view thirteen different exit signs through a smoke chamber.

The density of the white smoke used was increased in the chamber until the sign reached two threshold criteria, readability (can just read the sign), and detectability (can just detect the presence of something). The 13 signs were representative of exit signs used in Canada and consisted of four internally illuminated signs equipped with either incandescent or fluorescent lamps, three externally illuminated exit signs lit by incandescent lamps, and two self-luminous signs. Figure 9.8 shows the critical smoke densities, i.e. the smoke density required to bring each sign to the threshold criterion, for each of the 13 signs, plotted against the general luminance of the sign. The smoke density is defined as the optical density of the smoke per meter of path length. Optical density is defined as the logarithm of the reciprocal of the transmittance of the smoke. The general luminance of the sign is the average luminance of a circular area enclosing most of the letters and their immediate background. From Figure 9.8, it can be seen that the higher is the sign's general luminance, the greater is the smoke density through which the sign can be detected and read.

It is evident from the variation of critical smoke density for different signs in Figure 9.8 that general luminance is not the only important factor. Among other factors likely to be important are the color format of the sign, the polarity of the contrast, and the uniformity of the luminance of the letters. Among the signs used by Rea *et al.* (1985b) were those with red letters on

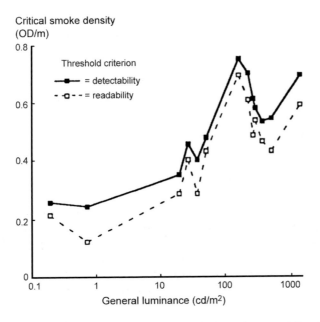

Figure 9.8 Smoke density required to make 13 different exit signs just detectable and just readable, plotted against general luminance of the signs (after Rea *et al.*, 1985b).

a white background, red letters on a black background, green letters on a white background, green letters on a black background, green letters on a red background, black letters on a green background, and white letters on a red background. There was no consistent effect of color format on the critical smoke density. This diversity of color combinations, which can be found in the United States as well as Canada, reflects the lack of information on the effects of exit sign color on sign visibility (but see Section 9.6.2).

Ouellette (1988) examined the visibility of signs with different polarity of contrast, i.e. bright letters on a dark background or dark letters on a bright background. He found that there were small but statistically significant differences in visibility due to contrast polarity. On average, signs with high luminance backgrounds needed a higher luminance to be seen through the same density of white smoke compared to those with a low luminance backgrounds. This effect can be explained by the scattering of the light emitted by the sign in the smoke. Light scattered from a high luminance background will tend to mask the lower luminance letters on the sign making the sign less visible. Collins *et al.* (1990) reached a similar conclusion using black smoke. Specifically, stencil signs (transilluminated letters and opaque background) were considered more visible than panel signs (transilluminated letters and background).

As for the effect of uniformity of luminance on visibility, so far this has not been methodically studied. It can be speculated that a very non-uniform luminance sign will be less readable, particularly in smoke, since light scattered from the areas of higher luminance would likely mask the areas of lower luminance and hence fragment the display. This is an aspect of sign design deserving of investigation.

Rea *et al.* (1985b) also examined the effects of having ambient illumination present in smoke, such as would occur when the normal lighting is on. The ambient illumination on a horizontal plane through the smoke chamber ranged from 170 to 1,200 lx. The illuminance falling on the face of the exit signs was 75 lx. Figure 9.9 shows the critical smoke density for readability for the 13 exit signs plotted against the sign general luminance, with the ambient lighting on and off. It is obvious that having the ambient lighting on makes all the signs less visible, although the reduction is of different magnitude for different signs.

Taken together, these studies provide a qualitative specification for conventional exit signs which will ensure they are effective in smoke. Ideally, the sign most easily read through smoke will be large in size, will be of the stencil type and will have a high letter luminance. However, there is one important question which needs to be considered: what is the smoke density that people can survive for a brief time? There is little point in making an exit sign visible through smoke that is so dense that anyone present is already dead. Unfortunately, there is no simple answer to this problem because it depends on the associated temperature and the constituents of the smoke. Death in fires can occur in three ways: by thermal collapse, by

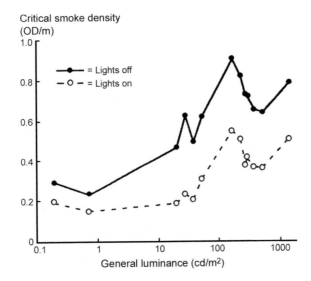

Figure 9.9 Smoke density required to make 13 different exit signs just readable, with the ambient lighting on and off, plotted against the general luminance of the signs. When the ambient lighting was on it
provided an illuminance on the face of the sign of 75 lx in the
absence of smoke (after Rea *et al.*, 1985b).

the inhalation of toxic gases, or by the inhalation of irritant gases. Thus, an
optical density which may be survivable in one type of smoke may not be
survivable in another. Newman and Kahn (1984) suggest that the critical
smoke density for short-term exposure is 0.22/m, while Gross (1986) and
Chittum and Rasmussen (1989) use a "just survivable" smoke density of
1.64/m.

It is interesting to consider the implications of these survivable smoke
densities. All the signs examined by Collins *et al.* (1990) at a distance of
19 m (62 ft) had disappeared in black smoke at smoke densities less than
0.17/m. The equation describing the relationship between smoke density,
viewing distance, and sign luminous intensity is

$$\log(I_0/I_s) = \mathrm{SD} \cdot d$$

where I_0 is the luminous intensity of the sign in a clear atmosphere (cd), I_s
the luminous intensity of the sign in smoke (cd), SD the smoke density equal
to the optical density per unit path length (m^{-1}), and d the path length (m).

Using this equation, the data from Collins *et al.* (1990) imply that the
value of the logarithm corresponding to the disappearance of all the signs
in black smoke is 3.23. For this ratio and a survivable smoke density of
0.22/m, the distance at which all the signs examined by Collins *et al.* (1990)

would have disappeared is 14.7 m (50 ft). If the survivable optical density is 1.64/m, the distance at which all the signs would have disappeared is 1.97 m (6.5 ft). Despite the uncertainty associated with survivable smoke densities, there can be little doubt that few commercially available exit signs will be visible at 30 m (100 ft) through survivable smoke densities, even in the absence of escape route lighting.

This raises the question of how the visibility of exit signs in smoke might be improved. The results discussed above indicate that the key to improving visibility in smoke is to increase the luminance of the sign. If the effect of smoke on light was limited to absorption, then increasing the luminance would be all that was required. However, smoke scatters as well as absorbs light, and scattered light will tend to mask the message carried by the sign. Therefore, the highest luminance of the sign should be generated by the part of the sign that carries the message. Gross (1986) describes an exit sign that uses LEDs to form a matrix spelling out the letters of the word "EXIT." Gross (1988) claims a marked increase in the distance at which such signs can be read in dense smoke compared to conventional, internally illuminated exit signs.

Rather than attempt to increase the luminance of exit signs sufficiently for them to be seen through a survivable smoke density at 30 m (100 ft), an alternative approach would be to use a low-mounted, path-marking system (see Section 9.5.2). Such systems have two advantages. The first is that any path-marking system provides information at much more frequent intervals than does the conventional exit sign/ceiling-mounted escape route lighting so the need to see information far away through smoke is eliminated. The second is the low level mounting position. This itself has two advantages. First, it places the light sources closer to the surface of the escape route. The second advantage is that the distribution of smoke is not always uniform. Hot smoke tends to accumulate at the ceiling, initially, and then gradually extend in layers down to the floor. This stratified structure will be evident until the smoke temperature falls or until sprinklers start operating, in which cases smoke will rapidly become evenly distributed throughout the space. A stratified smoke structure means that smoke close to the origin of the fire will be thinnest at floor level so the absorption and scattering of light originating close to the floor should be less. Chesterfield *et al.* (1981) compared the effectiveness of ceiling-mounted lighting and lighting mounted in armrests for the evacuation of an airliner in smoke. The results obtained showed that the lower mounted lighting allowed a 18 percent improvement in evacuation time compared with the ceiling-mounted lighting. Paulsen (1994) examined the time taken for people to cover a route simulating the interior of a ship and that involved moving along a corridor, up a flight of stairs, and along another corridor to a door giving access to an open deck. The whole interior was filled with white smoke. The escape lighting which achieved 100 percent correct evacuation in the shortest time (68 s) was a continuous, incandescent, low-mounted path-marking system with a mean luminance of 5.5 cd/m^2. An escape lighting system based on six exit signs positioned at head height and indicating changes of direction, produced a longer evacuation time, and allowed only two-thirds of the subjects to successfully find their way to the open deck.

Webber and Aizlewood (1993b) had people discover how far they had to be away before they could see the door at the end of a smoke-filled corridor, for five different exit signs, three different path-marking systems, and traditional ceiling-mounted escape route lighting. Also, they asked the subjects if they would be willing to move along the corridor in smoke with the lighting they had just seen and to rate how satisfactory the lighting was for an emergency smoke condition. The smoke density varied along the length of the corridor, being densest closest to the door, with an average value of about 0.4/m. The distribution of smoke was similar for all the lighting conditions. Figure 9.10 shows the distance from the door at which the door was first detected and then when it was confidently recognized. As would be expected from the discussion above, the distances increased approximately logarithmically with the luminance of the sign. The distances for the path-marking systems are intermediate in the range of exit signs. Figure 9.11 shows the percentage of people willing to start to move along the corridor. Clearly, a much

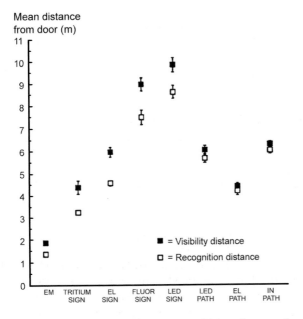

Figure 9.10 The mean distance at which a door at the end of a smoke-filled corridor could first be detected through a smoke density of 0.4/m. The lighting examined was radioluminescent (Tritium), electroluminescent (EL), fluorescent (Fluor), or LED exit signs alone; ceiling-mounted escape route lighting luminaires (EM); or incandescent (IN), electroluminescent (EL), or LED path-marking systems (after Webber and Aizlewood, 1993b).

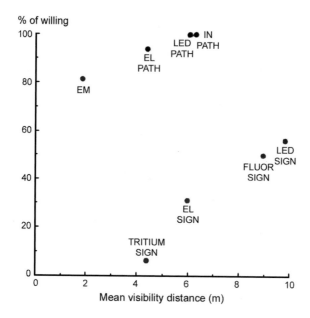

Figure 9.11 Percentage of subjects willing to start moving down a smoke-filled
corridor (smoke density = 0.4/m) when the corridor and the door at
the end were lit by radioluminescent (Tritium), electroluminescent
(EL), fluorescent (Fluor), or LED exit signs alone; ceiling-mounted
escape route lighting luminaires (EM); or incandescent (IN), electroluminescent (EL), or LED path-marking systems; plotted against
the mean distance at which the exit sign or the door marking could
first be detected (after Webber and Aizlewood, 1993b).

higher percentage of people would be willing to start to move along the corridor when the corridor was lit in some way, either by conventional,
ceiling-mounted escape route lighting or by path-marking systems, than
would when an exit sign alone was used. Figure 9.12 shows the mean rating
of how satisfactory for a smoke emergency condition the various systems
were. There can be little doubt that in smoke, the path-marking systems are
considered more satisfactory than the ceiling-mounted escape route lighting
and that in turn is considered more satisfactory than the exit signs alone.

Webber and Aizlewood (1994) offer another approach to assessing the
visibility of various means of providing emergency egress information in
smoke. For one observer, they measured the distance at which different
components in an escape lighting system, such as exit signs, marked door
frames, and path-marking, could just be seen along a corridor filled with
white smoke of various densities. They found that over the range of distances examined, the product of the smoke density and the viewing distance
of the observer, i.e. the optical density of the smoke, was a constant,

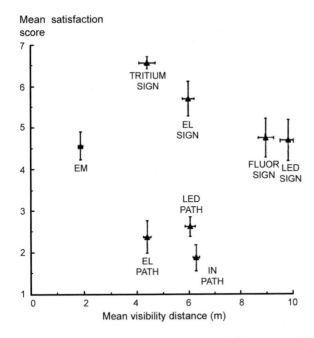

Figure 9.12 Mean satisfaction ratings, and the associated standard errors, for the lighting of a corridor and door in an emergency smoke condition, plotted against the mean distance at which the signs or marking identifying the door at the end of the corridor could be recognized through a smoke density of 0.4/m. The satisfaction ratings were given on a seven point scale with 1 = very satisfactory, 7 = very unsatisfactory. The escape route lighting consisted of radioluminescent (Tritium), electroluminescent (EL), fluorescent (Fluor), or LED exit signs alone; ceiling-mounted escape route lighting luminaires (EM); or incandescent (IN), electroluminescent (EL), or LED path-marking systems (after Webber and Aizlewood, 1993b).

although there was a different constant for each component. Table 9.3 gives the mean optical density at which each component was just visible. The larger is the mean optical density, the more visible is the component. Table 9.3 also gives the mean luminance of the letters in the exit signs, the door frame markings, and the path markings. There is clearly a broad relationship between the luminance of the component and the associated optical density; the higher the luminance, the higher the mean optical density. The mean optical density given in Table 9.3 is valuable because, given that it is constant for a given component, it can be used to predict the smoke density before that component becomes invisible for a fixed viewing distance, or, for a constant smoke density, how far away the observer can be before the component disappears. Figure 9.13 is derived from the smoke densities for a range of exit signs and a number of door frame markings using the same materials as the

Table 9.3 Mean optical density of smoke when various components of escape lighting systems were just visible and the mean luminance of that component
(after Webber and Aizlewood, 1994)

Component	Mean optical density	Component luminance (cd/m^2)
Photoluminescent exit sign after 1 min	0.84	0.042
Photoluminescent door frame marking after 1 min	1.60	0.042
Radioluminescent door frame marking	1.65	0.61
Radioluminescent exit sign	2.13	0.51
Electroluminescent exit sign	2.61	0.33
Electroluminescent door frame marking	2.61	7.32
Ceiling-mounted escape route lighting and a fluorescent pictogram sign	3.00	935
LED door frame marking	3.01	562
Fluorescent pictogram exit sign	3.19	935
Miniature incandescent door frame marking	3.23	1,610
Low-mounted LED exit sign	3.60	1,890
LED exit sign	4.01	3,280
LED pictogram exit sign	4.15	2,320

path-marking. It is clear from Figure 9.13 that any smoke density above
about 0.5/m severely restricts the distance at which any of these components are visible. This strong obscuring effect of smoke suggests that a
well-planned path-marking system, that provides information on the direction to go at closely spaced intervals, will be a better choice for escape route
lighting where the presence of dense but survivable smoke is considered
a possibility.

While there can be little doubt that escape route marking located close to
the floor is more effective in guiding people along an escape route in smoke
than conventional, ceiling-mounted escape route lighting, it is important to
remember why conventional escape route lighting and exit signs are
mounted above head height. The reason is to reduce the likelihood of the
route marking being obstructed by people, furniture, and equipment. This
implies that the low level escape route marking is, as its name implies,
strictly of value for marking defined escape routes, which are kept clear of
obstructions. This leaves open the question of what is the best way to guide
people from an occupied space to the escape route.

One complicating aspect of visibility in smoke that has not been studied
in the experiments discussed above is the effect of the smoke on the eye. Jin
(1978) measured the distance at which an exit sign could be read by people
walking down a corridor in the presence of irritant and non-irritant smoke of
a known density. The results showed that the irritant smoke reduced the

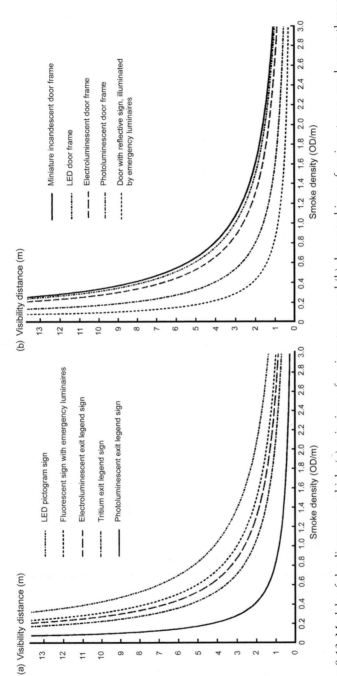

(a) Visibility distance (m)

Smoke density (OD/m)

.... LED pictogram sign
.... Fluorescent sign with emergency luminaires
– – – Electroluminescent exit legend sign
—··— Tritium exit legend sign
——— Photoluminescent exit legend sign

(b) Visibility distance (m)

Smoke density (OD/m)

——— Miniature incandescent door frame
.... LED door frame
·········· Electroluminescent door frame
—··— Photoluminescent door frame
– – – Door with reflective sign, illuminated by emergency luminaires

Figure 9.13 Models of the distance at which: (a) exit signs of various types; and (b) door markings of various types, can be seen through different smoke densities (after Webber and Aizlewood, 1994).

visibility distance markedly because the subjects' eyes watered profusely. Jin (1978) also examined peoples' walking speeds through both irritant and non-irritant smoke. Walking speeds were much reduced in irritant smoke, and providing more light was ineffective.

9.6.2   People with color defective vision

Given that color is an intrinsic component in exit signs and is important for the identification of a sign as an exit sign, it seems reasonable to ask how effective various colors of exit sign would be in conveying information to individuals with defective color vision. Eklund (1999) examined this question, using people with normal color vision and deutan and protan observers (see Section 2.2.7 for a description of the various forms of defective color vision). The apparatus used provided independent control of the letter and background color and luminance, for the word "EXIT." The word "EXIT" could appear normal or reversed and was sized to correspond to an exit sign conforming to the *Life Safety Code* seen from 30 m (100 ft). LEDs, with different peak wavelengths, were used to provide the light for the letters and background of the exit sign. The observer's task was simply to recognize the orientation of the sign. Because there are only two possible orientations, the recognition performance for the orientation of the exit sign can range from 100 to 50 percent, the latter being achieved by guessing. Figure 9.14 shows the recognition performance for color normal,

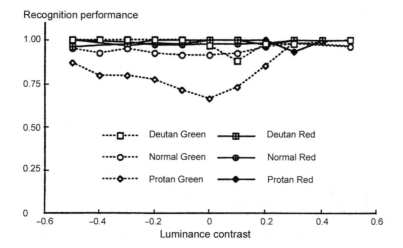

Figure 9.14 Recognition performance expressed as the proportion of correct identifications of the orientation of an exit sign, plotted against negative and positive luminance contrast, for color normal, deutan, and protan subjects. The two types of exit signs use green LEDs (peak wavelength = 530 nm) or red LEDs (peak wavelength = 660 nm) against a white background (after Eklund, 1999).

deutan, and protan observers; for green (peak wavelength = 530 nm) and red (peak wavelength = 660 nm) letters seen against a white background, plotted against luminance contrast. The results in Figure 9.14 show that the only condition in which recognition performance is much reduced is for the green letters on the white background, seen by protans. This result raises an interesting question. Why is the recognition performance of protans good with the red letters and relatively bad with the green letters? Eklund (1999) suggests that the explanation lies in the spectral sensitivity of the protans' visual system. Specifically, protans do not have long-wavelength-sensitive cones and consequently have a reduced sensitivity in the long-wavelength region of the visible spectrum. This alters the luminance contrast provided by the sign for protans. Figure 9.15 gives estimates of the "protan equivalent luminance contrast" matched to the luminance contrast for people with normal color vision, based on the spectral sensitivity curve for protanomolous and normal color vision observers (Wyszecki and Stiles, 1982).

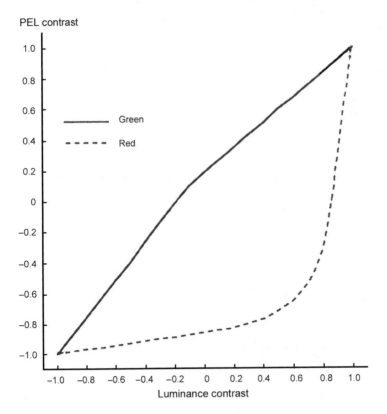

Figure 9.15 Protan equivalent luminance (PEL) contrast plotted against luminance contrast, for green and red LED signals The peak wavelengths for the LEDs are green = 530 nm and red = 660 nm (after Eklund, 1999).

From Figure 9.15 it can be seen that for the red LEDs, the protan equivalent luminance contrast is highly negative over the range of luminance contrasts examined (−0.5 to +0.5), i.e. the red letters on a white background appear to the protans as black on white. For the green LEDs, the protan equivalent luminance contrast is low for the luminance contrasts that are low, with a slight bias towards the negative, which explains the pattern of recognition performance in Figure 9.14.

These results are applicable to panel exit signs, where both letters and background have luminances above zero, even in the no-power condition. But what happens when a stencil sign is used, so that in the no-power condition, the background luminance is zero and the luminance contrast is very high? Figure 9.16 answers this question. It shows the recognition performance for color normals, deutans, and protans, for the four different LED colors. In this condition, protans show worse recognition performance than color normals or deutans for the red LEDs but not for the green LED. Again, the worse recognition performance for the red LEDs by the protans can be explained by the reduced sensitivity of protans at long wavelengths. The reduced sensitivity means the equivalent luminance of the red letters is less for protans, which, when combined with the black background, leads to a lower protan equivalent luminance contrast signal.

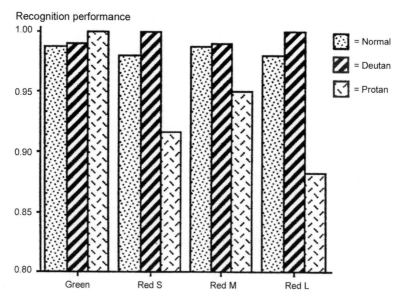

Figure 9.16 Recognition performance expressed as the proportion of correct identifications of the orientation of an exit sign for signs using green and red LEDs, by color normal, deutan, and protan subjects, on a dark background. The peak wavelengths for the LEDs are green = 530 nm, Red S = 622 nm, Red M = 632 nm, and Red L = 660 nm (after Eklund, 1999).

Two conclusions and one implication can be drawn from these results. The first conclusion is that what matters for recognition performance is the contrast between the letters of the sign and the background. The second conclusion is conditions that may be very good for color normals may be poor for color defectives. For example, Figure 9.15 suggests that luminance contrasts in the range +0.6 to +0.8, which would ensure a high level of recognition performance for color normals would lead to poor recognition performance by protans for red letters on a white background. The implication is that the most effective exit sign for color normals and for deutans and protans is a green stencil sign, i.e. green letters on a dark background. This format provides a high level of recognition performance regardless of ambient illumination (see Figures 9.14 and 9.16). This, at least, is a rational reason for choosing between exit sign colors.

9.6.3   People with low vision

All the studies discussed above have used people with normal visual acuities and visual field sizes. However, buildings are also used by people with limited visual capabilities. Age is by far the most common cause of limited visual capability. By the fourth decade of life, most people experience limits on focusing distance; by the sixth decade, the prevalence of pathological conditions in the optic media of the eye starts to increase; and by the seventh decade, the prevalence of pathological conditions in the retina increases rapidly (see Chapter 12). These changes reduce the ability to resolve detail, to discriminate colors, and to adapt to a sudden change in illumination and increase sensitivity to glare. Many of these changes can be expected to influence how people with limited visual capabilities can use the information provided by escape lighting.

Pasini and Proulx (1988) studied the manner in which visually impaired people moved through a building under normal conditions. They concluded that the visually impaired navigate through a building by wayfaring, i.e. by making a series of decisions at frequent intervals. They suggest that the visually impaired would benefit from regularly spaced information which is easily perceived and has a distinctive identity when attempting to move around a building under normal conditions. This implies that the path-marking approach discussed above, particularly if it could also have some tactile characteristic, would be of use to the visually impaired seeking to leave a building under emergency conditions.

Wright *et al.* (1999) report a study in which groups of 30 people with different forms of partial sight were asked to move over an escape route, involving both a corridor and some stairs. A comparison group of people with normal vision was also used. The escape route was lit by different combinations of ceiling-mounted lighting and path-marking. Measurements of speed of movement showed that the walking speeds of visually impaired people on the escape route was generally about 45–70 percent that of normally sighted people in the corridor and about 75–80 percent on the stairs. Figure 9.17 shows the movement speeds of the partially sighted in the corridor and on the stairs, plotted against the mean illuminance provided on the escape route, including that provided by the normal lighting. As would be expected, the higher is the illuminance, the faster is the speed of movement. Figure 9.18 shows the mean ratings of how difficult it was to see where to go plotted against the mean illuminance. Again, the mean rating shows a steady improvement with increasing illuminance. As for the different systems, the photoluminescent path-marking system produces both the slowest movement speed and the ratings of greatest difficulty; the ceiling-mounted escape lighting is worse than the powered path-marking systems at similar illuminances and the normal lighting provides the fastest speed and least difficulty of all. Cook *et al.* (1999) did a similar study, with similar results, although increasing the mean illuminance provided on the escape route by the ceiling-mounted system from about 1.9 lx to about 6.4 lx gave the ceiling-mounted system the lowest difficulty rating.

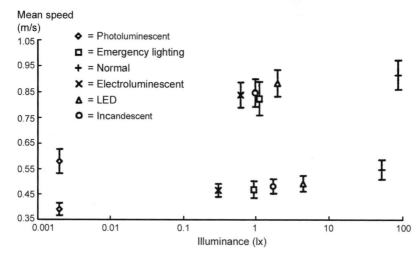

Figure 9.17 Mean speed of movement (and the standard error of the mean) of partially sighted people down a stair and along a corridor plotted against
the mean illuminance (lx) on the escape route. The upper, faster, set of
speeds are for the corridor. The lower, slower, set of speeds are for the
stairs. The escape route lighting consists of a photoluminescent path-
marking system, normal ceiling-mounted lighting providing 70 lx on
the floor, a LED path-marking system, a ceiling-mounted emergency
lighting system, an electroluminescent path-marking system, and a
miniature incandescent path-marking system. The 30 partially sighted
subjects included nine with retinitis pigmentosa, eight with macular
degeneration, four with cataract, three with glaucoma, two with diabetic retinopathy, and four with other causes of partial sight (after
Wright *et al.*, 1999).

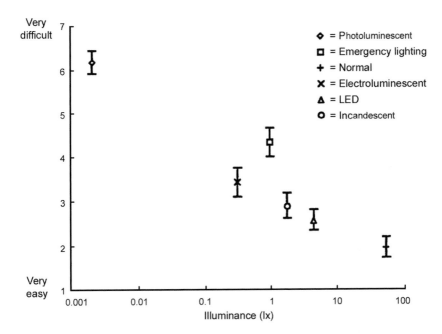

Figure 9.18 Mean ratings of difficulty of seeing where to go (and the standard error of the mean) given by partially sighted people moving down a stair and along a corridor plotted against the mean illuminance (lx) on the escape route. The escape route lighting consists of a photoluminescent path-marking system, normal ceiling-mounted lighting providing 70 lx on the floor, a LED path-marking system, a ceiling-mounted emergency lighting system, an electroluminescent path-marking system, and a miniature incandescent path-marking system. The 30 partially sighted subjects included nine with retinitis pigmentosa, eight with macular degeneration, four with cataract, three with glaucoma, two with diabetic retinopathy and four with other causes of partial sight (after Wright *et al.*, 1999).

This, again, emphasizes that the important factor is not the particular technology used, nor the location of the lighting, but rather the amount of light produced on the escape route.

9.6.4   *People with limited mobility*

Another special situation deserving consideration is an emergency requiring evacuation in a building where there are people of limited mobility. Typical buildings would be hospitals and nursing homes. Studies of the ease of evacuation of such buildings have been made (see Canter, 1980), but the role of emergency egress information is not mentioned. This is probably because people of limited mobility need assistance to move over the escape route, and it is assumed that the helpers will have normal visual capabilities. This is a reasonable assumption, but what it implies is that evacuation times will be much longer than for a normally mobile population. In this circumstance, the emergency egress information may need to be available for a much longer time period than is usually specified in the event of a power failure.

9.7 Escape lighting in practice

The research discussed above has been concerned primarily with the specification of escape lighting. While this is important, it is also necessary to consider how well any specification is implemented in practice. Ouellette *et al.* (1993) carried out a careful evaluation of the escape lighting system in seven, large, 20–30-year-old office buildings in Canada. The escape lighting systems in six of the buildings used conventional luminaires connected to a central generator that started when power failed. The other building used wall-mounted, incandescent floodlights powered by local rechargeable batteries. A visual inspection of the escape lighting in these buildings resulted in some disturbing findings. In many open-plan areas, discrepancies were found between the locations of the escape lighting and the defined escape routes. Often, the lighting of escape routes was shaded by high cubicle dividers. These failings are probably due to changes in the layout of office furniture, without changing the escape routes or the escape lighting. Given the churn rate in office buildings today, it is likely that this is a common situation. To make matters worse, exit signs were not always consistent with escape routes. In the worse cases, exit signs led to locked doors or cul-de-sacs. Measurements of illuminance were made on a horizontal plane, 1 m above floor level, at 1 m intervals along the center line of escape routes. These measurements revealed a very wide range of values along the escape route. Maximum illuminances in every building were more than 100 lx and the minimums where all less than 0.4 lx, many being less than 0.1 lx, the minimum measurable with the illuminance meter used. The picture that emerges from these measurements is one of pools of light separated by areas of darkness. The authors comment that "in some office areas, especially at night, it would be difficult for people to find their way to safety with the emergency lighting systems now in place."

Given that similar situations occur in other cities and countries, and there is no reason to suspect that they do not, the question that needs to be addressed is how to improve this situation. The most obvious answer is more frequent inspection of emergency egress systems, including operation of all parts of the system. In fact, the *Life Safety Code* (NFPA, 1997) does call for testing of emergency lighting equipment for function every 30 days, each test to last 30 s, and for a test lasting 90 min, annually. Such a testing schedule would catch any mismatches between the escape lighting and the escape routes and improve the reliability of the system components, particularly the battery-powered components. However, such inspections need not only to observe whether the escape lighting comes on but also to measure the illuminances provided on the escape route. The illuminance measurements made by Ouellette *et al.* (1993) form a multi-modal distribution. Given such a distribution, mean illuminance is likely to be misleading, particularly if the number of measurement points is limited. Ouellette *et al.* (1993) recommend that inspectors should look for the dimmest location on the escape route and measure the illuminance there, i.e. measure the minimum illuminance on the escape route. They support a minimum illuminance of 0.5 lx. This procedure has the advantage of being simple and quick, but a measurement is of little use unless there is a specification to compare it with. Fortunately, both the *Life Safety Code* (NFPA, 1997) and the *European Standard 1838* (CEN, 1999) include minimum illuminance specifications for the lighting of escape routes. One other possibility to enhance the quality of escape lighting in practice is to supplement powered exit signs and escape route lighting with photoluminescent path-marking. Although not as effective as powered exit signs and path-marking, it will operate even when the generator refuses to start or the battery in the exit sign is flat. It can also be easily moved when the furniture is rearranged.

9.8 Summary

The provision of some means of escape from a building is part of the legal framework of most countries. This provision usually involves defined escape routes and a means of telling the occupants when to leave. Lighting designed to tell occupants which way to go and to illuminate the escape route so that people can move quickly and safely along it when electrical power is absent and/or smoke is present, is an important component of emergency egress systems.

Informing occupants where to go to escape from the building is the role of the exit sign. These signs can consist of words or pictograms. In either case, the specification of the exit sign is based on the need for the sign to be visible and conspicuous at a specified distance. The specifications for exit signs usually define the minimum size of the elements of the sign, their luminances and luminance uniformity, and the luminance contrast between elements carrying the message and the background. Exit signs of different colors are used in different parts of the world. The value of having an exit sign of color is that it enhances the conspicuity of the sign relative to other nominally white luminaires that may be operating at the same time. Measurements of exit sign recognition by color normal and color defective individuals suggest that green, luminous letters or symbols on a black, opaque background is the most effective color and format for an exit sign. This stencil sign format will also be more effective in smoke than other formats, although very few commercially available exit signs will be visible at the maximum distance in a survivable smoke density.

The other part of an escape lighting system is the lighting of the escape route. This can be provided either by specially powered ceiling- or wall-mounted luminaires or some form of low-mounted path-marking system. There is little to choose between these systems in clear atmospheres. How fast people can move over the escape route and how often they make contact with obstacles depends on the illuminance produced on the escape route. A minimum illuminance of about 0.5 lx is sufficient to ensure that people will be able to avoid obstacles. Higher illuminances allow for faster movement speeds following a compressive function. Where a difference between the systems does emerge is in smoke. In smoke, the low-mounted path-marking systems are more effective in facilitating movement along an escape route than the wall- or ceiling-mounted systems, provided the path-marking system has sufficient light output to provide an illuminance on the escape route of at least 0.1 lx.

While what is needed to provide good quality escape lighting is fairly well understood, what is offered in practice often falls far short. This is for two reasons. The first is the shameful willingness of some of the organizations responsible for the escape lighting recommendations to pretend that while smoke can occur in buildings, it has no impact on the effectiveness of escape route lighting. This means that escape route lighting meeting many of the legal requirements will not be effective in the one situation where rapid escape is essential. The second is that many escape lighting systems are poorly maintained and/or not modified when the interior of the building is changed. This is because there is a widespread reluctance to provide the resources to enforce the legal requirements relating to escape lighting. In many ways, current escape lighting practice provides a false sense of security.

10 Lighting for driving

10.1 Introduction

In terms of the visual, cognitive, and motor components of tasks discussed in Chapter 4, the driver's task is to extract information from the environment, to determine what changes in behaviour are necessary, and to maneuver the vehicle appropriately. This is a continuous process, involving many interactions between the task components. Lighting is used either to provide information directly to the driver by day and night, e.g. traffic lights at intersections and brake lights on vehicles, or to help with the extraction of visual information from the environment after dark, e.g. headlights on vehicles and road lighting. For the driver, the value of lighting can be gauged by the extent to which it makes a journey safer, quicker, and more comfortable. Little is known about the effect of lighting on journey times but the effects on safety and comfort have been extensively studied. It is these aspects of lighting for driving that will be considered here.

10.2 The value of light

Figure 10.1 shows the average number of fatal vehicular accidents per hour occurring in the US in 1999, plotted against the time of day, for weekdays and weekends. These data are taken from the Fatality Analysis Reporting System (FARS) of the National Highway Traffic Safety Administration (NHTSA, 1992). For weekdays, the pattern of fatal accidents shows a minimum at about 4 a.m. followed by a rise as traffic flow increases during the morning commute, followed by a slight decrease and then a steady increase reaching a maximum in the evening commute, followed by slow decline to the minimum at 4 a.m. At the weekend, the pattern is somewhat different, the minimum occurring about 8 a.m. followed by an increase to a sustained maximum between 5 p.m. and 2 a.m. The difference between weekdays and weekends reflects the different activities that take place at these times.

Overall, the total number of fatal vehicular accidents is about the same in restricted lighting conditions as it is in daytime (Table 10.1). Given that traffic volumes are smaller at night than during the day, this pattern would

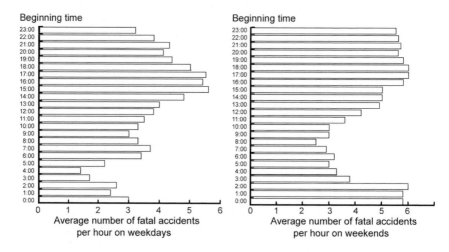

Figure 10.1 Average number of fatal vehicular accidents per hour in the US in 1999, plotted against time of day, for weekdays and weekends (after NHTSA, 2000).

Table 10.1 Number of vehicular crashes in the US in 1999, by lighting condition (after NHTSA, 2000)

Crash severity	Daylight	Dark but lit	Dark	Dawn or dusk	Total
Fatal	18,790	5,536	11,102	1,579	36,917
Personal injury	1,456,000	311,000	213,000	75,000	2,054,000
Property damage	2,985,000	579,000	465,000	158,000	4,188,000
Total	4,460,000	895,000	689,000	235,000	6,279,000

seem to imply that the absence of light makes the roads more dangerous, but such a conclusion would be premature. There are many differences between the driving conditions during day and night other than the presence of light. For example, the density of traffic is less, the incidence of driver fatigue and intoxication is higher, and the demographics of drivers are different at night. Before being able to conclude that the absence of light is the cause of the high level of fatal accidents at night, it is necessary to find a way to control these other differences.

An elegant solution to this problem is to use the change in lighting associated with the introduction of daylight saving time (Tanner and Harris, 1956; Ferguson *et al.*, 1995; Whittaker, 1996). In a daylight saving time system, the clock is moved forward by 1 h in Spring and back 1 h in Autumn. On both occasions, the effect is to suddenly change a period of driving from light to dark or vice versa. If it is assumed that activity and traffic patterns are governed by clock time, then it is likely that traffic density, driver fatigue and intoxication levels, and driver demographics are not likely to change substantially shortly before and shortly after the change in the clock, so any change in accidents can be safely ascribed to the change in lighting conditions. Sullivan and Flannagan (1999) used the FARS database to determine the total number of fatal collisions involving pedestrians for 46 of the 50 states in America, for 1 h about the civil twilight time, over the years 1987–97 (Arizona, Hawaii, and Indiana were excluded because they do not have daylight saving time, and Alaska was excluded because its solar cycle is markedly different from the other states). Civil twilight is defined as occurring when the sun is 6° below the horizon. The effect of the daylight saving time change on a Spring morning is to move the lighting conditions from twilight to night and then back through twilight and day, as day length increases. Figure 10.2(a) shows the total number of fatal pedestrian accidents occurring at twilight, for the morning transition, in the 9 weeks before and after the Spring daylight saving change. It can be seen that in the weeks before the change, there is a steady decrease in the number of fatal accidents but at the daylight saving change, there is a rapid return to a high level of accidents, a level than then reduces with the increasing day length. Figure 10.2(b) shows analogous data for the Spring evening twilight, for the 9 weeks before and after the daylight saving change. For the evening, the effect of the daylight saving change is to change driving conditions from night to day. The dramatic decrease in number of fatal pedestrian accidents with this transition is obvious.

The data presented in Figure 10.2 show the value of light to pedestrian safety but this should not be taken to mean that light is of value for all types of accidents. Sullivan and Flannagan (1999) carried out a similar analysis for fatal accidents involving a single vehicle leaving the road on curved, rural, high-speed roads. For this type of fatal accident, the change in lighting conditions produced by daylight savings time showed very little change in number of fatal accidents. The difference in the results for the two accident types reflects the underlying causes of the accidents. Pedestrian fatalities are commonly caused by the failure of the driver to see the pedestrian. Single vehicles leaving the road are much more likely to occur because of either fatigue or intoxication. Providing lighting will help make pedestrians more visible but will do little to help a driver who is fatigued or intoxicated.

This observation raises the question of what proportion of accidents are related to poor visibility. Figure 10.3 shows the percentage of accidents in which, in the opinion of the investigating team, each of the listed factors was involved (Sabey and Staughton, 1975; Hills, 1980). These percentages were based on on-the-spot investigations of 2,036 road accidents in a rural area of the UK. The percentages do not sum to 100 percent because road accidents rarely have a single cause, usually a combination of causes is involved. Even so, Figure 10.3 declares that in 95 percent of the accidents studied, the road user contributed to the accident. Further, 44 percent of the accidents involved what is termed perceptual error. Included under

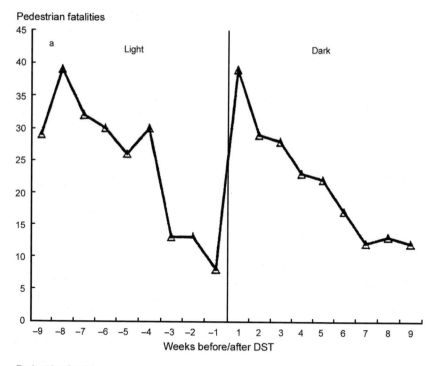

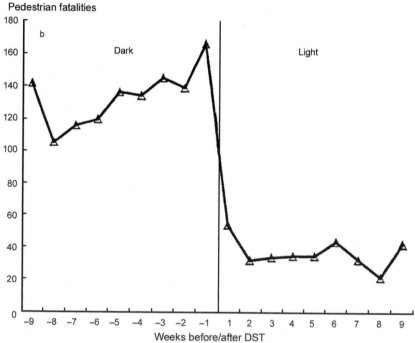

Figure 10.2 Number of pedestrian fatalities in 46 states of the US, over the years 1987–97, during twilight, for the 9 weeks before and after the Spring daylight saving time change: (a) for morning; (b) for evening (after Sullivan and Flannagan, 1999).

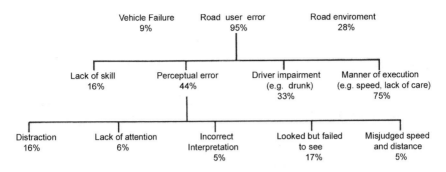

Figure 10.3 Apparent contributory factors in road accidents (after Hills, 1980).

perceptual error are distraction and lack of attention, incorrect interpretation, misjudgment of speed and distance, and a category intriguingly called "looked but failed to see." Not included under perceptual error are a lack of skill or care on the part of the driver and any reduction in the driver's ability, such as can be produced by the ingestion of drugs or alcohol.

A more detailed examination of the class of perceptual error can be used to show the visual requirements of driving. The first two contributory causes of accidents, distraction and lack of attention, both stem from the need to make decisions rapidly. Frequently, drivers have only one or two seconds in which to decide whether to make a particular maneuver. This time limit would matter less if all parts of the retina had the same capability for discrimination but they do not (see Section 2.4). Discrimination of detail is best, and hence the identification of objects is easiest, in the fovea, but declines rapidly off-axis. In the few seconds available, the driver will only be able to fixate and accommodate on a few places. But how does a driver know where the critical locations are before looking at them? Obviously, experience of both driving in general, and a particular piece of road is very important. There is certainly evidence that eye movement patterns are different for experienced and inexperienced drivers (Mourant and Rockwell, 1972). But experience is not the only factor. Another important feature is the conspicuity of any particular object. The more conspicuous a feature is the more likely it is to be detected in peripheral vision and then to be examined foveally. High conspicuity can be produced by relative motion against the background, and can be facilitated by having a high luminance contrast against the background (e.g. a motorcycle ridden with its headlight on during the day), or by rapidly flashing lights (as on emergency vehicles). However, it should be noted that although conspicuous objects are, by definition, highly visible, visible objects are not necessarily conspicuous. Conspicuity is also affected by the complexity of the background against which the object is seen. If all motorists drove with headlights on during daytime, they would all be visible but in dense traffic, none would be conspicuous.

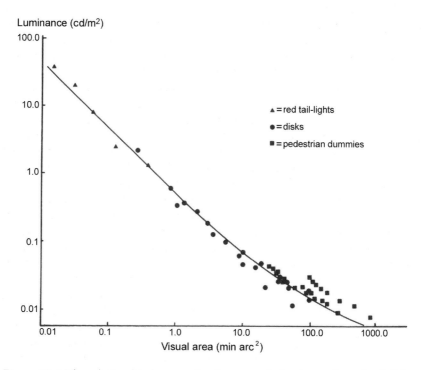

Figure 10.4 The relationship between luminance and visual area for red tail-lights, disks, and pedestrian dummies to be "just visible" under no road lighting/no glare conditions (after Hills, 1976).

When attention is directed to a location by the conspicuity of the object there, it is unlikely that any difficulty will be found in identifying the object. However, when experience suggests that a driver examine a particular location, then the visibility of any particular object there needs to be considered. As discussed in Chapter 4, the visibility of an object will depend on such attributes as its luminance, size, shape, color and movement, the background against which it is seen, and the state of adaptation of the driver. Much effort has been put into measuring the minimum values of some of these factors that are visible under different conditions. Figure 10.4 shows the relationship between luminance and size for red rear-lights, disk obstacles, and pedestrian dummies to be "just visible," i.e. at threshold, under no road lighting and "no glare" conditions (Hills, 1976). It can be seen that log luminance plotted against log visual area gives a nearly straight line; the smaller the obstacle, the greater the obstacle luminance has to be before it is just visible. This is just what would be expected from the spatial summation of the visual system for small targets (see Section 2.4.3).

Given that driving sometimes calls for the performance of tasks at or close to threshold, it seem likely that people with poor visual capabilities would be expected to have more accidents. Burg (1967) measured the static and dynamic visual acuity, the field of vision, the extent of misalignment of the two eyes and the rate of recovery from glare for 17,500 drivers who between them had had over 5,200 accidents in the previous 3 years. He was able to find only very weak correlations between the measured visual capabilities and the drivers' accident records. This failure can be explained in two ways. First, it is possible that drivers with worse visual capabilities are aware of their abilities and drive within them. Certainly, there is evidence that elderly people avoid driving at night long before they give up driving altogether. Second, there is the point that the visual capabilities measured were very simple. Davidson (1978) has confirmed that simple visual capabilities are at best weakly correlated with accident occurrences. This is to be expected because driving involves much higher levels of perception. The driver often has to detect the presence and/or movements of a number of different obstacles, recognize them and their characteristic patterns of movement, and then relate them to each other, all the while maintaining an understanding of the more slowly changing features of the road, and all in a short time before making a manoeuvre. It is the failure of this coordinating aspect of the drivers' task that appears, in Figure 10.3, as a cause of accidents under the heading "incorrect interpretation" (CIE, 1992a).

It is clear from this discussion that for people with sufficient visual capabilities to drive at all, their likelihood of an accident-free passage is strongly influenced by cognitive factors as well as visual factors (Wierda, 1996). This conclusion is supported by the existence of a "looked but failed to see" category of accident cause, a situation that others have found to be frequently involved in accidents in urban areas (Cairney and Catchpole, 1996). A classic example of looking but failing to see is a motorist who pulls out from a side road straight into the path of a motorcyclist on the main road. The car driver looked along the main road but failed to see the motorcyclist. It is not simply that the motorcyclist has a low visibility, if the driver had known the motorcyclist was there, he could have detected him easily. The problem is one of expectancy. The driver is searching for a car or a truck, he is not expecting a motorcycle and therefore does not see one. To overcome this phenomenon of expectancy, strong signals are necessary, i.e. conspicuity has to be increased. Using a headlight at all times on a motorcycle is an effective way to increase its conspicuity and hence its probability of detection, but such behavior also changes expectations. After exposure to motorcyclists that consistently use headlights at all times, the speed of detection of motorcycles without the headlight lit is slower (Hole and Tyrrell, 1996). Clearly, reducing the contribution of "looked but failed to see" to the prevalence of accidents is a matter of changing both visual and cognitive aspects of the situation.

The final category of perceptual contributions to accidents is misjudgment of speed and distance. These judgments are both involved in two maneouvres that produce many serious accidents, turning across traffic and overtaking, these maneuvers requiring the driver to move from the correct side of the road and go across or into the opposing traffic (Caird and Hancock, 2002). To do this safely, it is necessary to make accurate judgments of how far off an approaching vehicle is and at what speed it is approaching. Experimental evidence (Jones and Heimstra, 1964) suggests that people are not very good at making either of these judgments, particularly when the judgment has to be made at a long distance and the approaching vehicle is moving directly toward the driver, so the change in size and the perceived relative motion are small.

By now it will be appreciated that the drivers' task is a complex one. Within a very limited time the driver has to interpret what is likely to happen on the road ahead. To do this the driver has developed a series of expectations of other drivers' behavior and of what are the appropriate locations to examine. The driver will be faced with objects of different degrees of visibility and conspicuity and will have to make judgments for which the visual system is not well suited. It is this context that vehicle and roadway lighting has to operate.

10.3   Vehicle lighting

Vehicle lighting can be conveniently divided into two types: (a) lighting designed primarily to indicate the presence or give information about the movement of a vehicle; and (b) lighting designed to enable the driver to see after dark. The former category, known as signaling and marking lighting and including tail lights, side lights, brake lights, direction indicator lights, daytime running lights, emergency flashing lights, and license plate illuminators, are usually small in size and of limited luminous intensity. The latter category, known as forward lighting and including headlights, fog lights, auxiliary driving lights, and cornering lights, are somewhat bigger and some are of such a high luminous intensity that different aiming positions are used to avoid disability glare when two vehicles meet. Signal and marking lights are there to be detected. Forward lighting is a means of making the area in front of the vehicle more visible. One exception to this crude classification is reversing lights. Reversing lights provide both visibility to the rear and a signal to others around the vehicle.

10.3.1   *Signal and marking lighting*

Signal and marking lights on vehicles come in many different shapes and sizes, depending on the technology available and the design styling in fashion. Some signal and marking lights, such as side lights, are used only at night or in conditions of poor visibility in daytime while others, such as direction indicators and brake lights, have to be visible at all times, both day and night. The conditions necessary for such lights to be visible on roads under different conditions have been extensively studied (Dunbar,

1938; De Boer, 1951; Moore, 1952; Hills, 1975a,b; Sivak *et al.*, 1998). Using data similar to that shown in Figure 10.4, Hills (1976) produced a predictive model of the relationship between luminance increment and area for small objects to be just visible for a wide range of background luminances (Figure 10.5). A small object in this model is one for which spatial summation occurs in the visual system. For foveal vision, spatial summation is complete within a circle of diameter subtending about 6 min arc. For targets that occur 5° off-axis, spatial summation occurs over a circle of diameter about 0.5° (Boff and Lincoln, 1988). As shown in Figure 10.4, given sufficient viewing distance, these limits can include objects such as rear lights and pedestrians. The ordinate in Figure 10.5 is the logarithm of the increment of the object luminance necessary for it to be just visible against the background luminance. Different values of background luminance enable the effects of different lighting conditions to be estimated, from starlight, through road lighting, to daylight. Hills (1976) also shows that by using such curves he can plausibly predict the field results of Dunbar (1938) and Moore (1952). Such information forms the background for widely used standards for signal and marker lighting (SAE, 2001). These standards

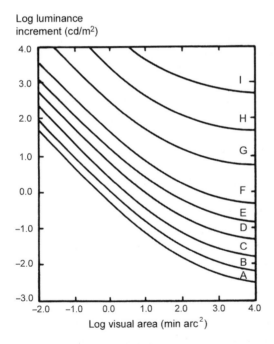

Figure 10.5 Relationships between log luminance increment and log visual area at different background luminances, for small targets to be just visible. Each curve is for one background luminance as follows: A = 0.01 cd/m², B = 0.1 cd/m², C = 0.32 cd/m², D = 1.0 cd/m², E = 3.2 cd/m², F = 10 cd/m², G = 100 cd/m², H = 1,000 cd/m², I = 10,000 cd/m² (after Hills, 1976).

cover the minimum luminous intensities that should be provided in different directions, the color of the signal or marker light, what colors and where each color can be used and, if flashing is called for, what the frequency of flashing should be. Of course, these heavily regulated aspects of vehicle lighting are usually measured by the manufacturers when the lights are new and clean. Cobb (1989) carried out a roadside survey of vehicle lighting and found that dirt typically reduced the luminous intensity of vehicle lighting by 30–50 percent.

Reductions in luminous intensity of a signal or marker light will reduce its visibility and conspicuity, especially during the day. Reducing visibility and conspicuity can be expected to increase reaction time to the signal. One signal light where this is likely to be important is the brake light, because of the close following distances adopted by many drivers. Another factor that may effect reaction time to the onset of brake lights, but one that is not considered in the standards, is the rise time to full light output. The rise time of a brake light is determined by the technology used to generate the light and the voltage applied. Incandescent lamps have long been the light source used in brake lights, and in all other forms of vehicle lighting, but recently, LEDs have begun to replace them. Incandescent lamps have a slow rise time, of the order of 200 ms, but LEDs have the potential to show a very fast rise time, of the order of nanoseconds. As for the effect of voltage applied, Sivak *et al.* (1994a) found that voltage drops from the nominal 12.8 V were common on large trucks, indicating an even longer rise time for incandescent brake lights on these vehicles. The effect of rise time in light output on reaction time to the brake light can be understood by considering the reaction time as the sum of two components, a visual reaction time and a non-visual reaction time. The visual reaction time is the time it takes for the light received from the brake light to be transformed to an electrical signal in the retina and transmitted up the optic nerve to the visual cortex. The non-visual component includes the time required for information perceived at the visual cortex to be processed and for neural signals to be sent to the muscles that make the response. Differences in rise time of light output for the brake light can be expected to influence the visual reaction time but not the non-visual reaction time. The effect of different rise times in light output can be estimated from the fact that to see the signal a constant level of energy, i.e. luminance integrated over time, is required to reach the retina (Teichner and Krebs, 1972). In other words, the visual reaction time is determined by the temporal summation properties of the visual system. Given a fixed maximum luminous intensity of the brake light, the shorter the rise time of the light output, the shorter the reaction time. This suggests that a fast rise brake light would be useful means for reducing rear-ending accidents (Sivak and Flannagan, 1993).

One final area of signal and marking lighting to consider is where the use of such lights interferes with the other objective of vehicle lighting, namely, for the driver to see what is happening around the vehicle. This is most common on emergency or road work vehicles where flashing lights are used to identify the purpose of the vehicle. Bullough *et al.* (2001b) measured how quickly drivers following a snowplough at night, in snow, could detect that the speed of the snowplough had changed, so that they were closing on it. They found that when the rear of the snowplough was fitted with two vertical, constantly illuminated LED bars, the change in speed of the snowplough could be detected 20 percent faster than when it was fitted with two flashing amber lights. Flashing marker lights are undeniably effective in attracting attention to the vehicle, particularly during daytime when there is much other visual information competing with the emergency vehicle for attention. However, at night, flashing lights make it more difficult to estimate relative speed, distance, and closure (Croft, 1971; Hanscom and Pain, 1990). Anyone who has come across an accident, on an unlit road, at night, being attended by three police vehicles, one ambulance and one fire engine, all equipped with flashing lights, has experienced the difficulty in extracting visual information. There is no doubt that the scene should be approached with caution, but what is expected of the approaching driver is unclear. This is because the flashing lights are often the only illumination of the scene and every flashing light produces glare and operates at a different phase, so the scene is continually changing in appearance. One approach to solving this problem would be to limit the number of flashing lights on any emergency vehicle and to reduce the percentage modulation of each flashing light. It seems likely that one or two flashing lights visible from all angles are sufficient to attract attention to a vehicle at night and that adding more flashing lights will do little other than cause confusion.

10.3.2   Forward lighting

The most common form of forward lighting is that provided by headlights. Like signal and marking lights, headlight performance and placement is strictly regulated (SAE, 2001). Headlight regulation has always involved a compromise between forward visibility and minimizing glare to approaching drivers. This is the reason why headlights have two luminous intensity distributions, represented by high and low beams. High-beam headlights are for use when there is no approaching vehicle so the need to limit glare is reduced. Low-beam headlights are for use when there is an approaching vehicle and it is necessary to limit the luminous intensity in the direction of the driver of the approaching vehicle so as not to compromise that driver's vision. The SAE recommended practice for headlights sets minimum and maximum luminous intensity values, for both low and high beams, for a number of relevant directions (see Table 10.2).

The compromise between forward visibility and limiting disability glare has resulted in different regulations for low-beam luminous intensity distributions in the US and Europe. The major difference occurs in the luminous intensity towards the approaching driver. The European low beam has

Table 10.2 Minimum and maximum luminous intensities in different directions recommended by the Society of Automotive Engineers for high- and low-beam headlights. The directions are specified by the angle, in degrees, up and down, and left to right, from an axis formed by the horizontal and vertical planes which intersect the principal filament of the headlight (from SAE, 2001)

Vertical direction; U = up, D = down	Horizontal direction; L = left, R = right	Minimum luminous intensity (cd)	Maximum luminous intensity (cd)
Low-beam			
10–90U	45L–45R	—	125
0.0–4.0U	8.0L–8.0R	64	—
0.0–2.0U	4.0L–4.0R	125	—
1.0U	1.5L	—	700
0.5U	1.5L	—	1,000
0.5D	1.5L	—	3,000
1.5U	1.0R	—	1,400
0.5U	1.0R, 2.0R, 3.0R	—	2,700
0.5D	1.5R	8,000	20,000
1.0D	6.0L	750	—
1.5D	2.0R	15,000	—
1.5D	9.0L and 9.0R	750	—
2.0D	15.0L and 15.0R	700	—
4.0D	4.0R	—	8,000
High-beam			
2.0U	0.0	1,500	—
1.0U	3.0L and 3.0R	5,000	—
0.0	0.0	20,000	75,000
0.0	3.0L and 3.0R	10,000	—
0.0	6.0L and 6.0R	3,250	—
0.0	9.0L and 9.0R	2,000	—
0.0	12.0L and 12.0R	500	—
1.5D	0.0	5,000	—
1.5D	9.0L and 9.0R	1,500	—
2.5D	0.0	2,000	—
2.5D	12.0L and 12.0R	750	—
4.0D	0.0	—	12,500

a much sharper cut-off than the American low beam, indicating the greater emphasis given to controlling disability glare in Europe. Over the last few years, there have been moves to resolve this difference by the production of harmonized vehicle headlight specification (SAE, 1995a; GTB, 1999). Sivak *et al.* (2001) have carried out an evaluation of the proposed harmonized low-beam luminous intensity distribution, in terms of the change in luminous intensities in directions where significant elements in the road environment are expected to be found, e.g. pedestrians at the edge of the road, road edge delineations, retroreflective traffic signs, glare towards oncoming drivers, glare towards the mirrors of the car immediately ahead, and glare reflected from wet road surface towards oncoming drivers. As might be expected from a luminous intensity distribution based on a compromise between the American approach, which emphasizes forward visibility, and the European approach, which emphasizes control of glare, headlights meeting the harmonized specification provide less light for traffic signs, and vehicle reflectors, and less light towards mirrors on the vehicle ahead than headlights meeting the American luminous intensity requirements. Conversely, headlights meeting the harmonized specification provide more light for pedestrians, traffic signs, and vehicle reflectors, more glare to oncoming drivers and more light towards mirrors on the vehicle ahead than headlights meeting the European luminous intensity requirements. The harmonized specification also provides more foreground illumination and more glare from a wet road towards oncoming drivers, than for headlights that conform to either the American or the European requirements. This process of harmonization has been driven by the globalization of the vehicle manufacturing industry but that should not be seen as a reason for rejecting the compromise. The fact is the recommended luminous intensity distributions used in different parts of the world are all compromises between the need for visibility and the need to control glare. Further, there is a large difference between the luminous intensity distribution of a new headlight in a laboratory and a headlight in a vehicle on the road. Headlights in a vehicle on the road may produce a different luminous intensities in important directions because the vehicle may not be level, or the headlight is incorrectly aimed, or the headlight is dirty. Yerrel (1971) reported a set of roadside measurements of headlight luminous intensities in Europe and found a very large range of luminous intensities for the same direction. Alferdinck and Padmos (1988) found similar results from roadside measurements in The Netherlands. They also examined the importance of aiming, dirt, and lamp age on the luminous intensity in a series of laboratory measurements. Figure 10.6 shows the cumulative frequency distributions of luminous intensity in a direction important for glare to an oncoming driver and in a direction important for forward visibility, for 50 cars taken from a parking lot. The luminous intensity measurements of the headlights, as found, but taken in the laboratory, agreed with measurements taken at the roadside. From Figure 10.6 it can be seen that the headlights, as found, tend to produce more glare and less forward visibility than new headlights. The forward visibility is most improved by correcting the aiming. Cleaning the headlights and operating them at 12 V increases the luminous intensity for forward visibility a little and brings it closer to that of new headlights. For the direction important for glare, correcting the aiming makes things slightly worse, but cleaning the headlights reduces the luminous intensity causing glare and again brings it close to that of new headlights. The ranges of luminous intensities shown in Figure 10.6 suggest that fine differences between the recommended headlight luminous intensity

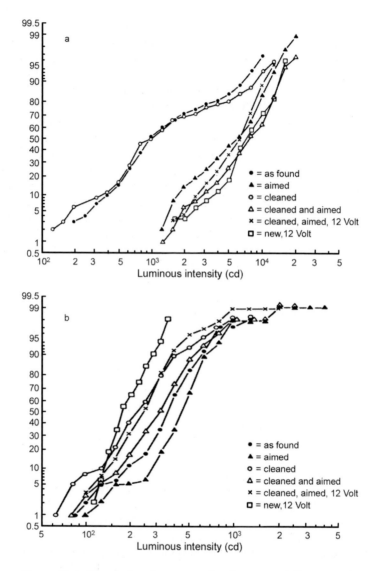

Figure 10.6 Cumulative frequency distributions of luminous intensities in directions (a) important for visibility of the nearest edge of the road and (b) important for glare to oncoming drivers, for headlights on 50 cars as found; aimed; cleaned; cleaned and aimed; cleaned, aimed, and operated at 12 V; and for new headlights (after Alferdinck and Padmos, 1988).

distributions used in America and in Europe are trivial compared to the differences that occur in practice, due to aiming, voltage, and dirt.

The most dramatic change in headlights over the last decade has been the introduction of the high-intensity discharge (HID) headlight. These headlights, identifiable by their blue–white color appearance, are essentially a metal halide discharge with xenon added to allow for rapid rise and restrike times. HID headlights differ from conventional tungsten or tungsten–halogen headlights in three respects; the size of the light source, the luminous intensity distribution from the headlight, and the spectral power distribution of the light emitted. The arc tube of the HID light source is smaller than the filament of tungsten and tungsten–halogen light sources, with the result that headlights using HID light sources can be smaller than those using other light sources. Initially, concern was expressed that the increased luminance of the headlight associated with a higher light output being produced from a smaller area would lead to greater discomfort glare. However, studies have shown that headlights subtend such a small solid angle from the normal viewing distance that headlight area has only a small effect on discomfort from headlights. The perception of discomfort from headlights is dominated by the illuminance at the eye (Sivak *et al.*, 1990; Alferdinck, 1996), although both light spectrum and duration of exposure also influence discomfort. Specifically, HID headlights giving the same photopic illuminance at the eye produce greater discomfort than tungsten–halogen headlights (Flannagan *et al.*, 1992; Flannagan, 1999) and longer durations of exposure produce greater discomfort (Sivak *et al.*, 1999). Schmidt-Clausen and Bindels (1974) have produced an equation relating the illuminance at the eye to the level of discomfort produced, expressed on the De Boer scale (De Boer, 1967) (1 = unbearable, > = unnoticeable). The equation is

$$W = 5.0 - 2\log(E/0.003(1 + \sqrt{(L/0.04)} * \phi^{0.46})$$

where W is the discomfort glare rating on the De Boer scale, E the illuminance at the eye (lx), L the adaptation luminance (cd/m^2), and ϕ the angle between line of sight and glare source (min arc).

As for the luminous intensity distribution of HID headlights, the recommended minimum and maximum luminous intensities used in different parts of the world apply regardless of the light source used, so HID headlights are designed to meet these requirements. However, the HID light source has a much higher luminous efficacy than tungsten or tungsten–halogen light sources so HID headlights typically have a higher maximum luminous intensity than tungsten and tungsten–halogen headlights, as well as putting more light to the sides of the vehicle in areas that are not controlled by the current regulations. These differences in the amount and distribution of light from HID headlights, together with the variability introduced by aiming, dirt, and the different geometries that can occur between two approaching vehicles, are probably enough to explain the widespread anecdotal complaints of disability and discomfort glare from drivers meeting vehicles equipped with HID headlights. It also explains the reason why people driving vehicles equipped with HID headlights like them. Figure 10.7 shows contours for the detection of a 40 cm side target of reflectance 0.1 by drivers using either HID headlights or tungsten–halogen

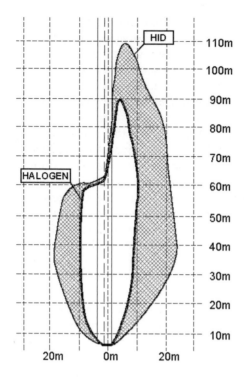

Figure 10.7 Contours for the distances at which a square target of 40 cm side, with a reflectance of 0.1, is detected by drivers for either HID headlights or tungsten–halogen headlights, both on low beam (after Rosenhahn and Hamm, 2001).

headlights, on a low-beam setting (Rosenhahn and Hamm, 2001). Clearly, the HID headlights conforming to the same regulations allow objects to be detected at greater distances and over a wider range of angles than the tungsten–halogen headlights. It is also worth noting that the locations where there is close agreement in detection distances for the two headlight types are the locations where the maximum luminous intensities are specified in regulations.

The spectral power distribution of the HID headlight is similar to that of the metal halide lamp shown in Figure 1.14 and is very different from that of the tungsten or tungsten–halogen headlight. At first there was some concern that the limited amount of power emitted at the long-wavelength end of the spectrum by the HID light source, would make it difficult to identify signs where color is an important part of the sign, e.g. the stop sign used in the US, which is red with white letters. However, Sivak *et al.* (1994b) showed that these fears were unfounded. Nonetheless, current SAE recommendations (SAE, 1995b) demand that any HID headlight shall have a CIE general CRI of 60 or higher. While concern about the ability of HID headlights to render road sign colors correctly represent the negative side of the effect of spectral power distribution, there is a positive side. There is evidence that at low light levels, i.e. in mesopic vision, light sources that provide greater stimulation to the rod photoreceptors allow faster reaction times off-axis than light sources that do not (He *et al.*, 1997). This finding suggests that HID headlights should allow faster reaction times for the detection of objects, such as pedestrians and animals, at the side of the road, than tungsten–halogen headlights, even when both provide the same photopic illuminance. So far, this possibility has not been tested for headlights alone but has been successfully demonstrated for road lighting (see Section 10.4).

Given the tendency for expensive options first introduced in up-market vehicles, to gradually spread into cheaper vehicles, it seems likely that HID headlights will soon become much more widely used, replacing tungsten–halogen headlights, just as they replaced tungsten headlights. This is also likely because the higher luminous intensities available with HID headlights and the smaller possible headlight sizes place fewer constraints on automobile design. The problems of disability and discomfort glare, currently associated with HID headlights, can be solved by regulating the maximum luminous intensity allowed in more directions relevant for glare, by more precise optical design, and by greater attention to the aiming and cleanliness of headlights when installed in vehicles. Already, some countries require that a vehicle equipped with HID headlights be fitted with a self-leveling suspension, while others require the installation of headlamp cleaning systems. It is ironic that, after many years spent arguing about the advantages or otherwise of yellow headlights (Reading, 1966; Schreuder, 1976), the latest development in headlight technology should be moving the color appearance of headlights away from the yellow and towards the blue.

Given that the purpose of headlights is to reveal the road and objects on and around it to the driver, it is appropriate to ask at what distances objects are visible when driving with headlights. Hills (1976) has developed a model of the luminance increment required for small objects of various sizes to be detected against different background luminances (Figure 10.5). Using this model, Hills was able to plausibly match the results of field measurements of the visibility of objects on a lit road (Dunbar, 1938) and an unlit road (Helmers and Rumar, 1975). This model is valuable. For known luminances of background and object, the area necessary to make the object visible at a required distance can be obtained. By juggling the visibility distance thought necessary for safety and the background luminance against which the object will be seen, it is possible to specify the luminance required for an object of a known size to be seen. Kosmatka (1995) discusses this process in detail. One limit of Hills' model is that it is based on the judgment of observers expecting an obstacle, when the obstacle was just visible. Roper and Howard (1938) have shown that an unexpected obstacle is seen at about half the distance that an expected obstacle is seen. As for the just visible criterion, Hills (1976) suggests a series of multiplying factors that can change the observer's criterion from just visible to just understandable to just obvious to just brilliant.

While Hills' model is undoubtedly valuable, it should be appreciated that it is frequently a simplification of reality. Obstacles seen on the road rarely have a single reflectance, sometimes have elements with both diffuse and specular reflection characteristics, can occur in different positions and hence have backgrounds that may be adjacent to the obstacle or much further away, and, when illuminated by headlights, can cast shadows on the background, which themselves can vary in complexity. Despite all this variation, there can be little doubt about how such changes will affect visibility. Any effect that takes the luminance contrast of the object further from zero will tend to increase its visibility and any effect that brings its luminance contrast closer to zero will decrease its visibility. How far the luminance contrast has to be from zero to ensure detection depends on where in the visual field the object is and the complexity of the object and its background (Paulmier *et al.*, 2001).

One effect that will always decrease object visibility at night is disability glare, caused by light from the headlights of approaching vehicles being scattered in the eye. The magnitude of disability glare is quantified by the equivalent veiling luminance, which is the luminance of a uniform veil that changes the luminance difference threshold as much as the actual lighting does. Light spectrum does not affect disability glare (Flannagan, 1999). The effect of disability glare can be taken into Hills' model by adding the equivalent veiling luminance to the background luminance and adjusting the luminance increment and size appropriately. But before doing this, it is important to note that the formula for equivalent veiling luminance given in Section 5.4.2.1 does not hold for small deviations from the line of sight, and such small deviations are common when driving against oncoming vehicles. Hartmann and Moser (1968) have shown that for angles less the 100 min arc from the line of sight, the loss of visibility associated with disability glare is much greater than would be predicted from the simple disability glare formula, probably because of neural interactions occurring in the retina in addition to scatter in the optic media. For such small angles, the CIE suggest a modified version of the age-adjusted disability glare formula given in Section 12.5 (CIE, 2002b). The formula is for use in the angular range 0.1–30° and takes the form:

$$L_\mathrm{v} = \sum [10E_n/\Theta_n^3 + (1 + (A/62.5)^4)\, 5E_n/\Theta_n^2]$$

where L_v is the equivalent veiling luminance (cd/m^2), A the age (years), E_n the illuminance at the eye from the nth glare source (lx), and Θ_n the angle of the nth glare source from the line of sight (degrees).

Theeuwes and Alferdinck (1996) have examined the impact of the presence of a glare source on drivers behavior by having people drive over urban, residential, and rural roads at night, with a glare source simulating the headlights of an approaching vehicle mounted on the bonnet of the car. They found that people drove more slowly when the glare source was on, particularly on dark winding roads where lane keeping was a problem. Older subjects showed the largest speed reduction. The presence of glare also caused the drivers to miss many roadside targets, particularly older drivers.

Given that the presence of headlights in the driver's field of view make it more difficult to detect targets beside the road it is interesting to consider how the distance at which an obstacle can be detected as two vehicles approach and pass each other varies with the luminous intensity distributions of the headlights of the two vehicles, the relative position of the two vehicles and the presence or absence of road lighting. Mortimer and Becker (1973), using both computer simulation and field measurements, have shown that the visibility distance for two opposing cars diminishes as the cars close and then starts to increase rapidly (Figure 10.8). The separation at which the visibility distance is a minimum depends on the relative luminous intensity distribution of the headlights, the relative positions of the two vehicles, the obstacles to be seen, and the physical characteristics of the obstacle. Helmers and Rumar (1975) measured visibility distances for flat gray 1.0 m by 0.4 m rectangles with a reflectance of 0.045. Observers were driven towards a parked car with its headlights on and asked to indicate

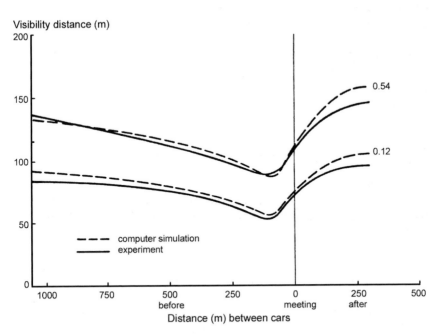

Figure 10.8 Visibility distance for targets of reflectance 0.54 and 0.12, plotted against the distance between two vehicles approaching each other, with headlights of equal luminous intensity (after Mortimer and Becker, 1973).

when they saw the obstacles. It was found that when both the headlights in the observer's vehicle and the opposing vehicle had about the same maximum luminous intensity, there was little increase in visibility distance above a luminous intensity of 50,000 cd, the visibility distance being about 60 m. However, when the opposing car had a maximum luminous intensity three or more times greater than the observer's vehicle, then the visibility distance was reduced by about one-third. Figure 10.9 shows these results in terms of visibility distances plotted against the distance between the two vehicles, for different combinations of luminous intensity for the observer's car and the opposing car. Helmers and Rumar (1975) show that for their small dark-gray obstacle, a headlight system with a maximum high-beam luminous intensity gives a visibility distance of about 220 m when no opposing vehicle is present. This is the same as the stopping distance for a vehicle moving at 110 km/h (65 miles/h), which is about 220 m on wet roads (AASHTO, 1990). However, when two opposing vehicles have equal luminous intensity headlights, the visibility distance is reduced to about 60–80 m, which is much less than the stopping distance and when the opposing vehicle had a luminous intensity about three times more than the observer's vehicle, the visibility distance is reduced to about 40–60 m. It is clear that driving at high speeds against opposing traffic at night approaches an act of faith.

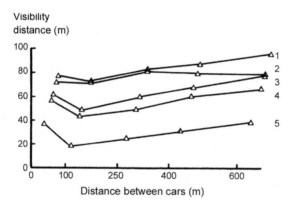

High beam luminous intensities (cd)

	Observer's car	Opposing car
1 -	260,000	220,000
2 -	low beam	low beam
3 -	130,000	220,000
4 -	87,000	220,000
5 -	29,000	220,000

Figure 10.9 The visibility distance for detecting the presence of 1.0 m by 0.4 m rectangle of reflectance 0.045 in the presence of opposing headlights, plotted against the distance between the observer's car and the opposing car, for different combinations of headlight luminous intensities (after Helmers and Rumar, 1975).

As for driving on low beams without any opposing vehicle, Olson and Sivak (1983) measured the distance at which observers could detect a pedestrian wearing a dark or light top while being driven along an unlit road in a car using low-beam headlights. They then calculated the stopping distance for a vehicle moving at 88 km/h (55 miles/h). The percentage of trials in which the detection distance was less than the stopping distance ranged from 3 percent, which was for young observers with the pedestrian wearing the light top to 83 percent for old observers with the pedestrian wearing the dark top. It is likely that these percentages are underestimates because the observers were told to look for the targets and were not distracted from that task by having to drive. Similar estimates of the percentages of people able to stop within the distance at which they can detect low-contrast targets have been made by Yerrel (1976).

These estimates of visibility distances go some way to explain the increase in pedestrian deaths around the change in daylight saving time discussed earlier. They also support the advice to pedestrians to wear high-reflectance clothing or retroreflective markers when out walking at night and add yet another reason for the groundswell of complaint about the glare from HID headlights from those driving vehicles equipped with tungsten–halogen headlights. Clearly, striking the right balance between enhancing visibility and controlling glare in the design of headlights is difficult and is likely to remain so.

Fortunately, there are a number of potential solutions to the problem of maximizing visibility while minimizing glare (Mace *et al.*, 2001). One is the use of polarization. In this system, headlights are designed to emit polarized light and the driver viewing the headlights sees them through a polarizing filter at an opposing orientation. Polarized headlights have been technically possible for more than 50 years (Land, 1948) but have never got beyond the stage of an interesting idea, partly because much higher luminous intensities are required to compensate for the absorption in the polarizing and analyzing components and because polarizing headlights will only be fully effective when all vehicles are fitted with the such headlights and this is inevitably a very gradual process (Yerrell, 1976).

A more recent solution to the problem of how to ensure visibility without glare to oncoming drivers has been the development of an ultraviolet (UV) emitting headlight. These headlights emit radiation in the wavelength range 320–380 nm and are designed to supplement conventional low-beam headlights. When these UV headlights are combined with fluorescing materials in road markings, the distances at which the road markings can be seen increase dramatically, as do the distances at which pedestrians who are wearing clothing washed in detergents containing a fluorescent whitening agent can be detected (Turner *et al.*, 1998). The UV radiation emitted by such headlights does not pose a health threat to people exposed to it (Sliney *et al.*, 1995). The combination of UV headlights with fluorescing materials is considered a very cost-effective approach to improving the safety of driving at night (Lestina *et al.*, 1999). The main problem associated with the introduction of a system based on UV headlights and fluorescing materials is a political one. The system requires the organizations who manufacture vehicles and who design and construct the road system to work together, something which past history suggests is not easily achieved.

A variation on this approach, that does not require action by anyone other than the vehicle manufacturer, is to use a headlight emitting infrared (IR) radiation at about 800 nm coupled to a video camera sensitive to this wavelength (Holz and Weidel, 1998). Many materials which have a low reflectance in the visible wavelengths have a much higher reflection in the near-IR. The result is that the video image shows a much brighter road and surroundings than can be seen with the naked eye.

Another solution to the problem of enhancing visibility without increasing glare is to use the night-vision technology developed for military uses. Such technology, based on far-IR imaging, is now available as an option on some up-market vehicles, the image being projected onto the windshield. This is a completely passive system in that it simply detects IR radiation emitted by surfaces at different temperatures. There is no additional radiation emitted from the vehicle. Such a system should be effective in detecting objects with a distinct temperature difference from the background, such as pedestrians and other animals, but how effective it would be in revealing road marking is open to question.

Both the UV headlights and the IR-imaging systems solve the problem of enhancing visibility without enhancing glare by using radiation outside the visible range and then converting some important elements in the scene to visible radiation. An alternative adopted by headlight manufacturers is to continue to use visible radiation but to alter the amount and distribution of the light emitted by the headlights according to the situation ahead of the driver. Such adaptive forward lighting systems, as they are called, are a topic of considerable interest amongst vehicle lighting manufacturers (Hamm and Rosenhahn, 2001). Adaptive forward lighting is certainly something that can be introduced by vehicle manufacturers alone but whether customers would be willing to pay the increased cost remains to be seen. It will be interesting to see which, if any, of these solutions to the problem of enhancing visibility without enhancing glare is the eventual winner.

10.4   Road lighting

Road lighting is provided in urban areas and on traffic routes mainly to enable a driver to see further and better than by vehicle forward lighting alone. Lip service is sometimes paid to the needs of the pedestrian and the appearance of the environment but it only takes a look at the typical road lighting installation layouts and the luminous intensity distributions of the luminaires used to appreciate that the main concern is with the driver. The discussion of the effects of the change to daylight saving time on accidents of different types presented at the start of this chapter indicates that an increase in the level of ambient lighting, such as may be provided by road lighting, can be a valuable accident countermeasure. This conclusion is supported by a meta-analysis of the safety benefits of introducing road lighting to previously unlit roads (Elvik, 1995). Based on results from 37 studies in which the effects had been measured in terms of the change in either number of night-time accidents or number of night-time accidents per million vehicle kilometers of travel, it was estimated that introducing road lighting would lead to a 65 percent reduction in night-time fatal accidents, a 30 percent reduction in night-time injury accidents, and a 15 percent reduction in night-time property damage accidents.

Given that the introduction of road lighting can be an effective accident countermeasure, the problem now is to identify what is the cost-effective level of road lighting. The costs of road lighting are fairly easy to establish, it is the benefits of different levels of lighting that are difficult to determine. The most comprehensive study of the effect of different lighting conditions on accident rate is that of Scott (1980). In this study, photometric measurements were taken of the lighting conditions on up to 89 different sites in the UK, using a mobile laboratory (Green and Hargroves, 1979). The sites were all at least 1 km long with homogeneous lighting conditions and both the lighting and the road features had been unchanged for at least 3 years. The sites were all two-way urban roads with a 48 km/h (30 miles/h) speed limit. The photometric measurements were made with the road dry and the accidents considered were only those that occurred when the roads were dry. Multiple regression analysis was used to determine the importance of various characteristics of the lighting on the night/day accident ratio. The average road surface luminance was found to be the best predictor of the effect of the lighting on the night/day accident ratio. Figure 10.10 shows the night/day accident ratio for the sites plotted against the average road surface luminance. The best-fitting exponential curve through the data is shown, the night/day accident ratios being weighted to give greater importance to those sites where accidents occurred most frequently. The equation for the curve is

$$N_R = 0.66\,e^{-0.42L}$$

where N_R is the night/day accident ratio and L the average road surface luminance (cd/m^2).

It is clear from Figure 10.10 that average road surface luminance does contribute to a reduction in accidents at night but the wide scatter in the individual points also indicates that there are many factors other than road surface luminance that matter.

One factor that may have contributed to the scatter in Figure 10.10 is risk compensation. The idea behind risk compensation is that drivers will adjust their behaviour to a constant level of risk. This implies that if road lighting

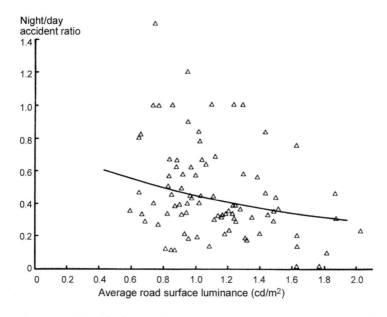

Figure 10.10 Night/day accident ratios plotted against average road surface luminance. The curve is the best-fitting exponential through the data, after weighting each ratio for the number of accidents to which it relates (after Hargroves and Scott, 1979).

is installed, the risk to drivers is reduced so drivers will increase their speed. Assum *et al.* (1999) found that drivers in Norway did indeed increase their speed at night after road lighting was installed, by about 5 percent on straight roads and about 0.7 percent on curves. Presumably, the reason for the difference between straight and curved roads is that road lighting is more effective in enabling the driver to see further ahead on straight roads than on curved roads. Despite this finding, the results from the meta-analysis discussed above (Elvik, 1995) suggest that such increases in speed are not enough to completely nullify the value of road lighting as an accident countermeasure, but it is undeniable that they may go some way to diminish its impact.

An alternative approach for determining desirable road lighting conditions is to measure driver performance under different lighting conditions. Gallagher *et al.* (1974) placed a human size obstacle in one lane of a three-lane urban road that could be lit to different illuminances and illuminance uniformities. Unsuspecting drivers were observed and the distances at which they first took avoiding action was recorded, as was the speed at which they were traveling. The speeds and distances were combined to form a measure of the time to target when avoiding action was taken. The mean times to target are plotted against the illuminance on the road at night and

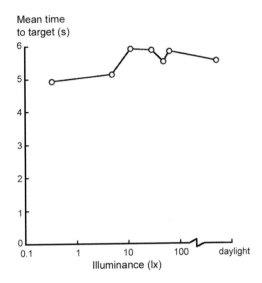

Figure 10.11 Mean time to target for unwarned drivers detecting a pedestrian-sized obstacle on the road, plotted against the average illuminance on the road provided by road lighting, and in daylight (after Gallagher *et al.*, 1974).

for daylight in Figure 10.11. It can be seen that time to target approaches the level reached in daylight at an illuminance on the road of about 10 lx. Fisher and Hall (1976) measured reaction times for the detection of a vehicle ahead accelerating or decelerating, for different luminances. They summarized their results as a nomogram from which the road surface luminance necessary to achieve detection of a vehicle ahead accelerating or decelerating within 10 percent of the fastest reaction time could be estimated. The range of road surface luminances to achieve this response varied with the headway of the vehicle and the acceleration/deceleration. The highest required road surface luminance of 5 cd/m^2 occurred for a headway of 80 m, regardless of acceleration/deceleration. The lowest luminance of 0.2 cd/m^2 occurred for a headway of 10 m, for an acceleration/deceleration of 4 m/s^2.

There can be little doubt that road lighting is a useful accident countermeasure and that average road surface luminance is a useful metric. However, what the optimum road surface luminance is remains open to argument because of the diversity of tasks to be undertaken. Fortunately, the CIE has produced a set of recommendations based on average road surface luminance that has been widely adopted (CIE, 1995c). Specifically, the CIE recommends a minimum average road surface luminance that should be maintained throughout the life of the installation. The values range from

0.5 cd/m^2 for local distributor roads to 2.0 cd/m^2 for high speed roads with separate carriageways.

It should not be surprising that average road surface luminance is an important factor in making driving at night easier and safer. The aim of road lighting is to produce a uniform luminance on the road, so that unlit objects can be seen against the road surface in silhouette. In other words, the aim is to light the road so that obstacles on the road have a high luminance contrast when seen against the road. Given this aim, it seems reasonable to assume that the uniformity of the luminance of the road surface should also be important. Gallagher *et al.* (1974) failed to find any effect of illuminance uniformity on time to target but the average illuminances for which uniformity was studied were all at 10 lx or greater. Scott (1980), in his study of the relationship between night/day accident ratio and road lighting, did find an effect of luminance uniformity, at least for accidents not involving pedestrians. Specifically, the equation predicting the night/day accident ratio was of the form

$$N_R = 0.66 \, e^{(1.49U - 0.75L)}$$

where N_R is the night/day accident ratio, U the ratio of minimum to average road surface luminance, and L the average road surface luminance (cd/m^2).

In this equation, the effect of greater uniformity is to increase the night/day accident ratio, the increase in night/day accident ratio being less as the average road surface luminance increases. At first, this link between greater luminance uniformity and greater night/day accident ratios seems counterintuitive. However, it may be explained by the importance of luminance contrast for detecting obstacles. In principle, if the road surface luminance were perfectly uniform, then any object with the same luminance as the road would be invisible at all positions because it would have zero luminance contrast. If the road surface luminance was variable, then the same object would be visible in some positions. Given the variations in reflectance and shape of objects that occur on roads, it is possible that at least some would be difficult to see against a completely uniform luminance road. This speculation suggests that some luminance non-uniformity is useful, but not too much. If non-uniformity is large enough to produce large areas of low luminance on the road, then it will be difficult to detect low reflectance objects. Narisada (1971) used a simulated roadway lighting scene to measure the ability to detect a square shape of side 0.2 m, with a minimum luminance contrast of 0.25. He found that for a given road surface luminance, the greater the uniformity, the higher the probability of detecting the obstacle. He also reported that in order to maintain a given probability of detection, the average road surface luminance should be increased if the uniformity is decreased, the relationship being of the form

$$L = L_p/U$$

where L is the average road surface luminance (cd/m^2), L_p the average road surface luminance for the set probability of detection when $U = 1.0$ (cd/m^2), and U the minimum/average road surface luminance.

Taken together, these results suggest that road surface luminance uniformity is a mixed blessing. Complete uniformity ensures that some obstacles will always be visible and others will never be visible. Very non-uniform road surface luminances implies that the same object will vary in visibility depending on what part of the road the object is seen against. CIE (1995c) recommends a minimum to average road surface luminance of at least 0.4.

As if this consideration of the effect of luminance uniformity were not complicated enough, it is important to appreciate that it still represents a very simplified version of reality. Objects on the road rarely consist of a single surface of one reflectance on a single plane. Often, they are composed of surfaces on different planes and the reflectances of the surfaces differ in both kind and degree, i.e. they can be both diffusely and specularly reflecting to different extents. Further, because these surfaces are in different orientations to the observer, the illumination falling on each surface will be different. All this means that many objects found on roads will have a wide range of luminances and hence will have multiple luminance contrasts. And this is without considering the interaction between road lighting and vehicle forward lighting. Road lighting is designed to illuminate a horizontal surface with the expectation that objects on the road will be seen against the road in silhouette. Vehicle forward lighting is effective in lighting the vertical surfaces of objects. Depending on the relative contribution of road lighting and forward lighting, the object on the road may be seen as darker than the road surface, may blend into the road surface or be brighter than the road surface. Ketvirtis (1977) measured the visibility distance of three different objects on lit dual carriageway roads with the test vehicle using low-beam headlights and experiencing opposing traffic, for different road surface luminances produced by road lighting alone (Figure 10.12). The visibility distance shows a deflection in the trend with increasing average road surface luminance. As average road surface luminance approaches 0.2 cd/m^2 there is decrease in visibility distance followed by an increase, that can be ascribed to the luminance contrasts of the object passing through zero, i.e. changing from positive to negative polarity.

Another factor that may influence the visibility of objects on the road is disability glare from the road lighting luminaires. Disability glare, as its name implies, produces a measurable change in visibility because light scattered in the eye reduces the luminance contrasts in the retinal image. The CIE (1995c) recommendations for road lighting limits disability glare by restricting the allowed percentage increase in luminance difference threshold occurring under the roadway lighting when the equivalent veiling luminance is allowed for. This criterion is known as the threshold increment, limiting values varying from 10 to 15 percent, depending on the class of

Visibility distance (m)

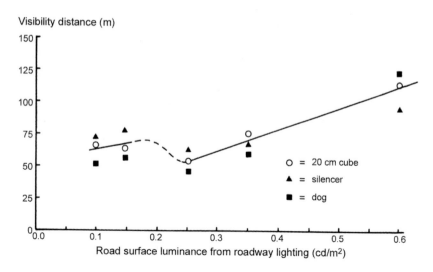

Figure 10.12 Visibility distances for a 20 cm side cube (reflectance = 0.05), a car silencer (reflectance = 0.12), and a poodle dog (reflectance = 0.04), plotted against the road surface luminance produced by roadway lighting. The visibility distances were obtained from a vehicle using low beam headlights and facing opposing traffic (after Ketvirtis, 1977).

road. The percentage threshold increment (TI) can be obtained from the formula

$$TI = 65 \ (L_v/L^{0.8})$$

where L_v is the equivalent veiling luminance (cd/m^2) (see Section 10.3.2) and L the average road surface luminance (cd/m^2).

The importance of disability glare diminishes rapidly as the angle between the line of sight and the luminaire increases. Scott (1980) failed to find any relationship between percentage threshold increment and accident frequency, probably because disability glare will vary widely depending on the relative positions of the vehicle and the luminaire as the former moves along the road. In general, disability glare from road lighting luminaires is a minor matter compared to the disability glare produced by vehicle headlights.

Another aspect of glare from road lighting luminaires that needs to be considered is discomfort glare. The study of discomfort glare from road lighting has a long and tortuous history, starting with Hopkinson (1940), and continuing until the present day, but without reaching any very definite conclusion. Three alternative metrics have been developed at various times; the luminaire luminance corresponding to the boundary between comfort and discomfort (De Boer *et al.*, 1960), the Glare Mark (Adrian and Schreuder, 1970), and the Cumulative Brightness Evaluation (IESNA, 1980). Adrian

(1991) examined each of these formulations and showed that despite their apparent differences they actually had a very similar form, and a form that would be expected from the work on discomfort glare in interiors. Specifically, discomfort glare from road lighting increased with increasing luminaire luminance and solid angle subtended at the driver's eye, and decreased with increasing road surface luminance and increasing angle between the road lighting luminaire and the line of sight. There were differences in the exponents used in the three metrics but Adrian concluded that they all had the same characteristics and would yield comparable results when used to evaluate different road lighting installations. Despite this conclusion, the latest CIE recommendations for road lighting (CIE, 1995c) have dropped the previously used Glare Mark system and replaced it with a simple assertion that experience suggests that road lighting installations designed to limit threshold increments to less than 10–15 percent are generally acceptable as regards discomfort glare.

As with all lighting installations, the designer of road lighting has three aspects to manipulate; the amount of light, the distribution of the light, and the color of the light. So far this discussion of road lighting has concentrated on the amount and distribution of light. It is now necessary to consider the last element, the color of the lighting. Road lighting varies greatly in the color of light produced, from the monochromatic yellow of low-pressure sodium discharge lamps, through the orange of high-pressure sodium discharge lamps, to the white of metal halide discharge lamps. Occasionally, other colors produced by relics of the past can be found in the form of road lighting installations using incandescent, fluorescent, or mercury vapor lamps. Several different studies have been made of the effectiveness of these light sources for making largely achromatic objects on the carriageway visible, without any clear conclusions, suggesting that any effects are small (Eastman and McNelis, 1963; De Boer, 1974; Buck *et al.*, 1975). One common feature of these evaluations is that all the measurements were taken directly viewing the object, i.e. the retinal image fell on the fovea of the retina. More recent measurements of the effect of light spectrum on the detection of off-axis targets, in mesopic conditions, suggest that there may be an effect of light color relevant to road lighting. Specifically, He *et al.* (1997) carried out a laboratory experiment in which high-pressure sodium and metal halide lamps were compared for their effects on the reaction time to the onset of an achromatic 2° disk, either on-axis or 15° off-axis, for a range of photopic luminances from 0.003 to 10 cd/m². The luminance contrast of the disc against the background was constant at 0.7. The same lamp was used to produce both the background luminance and the stimulus. Figure 10.13 shows the median reaction time to the onset of the stimulus, on- and off-axis, for a range of photopic luminances, for two practiced subjects. From Figure 10.13, it is evident that reaction time increases as photopic luminance decreases from the photopic to the mesopic state, for both on- and off-axis detection. There is no difference between the

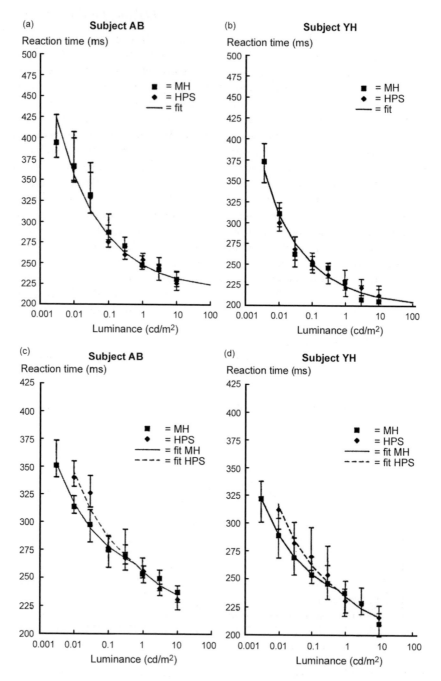

Figure 10.13 Median reaction times, and the associated interquartile ranges, to the onset of a 2°, high-contrast target seen either (a and b) on-axis or (c and d) 15° off-axis, and illuminated using either high-pressure sodium (HPS) or metal halide (MH) light sources, for photopic luminances in the range 0.003–10 cd/m² (after He *et al.*, 1997).

two light sources in the change of reaction time with luminance for on-axis detection, but for off-axis detection, the reaction times for the two light sources begin to diverge as vision enters the mesopic region. Specifically, the reaction time is shorter for the metal halide lamp at the same photopic luminance, and the magnitude of the divergence between the two sources, increases as the photopic luminance decreases. These findings can be explained by the structure of the human visual system. The fovea, which is what is used for on-axis vision, contains only cone photoreceptors, so its spectral sensitivity does not change as adaptation luminance decreases until the scotopic state is reached, at which point the fovea is effectively blind. The rest of the retina contains both cone and rod photoreceptors. In the photopic state, the cones are dominant but as the mesopic state is reached the rods begin to have an impact on spectral sensitivity until in the scotopic state, the rods are completely dominant (see Chapter 2). Given the different balances between rod and cone photoreceptors in different parts of the retina and under different amounts of light, it should not be surprising that the metal halide lamp produces shorter reaction times for off-axis detection than the high pressure sodium lamp in the mesopic range because it is better matched to the rod spectral sensitivity. It is also evident why there is no difference between the two light sources for on-axis reaction times.

Lewis (1999) has obtained similar results. Figure 10.14 shows the mean reaction time to correctly identify the vertical or horizontal orientation of a large, achromatic, high contrast, 13° by 10° grating, where the grating was lit by one of five different light sources; low-pressure sodium, high-pressure

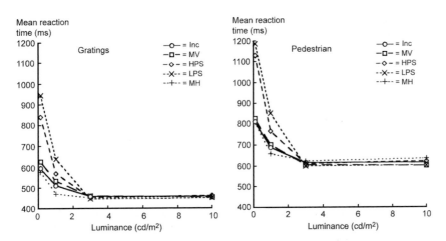

Figure 10.14 Mean time to correctly identify the vertical or horizontal orientation of a grating and the direction a pedestrian located adjacent to a roadway is facing, plotted against the photopic luminance produced by five different light sources (Inc = incandescent, MV = mercury vapor, HPS = high-pressure sodium, LPS = low-pressure sodium, MH = metal halide) (after Lewis, 1999).

sodium, mercury vapor, incandescent, and metal halide, plotted against the photopic luminance. As long as the visual system is in the photopic range, i.e. at 3 and $10 \, cd/m^2$, there is no difference between the different light sources provided they produce the same photopic luminance. However, when the visual system is in the mesopic state, i.e. at 1 and $0.1 \, cd/m^2$, then the different light sources produce different reaction times, the light sources that better stimulate the rod photoreceptors (incandescent, mercury vapor, and metal halide) giving shorter reaction times than the light sources that stimulate the rod photoreceptor less (low- and high-pressure sodium).

Such measurements of the time to detect the onset of abstract targets under different light sources may seem irrelevant to the task of driving but, in fact, driving often requires the visual system to extract information from the peripheral visual field. Lewis (1999) verified that the spectral power distribution of a light source does have an effect on the time taken to extract information of relevance to driving, by repeating the experiment described above but replacing the gratings with a transparency of a woman standing at the right side of a roadway in the presence of trees and a wooden fence. In one transparency, the woman was facing towards the road, in the other she was facing away from the road. The subject's task was to identify which way the woman was facing. Figure 10.14 also shows the mean reaction times for this task, under the different light sources, over a range of photopic luminances. Again, there is no difference between the light sources as long as the visual system is in the photopic state but once it reaches the mesopic state, the light sources that more effectively stimulate the rod photoreceptors produce faster reaction times.

Another approach to evaluating the effect of light spectrum in mesopic conditions measured the probability of detecting the presence of a target off-axis. Bullough and Rea (2000) used a simple driving simulator based on the projected image of a road, controlled through computer software. The subject could control the speed and direction of the vehicle along the road through a steering wheel and accelerator. The computer monitored the time taken to complete the course and the number of crashes occurring. Filters were applied to the projected image of the course to simulate the light spectrum of both high-pressure sodium and metal halide lighting and more extreme red and blue light for a range of luminances. Interestingly, there was no effect of light spectrum on the time taken to complete the course, i.e. on driving speed, but there was a marked effect on the ability to detect the presence of a target near the edge of the roadway. The light spectra that stimulates the rod photoreceptors more (blue and metal halide) led to a greater probability of detection than light spectra that did not stimulate the rod photoreceptors so effectively (red and high-pressure sodium).

Of course, the consequences of making an error while driving a simulator is not quite the same as driving a real vehicle on the road and so may lead subjects to pay less attention to what they are doing. Fortunately,

Akashi and Rea (2001a) have extended this work into the field, having people drive a car along a short road while measuring their reaction time to the onset of targets 15° and 23° off-axis. The lighting of the road and the area around it was provided by either high-pressure sodium or metal halide road lighting, adjusted to give the same amount and distribution of light on the road, and seen with and without the vehicle's halogen headlights. There was a statistically significant difference between the high-pressure sodium and metal halide lighting conditions. Specifically, the mean reaction time to the onset of the targets was shorter for the metal halide lighting than for the high-pressure sodium lighting at both angular eccentricities (Figure 10.15). Using the same experimental site and equipment, Akashi and Rea (2001b) also examined the effect of disability glare, provided by halogen headlights from a stationary car in the adjacent lane, on the ability of a stationery driver to detect off-axis targets at 15° and 23° when the road lighting was provided by metal halide and high-pressure sodium lighting. Again, the mean reaction times to the onset of the targets were longer for the high-pressure sodium road lighting than for the metal halide road lighting. As might be expected, the mean reaction times were also longer when the headlights in the opposing vehicle were switched on.

Given the results discussed above there can be little doubt that light spectrum is a factor to be considered for road lighting. Currently, the most

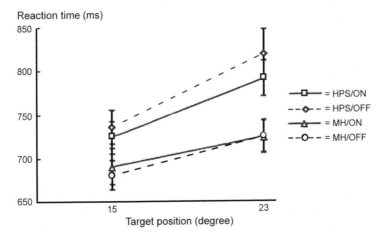

Figure 10.15 Mean reaction times (and the associated standard errors of the mean) to the onset of a target at 15° and 23° off-axis while driving, with high-pressure sodium (HPS) and metal halide (MH) road lighting, and with halogen headlights turned on and off. The road lighting using the two light sources was adjusted to give similar illuminances and light distributions. The rectangular target subtended 3.97×10^{-4} steradians for the 15° off-axis position and 3.60×10^{-4} steradians for 23° off-axis position. Both targets had a luminance contrast against the background of 2.77 (after Akashi and Rea, 2001a).

widely used light source for road lighting is HPS although the low-pressure sodium lamp is extensively used in some countries where its high luminous efficacy is valued enough to offset its complete lack of color rendering. In North America, the battle for the road lighting market is presently being fought between the high-pressure sodium lamp and the metal halide lamp. The results discussed above have been seized upon by advocates for the metal halide lamp as confirming the universal benefits of "white light" as opposed to "orange light" but, in reality, the implications of the results are rather more complex than such a simple statement suggests. Specifically, the benefit of choosing a light source that stimulates the rod photoreceptors more depends on the driver's adaptation luminance and the balance between on- and off-axis tasks. Provided the adaptation luminance is such that the visual system in operating in the photopic state, say $3 \, cd/m^2$ and above, there is no effect of light spectrum on off-axis reaction time so the choice can be made on the other factors such as lamp luminous efficacy and lamp life. If the adaptation luminance is in the high mesopic, say about $1 \, cd/m^2$, the effect of light spectrum is slight. It is only when the adaptation luminance is well below $1 \, cd/m^2$ that the choice of light source is likely to make a significant difference to off-axis visual performance. How often this occurs is open to question. Olson *et al.* (1990) have estimated that the adaptation luminance for a driver using low-beam headlights on an otherwise unlit road is about $1 \, cd/m^2$. Adding road lighting will increase this adaptation luminance, thereby reducing the impact of light spectrum. Of course, this argument depends on the assumption that the driver continues to fixate in the same positions, with and without the road lighting. This too seems unlikely. Mortimer and Jorgeson (1974) found that when driving at night, eye fixations tended to be confined to the lit area. The lit area is much larger in the presence of road lighting than with headlights alone, and the range of luminances will be much wider, in which case, what the adaptation luminance is remains open to question. Some measurements of the state of adaptation of the visual system while driving at night in different lighting conditions are urgently needed.

However, if it is assumed that there are conditions in which the visual system will be operating in the mesopic state while driving at night, it is then necessary to consider the balance between on- and off-axis tasks. This balance is important because it is sometimes argued that the metal halide lamp can be used at a lower road surface luminance than the high-pressure sodium without penalty. Certainly, the results discussed above suggest that, in mesopic conditions, the same reaction time to the onset of an off-axis target can be achieved at a lower adaptation luminance with a metal halide lamp (see Figure 10.13c and d). However, for an on-axis target, the same adaptation luminance is required for the same reaction time for both light sources (see Figure 10.13a and b). Thus, it is only if off-axis detection is assumed to be the most important task in driving that a reduction in road surface luminance for metal halide lamps can be justified. There cannot be many drivers who would be willing to deny the importance of both on- and off-axis vision for driving.

There is clearly some way to go before the full impact of this new understanding of the effect of light spectrum is incorporated into practice. Current recommendations for good road lighting confine themselves to the amount and distribution of light. The CIE (1995c) recommends a maintained average road surface luminance, a minimum overall and a longitudinal road surface luminance uniformity and a maximum initial threshold increment, for five different classes of road, and a surround illuminance ratio for roads with adjacent footpaths. The roadway lighting recommendations used in North America (IESNA, 2000b) take three forms, first based on the illuminance distribution on the roadway, second based on the pattern of road surface luminance and the disability glare from the luminaires, and third based on what is called small target visibility. This last form is an attempt to develop a metric that directly quantifies how effective the lighting is in achieving one of its objectives, namely, to make objects on the road visible. Small target visibility is measured as the weighted average of the visibility level of a flat, 18 cm side, square target with a diffuse reflectance of 0.5, mounted vertically, at an array of points on the road surface. The target is always taken to be 83 m ahead of the observer, with the observer-to-target sight line parallel to the centerline of the road. This distance and viewing direction convention maintains constant both target size and target/observer geometry. The visibility level of the target is defined as the ratio of the actual contrast of the target to the threshold contrast of the target; the higher the ratio, the more visible the target. The visibility level of the target at a given position on the road can be calculated from its components. Adrian (1989) has provided a quantitative model that allows the calculation of threshold luminance difference for different sizes of targets as a function of background luminance, for positive and negative contrast. The actual luminance difference that will occur under a proposed lighting installation can be obtained by calculating the road surface luminance and the target luminance, including, for the latter, both the light received on the target directly from the luminaires and after reflection from the road surface (Adrian *et al.*, 1993). Small target visibility studies have an extensive history starting with Gallagher *et al.* (1974), who was able to show a relationship between the visibility of a target cone in the road and the distance at which unwarned drivers took avoiding action, continuing through Janoff (1990), who showed that there was a clear relationship between the visibility of the target and the distance at which it could be detected and recognized, and Janoff (1992) who found a link between the visibility level of small targets and subjective ratings of their visibility.

The small target visibility approach is a creditable attempt to move the design of road lighting away from photometry to perception. Unfortunately, it is probably doomed to failure because the visual task of driving is multifaceted and small target visibility addresses only one facet and only one source of light, the road lighting. The driver certainly has to be able to detect small targets on the roadway but also needs to be able to detect movement at the edges of the roadway using peripheral vision, to keep the vehicle between lane markers; to judge the relative speeds and direction of movement of other vehicles and to identify many different signs and signals, and the driver has to do all this under lighting conditions created by a combination of road lighting and the forward lighting of vehicles. This last point is important because the different light distribution patterns of road lighting and vehicle forward lighting can produce dramatic differences in visibility depending on the balance between them (Oya *et al.*, 2000). Even if the driver's task were limited to the detection of obstacles on the road, it must be admitted that obstacles are not always small and they can have multiple contrasts, either within the obstacle or between the obstacle and the background. The existence of a number of different contrasts associated with a single target poses a problem for the concept of a single visibility level for that target, because the target then has multiple visibility levels, a difficulty experienced by Blackwell *et al.* (1964) in their early work on visibility measurement. A cynic could argue that the small, diffusely reflecting target placed on the road and used in the small target visibility approach, was chosen more because it eliminates the problem of multiple contrasts that for its relevance to driving. Until the complexity of real targets and the multi-faceted nature of the driver's task is recognized it seems likely that the visibility approach to determining road lighting will be barren. A more fruitful approach would be to first carry out a task analysis to identify the objects that are critical to safety in different driving situations. These objects are most likely not to carry any form of luminous marking, e.g. a pedestrian on a city street or a piece of tyre on a motorway. Second, to examine the range of contrasts that occur with representative samples of these critical objects viewed over an array of different positions under different combinations of road lighting and vehicle forward lighting. Third, to set up a system for grading the quality of different road lighting systems based on the proportion of contrasts that fall below a critical value. This is only an example of what might be done. In general, it is not until an effort is made to examine the combined effects of road lighting and vehicle forward lighting on the detection and recognition of realistic objects that progress in designing better road lighting systems will be made. If this were to be done, some innovative approaches to road lighting might appear (Oya *et al.*, 2000; Rea, 2001).

10.5 Tunnel lighting

One road lighting situation in which special care is required if good visibility is to be maintained is when the road passes through a tunnel. Tunnel lighting has to address two different problems. The first is the black-hole effect experienced by a driver approaching a tunnel. The second is the black-out effect caused by a lag in adaptation on entering the tunnel. Neither of these problems occurs at night, because then the road surface luminance inside the tunnel is recommended to be a minimum of 2.5 cd/m^2, a value likely to be similar to, if not more than, the road surface luminances outside the tunnel (IESNA, 1996a). By day, this is not the case. By day, the luminances around the tunnel portal will be much higher than those inside the tunnel so both the black-hole effect and the black-out effect may be experienced and driver safety may suffer (Ueki *et al.*, 1992).

The black-hole effect refers to the perception that from the distance at which a driver needs to be able to see an obstruction in the entrance to the tunnel, that entrance is seen as a black hole. The original studies of tunnel lighting assumed that misadaptation was the major cause of the black-hole effect (Schreuder, 1964). However, Narisada and Yoshikawa (1974) showed that drivers typically start to fixate on the tunnel entrance at about 150–200 m from the entrance. At a speed of 100 km/h (62 miles/h), this means the driver is fixated on the tunnel entrance for about 5–7 s before entering, which would allow time for foveal adaptation to occur, especially as the solid angle subtended by the tunnel entrance increases as the driver approaches the tunnel. This finding, together with awareness of the fast neural process of adaptation, led Adrian (1982) and Vos and Padmos (1983) to conclude that the major cause of the black-hole effect is the reduction in luminance contrast of the retinal image of the tunnel entrance caused by scattered light in the eye.

Given that the low luminance contrasts of vehicles and objects in the tunnel entrance is the problem drivers experience when approaching a tunnel in daytime, there are two approaches that can be used to alleviate it. The first is to reduce the luminance of the surroundings to the tunnel. Reducing the luminance of the surroundings reduces the amount of scattered light and hence the extent to which the luminance contrasts of the retinal image of vehicles and obstacles in the tunnel entrance are reduced. The luminance of the surroundings can be reduced given appropriate construction, e.g. by ensuring that the tunnel portal is of low reflectance, by shading the tunnel portal and the road close to the tunnel entrance with louvres designed to exclude sunlight, by using low-reflectance road surface materials outside the tunnel and by landscaping to shield the view of high-luminance sources, such as the sky. This last possibility is particularly valuable where the tunnel is oriented east–west so that the sun can be seen immediately above the tunnel entrance at some times of the day and year. The second is to increase the luminance contrast of vehicles and obstacles inside the tunnel entrance. This can be done by the choice of materials used in the tunnel entrance and by increasing the road surface luminance. The road surface inside the tunnel entrance should be of higher reflectance than that immediately outside and the walls of the tunnel, against which vehicles and objects in the tunnel are usually seen, should have a reflectance of at least 0.50. As for the road surface luminance, there are a number of ways to determine the luminance required in the tunnel entrance by calculating the equivalent veiling luminance corresponding to the scattered light at the fovea, given the luminances that occur in different parts of the visual field around the tunnel entrance, and then estimating the road surface luminance required to keep targets of different sizes and luminance contrasts above visual threshold (Adrian, 1982, 1987). There are also tabulated sets of recommendations of the road surface luminances required in tunnel entrances, in different settings, based on these methods (CIE, 1990b; IESNA, 1996a).

As the driver approaches the tunnel, the proportion of the visual field taken up by the tunnel entrance increases and the contribution of the scattered light to the luminance of the retinal image of the entrance is reduced until, at the tunnel entrance, the luminous veil produced by scatter in the eye is much diminished (Hartmann *et al.*, 1986). This does not mean the drivers problems are at an end. There is still the potential for a reduction in visual capabilities caused by misadaptation on entering the tunnel (Bourdy *et al.*, 1988). During the day, the driver approaching the tunnel will be adapted to the exterior conditions and on entering the tunnel will have to adapt to a much lower luminance. Whether there is enough time for adaptation to occur depends on the driver's speed and the distance allowed for the change of luminance. Given a slow enough speed and a long enough distance, the visual system is well able to look after itself so there would be no need for the tunnel lighting to be designed to allow for adaptation. However, reducing speed on the approach to a tunnel is not beneficial in terms of traffic flow and may be dangerous in heavy traffic, so such a behavioral solution is not often adopted. Rather, the lighting is designed to overcome the black-out problem. The approach used is to gradually decrease the road surface luminance in the tunnel, from a threshold zone starting at the tunnel portal, through a transition zone, to the interior zone. The length of these zones is determined by what is termed the safe stopping sight distance (SSSD) (AASHTO, 1990). This is the estimated distance in which a vehicle can stop on a straight and level but wet tunnel approach when traveling at or near the speed limit. The faster is the speed of the vehicle approaching the tunnel, the longer is the SSSD. Figure 10.16 shows the profile to be followed for grading the road surface luminance from the threshold zone, through the transition zone to the interior zone. Different points on Figure 10.16 are determined by different conditions. The recommended road surface luminances for the threshold zone in daytime, which correspond to 100 percent luminance in Figure 10.16, covers a luminance range of 140–370 cd/m^2 (IESNA, 1996a). The exact luminance recommended is determined by the surroundings of the tunnel portal, and adjusted for different traffic flow rates, tunnel lengths, tunnel surface reflectances, and daylight penetration. The length of the threshold zone is one SSSD less 15 m; because at 15 m from the tunnel portal, the tunnel fills a significant proportion of the visual field and adaptation is assumed to be occurring. The length of the transition zone is one SSSD, which in turn

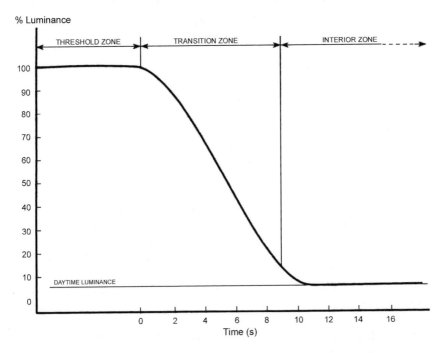

% Luminance

Figure 10.16 A road surface luminance profile from the threshold zone, through the transition zone to the interior zone of a tunnel, plotted against time from entering the transition zone. The length of the transition zone will depend on the assumed speed of vehicles using the tunnel (after IESNA, 1996a).

depends on the assumed vehicle speed. The road surface luminance of the interior zone in daytime depends on the speed and density of traffic in the tunnel and covers a range of 3–10 cd/m^2, the faster the speed and the higher the traffic density, the higher the average road surface luminance recommended in the interior zone (IESNA, 1996a). This interior zone luminance is continued up to the exit from the tunnel, no special treatment being needed at the exit because of the high luminance of the tunnel exit and the faster speed of adaptation from lower to higher luminances than from higher to lower luminances (see Section 2.3.1). This approach of grading the luminance over the threshold zone, transition zone, and interior zone is only used where the tunnel is longer than the SSSD or where the tunnel is curved. If the tunnel is shorter than the SSSD and the exit of the tunnel can be seen as the tunnel is approached, vehicles and objects in the tunnel entrance will be seen in silhouette against the tunnel exit and so will be highly visible during the day. For such short tunnels the same luminance is used throughout the tunnel, the luminance being that recommended for the threshold zone in longer tunnels.

As for the type of lighting used to provide the luminances in the tunnel, the light source most commonly used is one of the discharge sources, because of their high luminous efficacy, long life, and robustness. The luminaires used in tunnels have to be of rugged construction to deal with vibration, dirt, chemical corrosion, and washing with pressure jets. Three types of light distribution are used, symmetrical, counter-beam, and pro-beam lighting. Symmetrical light distributions produce uniform luminance lighting throughout the tunnel so vehicles of different reflectances will tend have either positive or negative luminance contrasts with the road. Counter-beam light distributions are those where the light is directed predominantly against the traffic flow. This gives a high pavement luminance so that vehicles tend to be seen in negative contrast, but there is some risk of the driver experiencing discomfort and disability glare. Pro-beam light distributions are those where the light is directed predominately in the direction of the traffic flow. This gives a low road-surface luminance but high luminances for vehicles so the vehicles tend to be seen in positive contrast. Various claims have been made about the benefits of these systems (Novellas and Perrier, 1985; Schreuder, 1993) but no consensus has been reached. About the only thing that can be said with confidence is that a lighting system that changes the luminance contrast of a vehicle from positive to negative or vice versa as the vehicle moves through the tunnel must produce at least one location where the luminance contrast is zero, something that will make the vehicle impossible to see.

Finally, it is necessary to consider the potential for flicker and the consequent discomfort and distraction to the driver. When tunnel lighting is provided by a series of regularly spaced, discrete luminaires, there is always a possibility of flicker being perceived. It is recommended that care be taken to avoid spacing individual luminaires so that drivers moving at representative speeds in the tunnel are not exposed to flicker in the range 5–10 Hz (IESNA, 1996a). Of course, flicker is only a consideration if the lighting is provided by discrete luminaires. An alternative system based on a continuous linear luminaire through the tunnel avoids any flicker problem and provides good visual guidance for the tunnel, a feature that is particularly valuable where the tunnel curves.

10.6 Signals, signs, and messages

A ubiquitous feature of roads in urban and suburban areas is the traffic signal. Traffic signals are placed at intersections to identify priorities for both vehicular and pedestrian traffic. The photometric characteristics of traffic signals are closely regulated in terms of their luminous intensity distributions and color (ITE, 1985, 1998; National Police Agency, Japan 1986; CIE, 1994; CEN, 1998). The recommendations are consensus decisions made by a committee, but those decisions is based, at least in part, on studies of the reaction time to the onset of the signals and the number of signals that are not detected under different conditions. Recently, Bullough *et al.* (2000) have reported an extensive series of measurements of reaction time and missed signals using a tracking task requiring continuous fixation and simulated traffic signals occurring a few degrees from the fixation point, the traffic signals being provided by both incandescent and LED light sources. Figure 10.17 shows

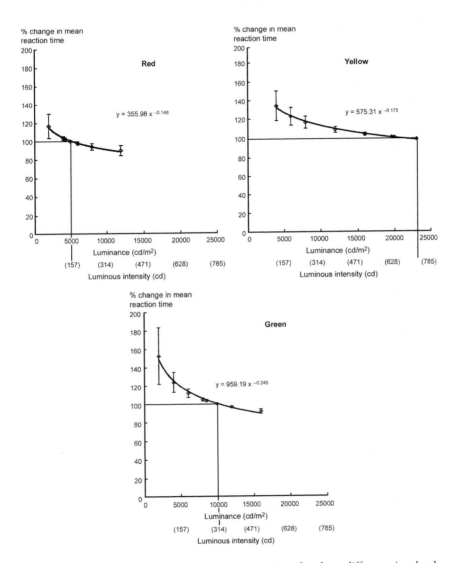

Figure 10.17 Percentage change in mean reaction time for three different signal colors provided by LEDs, plotted against signal luminance. The normalization for each color, i.e. 100 percent mean reaction time, is set at the signal luminance for 200 mm diameter signals recommended by the Institute of Transportation Engineers; 5,000 cd/m² for red, 23,121 cd/m² for yellow, and 10,000 cd/m² for green (ITE, 1985). The second horizontal axis is the luminous intensity corresponding to the signal luminance for a 200 mm diameter signal (after Bullough *et al.*, 2000).

the percentage change in mean reaction times for three LED traffic signal colors, over a range of signal luminances, seen against a 5,000 cd/m² large-area background, i.e. against a simulated daytime sky. The trend for reaction times to become shorter with increasing signal luminance is clear. Figure 10.18 shows the percentage of missed signals for the same signals plotted against signal luminance. A missed signal was one that was lit for more than 1 s without a response from the subject. Again, it is evident that increasing the signal luminance reduces the percentage of missed signals until a minimum level is reached.

The relevance of small changes in reaction time to how effective a traffic signal is may be open to question. After all, a vehicle traveling at 80 km/h (50 miles/h) only covers 2.2 m (7.3 ft) in 100 ms, so adding 100 ms to the reaction time will make little difference to whether a driver will be able to stop at

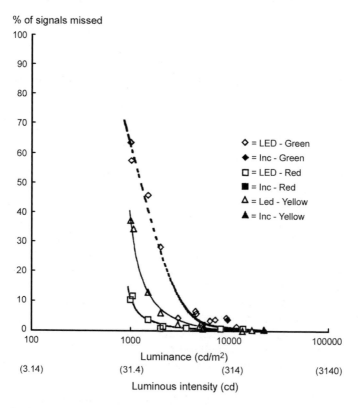

Figure 10.18 Percentage of signals missed for each signal color and luminance plotted against signal luminance. The signals were provided by either LED or filtered incandescent light sources. To be counted as a missed signal, the signal had to have been on for 1 s without a response from the subject. The second horizontal axis is the luminous intensity corresponding to the signal luminance for a 200 mm diameter signal (after Bullough *et al.*, 2000).

a traffic signal. The percentage of missed signals is much more relevant because if the signal is not seen the driver will not behave appropriately. The practical implication of Figure 10.18 is that different traffic signal colors need to have different luminances to achieve the same percentage of missed signals, unless the luminances are so high that virtually no signals are missed. This observation explains differences in practice in different countries. The US has different luminous intensities for different signal colors (ITE, 1985, 1998) while Japan and Europe recommend the same but higher luminous intensity for all traffic signal colors (National Police Agency, Japan 1986; CEN, 1998).

Another point of practical importance is that there is no difference in missed signals between LED and incandescent traffic signals of the same luminous intensity and the same nominal color (Bullough *et al.*, 2000). This suggests that either source can be used to create a visually effective traffic signal. However, it would be premature to reach such a conclusion without considering the ability of people to correctly identify the signal color. Boyce *et al.* (2000c) examined this question for both incandescent and LED signals in daytime, both light sources producing signals with chromaticity coordinates inside the ITE color boundaries for signal lights (ITE, 1985). They found that errors in color identification were rare but, when they did occur, were most likely to occur for off-axis viewing, at signal luminances below about 8,000 cd/m^2, for the incandescent light source and the yellow signal, the yellow signal being identified as red. These results suggest that, visually, the LED light source has a slight perceptual advantage over the incandescent light source for use in traffic signals

One general conclusion from the results discussed above is that the higher is the luminous intensity of the signal, the shorter is the reaction time, the fewer are the number of missed signals and the more likely it is that the signal color will be correctly identified. This suggests that the higher is the luminous intensity, the better is the signal, but there is a limit to how far the luminous intensity of a signal can be taken. A traffic signal has to be seen both day and night. A higher luminous intensity is of value during the day because it will tend to increase the conspicuity of the signal but by night a high luminous intensity can become a source of discomfort and even disability glare. Bullough *et al.* (2001a) measured the percentage of people considering traffic signals of different luminances uncomfortable when viewing them directly (Figure 10.19). Such data can be used to set desirable traffic signal maximum luminances at night, which might be different from the maximum allowed by day.

It is important to appreciate that all the findings discussed so far have been collected from people with normal color vision. Cole and Brown (1966) examined the difference in reaction time to a red signal for people with normal color vision and for protanopes, i.e. with the long-wavelength cone photoreceptor missing (see Section 2.2.7). They found that the protanopes had longer reaction times than color normals to the same red signal and were much more likely to miss the signal at lower signal luminances.

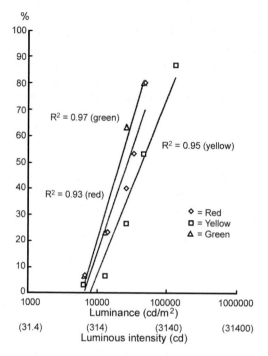

Figure 10.19 Percentage of subjects considering a signal uncomfortable for the three signal colors seen in darkness, plotted against signal luminance. The signal simulates a 200 mm diameter signal seen from a distance of 20 m. The second horizontal axis is the luminous intensity corresponding to the signal luminance for a 200 mm diameter signal (after Bullough *et al.*, 2001a).

Nathan *et al.* (1964) also found longer reaction times for people with color defective vision as well as more mistakes in identifying the signal color. That such observations can have behavioral consequences is shown by the finding that people with defective color vision had more accidents involving failure to halt at a red light than people with normal color vision (Neubauer *et al.*, 1978). There are three possible solutions to this problem. The most draconian would be to ban people with seriously defective color vision from driving. This is what is done for public service vehicles, but given the need for individual mobility in modern life and the long history of people with defective color vision driving, it seems unlikely that there would be public support for such a ban. An alternative approach would be to increase the minimum luminous intensities of traffic signals to compensate for the reduced sensitivity to some wavelengths shown by different forms of color defective vision and to more closely define the chromaticity coordinates of traffic signals so that the colors are more widely separated in color space. The former could be done but would pose a risk of greater discomfort for drivers with normal color vision, especially at night. The latter could be easily achieved by adopting the restricted color boundaries in the CIE signal color recommendations (CIE, 1994). These boundaries are designed to make signal colors easier to discriminate by protans and deutans. The third possibility is to modify traffic signals so that color is not the only difference between signals with different meanings. With present traffic signal designs, color defective people can use the relative positions of the red, yellow, and green signals to supplement the color information they find hard to extract. It would be interesting to incorporate other means of conveying the necessary information, such as the shape of the signal. Then, as long as the driver could tell which signal was lit, the meaning would be clear.

Another common feature of roads is signs giving information on directions, lane changes, speed limits, etc. The size, shape, and content of such signs has been extensively studied (Forbes, 1972) and have to conform to the recommendations of the authority responsible for the road, e.g. the *Manual for Uniform Traffic Control Devices* (FHA, 2000) in the US. The size, shape, and content of such signs is not of interest here but the question of whether such signs should be illuminated is. The decision on whether or not to illuminate a sign depends on the distance at which the sign needs to be detected and the distance at which it needs to be legible. These distances depend on the speed and density of traffic approaching the sign and whether the driver has to carry out some maneuver in response to the sign. High speed, dense traffic, and the need for a maneuver, all increase the distance at which the sign needs to be detected and legible. Olson *et al.* (1989) examined the effect of sign luminance on detection distance. There was a linear relationship between the logarithm of the sign luminance and detection distance, the higher the sign luminance, the greater the detection distance. As for legibility of the sign, this depends on both the luminance contrast of the elements of the sign, something that is determined by the combination of the reflectances of the elements and the luminances of the elements, something that is determined by the illumination of the sign. Allen *et al.* (1967) showed that decreasing the illuminance on the sign produced a decrease in legibility distance, following an accelerating function, as would be expected for printed material. Recommendations of the illuminance on externally illuminated signs and of the luminance of internally illuminated signs are available (IESNA, 2000b).

Providing the recommended illuminance on a sign poses no difficulty when the sign has its own lighting, as is often the case on major roads, where detection distances and legibility distances are long and overhead signs are frequently used. For this situation, the luminaire is placed close to the sign in a position where it does not obstruct the view of the sign from any meaningful direction and any specular reflections from the sign do not occur in the direction of the approaching traffic. Further, the light source is chosen to accurately reveal the colors of the sign, color often being an important element in identifying the sign before any printing is readable.

More problematic are signs on minor roads where the lighting of the sign at night is received from the forward lighting of approaching vehicles. The need to control the light distribution from forward lighting in order to limit glare to other vehicles also limits the amount of light that can be expected to reach a sign at the side of the road. To increase the amount of light reflected from the sign to the driver, these signs often contain retroreflective materials. Ideally, these materials have the optical characteristic of reflecting incident light back along its own path, i.e. light received at a sign from a headlight will be reflected back to the headlight, regardless of the angle of incidence on the sign. Of course, such material is not perfect so there will always be some spread in the reflected light distribution. What the luminance of the sign, as seen by the driver, will be will depend on the amount of incident light, the efficiency of the retroreflective material, and the angular separation of the driver from the headlights. The position of the driver relative to the headlights is not usually a problem for cars but for trucks it can be. Sivak *et al.* (1993) have shown that the luminance of retroreflective signs can be much less for truck drivers than for car drivers, and that such reduced sign luminances could have a major effect on the legibility of signs.

The other area of concern for signs is the background against which the sign is seen. The background can be important for two different reasons. The first is the presence of a very bright light source close to the sign. Such a source can produce enough disability glare to make the sign invisible. The classic example of this is a traffic signal that is seen with the setting sun immediately adjacent to it. This problem is usually solved by surrounding the traffic signal with a low reflectance screen that cuts off the view of the sun within a few degrees of the signal. The second reason is where the background against which the sign is seen is visually complex so that the sign is just one sign amongst many. This often occurs in city centers where there are a multitude of advertising signs of high luminance to compete with the road sign. Schwab and Mace (1987) examined the detection and legibility distances for signs seen against backgrounds of different complexity. They found that the more complex was the background, the shorter was the detection distance, but there was little effect on legibility distance. This is not surprising because legibility is primarily dependent on the details within the sign when the sign is fixated while a sign is usually first detected off-axis. The effectiveness of off-axis detection during visual search will be influenced by the presence of competing visual information. Schwab and Mace suggest that in the presence of a complex background either the luminance of the sign or its size should be increased, or an advanced warning sign should be used. An alternative would be to separate the sign from the background by using a large, low-reflectance surround. It would also be possible to attract attention to the sign by means of a simple strobe light.

Another form of sign that is found alongside the road with increasing frequency is the changeable message sign. These signs are used to provide information about temporary road conditions, such as the presence of work zones, and variable speed limits. Changeable message signs use a series of pixels to display a text message. The pixels may be light reflecting or light emitting. Light-reflecting pixels are usually of low reflectance on one side and high reflectance on the other, electromechanical means being used to turn them over as required. During the day, the reflectances are enough to ensure visibility of the message. At night, some illumination of the pixels is required, illumination that sometimes uses UV radiation to excite phosphors in the high-reflectance material. For the self-luminous pixels, the light sources are usually either miniature incandescent lamps or LEDs. The legibility and readability of such signs depends on many factors, including the pixel shape, letter width/height ratio, font, and letter separation. Collins and Hall (1992) showed that words with a regular pixel, a letter width/height approaching unity, an upper case font, and a letter separation of two pixels are most legible. Collins and Hall worked with light reflecting signs and so had no need to consider the luminance of the display. Padmos *et al.* (1988) carried out a series of field evaluations of self-luminous message signs, with a different number of pixels used to form the message. Figure 10.20 shows the mean message luminances set for three different displays of the number 5, for two different visibility criteria, plotted against the horizon luminance. The number 5 was viewed from 100 m. The message luminance is given by the equation

$$L_{\mathrm{mes}} = 10^6 \cdot I_{\mathrm{px}}/d^2$$

where L_{mes} is the message luminance (cd/m^2), I_{px} the pixel luminous intensity (cd), and d the distance between pixels (mm).

Figure 10.20 shows that the visibility of the message varies with the horizon luminance, the higher the horizon luminance the higher the message luminance required for the message to be visible. By using other visibility criteria, Padmos *et al.* (1988) were able to show that the message luminances necessary for a ratings of "optimum" on a bright day would be rated as "glaring" at night. This finding implies that some degree of luminance control is necessary to ensure comfortable and effective viewing of the message by day and night. Padmos *et al.* (1988) suggest that a sufficiently legible but not too bright message can be obtained by a two-step message luminance, 4,000 cd/m^2 by day and 100 cd/m^2 by night, although three steps (4,000, 400, and 40 cd/m^2) would be better.

One other application where lighting has been used to enhance drivers' awareness of a potential hazard is the pedestrian crossing. Such crossings are diligently marked by warning signs, by special road markings, and sometimes by continuously lit beacons. Despite these efforts, accidents involving pedestrians on crossings persist. One approach to increase the conspicuity of pedestrian crossings at night is to provide special lighting of the crossing (Janoff *et al.*, 1977). In this case, the lighting is designed to provide a bright stripe of light over the crossing, without glare, the light being

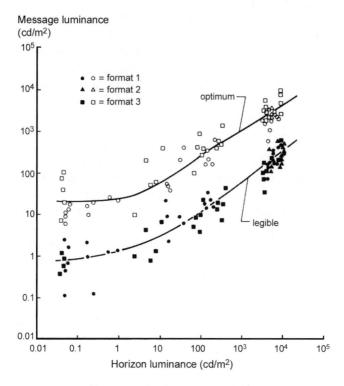

Figure 10.20 Message luminances set by individual subjects for the number 5 presented in three different formats on a self-luminous message sign, seen from 100 m, plotted against horizon luminance. The luminances were set to match two different visibility criteria at different times of day and night and hence for different horizon luminances. The criterion "optimum" is based on the perception that the display is conspicuous but not glaring. The criterion "legible" is based on the perception that the display is just recognizable (after Padmos *et al.*, 1988).

markedly different in colour to the lighting of the environment around it. Another approach, used in Switzerland, is to position luminaires to produce a vertical illuminance of 40 lx on a pedestrian on the crossing (Hasson *et al.*, 2002).

These approaches, while making the crossing more conspicuous at night, do not discriminate between an empty crossing and one in use and do nothing to increase the conspicuity of the crossing during the day. An approach that addresses both these limitations is the placing of flashing lights into the road surface at intervals across the road at the edge of the crossing (Whitlock and Weinberger Transportation, 1998; Boyce and Van Derlofske, 2002). These in-road lights, which are basically the same as airport runway marking lights, provide a narrow beam of light toward the driver. They can be made to flash, either by a simple manual control available to the pedestrian, or automatically, by the use of a system for detecting the presence of a pedestrian waiting to cross. Such systems have been shown to make the crossing more conspicuous to drivers by night and day, especially in adverse weather conditions, and to reduce drivers' approach speeds.

10.7   Rain, fog, and snow

The presence of rain, snow, or fog makes the driver's task more difficult, but in different ways. Rain, unless it is very heavy or the windscreen wipers are not working, has its main effect by changing the reflection properties of the road surface. Fog and heavy spray thrown up by other vehicles have their effect by absorbing and scattering light in the atmosphere, thereby reducing the luminance contrast of everything ahead of the driver. Snow also scatters incident light and in doing so creates a lot of visual noise in the driver's visual field as the snowflakes swirl about. In addition, if the snowfall is heavy enough, the snow will cover the road surface, obscuring lane markings and other information designed to aid the driver.

When rain covers a road surface, the effect is to make the road surface more of a specular reflector than a diffuse reflector. This is not usually a problem during the day because the meteorological conditions that tend to produce heavy rain also tend to produce diffuse illumination from the sky rather than direct illumination from the sun. However, at night, when the road is lit either by a combination of road lighting and vehicle forward lighting or by vehicle forward lighting alone, a wet road surface dramatically alters the effectiveness of that lighting. For road lighting, a wet road means that some parts of the road surface become much higher in luminance and others much darker, i.e. the road surface luminance becomes much more non-uniform. As the method by which objects are seen on the road is by silhouette against the road surface luminance, this increase in luminance non-uniformity means that individual objects will vary more in visibility as they are seen against different parts of the road. Objects that change in visibility from moment to moment are likely to be more difficult to identify. This situation is worsened by the presence of vehicle forward lighting. For a wet road, a higher proportion of the light from the vehicles forward lighting will be reflected away from the vehicle rather than back to the driver. This increased forward reflection will increase the disability glare to approaching drivers. The most successful approach to the problem of wet roads lies in the choice of road materials. A road constructed with a high proportion of aggregates will have a more coarse texture. Such a road will show less change in reflection properties between dry and wet conditions but will take longer to dry. Attention should also be paid to the permeability of the road, a road constructed of more permeable materials will reduce the probability of specular reflections occurring because water is less likely to collect on the surface. How much a road surface will change is reflection properties between dry and wet conditions and how it compares in this respect to roads constructed of other materials can be judged from the reflection classification system of Frederiksen and Sorensen (1976). The other approach to limiting the problems caused the driver by wet roads is to modify the road lighting. By changing the luminous intensity distribution of the luminaires and their position relative to the road surface, it is possible to limit the increase in road surface luminance non-uniformity when the road becomes wet. Where roads are likely to be wet for a considerable proportion of the time, the CIE (1995c) recommends that the road lighting should be designed to have a minimum luminance uniformity (minimum/average) of 0.15 for the most specular of the wet road surface reflectances in their classification system (CIE, 1979). How effective this approach is in maintaining visibility of objects on the road is open to question, the point being that a good luminance uniformity achieved at the cost of a low road-surface luminance may not represent an improvement in visibility (Lecocq, 1994).

Fog and spray consists of water droplets suspended in the atmosphere. Photons of light incident on a droplet are mainly forward scattered. If the density of the particles is high, which is what defines a fog, then a photon will have multiple impacts with droplets, scattering occurring each time. The outcome of this multiple scattering is a somewhat uniform veil of luminance covering the driver's visual field. The effect of this uniform luminance veil is to reduce the luminance contrast of all the things in front of the driver and hence to reduce their visibility. This general reduction in visibility has been shown to lead drivers reducing speed in fog (White and Jeffery, 1980), although not always enough to make the stopping distance of the vehicle less than the distance the driver can see (Sumner *et al.*, 1977). Another aspect of driving behavior that deteriorates in fog is the ability to maintain lateral position the road (Tenkink, 1988). Clearly, driving in fog is difficult and dangerous, particularly in the presence of many other vehicles.

There are a number of ways to help the driver in fog but none of them are a complete solution. One approach used in locations where fog is common is to inset retroreflective elements (also known as cats eyes) into the road, at regular intervals, as lane markers. When illuminated by the forward lighting of a vehicle, the retroreflectors have a high luminance and hence an increased contrast against the road. This helps the driver to keep the vehicle in the lane.

Another approach is to fit the vehicle with forward lighting intended for use in fog (SAE, 2001). Such foglights are usually mounted low on the vehicle, below the conventional forward lighting, and have a luminous intensity distribution that is both wide and flat. The low mounting position is advantageous because fog is usually thinner close to the road so less scatter occurs. There has been a long controversy associated with the best spectrum to be used for forward lighting in fog (Schreuder, 1976). For many years,

France required the use of yellow forward lighting, on the basis the yellow light provided better visibility in fog than white light. Whether this is true depends on the size of the water droplets forming the fog. If the water droplet size is less than the wavelength of the incident light, then scattering can occur that is wavelength dependent. However, many fogs have droplet sizes that are much larger in diameter than the wavelengths that stimulate the human visual system (Middleton, 1952). In these conditions, different degrees of scattering for different visible wavelengths does not occur. It is clear that any advantage of yellow forward lighting in fog has to rely on some factor other than decreased scattering (Reading, 1966).

Foglights are designed to help the driver see ahead. Another form of vehicle lighting used in fog is high-intensity rear lights (SAE, 2001). These are used to indicate the position of the driver's vehicle to other vehicles. They are effective because they increase the distance from which the vehicle can be detected. However, care is required to limit their use to foggy conditions because in a clear atmosphere their luminous intensity is high enough that they become a source of discomfort and sometimes a source of disability glare to following drivers.

Road lighting is sometimes installed in areas prone to fog in otherwise unit roads. How effective such road lighting will be in enhancing visibility depends on the density of the fog. This is because adding more light will increase both the luminance of the objects on the road and the luminance of the luminous veil caused by scatter. The balance between these two effects depends on the density of the fog. Of course, this assumes the road lighting is of the conventional type, mounted about 10 m above the road and therefore, light has to pass through a lot of fog before it reaches the road. Girasole *et al.* (1998) carried out a computer simulation, based on a mathematical model of multiple scattering, of the visibility of the back of a truck, a broken tyre on the road, and some lane markings, under two different forms of road lighting in fogs of different density. The two lighting systems were conventional road lighting mounted 9 m above the road and a low-mounted system (0.9 m above the road) placed on the central reservation, the light being aimed 45° across the road. Figure 10.21 shows the estimated veiling luminance from vehicle forward lighting, conventional road lighting, and the low-mounted road lighting, separately, plotted against fog concentration. Figure 10.22 shows the visibility distances for the truck and the tyre, under headlights and either the conventional road lighting or the low-mounted system, plotted against fog concentration. The low-mounted road lighting system produces much less veiling luminance than the conventional road lighting and hence increases visibility distances dramatically for low fog concentrations, but there is little difference at high fog concentrations. These results, and it must be remembered that they are based on a computer simulation rather than field measurement, suggest that new types of road lighting could be designed for use in areas where fog is common and would provide much more effective lighting than is currently available.

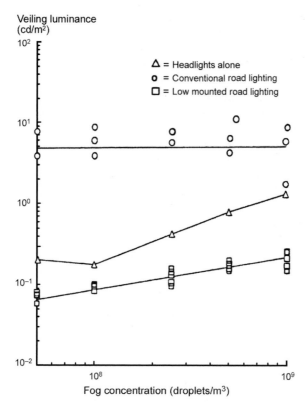

Figure 10.21 Veiling luminance plotted against fog concentration for head-lights alone, conventional road lighting and low-mounted road lighting separately (after Girasole *et al.*, 1998).

Finally, it is important to appreciate that there are other ways to improve the flow of visual information to the driver than simply improving the lighting. Nilsson and Alm (1996) examined the impact of a vision enhancement system on a driver's ability to drive safely in fog, using a driving simulator. The vision enhancement system produced a clear image of the road ahead as a small window in the scene. With the vision enhancement system drivers choose to drive in fog at a speed similar to the speed they used in a clear atmosphere and their reaction time and the distance they moved after an unexpected hazard appeared were similar to what they were without fog. However, the lateral position of the vehicle varied most with the vision enhancement system. Technological advances in sensors and computing power are making such a vision enhancement system a real possibility for everyday use in vehicles.

Snow consists of large, high-reflectance particles falling through the atmosphere. Light from vehicle forward lighting striking a snowflake will

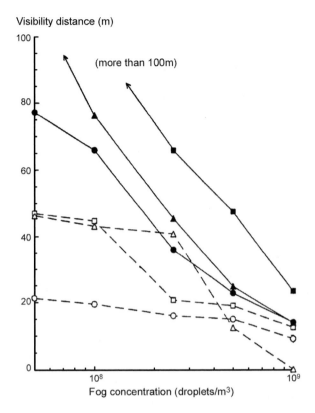

Figure 10.22 Visibility distance for the lower (square) and upper (triangle) parts of a truck and a broken tyre (circle), seen under conventional road lighting (open symbols) and under low-mounted road lighting (filled symbols), both with headlights, plotted against fog concentration (after Girasole *et al.*, 1998).

be back-scattered towards the vehicle. What the result of this is will depend on the density of snowflakes. If the density is high so there are no gaps between the snowflakes, the result is a continuous veil of high luminance that restricts visibility in all directions to very short distances. This is known as a whiteout. It is similar in its effects on driving as a dense fog. More usually, each snowflake is seen separately against a low luminance background. This would not be too much of a problem if the snowflakes were stationary and so fixed against the background. After all, in daylight the detailed structure of the background is always visible and causes no complaint. However, snowflakes are not fixed, and their movement against the background provides a strong signal to the peripheral visual system that is difficult to ignore, the peripheral visual system being designed specifically to detect changes in the visual world.

How distracting snowflakes are will depend on their luminance contrast, the higher the luminance contrast, the more distracting the snowflakes. Maximum luminance contrast will occur when the vehicle forward lighting is the only source of light. In this situation, the background against which the snowflakes are seen will be at a minimum. During daylight or at night when road lighting is present, the background against which the snowflakes are seen is higher in luminance and thus the snowflakes are at a lower luminance contrast.

There are three approaches used to reduce the distraction caused to the driver by snowflakes. The one most easily available to the driver is to use the forward lighting on low beam rather than on high beam. The high-beam condition increases the amount of light emitted by the forward lighting and increases the area covered by the beam, with the result that more snowflakes are seen at a higher luminance contrast. Another approach, but one that is only available to the vehicle designer, is to place the forward lighting as far away from the driver's usual line of sight as possible. This is effective because the intensity of back-scattered light in snow is highest when the angle between the incident light and drivers' line of sight is small and lowest when this angle is around 90° (Hutt *et al.*, 1992). Usually, vehicle forward lighting is mounted below the drivers line of sight but in some situations this is not possible, e.g. a snowplough. Field studies have shown that mounting the forward lighting on a snowplough as far away as possible from the driver's line of sight results in better visibility and reduced perceptions of glare in snow (Bajorski *et al.*, 1996; Bullough and Rea, 1997). The third approach is to modify the spectrum emitted by the forward lighting. As discussed in Section 10.4, there is evidence that light spectra that more effectively stimulate the rod photoreceptors of the retina ensure better off-axis detection in mesopic conditions, at the same photopic luminance. This is to the advantage of the driver in a clear atmosphere at night but when that atmosphere contains snowflakes, ensuring better off-axis detection may be detrimental because it will make the snowflakes more distracting rather than less. Bullough and Rea (2001) have demonstrated this effect of light spectrum for the performance of a tracking task viewed through a curtain of air bubbles moving through water. If this result could be confirmed for snow in realistic conditions, it would introduce another possibility for those developing adaptive forward lighting to consider.

10.8   Summary

Lighting for driving involves the consideration of several different components. The first component is vehicle lighting. Vehicle lighting takes two forms: (a) signal and marking lighting, designed primarily to indicate presence or give information about the movement of a vehicle by day and night; and (b) forward lighting, designed to enable the driver to see after dark. Signal and marking lights are there to be detected. Forward lighting is a means of making the area in front of the vehicle, and the objects in that area, more visible. Both types of vehicle lighting are closely regulated. For signal and marking lights the regulations are based on the visibility of the lights, which in turn is dependent on the luminous intensity, area, and color of the light. For forward lighting, the regulations reflect a balance between the desire to brightly illuminate whatever is ahead of the vehicle and the need to avoid blinding an approaching driver. The result of this compromise is often to limit the distance at which significant obstacles can be seen to less than that needed for safety. Various methods have been proposed to increase forward visibility without increasing glare, using polarization, UV, or IR headlights, IR sensors, and headlight systems that automatically adapt themselves to the prevailing conditions. None have yet been widely adopted.

Another component involved in lighting for driving is road lighting. The principle behind road lighting is to light the road surface brightly enough so that objects on the road are seen in silhouette against the road. The value of light as an accident countermeasure has been established for fatal accidents involving pedestrians at night, where the low visibility of the pedestrian is a contributory factor in the accident. There are well-recognized recommendations of road surface luminance, luminance uniformity, and disability glare for roads with different traffic speeds and densities, but there is no recommendation as regards light spectrum. Recent research has shown the light spectrum that effectively stimulates the rod photoreceptors of the retina will lead to shorter reaction times and fewer misses for off-axis targets than one that does not stimulate the rod photoreceptors, in mesopic conditions. This knowledge had yet to be incorporated into road lighting practice.

A special problem for road lighting is that of tunnels during the day. The problems posed are those of the black-hole effect, in which the tunnel entrance is seen as a black-hole by approaching drivers, and the black-out effect, in which drivers entering the tunnel find themselves misadapted. The black-hole effect is due to light scattered in the eye from the surroundings of the tunnel while the black-out effect is due to the driver having insufficient time to adapt to the much lower luminances in the tunnel. Both these problems can be overcome by appropriate construction of the tunnel entrance and by following the recommendations for grading the luminances from the entrance to the interior of the tunnel.

Another component in lighting for driving is the visibility of signs and signals. Signs giving information on speed limits, directions, etc. either have their own lighting, usually where they need to be seen from a long distance, or are illuminated by light from the vehicle's forward lighting. In the latter case, the signs are usually coated in some form of retroreflective material that reflects light back in the direction from which it came, regardless of the angle of incidence on the sign. These signs rely on reflected light to be seen. Other types of sign emit light. Probably the most ubiquitous are traffic signals designed to control traffic flow. The luminous characteristics of traffic signals are closely regulated, but in different ways, in different parts of the world. The basis of these characteristics is the reaction time and proportion of missed signals to the onset of the signals. Another type of sign increasingly used on roads is the changeable message sign. Both traffic signals and changeable message signs have to be designed so that they are bright enough to be conspicuous and legible by day, but not so bright that they become glare sources by night.

All these aspects of lighting for driving can be rendered less effective in the presence of rain, fog, and snow. Rain has its main effect by changing the reflection properties of the road surface from diffuse to specular, thereby making the luminance of the road surface provided by road lighting less uniform and the disability glare from vehicle forward lighting greater. The effect of rain can be minimized by careful selection of the materials used in the construction of the road, and by designing the road lighting on the basis of a wet road. Fog has its effect by multiple scattering of light, thereby producing a luminous veil that reduces the luminance contrast of everything ahead of the driver. Lane keeping in fog can be helped by the use of retroreflective markers set into the road. Vehicles foglights mounted low on the vehicle are also used. These have a luminous intensity distribution that is both wide and flat. These characteristics usually mean that less scattering of light occurs so the luminance of the luminous veil is reduced. Road lighting is also used to alleviate the effect of fog. Whether it does so depends on the density of fog and the mounting height of the lighting. Road lighting is most effective in thin fogs, when the lighting is at a low mounting height. Snow backscatters incident light and, provided each flake is seen as an individual item, creates a lot of visual noise in the driver's visual field. How distracting this visual noise is depends on the luminance contrast of the snowflakes. Road lighting will reduce the luminance contrast of the snowflakes and therefore helps to alleviate the distraction. As for vehicle forward lighting, the distraction can be reduced by using headlights on low beam and by placing the forward lighting as far from the line of sight of the driver as possible. One other possibility is to modify the light spectrum so as to reduce the stimulation of the rod photoreceptors.

The overall impression created by this chapter is that the components of the lighting systems designed to help the driver have been developed piecemeal, without regard to the variety of conditions that the driver may experience. This is most marked for road lighting and vehicle forward lighting. Vehicle forward lighting primarily lights the vertical surfaces of objects on the road while road lighting primarily lights the horizontal road surface. The combined effect can be to eliminate the contrast of the object against the road, yet only rarely is the combined effect of vehicle lighting and road lighting considered. The physics of what is required to make an object on or near the road visible is well understood. What appears to be missing is the will to consider all the components involved in making an object visible to the driver in all conditions.

11 Lighting and crime

11.1 Introduction

Crime began with Adam and Eve and has been with us ever since. Crime can take many different forms, some major, some petty; some against people and some against property. The consequences of crime for the victim can range from irritating to devasting. The consequences for society can also be dramatic, changing the ways in which people perceive each other and how they behave towards each other. But crime is not always a negative for society. Every so often in human history, what at the time was called widespread criminal activity has lead to advances in human rights and tolerance. Indeed, many nations owe their existence to such activity. This chapter is devoted to the role of lighting as a means of preventing and detecting criminal activity, regardless of whether that activity is later considered to be justified or not.

11.2 Lighting and the incidence of crime

Attempts to use lighting as a measure to combat crime on the streets have been made since the fifteenth century (Painter, 1999). In 1415, all owners of property in London rated at £10/year or more were ordered to hang out a lantern each winter evening between All Hallows (1 November) and Candlemas (2 February), except on the nights from 7 days before the full moon to 7 days after the full moon, when moonlight was considered to be sufficient. To fill in the gaps between this inevitably sparse provision, citizens who had to go out at night relied on linkmen, men and boys who carried flaming torches and escorted their clients through the darkened city. This combination of lighting provided by individual householders and the use of linkmen persisted until the eighteenth century, despite complaints about its inadequacy and the widespread belief that the linkmen where often hand in glove with footpads and highwaymen.

Paris followed a similar route to London but more rapidly developed a public lighting system. In the fifteenth century, it was decreed that "during the months of November, December, and January, a lantern is to be hung out under the level of the first floor window sills before 6 o'clock every night. It is to be placed in such a prominent position that the street receives sufficient light" (Schivelbusch, 1988). In 1667, the authorities in France decided to suspend lanterns on cables over the center of the streets rather than mount them on houses and thereby to create a public lighting system under the control of the police. This link between lighting and the police had the consequence of associating street lighting with the maintenance of public order under an absolutist monarchy and hence as an instrument of repression rather than as a friend of the citizen. The result was an enthusiasm for lantern smashing at times of political unrest.

As cities grew and the concepts associated with the social contract between the governing and the governed developed, there was increased demand for some form of public lighting at night. This demand was first widely met by the introduction of gas lighting, the gas being delivered to each lantern through a distribution system from a central source. In London, by 1823, the gas lighting system had grown to such an extent that 39,000 gas lamps provided lighting for 215 miles of road (Chandler, 1949). Gas lighting rapidly spread to other cities in Europe, although not without resistance. In Cologne, many citizens asserted that God had ordained that certain hours of darkness should prevail and that any attempt to illuminate the streets at night was an encroachment on the divinely established order of the universe (Roberts, 1997). Despite this fundamentalist view, public lighting powered by gas continued to spread until virtually every major city had such provision. Gas was the major source for exterior lighting at night for about 100 years, although the first exterior electric lighting installations, using arc lamps, were installed in the 1850s. The brightness of these light sources were believed to be of great benefit in the fight against crime. Indeed, the police chief of New York was quoted as saying "Every electric light erected means a policeman removed" (O'Dea, 1958). But arc lighting required continuous maintenance and so was never widely used, although Detroit had a system based on 50 m high towers that was used to light 54 km^2 of the city at the end of the nineteenth century. After 30 years, this system was dismantled, the criticism being that it provided a twilight glow over the whole area – but no effective lighting anywhere (Schivelbusch, 1988). It was not until the introduction of the incandescent lamp and the associated electrical distribution systems that gas gave way to electricity as the primary mean of providing light at night. Since then, the electric light sources for exterior lighting used have changed from incandescent, through a range of discharge sources such as low-pressure sodium, mercury vapor, and tubular fluorescent to today's most widely used light sources, high-pressure sodium and metal halide.

Strangely, once widespread provision of exterior lighting at night had been accomplished, interest in the question of its effects on crime diminished and attention switched to the most appropriate form of exterior lighting for driving (see Chapter 10). Interest in lighting's potential to affect crime resurfaced in the US in the 1960s, coincident with a dramatic rise in the incidence of crime. Municipalities across America improved their street lighting to combat crime and some encouraging results were reported (Wright *et al.*, 1974). However, in 1979, Tien *et al.* (1979) published an extensive review of the impact of lighting projects on crime in the US. Inclusion in the review was restricted to street lighting projects that had been installed with an effect on crime in mind and that were not clearly highway lighting. This latter criterion was adopted because it was assumed that road lighting was primarily concerned with vehicle safety and not pedestrian security. Applying these two criteria led to a total of 103 street lighting projects that had been implemented in the US, from 1953 to 1977. Two other criteria were then applied; projects were only considered further if they took place in cities with a population of at least 25,000 and after 1970. The population limit was adopted because of the desire to compare like situations. The date requirement was adopted because of the difficulty in recovering or collecting data from long completed projects. Applying these two criteria reduced the project pool to 45, Finally, a fifth criterion was applied; a project had to have some data available on the lighting installation used, people's attitudes or behavior, and the incidence of crime. The outcome was a set of 15 street lighting projects for detailed evaluation.

The detailed study revealed a pattern of partial information, inadequate or inappropriate measurements, limited control of relevant variables, and invalid statistical analyses, where any statistical analysis had been applied at all. The conclusion reached by Tien *et al.* (1979) was that there was no statistically significant evidence that improved street lighting influences the level of street crime. However, there was some indication that improved street lighting decreases the fear of crime.

Other studies of specific street lighting projects and specific types of crime conducted at the same time have confirmed the difficulty in obtaining unambiguous evidence of the effect of street lighting on street crime. In one case (Griswold, 1984), the influence of the improved lighting could not be disentangled from the effects of security surveys carried out at the same time. In two other studies (Krause, 1977; Lewis and Sullivan, 1979), the apparent reduction in crime against property following an improvement in street lighting could be seen as part of a continuing trend that started before the lighting was changed.

The effect of the Tien *et al.* review was to dampen enthusiasm for lighting as a means to combat crime for about a decade. Then, Painter (1988) reopened the question. She identified three aspects of the studies reviewed by Tien *et al.* (1979) that might be expected to limit their sensitivity. The first was the fact that virtually all the studies involved large areas. Large areas lead to averaging, which makes it very difficult to isolate the impact of lighting from that of all the other factors which may affect the level and type of crime and the fear of crime. The second was the use of police crime statistics. Crime statistics, as conventionally collected, are coarse measures that group together a wide range of offences. Further, not all crimes are reported to the police. This makes it difficult to know if improved street lighting generates a change in the level or the pattern of crime. The third was that there had been no examination of the effects of different types of lighting on the incidence of various types of crime. Different lighting installations are likely to have different impacts on different types of crime, so to group them together risks masking any effects.

Painter's response to this analysis was to conduct a field experiment in an outer city area of London (Painter, 1988). The project focused on the effect of lighting on particular crimes in a very localized area. Specifically, the area examined was one street and a tunnel under a railway. The street was heavily used because it served as a pedestrian route from a residential area to commercial, transport, and leisure facilities. The types of crime examined were those representative of common street crimes: violence against the person (robbery, theft, physical and sexual assault); auto crime (theft and damage); and incidents of harassment. The study was conducted as a "before and after" design. The street lighting "before" gave illuminances on the street in the range 0.6–4.5 lx from low-pressure sodium lanterns, while the street lighting "after" gave illuminances in the range 6–25 lx on the street from high-pressure sodium lanterns. Data were collected by a street survey carried out after dark. The information sought from people using the street was their experience of crime in the area, their fear of crime in the area, and any physical precautions they took. The interviewers carrying out the survey also recorded any crime and/or harassment they observed or experienced. Table 11.1 lists the number of crime incidents the 207 respondents had experienced on the street over a 6-week period before the lighting was changed and the number of crime incidents another 153 respondents had experienced on the street during the 6 weeks immediately

Table 11.1 Crime experienced on the street by respondents over six week periods before and after the change in lighting (after Painter, 1988)

Type of crime	Number of respondents experiencing	
	Before lighting change (n = 207)	After lighting change (n = 153)
Robbery	2	0
Sexual assault	1	0
Physical assault	2	1
Threats	4	0
Stolen automobile	4	1
Stolen motorcycle	4	0
Stolen bicycle	1	0
Automobile damage	2	1
Motorcycle damage	2	0
Total	22	3

Table 11.2 Changes in perceptions of 153 respondents following the change in lighting (after Painter, 1988)

Question	Percent answering positively		
	All	Men	Women
Noticed the change in the lighting of this street?	69	63	82
If you noticed the change in lighting in what way is the lighting different?			
Lighting brighter	99	—	—
Lighting makes it easier to recognize people	97	—	—
Lighting improved	96	—	—
Lighting better maintained	82	—	—
Lighting casts less shadows	65	—	—
Lighting more attractive	58	—	—
Lighting improved look of area	47	—	—
During the past six weeks, while walking on this road, would you say that your feelings of personal safety have			
Increased	62	61	63
Decreased	3	4	2
Remained the same	31	29	33
Do not know	4	5	2

after the change in lighting. Table 11.2 lists the percentage of respondents identifying the change in lighting conditions and, if they did, the nature of the change they had observed. It also gives the respondents' change in fear of crime following the change in lighting. In this study, at least, there is some evidence that improving street lighting does reduce the incidence of some types of crime and people's fear of crime.

Obviously, this study, with its limited area examined and brief time of exposure, cannot be said to prove conclusively that improved street lighting reduces the incidence of crime. Nonetheless, after the ambiguities of the macro-scale studies described by Tien *et al.* (1979), the results of this micro-scale study by Painter (1988) are at least clear. These findings lead to an outburst of similar studies in the UK, some done by Painter herself (Painter, 1989, 1991a, 1994), some done by others (Barr and Lawes, 1991; Burden and Murphy, 1991; Davidson and Goodey, 1991; Glasgow Crime Survey Team, 1991; Herbert and Moore, 1991; Nair *et al.*, 1993; Ditton and Nair, 1994; Cridland, 1995). The results were mixed. Nearly all these studies showed a reduction in fear of crime following improved street lighting, particularly for women and the elderly (Painter, 1991b). However, while most of the studies showed a reduction in the incidence of crime following improvements to the street lighting, others found no effect while yet others found an increase in some types of crime (Painter, 1996). Further, at least one follow-up study found that the marked reduction in level of crime that immediately followed the improvement in street lighting was not sustained

Table 11.3 Crime experienced by households on the estate before and after the lighting was improved (after Painter, 1991a)

Type of crime	Number of respondents experiencing	
	Before lighting change (n = 197)	After lighting change (n = 197)
Burglary with loss	46	25
Attempted burglary	53	51
Outside household theft	35	10
Theft from person	6	1
Street robbery	15	3
Public physical assault	9	19
Vandalism/home	25	27
Vehicle stolen	5	5
Theft from vehicles	5	15
Vandalism/vehicle	25	13
Pestered/insulted	81	64
Sexual assault/rape (women only)	2	0
Sexual harassment (women only)	42	35
Total	349	268

over time (Nair and Ditton, 1994). Obviously, there is no simple link between lighting conditions and the prevalence of crime. A study conducted in Ashton-under-Lyme, in northwest England, will serve as an example of the variability of the effect of improving street lighting (Painter, 1991a). This study took place on a public housing estate consisting of three tower blocks surrounded by maisonettes, and examined the incidence of crime over a period of 12 months. Table 11.3 shows the incidence of different types of crime before and after the estate lighting was improved, as reported by the households on the estate. It should be noted that these data indicate a much higher level of crime than was recorded in police statistics, both before and after the lighting was improved. Nonetheless, there is again evidence that, overall, the incidence of crime has been reduced following the improvement in the lighting, although some types of crime such as theft from vehicles and physical assault have increased. As for fear of crime, the improvements in the lighting produced a 41 percent reduction in those who felt unsafe on the estate after dark and lesser reductions in the number afraid of being robbed (25 percent reduction), afraid of vandalism (14 percent reduction), and afraid of sexual assault (10 percent reduction).

While this work was being done, another study covering a large urban area was undertaken in which the street lighting of the whole of Wandsworth, a borough of London, was improved. The level of crime reported to the police was monitored for 12 months before and after the lighting was changed (Atkins *et al.*, 1991). The conclusion of the authors of this study was that there was no effect of improved street lighting on the incidence of crime. However, a later review of this study (Pease, 1999) came to a different conclusion, pointing out that there was a 15 percent reduction in the level of crime following the improvement of the lighting, although the reduction occurred both day and night, 11 percent by day and 17 percent by night. The reason for the different interpretations of the data collected in Wandsworth lies in the question, can street lighting that only operates at night have an impact on crime during the day? If it is assumed that street lighting can only have an effect at night, then the interpretation of the original authors is correct, but if street lighting can affect crime during the day, then the conclusion of the revisionists is correct. This question will be discussed in Section 11.3.

While argument was raging over the Atkins *et al.* (1991) study, more data were accumulating. Probably the most sophisticated study is that under taken in Stoke-on-Trent in the North-Midlands of England (Painter and Farrington, 1999). Three areas of housing were identified, each with a stable population. One area was designated as the experimental area, the second as an adjacent area, and the third as a control area. Before the street lighting in the experimental area was improved, interviews were conducted with households in all three areas about their experience of crime in the last 12 months, their perceptions of the area, and their behavior. Then, the lighting of the experimental area was improved from widely spaced incandescent lighting on roads, and unlit footpaths, to more closely spaced high-pressure sodium lighting on both roads and footpaths. The lighting of the adjacent and control areas was unchanged. Twelve months after the lighting in the experimental area was improved, as many of the same households as were available in the three areas were again interviewed about their experience of crime in the preceding 12 months and about their perceptions and behavior. The experimental area was intended to directly measure the effect of lighting. The control area was intended to provide a baseline against which any changes in the level of crime in the city could be monitored. The adjacent area was intended to explore the possibility of diffusion occurring. In the case of lighting, this means that the provision of improved lighting in one area will lead to similar benefits for crime reduction in adjacent areas. Diffusion has been found to occur in relation to other crime prevention measures. For example, the installation of a closed circuit television (CCTV) system aimed at reducing thefts of cars from a university car park also produced a reduction in theft of vehicles from an adjacent car park not covered by the cameras (Poyner, 1991). Diffusion is the opposite of another possible effect of taking crime prevention measures in a given area, namely the displacement of the crime to other areas. Displacement, too, has been found following the introduction of crime prevention measures (Gabor, 1990) although it is by no means inevitable (Clarke, 1995), presumably because a certain amount of crime is opportunistic rather than systematic in nature.

Table 11.4 shows the incidence of crime, expressed as the percentage of households in the three areas that had been the victim of crime in the

Table 11.4 Percentage of households experiencing different categories of crime in three areas, before and after the lighting was improved in the experimental area (after Painter and Farrington, 1999). Percentages printed in italics represent the differences that are statistically significant ($p < 0.05$)

Crime category	Experimental area		Adjacent area		Control area	
	Before ($n = 317$) (%)	After ($n = 278$) (%)	Before ($n = 135$) (%)	After ($n = 121$) (%)	Before ($n = 88$) (%)	After ($n = 81$) (%)
Burglary	24	21	20	18	13	16
Outside theft/ vandalism	*21*	*12*	30	22	17	16
Vehicle crime	*26*	*16*	19	12	11	9
Personal crime	*13*	*6*	16	11	7	5

Table 11.5 Average number of victimizations per 100 households for four different categories of crime in three areas, before and after the lighting was improved in the experimental area (after Painter and Farrington, 1999). Percentages printed in italics represent the differences that are statistically significant ($p < 0.05$)

Crime category	Experimental area		Adjacent area		Control area	
	Before ($n = 317$)	After ($n = 278$)	Before ($n = 135$)	After ($n = 121$)	Before ($n = 88$)	After ($n = 81$)
Burglary	38.5	32.7	31.1	24.8	15.9	16.0
Outside theft/ vandalism	43.8	27.0	65.2	38.8	26.1	34.6
Vehicle crime	*47.6*	*25.5*	34.8	*18.2*	17.0	11.1
Personal crime	*43.8*	*14.0*	48.9	*16.5*	10.2	6.2

12 months before and after the relighting of the experimental area. Crime was divided into four categories; burglary, including attempts; theft from outside the home, vandalism of the home or bicycle theft; theft of or from vehicles or damage to vehicles; and personal crime against any member of the household, including street robbery, snatch theft, assault, threatening behavior, and sexual pestering of females. An examination of Table 11.4 shows marked reductions in the prevalence of crime in three of the four crime categories in the experimental area and no statistically significant changes in the adjacent and control areas for any crime category, following the improvement of the lighting in the experimental area.

Measuring the incidence of crime as the percentage of households experiencing a particular type of crime has the limitation that a household may experience the same type of crime more than once. Table 11.5 shows the incidence of crime, expressed as the average number of crimes of each category experienced per 100 households, in the three areas, following the improvement of the lighting in the experimental area. Again, there is a statistically significant reduction in two categories of crime in the experimental area following the improvement of the lighting. Interestingly, for this measure there are statistically significant reductions in the same crime categories for the adjacent area as well, suggesting that diffusion has occurred. There are no statistically significant changes in this measure for the control area, following the improvement of the lighting in the experimental area.

As for perceptions, there were statistically significant increases in the number of households in the experimental area who considered their estate well kept (39–57 percent) after the improvement in the lighting and that their quality of life had improved in the last year (4–23 percent). The improved lighting was certainly noted, only 4 percent of households saying the estate was badly lit after the lighting was improved (74 percent before). As for behavior, counting of the number of pedestrians on the streets after dark revealed a 70 percent increase in males in the experimental area, compared with 29 and 25 percent increases in the adjacent and control areas, respectively. For females, there was a 70 percent increase in the experimental area and 42 and 41 percent increase in the adjacent and control areas, respectively. Clearly, improving the lighting has lead to a greater use of the streets at night.

About the same time as the Stoke-on-Trent study, Painter and Farrington (1997) completed a similar study in Dudley, in the West Midlands of England. In this study, the lighting of a local authority housing estate was improved while the lighting of a comparable, nearby estate was not. What makes this study interesting is that details of the level of crime on the two estates, before and after improving the lighting of one, were obtained from two different sources. One source was similar to that used in Stoke-on-Trent, namely the adult residents of the two estates. Interviews with these adults revealed that following the lighting improvements, they experienced less crime on the estate with improved lighting (a 23 percent decrease) but not on the control estate, where the lighting was unchanged (a 3 percent decrease). The second source was self-reported delinquency data collected from young people in the age range 12–17 years and living on the two estates (Painter and Farrington, 2001a). These data showed that the admitted level of delinquency decreased more on the relighted estate than on the control estate after the lighting was improved.

Another relevant piece of work using a different method and done in a different country is that of Loomis *et al.* (2002). They carried out a thorough epidemiological case-control study of the impact of various safety measures on the incidence of workplace homicide in North Carolina. One hundred and five workplaces, where a worker had been murdered in the years between 1994 and 1998 formed the case group. The control group was an industry-matched random sample of 210 workplaces at risk during the same period. The safety measures considered were environmental ones, such as bright lighting inside and outside, surveillance cameras and cash drop boxes; and administrative measures such as limiting public access, pre-screening employees, and never having staff work alone. The results showed that strong and consistent reductions in the risk of a worker being murdered at work were associated with bright exterior lighting (odds ratio = 0.5) and with not having people working alone at night (odds ratio = 0.4).

11.3   The reason why

The evidence given in the studies considered above leave little doubt that lighting has a place to play in crime prevention. Improving lighting can lead to a decrease in crime, but it may not. There can be no guarantees. After all, if all that was necessary to prevent crime was to provide a lot of light, there would be no crime during daytime. This conclusion implies that there are circumstances in which lighting can be an effective crime countermeasure, either alone or in combination with other measures, and circumstances in which it will not. To determine what those circumstances are, it is necessary to consider the mechanisms by which lighting might impact crime.

Anderson (1981) asserts that virtually all human thought and behavior has multiple causes, the result of many co-acting factors. There seems little reason to suppose that criminal behavior and the fear of crime depart from Anderson's assertion. This means that lighting is only one among many factors that can influence the incidence of crime. The question then arises: Why might lighting be expected to reduce crime and the fear of crime?

The answer to this question can be framed in terms of the visibility and "message" routes by which lighting conditions affect human performance (see Section 4.2) Functionally, the most obvious and only certain effect better lighting can have is to change how well people can see, i.e. improve visibility. It is well established that increasing the adaptation luminance increases the speed of visual processing, improves the discrimination of detail and makes color judgments more accurate (see Section 2.4). Reducing the adaptation luminance has the reverse effects. Different types of street lighting generate different adaptation luminances. Different types of street lighting also produce different patterns of light distribution. If these patterns give rise to shadows and disability glare, the ability to see may be impaired.

Given that street lighting conditions can influence how well we see, the next question to ask is why that should be expected to influence the incidence of crime and fear of crime. After all, improved street lighting enhances visibility for both the criminal and the law-abiding equally. A plausible tactical answer is that better lighting increases the distance at which something suspicious can be seen. The greater the distance at which a threat can be detected and the finer the discrimination of detail possible, the greater is the time available to select and act out an appropriate response. For example, it may be possible, because facial expression and body language are visible at a distance, to recognize a threatening situation while there is still time to escape. Similarly, greater visibility at a greater distance may enable people behaving in a suspicious manner to be recognized or at least described. Such observations at a distance are a benefit to the law-abiding and can be a disadvantage to the criminal.

Strategically, improving lighting can be considered as one of several different contributors to situational crime prevention. Situational crime prevention involves the modification of environments so that crime requires more effort, poses more risk and produces lower rewards (Clarke, 1995; Pease, 1997). Better lighting may affect the perception of risk by increasing the ease of surveillance of the street, either formal, by the police in person or through a CCTV system, or informal, by members of the community. Better lighting may increase the effort required by enabling potential victims to take action at a distance and by limiting the locations where victims can be taken by surprise. However, better lighting may also decrease the effort required and increase the rewards by making it easier for a criminal to pick out an easy, valuable target. Thus, whether better lighting will reduce the prevalence of crime will depend on the criminal's perceived risk/reward ratio for the crime, whether the lighting helps or hinders the commissioning of the crime, and the likelihood that surveillance will be translated into action. Certainly, it will not eliminate crime. This is because improving street lighting is unlikely to deter the very experienced criminal. Weaver and Carroll (1985) took groups of experienced and novice shoplifters through retail stores and asked them to assess the opportunities. The results showed that experienced shoplifters considered conventional crime deterrents, such as store personnel and security devices, as obstacles to be overcome and were more strategic in their assessments. Novice shoplifters decided against shoplifting in the presence of any deterrent. Applying these data to street crime suggests that improved street lighting may deter the tyro street criminal, but is unlikely to deter the more experienced.

So far, this discussion of how better lighting might impact crime has been concerned with the direct effect of lighting conditions on visibility. The other mechanism by which lighting might impact crime is through changes of behavior and community confidence. Most of the studies of the impact of better street lighting have shown that improving street lighting tended to decrease peoples' fear of crime. These findings can be understood from the results of Fisher and Nasar (1992). These authors examined the fear of crime in relation to exterior site features on a college campus. Using three different approaches they found that fear of crime was highest in areas which offered places for criminals to hide and which had a restricted view, with few avenues for flight. At night, the effect of good lighting will tend to diminish the number of places where criminals can hide, increase the distance over which people can see and, possibly, reveal opportunities for flight. Thus, good street lighting can be expected to reduce fear of crime. The behavioral consequence of this is that more people use the streets at night (Painter, 1994; Painter and Farrington, 1999). More people on the street at night increases the number of pairs of eyes and hence the amount of informal surveillance, something that criminals consider increases the risks of their activities (Bennett and Wright, 1984).

Of course, improving street lighting will only enhance visibility after dark but this should not be taken to mean that providing good street lighting only effects crime after dark. The installation of improved street lighting is a highly noticeable activity that sends the "message" that someone cares about the neighborhood (Taylor and Gottfredson, 1986). Such perceptions can to lead to greater community confidence, cohesion, and informal social control and these in turn tend to lead to more surveillance by residents and a greater likelihood that such surveillance will be used to support the authorities against the criminally inclined. And this will occur by day as well as at night. It is important to note that for greater community cohesion, etc. to occur, it is necessary that the new lighting be seen as a marked improvement over what was there before and that the intention of the people providing the lighting is perceived to be to help the community and not simply to control it. It is also necessary to appreciate that the same lighting conditions may deliver different "messages" to different people. For the resident the "message" may be that with this new lighting I can see everything so it is safe to go out at night. For someone driving by the "message" may be that this must be a dangerous area or they would not need such bright lighting. The possibility of such mixed "messages" suggest that it is wise to identify the recipient of any "message" before predicting the impact of improved lighting.

To summarize, lighting does not have a direct effect on the level of crime. Rather, lighting can affect crime by two indirect mechanisms. The first is the obvious one of facilitating surveillance by the authorities and by the community after dark. If such increased surveillance is perceived by criminals as increasing the effort and risk and decreasing the reward for a criminal activity, then the level of crime is likely to be reduced. Where increased surveillance is perceived by the criminally inclined not to matter, then better lighting will not be effective. The second mechanism by which an investment in better lighting might affect the level of crime is by enhancing community confidence and hence increasing the degree of informal social control. This mechanism can be effective both day and night but is subject to many influences other than lighting.

11.4 Essential characteristics of lighting

One characteristic of many of the studies described in Section 11.2, is the very skimpy level of detail given about the lighting installations, either before or after improvement. Usually, the improvement in lighting is considered adequately described by a listing of the new lighting equipment used. Typically, the improved lighting involves the use of more light sources with higher light output and better colour rendering, more closely spaced.

Only rarely is any quantitative information given about the lighting conditions used. This lack of information about the lighting conditions is understandable because all the studies described above have been done by criminologists rather than lighting designers, and we all tend to emphasize what we know about. Criminologists know a lot about crime but little about lighting. While it is understandable, the lack of information about the lighting used in the above studies is nonetheless disappointing because having established that improved lighting can have an effect on the level of crime, it is now necessary to turn to the question of what are the essential aspects of lighting needed to help reduce crime. Given that one mechanism by which lighting might help in the fight against crime is through the facilitation of surveillance, the aspects of lighting that are likely to be important are the average illuminance, the illuminance uniformity, the glare, and the light source color properties.

11.4.1 *Illuminance*

As in most lighting questions, the measure that immediately springs to mind when considering improvements in visual capabilities is the luminance to which the visual system is adapted. This luminance is determined by the luminances of the surfaces forming the scene, which in turn are determined by the illuminances on the surrounding surfaces and the reflectance of those surfaces. The usual practice in street lighting is to ignore the inter-reflected light and only consider the light directly incident. This is reasonable for practice given the great diversity of situations in which street lighting is installed, but, if there are any nearby surfaces, it is worth remembering that high reflectance, diffusely reflecting surfaces produce much more diffuse lighting. This will increase the adaptation luminance, reduce the strength of any shadows, and diminish the impact of disability glare.

But what adaptation luminance, or more practically, illuminance, should be provided to facilitate surveillance? There are two approaches by which quantitative knowledge might be obtained. The first approach is to carry out practical tests of how far away it is possible to see various levels of detail under different illuminances. The second is to measure how safe people perceive a location to be under different illuminances.

Rombauts *et al.* (1989) studied the ability to recognize a face from various distances. Following the work of Caminada and van Bommel (1980), semi-cylindrical illuminance was used as a measure of the lighting conditions. Semi-cylindrical illuminance is the average illuminance on the surface of an upright half-cylinder. Figure 11.1 shows the relationship between the distance at which the observers were completely confident that they recognized the person they were approaching and the semi-cylindrical illuminance on the face of the person being approached. As far as surveillance is concerned, the further the distance at which people can be recognized the better. Rombauts *et al.* claim that confident face recognition is not possible

Semi-cylindrical
Illuminance (lx)

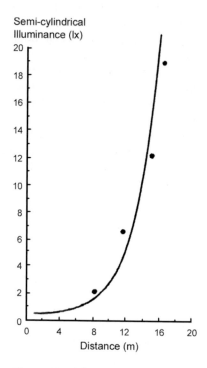

Figure 11.1 The semi-cylindrical illuminance on the face necessary for completely confident recognition plotted against distance (after
Rombauts *et al.*, 1989).

beyond 17 m and that a semi-cylindrical illuminance on the face of 25 lx is
sufficient to give confident identification at this distance. Obviously lower
semi-cylindrical illuminances can be used if the confident recognition at
shorter distances is acceptable. Hall (1966) claims that 4 m is close to the
boundary of what is called the public space surrounding an individual and
that anyone unexpectedly approaching closer will cause alarm. Rombauts
et al. identify a minimum semi-cylindrical illuminance of 0.6 lx on the face
as necessary to ensure confident identification at 4 m.

This is all very interesting but it is of little use because semi-cyindrical
illuminace is so rarely used. Fortunately, Rombauts *et al.* also found that
people considered the lighting of facial features to be well-balanced when
the vertical/semi-cylindrical illuminance ratio was in the range 1.1–1.5.
Assuming a desirable vertical/semi-cylindrical illuminance ratio of 1.3, their
results convert to a vertical illuminance of 33 lx for confident face recognition at 17 m, and 0.8 lx at 4 m.

Boyce and Rea (1990) examined the effects of different perimeter security
lighting installations on people's ability to detect someone walking towards
them and then to recognize them from a selection of four photographs.

The results obtained showed that the probability of detecting someone approaching reached 90 percent at a vertical illuminance on the person of 4–10 lx, the lower illuminance occurring when the person was approaching along a known path while the higher illuminance occurred when the person could come from anywhere ahead of the observer. Higher illuminances are needed to approach 100 percent detection.

Being able to detect the presence of another person is necessary for a sense of security but it may not be sufficient. It is also desirable to be able to recognize the person's dress, facial features and any mannerisms. Such information is valuable in deciding if the approaching person represents a threat, and, if the worse occurs, for giving an accurate description to the police. Boyce and Rea (1990) found a vertical illuminance of about 10 lx to be sufficient to obtain 90 percent correct recognition of an approaching person. A higher vertical illuminance will allow a higher probability of recognition, but the possibility for improvement is limited.

These two experiments used the ability to detect and recognize people approaching as a criterion to determine the suitable illuminance to facilitate surveillance. Another approach to determining how much light is needed to facilitate surveillance is to ask people when they think they can see well enough. Simons *et al.* (1987) carried out field appraisals of 12 different street lighting installations in London using a panel of observers experienced in street lighting, the assessments being made from the pedestrian viewpoint. The results obtained showed that mean horizontal illuminance was slightly more closely related to the observers' assessments of the installations than semi-cylindrical illuminance, so there is no advantage in using the latter rather than the former. It was also established that a horizontal illuminance of 5 lx was considered adequate and about 11 lx was seen as good. Assessments of street lighting in a much smaller city and using less experienced observers again revealed that an average horizontal illuminance of 5 lx was again considered adequate and 10 lx was considered good. In order to compare these results with those discussed earlier, it is necessary to convert horizontal illuminance to vertical illuminance. Unfortunately, there is no fixed relationship between the horizontal and the vertical illuminance produced by a street lighting lantern. It is a function of the angle from the line joining the light source to the measurement point, relative to the downward vertical, and the spacing of the lanterns. Given the variability inherent for different angles, it is not possible to make an exact comparison between the results of Simons *et al.* and those discussed earlier. The best that can be done is to note that the lighting considered good by their panel has an average horizontal illuminance of the same order as the vertical illuminances found by other investigators. This difficulty experienced in comparing results emphasizes the value of giving a comprehensive photometric description whenever a lighting installation is the subject of an evaluation.

The most extensive series of evaluations of street lighting as regards perceptions of safety are described in Boyce *et al.* (2000b). In the first study,

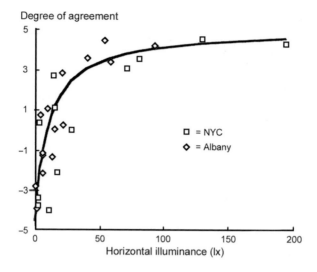

Degree of agreement

Figure 11.2 Mean levels of agreement with the statement "This is a good example of security lighting" plotted against horizontal illuminance, for sites in New York City and Albany, NY. A value of +5 indicates strong agreement and −5 indicates strong disagreement (after Boyce *et al.*, 2000b).

two field surveys were carried out; one in New York City and one in Albany, the New York State capital. In both cities, a number of exterior areas around multi-family housing projects, commercial strip developments, and industrial buildings that were accessible to the public were selected. Each site was visited at night by a panel of people, most of whom had no knowledge of lighting. Figure 11.2 shows the mean agreement with the statement "This is a good example of security lighting," plotted against the horizontal illuminance at the viewing position, for each site in New York City and in Albany. The degree of agreement with the statement "This is a good example of security lighting" was highly correlated with degree of agreement with the statements "I can see clearly around me" ($r = 0.90$) and "I can see far enough ahead" ($r = 0.89$). Figure 11.3 shows the mean agreement with the statement "This is a good example of security lighting," plotted against horizontal illuminance at the viewing position, for the male and female subjects separately. It is clear that females require a higher illuminance for the same perception of good security lighting than the males. From Figures 11.2 and 11.3 it is possible to determine an illuminance necessary to achieve a perception of good security lighting. Assuming that the objective should be to have a mean agreement level of + 3, i.e. that the average person should moderately agree that this is a good example of security lighting, Figure 11.2 suggests an illuminance of 40 lx is required. For the

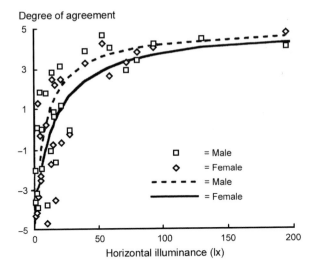

Degree of agreement

□ = Male
◇ = Female
- - - - = Male
—— = Female

Horizontal illuminance (lx)

Figure 11.3 Mean levels of agreement with the statement "This is a good example of security lighting" plotted against horizontal illuminance, for
sites in New York City and Albany, NY, for male and female subjects separately. A value of +5 indicates strong agreement and −5
indicates strong disagreement (after Boyce *et al.*, 2000b).

same criterion, Figure 11.3 shows that an illuminance of 35 lx is required
for males and 60 lx for females.

These field studies were both conducted in urban areas. The relationship
between illuminance and the perception of good security lighting found in
urban areas may not hold for suburban areas, where the risk of crime is less
(USDOJ, 1998) and the ambient illumination, which influences the perception of brightness, is less. Another field study was undertaken to answer this
question, using parking lots in an urban and an adjacent suburban area.
The same methodology was used as in the previous study, with the exception that the parking lots were also visited in daytime

Figure 11.4 shows the mean agreement with the statement "This is a
good example of security lighting," plotted against the median illuminance
on the pavement for the urban and suburban parking lots. The results
shown in Figure 11.4 imply that a lower illuminance could be used in suburban parking lots to produce the same perception of goodness of security
lighting as a higher illuminance in urban parking lots. However, there is an
alternative way of looking at the data collected and that is to look at the
differences on the same question, for the same parking lot, for day and
night. The question that was asked about the parking lots both day and
night concerned the perceived safety of walking alone in the parking lot.
Figure 11.5 shows the mean ratings on this question for the urban and suburban parking lots, for day and night. It is evident from Figure 11.5 that the

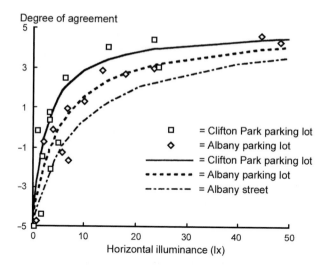

Degree of agreement

□ = Clifton Park parking lot
◇ = Albany parking lot
—— = Clifton Park parking lot
••••••• = Albany parking lot
—•——•— = Albany street

Horizontal illuminance (lx)

Figure 11.4 Mean levels of agreement with the statement "This is a good example of security lighting" plotted against median pavement illuminance, for parking lots in Albany, NY (urban) and Clifton Park, NY (suburban). A value of +5 indicates strong agreement and −5 indicates strong disagreement. Hyperbolic functions are fitted through the data for Albany and Clifton Park separately. Also shown is the hyperbolic function that fits the equivalent data for streets in Albany, NY (after Boyce *et al.*, 2000b).

perceived safety of walking alone in the parking lots during the day is higher in the suburban area than in the urban area. There are only two parking lots in urban areas which approach the level of perceived safety of the suburban areas by day, and these are the only two urban parking lots which have attendants. As for perceived safety when walking alone at night, Figure 11.5 shows that, for both urban and suburban parking lots, lighting can bring that perception close to what it is during the day but cannot exceed it. The interesting question now is how does illuminance relate to how close the perception of safety at night can be brought to what it is by day. Figure 11.6 shows the difference in ratings of safety when walking alone by day and night plotted against the median illuminance in the parking lot at night, for the urban and suburban parking lots. These results suggest that at a high enough illuminance the difference in ratings of safety for day and night approach zero. However, the approach to zero difference is asymptotic. Above 10 lx, the difference is less than one scale unit, and above 30 lx, the difference is less than half a scale unit on a seven-point scale.

It is clear from Figures 11.5 and 11.6 that the conclusion reached above about suburban areas needs to be qualified. The results shown in Figure 11.4 suggest that a lower illuminance can be used in a suburban context to get the

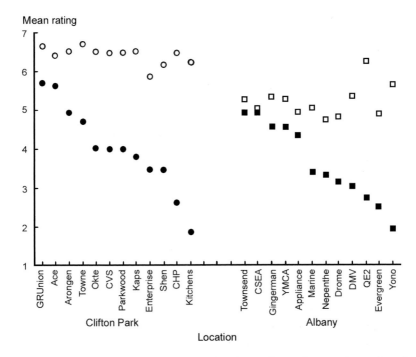

Mean rating

Figure 11.5 Mean ratings of perceived safety for walking alone in the parking lot, by day and night, for the parking lots in Albany, NY (urban) and Clifton Park, NY (suburban). The parking lots are presented in order of decreasing perceived safety at night (1 = very dangerous; 7 = very safe; ●, ■ = night, ○, □ = day) (after Boyce *et al.*, 2000b).

same rating of "good security lighting" as an urban context. But, Figure 11.5 suggests that the same level of perceived safety at night in a suburban context is further from what it is during the day than is the case for an urban context. What this implies is that while you cannot produce a suburban level of perceived safety at night in an urban parking lot by increasing the illuminance, you can reduce the suburban level of perceived safety at night to that of an urban parking lot by lowering the illuminance. Whether you would want to do that must be a matter of judgment.

These field studies were undertaken by different people, in different locations, using similar methods. Despite this diversity, the field studies show similar results. Specifically, they all show a non-linear change in the explicit or implicit perceptions of safety at night with increasing illuminance. For illuminances in the range 0–10 lx, small increases in illuminance produce a large increase in perceived safety. For illuminances above 50 lx, increases in illuminance make little difference to perceived safety. For illuminances in the range 10–50 lx, increases in illuminance show a law of diminishing returns. These results suggest an illuminance around 30 lx is what is

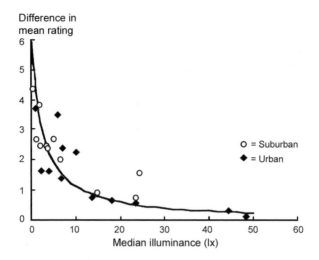

Figure 11.6 Difference in mean ratings of perceived safety for walking alone in the parking lot, by day and night (day – night) plotted against median pavement illuminance for the parking lots in Albany, NY (urban) and Clifton Park, NY (suburban) (after Boyce *et al.*, 2000b).

required for an exterior lighting installation to be seen as an example of good security lighting and for the site to be one where it would be considered to be as safe to walk alone in at night as it is during the day. Further, these conclusions are not wildly out of line with the conclusions reached from the studies on the ability to detect and recognize someone approaching.

11.4.2   Illuminance uniformity

This extensive discussion of illuminance should not be interpreted as meaning that illuminance is the only aspect of lighting that matters to the perception of safety at night. The variability in perceived safety at night, particularly at low illuminances (see Figures 11.2–11.4), suggests that there are other factors operating in addition to illuminance. For example, it is widely believed that uniformity of illuminance and the presence of disability glare also matter to perceptions of safety. Concern with illuminance uniformity is reasonable because if lighting is to be effective because it allows better surveillance, then the presence of areas lit to a much lower illuminance in which criminals can lurk without being seen is likely to undermine the effectiveness of the lighting. Despite this argument, evidence for a uniformity criterion is notable by its absence. However, illuminance uniformity recommendations are still made, based on experience. For example, in North America it is recommended that parking lots be lit so that the minimum illuminance anywhere on the pavement is 2 lx, with a maximum to minimum illuminance ratio of 20 : 1. Further, the minimum vertical illuminance of 1 lx is recommended. Where crime is likely, the recommended minimum horizontal illuminance shifts to 5 lx, the uniformity ratio improves to 15 : 1, and the minimum vertical illuminance increases to 2.5 lx (IESNA, 2000a). The main cause of poor illuminance uniformity is the over-spacing of luminaires and the shadows cast by buildings and vegetation. Shadows and vegetation can be dealt with by careful layout of structures and regular maintenance of the area, but over-spacing is a matter of lighting design. Reputable manufacturers of lighting equipment specify a maximum spacing for a given mounting height. These limits should be strictly observed.

11.4.3 Glare

There is no direct information on the effect of glare from street lighting on people's ability to detect and recognize people approaching them. Rombauts *et al.* (1989) calculated the effect of glare from street lighting using the measure of disability glare commonly used for drivers, the threshold increment (see Section 10.4). Using this approach they found that the minimum semi-cylindrical illuminance for confident facial recognition at 4 m, when the street lighting lanterns produced a threshold increment of 15 percent, increased from 0.4 to 0.6 lx. This very limited evidence suggests that disability glare is unlikely to be a problem for law-abiding pedestrians as long as street lighting is designed to existing criteria. Against this, Simons *et al.* (1987) doubt whether threshold increment is a suitable measure of the effects of disability glare on pedestrians because the state of adaptation of the driver might be different from that of the pedestrian and the restriction of the driver's field of view imposed by the roof of the car limits the view of some of the lanterns. Simons *et al.* recommend limiting disability glare by restricting the luminous intensity distribution of street lighting lanterns to maxima of 175 and 100 cd/klm at 80 and 90° from the downward vertical, respectively. How effective these limits are will depend on the mounting height of the lanterns. More generally, Boyce and Eklund (1998) have shown that the ratio of the illuminance received at the eye from above and below a horizontal plane running through the eye can be related to the perception of glare, although whether the glare is disability or discomfort glare is unclear. Field measurements have shown that when this glare ratio is below four, there will be little perception of glare. By keeping below this limit, any lighting system that has lanterns above the horizontal plane through the eye can be designed to be largely free from glare.

11.4.4 Light source color

There are three reasons why the color properties of the light source used might be important to the effectiveness of lighting as a countermeasure to crime. The first is that in mesopic conditions, light sources that more effectively stimulate the rod photoreceptors will make off-axis visual detection better (see Section 10.4). The second is that where there are colors in the scene, light sources with better color rendering properties will create larger color differences. Such color differences are an important part of the stimulus to the visual system, particularly when luminance contrast is low, and can be expected to produce a perception of greater brightness. The third is the fact that color is an important element in witness descriptions. Light sources with good color rendering properties will allow more accurate color naming.

While these arguments suggest that light source color properties ought to be important, evidence that such properties make a difference is difficult to come by. For example, Boyce and Rea (1990) showed that the probability of detecting a person approaching along a known path and the recognition of their face is the same under both low- and high-pressure sodium discharge lighting of the same vertical illuminance. Low-pressure sodium light sources are essentially monochromatic and so give no color information. High-pressure sodium light sources, while far from perfect, do give much clearer color perception. These results would seem to imply that light source color is not important for the detection and recognition of people approaching.

Another study that examined the effect of light source color on the perception of safety under different light sources was that of Boyce and Bruno (1999). This study involved the performance of a number of tasks and the collection of opinions on the lighting of a large, rectangular parking lot that could be divided into three approximately $1,000\,m^2$ areas. Each area was lit by the same number of new lanterns fitted with either high-pressure sodium or metal halide lamps. At two locations in each bay, the subjects' visual acuity and contrast threshold were measured using charts with Landolt rings of a fixed contrast and reducing size, and letters of fixed size but decreasing contrast, respectively. In addition, the subjects gave their opinions of the lighting on various dimensions. While doing the tasks and answering these questions, the subjects were seated in a car looking down the length of one of the driving aisles, wearing and not wearing gray wrap-around glasses with a transmittance of 0.10. When the glasses were worn, the subjects' state of adaptation was mesopic.

Figure 11.7 shows how the number of Landolt rings whose orientation was correctly identified and the number of letters correctly read vary with the different lighting conditions. Clearly, the dominant factor is the luminance on the charts. Light spectrum has no effect. Eloholma *et al.* (1999) have found similar results for high and low contrast visual acuity over a luminance range of 0.19–$5.2\,cd/m^2$. Of course, both visual acuity and contrast sensitivity measurements require the use of foveal vision and do not involve any color differences.

Figure 11.8 shows the result of a more realistic task, the identification of objects carried by a person in the parking lot. Under each lighting condition, the subjects were asked to identify whether a person about $10\,m$ away was carrying a metal ruler, a hammer, a spanner, a spray can, a screwdriver, a torch, a beer bottle, a gun, an umbrella, a knife, or a pair of scissors. The mean number of objects correctly identified out of a maximum possible of five, is closely related to the illuminance in the parking lot, independent of light spectrum.

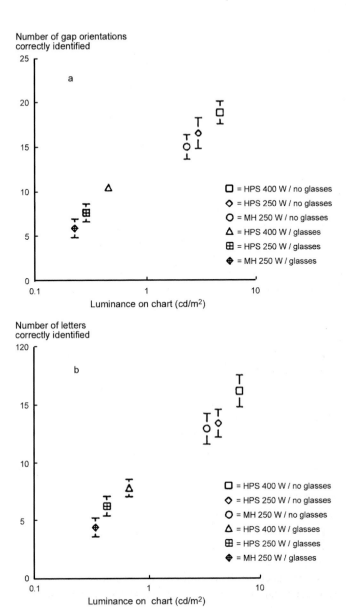

Figure 11.7 The mean number of gap orientations correctly identified plotted against the luminance of the visual acuity chart background. The mean number of letters correctly identified on the contrast threshold task plotted against the luminance of the chart background. The error bars in both graphs are standard errors of the mean. Data is given for different combinations of high-pressure sodium (HPS) and metal halide (MH) lighting seen with the naked eye (no glasses) and through low-transmittance glasses (glasses) (after Boyce and Bruno, 1999).

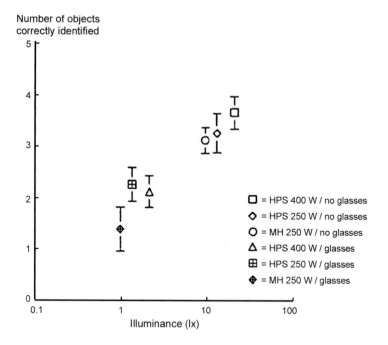

Figure 11.8 Mean number of objects correctly identified plotted against the mean illuminance on the pavement. The error bars are standard errors of the mean. Data is given for different combinations of high-pressure sodium (HPS) and metal halide (MH) lighting seen with the naked eye (no glasses) and through low-transmittance glasses (glasses) (after Boyce and Bruno, 1999).

The only task that did show a clear effect of light spectrum is that of color naming. The subjects were shown nine matte Munsell color plates, the nine colors being the basic colors identified by Boynton and Olson (1987). Figure 11.9 shows the mean percentage of colors correctly identified. The metal halide light source produces a higher percentage correct naming than the high-pressure sodium light sources, even though the former produces a lower illuminance on the colors. However, it is worth noting that increasing the illuminance does improve the percentage correct color naming for the high-pressure sodium light sources, so correct color naming at low light levels is a matter of both light spectrum and illuminance.

While the performance of tasks is important for quantifying how well a person can see, if lighting is to affect crime by increasing the use of a space after dark, then whether people perceive the space to be safe is also important. Figure 11.10 shows the mean rating of safety in the parking bay plotted against the mean illuminance on the pavement. The perceived safety is clearly linked to the average illuminance on the pavement, regardless of

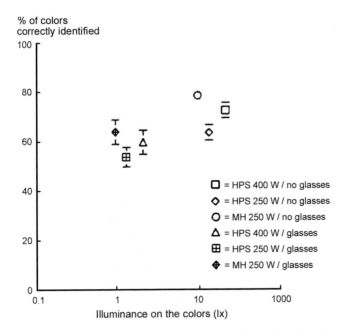

% of colors
correctly identified

Figure 11.9 Mean percentage of colors correctly identified plotted against the mean illuminance on the colors. The error bars are standard errors of the mean. Data is given for different combinations of high-pressure sodium (HPS) and metal halide (MH) lighting seen with the naked eye (no glasses) and through low-transmittance glasses (glasses) (after Boyce and Bruno, 1999).

lamp spectrum. Incidentally, it is interesting to note that an illuminance of about 30 lx produces a "very safe" rating which is consistent with the discussion in Section 11.4.1.

The implication of these results is clear; relative to illuminance the light spectrum is of minor importance for the perception of safety and the ability to detect and recognize people approaching. No doubt, given the same illuminance, people would prefer the better color rendering lamp but if that is not possible, these results suggest people would rather have the more illuminance at the cost of worse color rendering rather than better color rendering at the cost of a lower illuminance. Certainly, it would be a foolish designer who significantly reduced the illuminance simply because a better color rendering lamp was to be used.

This conclusion should not be taken to mean that the impacts of light spectrum on off-axis detection and brightness never occur. Rather, it should be taken to imply that the conditions commonly occurring in parking lots and on streets are not conducive to such effects. For example, for the improvement in off-axis detection with light spectrum to occur, the visual

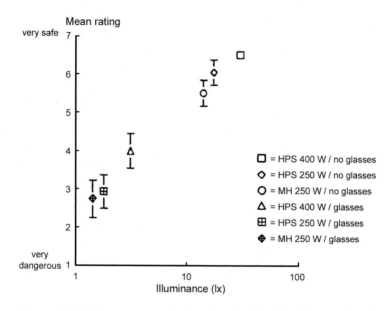

Figure 11.10 Mean ratings of safety provided by the lighting plotted against
the mean illuminance on the pavement. The error bars are standard errors of the mean. Data is given for different combinations
of high-pressure sodium (HPS) and metal halide (MH) lighting seen with the naked eye (no glasses) and through low-
transmittance glasses (glasses) (after Boyce and Bruno, 1999).

system has to be operating in the mesopic state. When parking lots are
lit to the recommended illuminances the visual system is operating at the
photopic/mesopic border so any enhancement of off-axis detection by light
spectrum will be slight. As for brightness, most streets and parking lots
are largely achromatic so there are few large differences in colors to be
enhanced. It would be interesting to discover if there would be a more pro-
nounced effect of light spectrum on brightness in an area such as a park
where the different colors of the vegetation would form a palette for the
better color rendering lamps to work on. Support for this possibility comes
from Rea (1996), who reports a study using a colorful diorama that could
be lit by either metal halide or high-pressure sodium lamps. At a background luminance of $1 \, cd/m^2$, a 40 percent higher luminance was needed
for lighting provided by the high-pressure sodium lamps to appear as bright
as lighting provided by metal halide lamps.

11.4.5 Design approaches

The above discussion makes it possible to specify what constitutes good
lighting for facilitating surveillance in photometric terms, although some

parts of the specification are more soundly based and more important than others. The mean illuminance on the pavement should be in the range of 10–50 lx, the illuminance uniformity should be in less than 15/1, the glare ratio should be less than four and a good color rendering source should be used, especially if a wide range of colors is present. Different design approaches can be used to meet this specification but not all will be effective. One to be avoided is the use of bollards. These luminaires are typically about 1 m high and direct light onto the ground. This is appropriate where the objective of the lighting is to light the path so that people do not trip and fall but for a perception of safety it is necessary to light the face of anyone approaching and not just their knees. A much more common approach in pedestrianized areas is to use post-top lanterns. These typically have the light source at a height of 3–6 m. Post-top lanterns can be very effective but care has to be taken to limit the luminance of the light source so glare is avoided, and the lanterns have to be spaced close enough so that a person standing under one lantern also receives light from others. If this is not done, as a person walks past a lantern, their face goes from being well illuminated when approaching the lantern to completely shadowed as they pass the lantern.

Another approach often used around buildings is the ubiquitous wall-pack. Typically, these are mounted on the wall of the building at 3–6 m. This low mounting height makes the wall-pack a potential glare source. Glare can be avoided by choosing a wall-pack which does not allow a direct view of the light source. Similar advice applies to post-top luminaires. Finally, there are an array of parking lot luminaires that are typically mounted at 6–15 m. The higher is the mounting height the less likely it is that glare will occur and the more likely that the desired illuminance uniformity will be achieved. For lower mounting heights care must be taken to limit the luminous intensity distribution at high angles from the downward vertical, i.e. use a full cut-off luminaire. Details of exterior lighting installations designed to combat crime and that are suitable for many different contexts can be found in CIBSE (1992), Leslie and Rodgers (1996), and IESNA (2000a). These designs also demonstrate that lighting that can be effective in reducing crime need not be ugly. Indeed, it can be attractive.

11.5 Special situations

All the above discussion has been focussed on the impact of lighting on areas to which the public has access and where the main means of detecting crime is the human visual system. It is now necessary to consider how lighting might act as a counter-measure against crime in protected locations and where surveillance through CCTV is used.

11.5.1 Fenced areas

It is common practice to protect valuable property by fences and to patrol the fence either from inside or outside the site. Lighting is provided for the people patrolling the fence. If the patrolling is done from inside the fence, the lighting is designed to light both sides of the fence so that anyone approaching the fence or any damage to the fence can be seen. The rest of the site may be lit or it may not. If the patrolling is done by the police from the outside, then the whole of the site, including the fence, is usually lit. The problem of lighting the fence is how to make it easy for the patrolling guard to see through the fence. The fence is usually a lot closer to the light source than the area that the guard needs to see. This means that there is a risk that the luminance of the fence will be higher than the luminance of the area seen through it. Boyce (1979b) showed that the ability to detect someone through a fence was greatest when the luminance of the fence matched or was less than the luminance of the surface being viewed through the fence. Fence luminances higher than the luminance of the area outside the fence reduce the visibility through the fence, and this reduction is greater for smaller fence mesh sizes. This understanding can be used to either increase or decrease visibility through a fence. Where it is desired to increase visibility through the fence, the luminance of the fence should be kept low. This can most easily be achieved by using very low reflectance materials for the fence. Where it is desired to make the visibility through a fence difficult, high reflectance materials should be used for the fence.

11.5.2 Gatehouses

Every fenced area has a means of access, usually protected by a guard in a gatehouse. The role of the guard is to check people and vehicles arriving and departing and to make sure nobody gets in who should not. Lighting designed to help with the inspection of people and vehicles, including the underside, is a common feature of gatehouse lighting (Lyons, 1980). The average illuminance recommended for the area immediately outside a gatehouse is much higher than for the rest of the site, 100 lx being typical (IESNA, 2000a). The most common failing of gatehouse lighting is the excessive amount of light inside the gatehouse after dark. This enables any would-be offender to see into the gatehouse and determine what the guard is doing. To avoid giving the game away, it is necessary to use only the minimum amount of light in the gatehouse after dark. It would also be possible to make seeing into the gatehouse difficult by covering the windows with a mesh of high reflectance on the outer surface and then lighting that surface. The inner surface should have a low reflectance.

11.5.3 Unfenced areas

Sometimes it is desired to protect a large open area but the cost of fencing the area is prohibitive. One approach to solving this problem is to use glare lighting (Lyons, 1980). Glare lighting is designed to provide the maximum amount of disability glare to anyone approaching the line of glare sources.

Therefore, the luminaires are usually mounted at eye height and aimed so that there maximum luminous intensity is horizontally out from the protected area. Glare lighting is rarely used because it is not popular with the neighbors and it is only effective when the site behind the line of glare luminaires is completely dark and there is some possibility that there might be a guard who cannot be seen behind the line of glare sources.

11.5.4   Facade lighting

Buildings on both fenced and unfenced sites are usually protected by means of locks on doors, bars on windows and alarms. Lighting of the building's facade is sometimes used as part of this system of protection. The idea of facade lighting is that it enables anyone tampering with the doors or windows to be seen from a distance. Facade lighting will only be effective if it is comprehensive, i.e. it covers all the facade uniformly, without glare. In this situation, anyone attempting to break into the building can be seen in silhouette against the wall. The completely opposite approach is sometimes used, namely, to eliminate all lighting in and around the building. This has the effect of making the building inconspicuous and the presence of any lighting indicates illegal activity. The problem with this approach is that some illegal activities can be undertaken with very little light. There is no right and wrong answer to this problem. The designer has to make a choice on the best approach to use depending on the type of illegal activity expected, the level of risk and the system of protection proposed.

11.5.5   Closed circuit television

CCTV has been used for many years in shops and banks as a means of deterring shoplifters and robbers, or if deterrence does not work, then of identifying and convicting them when caught. More recently, CCTV has spread to the public domain with the installation of CCTV cameras overlooking city centers and linked to a police control room. As with improved lighting, the effect of such CCTV systems in reducing crime has been mixed, depending on the circumstances and the increase in risk perceived by the criminal (Phillips, 1999). The role of lighting in CCTV surveillance is to allow the camera to provide clear pictures. Exactly how much light is needed and what is the ideal light spectrum depends on the characteristics of the camera. Available CCTV devices cover a large range of sensitivities, from a minimum illuminance of 10 lx to the very low illuminance provided by starlight alone. As for spectral sensitivity, most CCTV cameras do not have the same spectral sensitivity as the human visual system, usually being much more sensitive to IR radiation. The minimum illuminance necessary for the camera to operate given by manufacturers usually assumes an incandescent light source. If other light sources are to be used, which will usually be the case, it is always a good idea to check that the proposed light source will provide enough radiation for the camera to operate successfully.

Having determined the amount and spectral content of the lighting to be used with a specific camera it is then necessary to decide on the light distribution. Care has to be taken with distribution because the one thing all CCTV cameras have in common is a limited dynamic range. This means that too large a range of luminance will lead to areas of the image being black while other areas are white. In both black and white areas, no detail can be seen. The first rule to limit the range of luminances is to keep all light sources out of the field of view of the camera. For exteriors, this means keeping the sun and any luminaires out of the field of view. For interiors, this means keeping windows and luminaires out of the field of view. The second is to provide lighting that is uniform and avoids shadows on faces. Hargroves *et al.* (1996) examined the impact of different light distributions on the CCTV image of a face, the light distributions being characterized by a series of illuminance and luminance ratios. They identified two critical ratios for an acceptable appearance of a CCTV image of a face. The first was the ratio of the illuminance on the top of the head to the illuminance on a plane containing the face, the normal to the plane being in the direction of the camera. The maximum illuminance ratio for acceptable CCTV images was 5.0. Illuminance ratios larger than 5.0 tend to produce strong shadows under the eyes, nose, mouth, and chin which distort the appearance of the face. The second ratio was the average luminance of the face to the average luminance of the background against which the camera sees the face. The range of values of this ratio for acceptable CCTV images was from 0.3 to 3.0. When the luminance of the background against which the face is seen is too high, so that the luminance ratio is less than 0.3, the image of the face will be very dark. If the luminance of the background is too low, so that the ratio is more then 3.0, the image of the face will be washed out. For interiors, the simplest way to meet these two ratios is to use indirect lighting and to position the camera so that it does not have a window in its field of view. The likelihood of getting a good CCTV image is further increased by using medium reflectance wall finishes and a floor reflectance of 0.20. Lighting installations using direct lighting luminaires with a narrow luminous intensity distribution in a low reflectance room are guaranteed to produce poor CCTV images. For exteriors, indirect lighting is not possible, but the same criteria apply. Fortunately, lighting that facilitates visual surveillance by people in the area should also be effective in meeting the criteria for a good CCTV image, namely, lighting that provides a uniform illuminance over a large area, without glare.

11.6   Generalization and value

One feature of the study of the effects of lighting on the prevalence of crime that marks it out as different from other areas of lighting research is the fact the most of the recent research has been done in the UK and USA. Given this situation, the question naturally arises as to whether the conclusions reached can be generalized to other countries where the conditions and culture are different. The answer is a definite maybe. It is not possible to be more definite without studies done in other countries but the conditions for successful generalization can be defined. The important point is that lighting, per se, has no direct effect on crime. Rather, it has an indirect effect by facilitating surveillance, community confidence and social control. In countries or communities where criminals consider increased surveillance makes criminal activity more risky and less rewarding, and where public lighting is inadequate for good surveillance, improving lighting sufficiently to ensure good surveillance can be expected to reduce criminal activity. In countries where criminals are not bothered about surveillance, either because the community is intimidated by or supports the criminals, or there is little prospect of action by the authorities, improving the lighting to enhance surveillance will be ineffective.

One other point to consider is the value of improving lighting in terms of its cost-effectiveness. Painter and Farrington (2001b) consider this question, using estimates of the cost of individual crimes to the victims and to the authorities and the costs of improving the lighting. Based on their studies in Dudley (Painter and Farrington, 1997) and Stoke-on-Trent (Painter and Farrington, 1999) they conclude that the financial benefits of better street lighting due to the reduction in crime can enormously outweigh the financial costs of providing the lighting. Specifically, they estimate that the financial benefits of reduced crime are enough to cover the capital costs of improving the lighting within 1 year, even when only tangible costs are considered. Of course, this conclusion is based on British costs but, in general, such a finding is good news for all those who believe in the value of lighting.

11.7   Summary

Attempts to use lighting as a means to reduce or at least limit criminal activity have a long history. Starting in the fifteenth century, major cities in Europe attempted to provide some form of exterior lighting at night, either by requiring householders to provide lanterns on their property or by a developing a system of public lighting controlled by the authorities. Since that time, the provision of public lighting has become more sophisticated, more widespread, and more centralized until today virtually all cities, towns and villages in the developed world have some form of public lighting. This public lighting can fulfill many roles. It can provide for safe movement of pedestrians and vehicles. It can enhance the appearance of an area and hence encourage its use after dark. It can act as a crime prevention measure. It is this last objective that is considered in this chapter, although the same lighting installation can fulfill all three roles.

A series of studies of increasing sophistication, done in the UK over the last decade, leave little doubt that lighting has a place to play in crime prevention. Improving street lighting can lead to a decrease in crime, but it may not. There can be no guarantees. This is because lighting, per se, does not have a direct effect on the level of crime. Rather, lighting can affect crime by two indirect mechanisms. The first is the obvious one of facilitating surveillance by people on the street after dark, by the community in general and by the authorities. If such increased surveillance is perceived by criminals as increasing the effort and risk and decreasing the reward for a criminal activity, then the incidence of crime is likely to be reduced. Where increased surveillance is perceived by the criminally inclined not to matter, then better lighting will not be effective. The second indirect mechanism by which an investment in better lighting might affect the level of crime is by enhancing community confidence and hence increasing the degree of informal social control. This mechanism can be effective both day and night but is subject to many influences other than lighting.

Unfortunately, the studies that demonstrate the value of better lighting as regards reduced levels of crime contain few details of the characteristics of the lighting necessary to achieve the desired effect. However, in these studies better lighting usually involves the use of more light sources with higher light output and better color rendering, more closely spaced. From such information and basic knowledge of how to make it easier to see details at night, it can be concluded that the important factors are the illuminance provided, the illuminance uniformity, the control of glare and the light spectrum. From a combination of experimental studies and practical experience, it is possible to specify what constitutes good lighting for facilitating surveillance, although some parts of the specification are more soundly based and more important than others. In public pedestrianized areas the mean illuminance on the pavement should be in the range of 10–50 lx, the illuminance uniformity should be less than 15/1, the glare ratio should be less than 4 and a good color rendering source should be used, especially if a wide range of colors is present. Lighting meeting this specification should allow anyone on the street to detect and recognize a threatening situation while there is still time to do something about it, and any witnesses to provide accurate information about the perpetrators.

While most of this chapter is devoted to the lighting of areas to which the public has access, lighting can also be used protect private areas. For example, lighting can be used to increase or decrease visibility through a fence. Visibility will be enhanced when the luminance of the fence is the same as the luminance of the area seen through the fence. Visibility will be reduced when the luminance of the fence is much higher than the luminance of the area being viewed through the fence. Lighting can also be used to enhance the performance of remote surveillance based on a CCTV system. The amount and spectrum of the light needed will depend on the characteristics of the CCTV camera used, but one factor that requires attention for all CCTV systems is the light distribution. CCTV cameras all have a limited dynamic range so they provide the best images when the range of luminances in their field of view is limited and shadows are avoided on significant features, such as faces.

There remains much to be understood about what lighting conditions are most effective in reducing crime in different countries, especially as regards limits on glare and illuminance uniformity, but one thing at least is clear; it can be done.

12 Lighting for the elderly

12.1 Introduction

Given the alternative, everyone should look forward to being old. With increasing age comes knowledge and, in some cases, wisdom. Unfortunately, knowledge and wisdom have companions in the form of physical and mental decline, ultimately leading to loss of independence, dementia, and death. This chapter examines the changes that occur in the visual system and circadian system with increasing age, the consequences these changes have for visual capabilities, and how lighting can be used to offset some of these changes so that the quality of life of the elderly may be sustained.

12.2 Optical changes with age

The human visual system can be considered as an image-processing system. Like all such systems, the visual system is most effective when it is operating at an appropriate sensitivity with a clear retinal image to process. The factors that determine the operating state of the visual system are the amount of light that reaches the retina and the wavelengths from which it is constituted. The factors that determine the clarity of the retinal image are the ability to focus the image of the external object on the retina; the extent to which light is forward scattered as it passes through the eye; and the presence of stray light produced by back-reflection from the components of the eye, transmittance through the eye wall, and fluorescence in the lens of the eye (Boynton and Clarke, 1964; Van den Berg et al., 1991; Van den Berg, 1993). Virtually all these characteristics change with age (Werner et al., 1990; Weale, 1992).

In simple optical terms, the eye has a fixed image distance and a variable object distance. To bring objects at different distances to focus on the retina, the optical power of the eye has to change. The optical power of the eye is determined by the curvature of the cornea, which is fixed, and the thickness of the lens, which is variable. If there is a mismatch between the distance of the retina from the lens and the combined optical power of the cornea and lens, the image of the outside world will not be in focus on the retina so the resulting retinal image will be blurred. Blur has been shown to be a potent cause of reduced visual performance (Johnson and Casson, 1995). The range of object distances that can be brought to focus on the retina decreases with age, because of increasing rigidity of the lens. After about 60 years of age, the eye is virtually a fixed focus optical system (Figure 12.1). Spectacles or contact lenses are used to modify the optical power of the eye, the prescription of the spectacles or contact lens changing over the years as the lens becomes increasingly rigid.

The optical factors determining the amount of light reaching the retina are the pupil size and the spectral absorption of the components of the eye. The area of the pupil varies as the amount of light available changes, the pupil opening to admit more light when there is little and closing when there is plenty. The ratio of maximum to minimum pupil area decreases with age, the maximum decreasing much more than the minimum (Figure 12.2). This means the elderly are much less able to compensate for low light levels by opening their pupils than are young people.

As for the spectral absorption of the eye, the majority of absorption takes place on passage through the lens (Murata, 1987). The absorbance of the

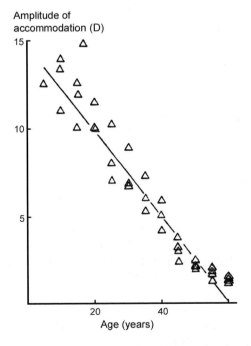

Figure 12.1 The variation of the amplitude of accommodation with age. The amplitude of accommodation is measured in dioptres, the difference between the reciprocals of the shortest and longest distances from the eye at which a sharp retinal image can be achieved, the distances being measured in meters (after Weale, 1990).

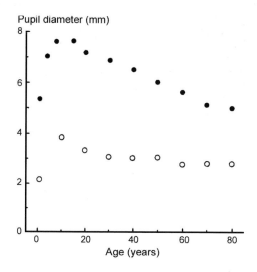

Figure 12.2 Maximum and minimum pupil diameters as a function of age (after Weale, 1982).

human lens increases exponentially from birth, following the formula (Weale, 1992):

$$D = D_0\, e^{\beta A}$$

where D is the absorbance, D_0 the absorbance at birth, β a constant that varies with wavelength, and A the age in years.

Using this formula and the values for D_0 and β given in Weale (1988), it is possible to calculate the absorbance of the lens over a range of visible wavelengths, for different ages (Figure 12.3). From Figure 12.3 it can be seen that the absorbance at short wavelengths increases dramatically with age. This goes some way to explain the diminished color vision capabilities of elderly people. Investigation of the causes of this increased absorbance with age have demonstrated that the change occurs primarily in the nucleus of the lens (Mellerio, 1987). This implies that the spectral absorbance of the lens will also vary with pupil size, smaller pupil sizes leading to greater absorbance (Weale, 1991). There can be little doubt that the reduction in pupil size and the increased absorption of light during its passage through the lens reduce the retinal illumination of older people, particularly at short wavelengths.

In addition to absorbing light, transmission through the lens and the other optical components of the eye scatters light. This is important because, whereas increased absorption does not degrade the retinal image and can be compensated by providing more light, scattered light degrades

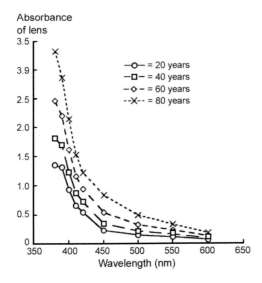

Figure 12.3 Spectral absorbance of the lens plotted against wavelength, for different ages (after Weale, 1988).

the retinal image and providing more light does not help. Scattered light degrades the retinal image by reducing the difference in luminance either side of an edge, thereby reducing the magnitude of its higher spatial frequencies. Scattered light also degrades the retinal image in terms of color by adding wavelengths from one area onto another, thereby reducing the color difference at the edge. The scattering occurring in the eye is primarily large-particle scattering, so is largely independent of wavelength. Measurements have shown that about 30 percent of scattering occurs at the cornea (Vos and Boogaard, 1963), with most of the rest occurring at the lens, vitreous humor, and fundus (Boettner and Wolter, 1962). The amount of scatter increases with age, due mainly to changes in the lens (Wolf and Gardiner, 1965).

Scatter can be quantified by a point spread function, which typically shows that the amount of scattered light decreases with increasing deviation from the beam of light being scattered (Vos and Boogaard, 1963). Straylight is characterized by a homogenous distribution of luminance over the whole retinal image. Straylight within the eye is caused by light back-reflected from the retina and pigment epithelium, by transmission of light through the iris and the eye-wall, and by lens fluorescence. Straylight matters because it falls uniformly across the retinal image, thereby reducing the luminance contrast of all edges and the saturation of all colors in the image. The amount of straylight generated by these causes increases with age. This is particularly so for lens fluorescence. The effect of lens florescence is negligible in young eyes, but as aging continues, the luminance of straylight due to fluorescence increases and fluorphores with emission wavelengths in the most sensitive part of the visual spectrum emerge. The effect of lens fluorescence can be seen as a haze over the visual scene (Jacobs and Krohn, 1976; Weale, 1985).

12.3 Neural changes with age

The optical changes that occur with age affect the retinal image, but for the visual system to be effective, the retinal image has to be processed by the retina and the visual cortex. There is no reason to suppose that aging is limited to only the optical elements of the visual system. Indeed, morphological changes have been reported in rod and cone photoreceptors in older people (Marshall *et al.*, 1979); the density of rods and cone photoreceptors have been shown to decrease in extreme old age (Feeny-Burns *et al.*, 1990); and the number of ganglion cells in the retina and the number of neurons in the visual cortex both decrease with increasing age (Devaney and Johnson, 1980; Balazsi *et al.*, 1984). Weale (1992) provides a useful review of these neural changes and their possible causes. The fact that the neural elements of the visual system also show changes with age is important because it implies that the compensation for visual system aging that can be provided by lighting is inevitably limited. Returning the optical characteristics of the eye to what they were before aging will not restore vision to its pristine state.

12.4 Low vision

Both the optical and neural changes discussed above are part of the normal process of aging. Everyone who lives long enough will experience these changes, but with increasing age there is also an increased probability of pathological change occurring in the eye. These pathological changes can lead to low vision and, ultimately, blindness. Before discussing what these pathological changes are, it is necessary to define what is meant by low vision and blindness. The accepted international definition of these terms is based on a classification of vision developed by the World Health Organization (WHO, 1977) (Table 12.1). This classification system uses the visual acuity of the better eye, with optical correction, and the size of the visual field, to discriminate between different levels of visual capability. Visual acuity is expressed as the ratio of two distances, e.g. 20/200. The numerator is always 20 and refers to the distance, in feet, from which the person being tested looks at a test chart and determines the smallest size of target where the detail can be resolved, e.g. a letter can be correctly identified. The denominator (200 in this example) is the distance at which a person with normal vision can be expected to resolve the same detail (in the metric system this ratio will be given as 6/60). A person with a visual acuity of 20/200 has low vision. A person with a visual acuity of 20/20 has

Table 12.1 The World Health Organization's classification of vision (WHO, 1977)

Category	Grade	Criteria
Normal vision	0	20/25 or better
Near normal vision	0	20/30 to 20/60
Low vision		
Moderate visual impairment	1	20/70 to 20/160
Severe visual impairment	2	20/200 to 20/400
Blindness		
Profound visual impairment	3	20/500 to 20/1,000 or a visual field less than 10°
Near-total visual impairment	4	Worse than 20/1,000 or a visual field less than 5°
Total visual impairment	5	No light perception

normal vision. Using this classification system, the WHO has defined blindness as a visual acuity of worse than 20/400 and a visual field diameter of less than 10° in the widest meridian of the better eye. Despite this international definition, there are still national variations in the criteria for what constitutes low vision and blindness. For example, in the US the legal definition of blindness is a visual acuity of 20/200 and a visual field of less than 20°. The important point to note from this discussion is that normal vision, low vision, and blindness are not discrete states, but rather a continuum, and the borders between these states are somewhat arbitrary.

There have been many attempts to quantify the prevalence of low vision and blindness in different populations (Tielsch, 2000). Probably the most interesting one for the purposes of this chapter is the Baltimore Eye Survey. This survey examined 5,308 residents of 40 years or older in an urban area in the US. Table 12.2 shows the measured prevalence of blindness and low vision for different age and racial groups. From Table 12.2, it can be seen that the prevalence of blindness and low vision is strongly related to age and more loosely linked to race. Specifically, the prevalence of low vision increases dramatically after about 70 years of age, and that increase seems to occur earlier for blacks than whites. As for the causes of blindness and low vision, Table 12.3 shows the percentage of people of different races classified as blind and with low vision due to various pathological conditions. From Table 12.3, it can be seen that the most common causes of blindness and low vision are cataract, macular degeneration, glaucoma, and diabetic retinopathy, although the most common causes are different for blacks and whites. Cataract and glaucoma are the most common causes of blindness and low vision amongst blacks while macular degeneration is much more common amongst whites. To what extent these differences between races are caused by physiology or by differences in access to health care remains an open question. What can be said is that cataract can be

Table 12.2 Prevalence per 100 people of blindness and low vision for different age groups and races (after Tielsch *et al.*, 1990). Blindness is defined as a visual acuity of 20/200 or worse. Low vision is defined as a visual acuity of from 20/40 to 20/200

Age range (years)	Blindness		Low vision	
	Whites	Blacks	Whites	Blacks
40–49	0.6	0.6	0.2	0.6
50–59	0.5	0.7	0.7	1.3
60–69	0.2	1.6	1.1	3.4
70–79	0.6	2.9	5.2	8.1
80+	7.3	8.0	14.6	18.0

Table 12.3 Percentage of people of different races classified as blind or with low vision according to the Baltimore Eye Survey (Sommer *et al.*, 1991; Rahmani *et al.*, 1996), with various causes of blindness and low vision. Blindness is defined as a visual acuity of 20/200 or worse. Low vision is defined as a visual acuity from 20/40 to 20/200

Cause	Blindness		Low vision	
	Whites	Blacks	Whites	Blacks
Cataract	13	27	38	34
Macular degeneration	30	0	22	6
Diabetic retinopathy	6	5	3	11
Other retinal disorder	7	15	10	5
Glaucoma	11	26	3	7
Optic neuropathy	2	5	3	7
Other	28	22	10	16
Unknown	4	0	13	15

dealt with and the progress of glaucoma can be slowed and sometimes stopped, given the appropriate medical intervention. This pattern of causal factors is typical of the developed world. In the Third World, the situation is very different with cataract accounting for between 30 and 60 percent of blindness (Thylefors *et al.*, 1995) and infections that can lead to blindness, such as trachoma and onchocerciasis, being common.

It is now necessary to consider the nature of each of the more common causes of blindness and low vision. Cataract is an opacity developing in the lens. In fact there are four main types of cataract; cortical, posterior sub-capsular, nuclear, and mixed, i.e. some combination of the other three (Chylack, 2000). The effect of all these types of cataract is to absorb and scatter more light as the light passes through the lens. This increased absorption and scattering results in reduced visual acuity and reduced contrast sensitivity over the entire visual field, as well as diminished color discrimination and greater sensitivity to glare. The extent to which more light can help a person with cataract depends on the balance between absorption and scattering. More light will help overcome the increased absorption but if scattering is high, the consequent deterioration in the luminance contrasts in the retinal image will reduce visual capabilities. The usual treatment for cataract in the developed world is the surgical removal of the cataract and its replacement with a fixed-focus plastic lens. Of course, this is only done after the cataract has fully developed, a process that takes between 4 and 7 years. Studies are currently in progress to determine if some form of multi-vitamin mineral supplement can be used to slow the development of cataracts.

Macular degeneration is detected when the macular of the retina, which is a circular, yellow-pigmented area of the retina, 2–3 mm in diameter and centered on the fovea, changes in appearance. These changes are evident to the ophthalmologist as the presence of drusen (whitish yellow spots) in the macula, hypo- or hyperpigmentation of the retinal pigment epithelium, distortion of the retinal pigment epithelium, or new blood vessels, blood, and scars in the macula (Schwartz, 2000). The presence of any of these features implies damage in and around the fovea which, in turn, implies a serious decline in central vision, ultimately making such everyday activities as reading and seeing faces impossible. However, peripheral vision outside the macular is unaffected so the ability to orient oneself in space and to find one's way around is little changed. Providing more light, usually by way of a task light, will help people in the early stage of macular deregulation to read, although as the deterioration progresses additional light will be less effective. Increasing the size of the retinal image by magnification or by getting closer is helpful at all stages. Macular degeneration is a leading cause of blindness in the developed world and is strongly linked to age. Klein *et al.* (1997) found that 30 percent of whites of 75 years and older had early signs of macular degeneration. These early signs are strongly predictive of the later, visually catastrophic consequences of continuing macular degeneration. Unfortunately, the only available treatment for macular degeneration, photocoagulation, is of limited effectiveness, being able to do little more than slow the rate of vision loss in less than 10 percent of cases (Schwartz, 2000). Given the increasing number of elderly people present in the populations of developed nations, it is to be fervently hoped that some of the effort now being put into finding treatments for macular degeneration bears fruit.

Glaucoma is best thought of as the ultimate outcome of a number of diseases that affect the eye, that outcome being progressive visual field loss (Ritch, 2000). Most forms of glaucoma follow the pattern of an event that alters the outflow from the aqueous humor, leading to elevated intra-ocular pressure, that produces damage to the optic nerve head and hence progressive visual field loss, leading ultimately to blindness (Shields *et al.*, 1996). As glaucoma develops, it leads to reduced contrast sensitivity, poor night vision, and slowed transient adaptation but the resolution of detail seen on-axis is unaffected until the final stage. Modifying lighting is of little value in helping people who show symptoms of glaucoma, because where damage has occurred the retina has been destroyed. The incidence of glaucoma is strongly related to age. The treatment of glaucoma is based on reducing intra-ocular pressure, either by pharmaceuticals or by surgery.

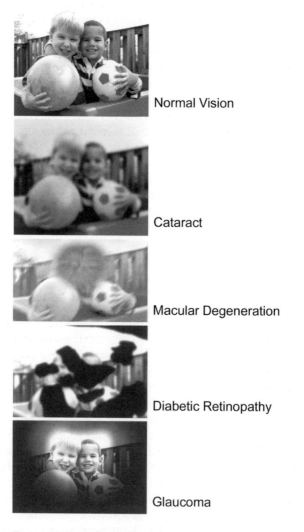

Normal Vision

Cataract

Macular Degeneration

Diabetic Retinopathy

Glaucoma

Figure 12.4 An illustration of a scene as it might appear to people with normal vision and with cataract, macular degeneration, diabetic retinopathy, and glaucoma (from the National Eye Institute, National Institutes of Health).

Diabetic retinopathy is a consequence of chronic diabetes mellitus (Leonard and Charles, 2000). Chronic diabetes mellitus effectively destroys parts of the retina through the changes it produces in the vascular system that supplies the retina. Specifically, diabetic retinopathy is identified by the presence of microaneurysms, hemorrhages, hard extrudates, changes in retinal arteries and veins and, sometimes, neovascularization. The effect these changes have on visual capabilities depends on where on the retina the hemorrhages, exudates, etc. occur and the rate at which they progress. Despite this uncertainty, the endpoint of diabetic retinopathy is clear. It is blindness. Blindness occurs 25 times more commonly in the diabetic than in the non-diabetic population (Ferris, 1993). The medical treatment of diabetic retinopathy is based on close control of blood glucose, and damage control using laser photocoagulation and vitreous surgery.

Although cataract, macular degeneration, glaucoma, and diabetic retinopathy have been discussed separately, it is important to appreciate that having any one of them does not confer immunity to the others. In fact, the older the individual, the more likely it is that more than one of these causes of low vision will occur. To give an impression of what it is like to have one of these conditions, Figure 12.4 shows a simulation of a scene as it would appear to people with normal vision and with cataract, macular degeneration, glaucoma, and diabetic retinopathy. The difficulties that must be experienced by people with any of these conditions in carrying out everyday tasks is obvious.

12.5 The aging of the circadian system

The output of the human circadian system can be characterized by its amplitude, period, and phase. With increasing age, the amplitudes of the many circadian rhythms, including core body temperature and melatonin concentration, have been shown to diminish (Brock, 1991; Copinschi and van Cauter, 1995). There is also evidence of a shortening of the period and a phase advance with increasing age (Czeisler *et al.*, 1986; Renfrew *et al.*, 1987). How much this change is due to the degeneration of the physiology of the circadian system (Swaab *et al.*, 1985) and how much it is related to the reduced amount of light reaching the retinas of people as they age is an open question. What is not open to question is that such changes may have an effect on the health and well-being of the elderly. Two groups of people whose circadian systems might be expected to be severely compromised are the blind and those with advanced cataract, glaucoma, and diabetic retinopathy. Some blind people do indeed suffer from a lack of entrainment of their circadian rhythms and consequent sleep disturbances (Sack *et al.*, 1992b) but some do not. Czeisler *et al.* (1995) found blind people with and without melatonin suppression by light. Those who showed melatonin suppression by light had no history of sleep disturbance while those who did not show melatonin suppression by light did. Whether the blind have a functioning circadian system or not depends on the cause of their blindness. Causes of blindness that leave the retina intact may also leave the circadian system intact. Apart from these investigations of the blind, primarily undertaken to better understand the circadian system, there has not been a systematic investigation of the use of light to compensate for an aged circadian system. What there have been are investigations of the use of light to treat specific health problems. These are discussed in Section 13.4. The rest of this chapter is devoted to compensating for the effects of age on the human visual system.

12.6   The effects of age on visual performance

As might be expected, the changes in the optical and neural characteristics of the visual system that occur with increasing age have an impact on what the visual system is capable of doing. The most likely place to find such effects is at threshold, where the visual system is operating at its limits. A consideration of the effects of age on threshold performance reveals that the effect of increasing age is almost always negative, in the sense that the visual system becomes less discriminating, and more sensitive to adverse conditions. Specifically, older people tend to show reduced visual field size, increased absolute threshold luminance, reduced visual acuity, reduced contrast sensitivity, increased sensitivity to glare, and poorer color discrimination. Figure 12.5 shows the functional visual field of one eye for a 24- and a 75-year-old (Williams, 1983). The boundaries of the fields are formed by

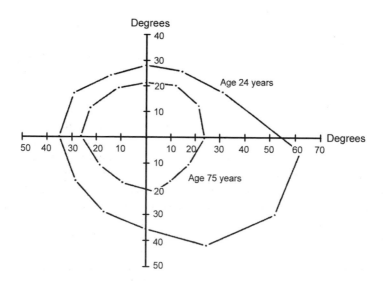

Figure 12.5 The functional visual fields for one eye of a 24- and a 75-year-old. The fields are defined by contours of equal detection performance (after Williams, 1983).

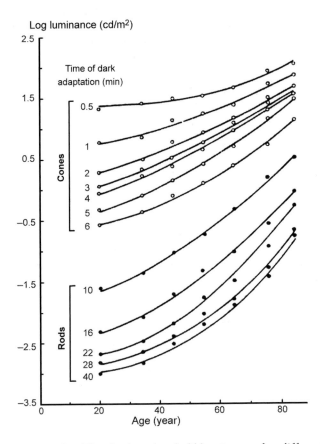

Figure 12.6 The absolute threshold luminance after different durations of dark
adaptation following exposure to a luminance of 5,100 cd/m², for
people of different ages (after Domey and McFarland, 1961).

contours of equal detection performance. The decrease in functional visual
field size with age is clear. Figure 12.6 shows the absolute threshold luminance after different times of dark adaptation plotted against the age of the
observers (Domey and McFarland, 1961). It is readily apparent that as the
visual system ages, the absolute luminance threshold increases. Figure 12.7
shows the change in visual acuity plotted against age (Adrian, 1995). The
data is taken from four different studies, all using different methods for
measuring visual acuity. Despite these differences, the trend with age is
consistent, increasing age leads to worse visual acuity.

Figure 2.29 shows the threshold contrast achieved by different people of
different ages for the detection of a 4 min arc disk presented against three
different background luminances (Blackwell and Blackwell, 1971). These
results show the expected increase in threshold contrast with decreasing

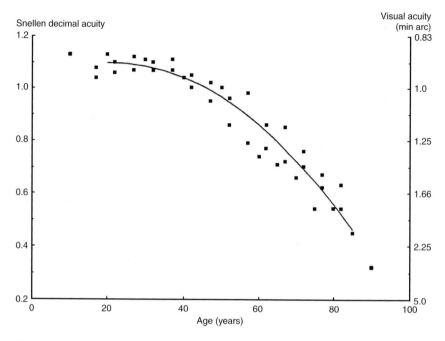

Figure 12.7 Visual acuity, expressed as Snellen decimals and as angle subtended, as a function of age (after Adrian, 1995).

adaptation luminance but there is also a trend of increasing threshold contrast with increasing age, at all three adaptation luminances.

Figure 12.8 shows contrast sensitivity functions for four people of four different ages (McGrath and Morrison, 1981). Examination shows that the effect of age is to decrease the maximum contrast sensitivity and decrease the range of spatial frequencies over which resolution can occur. More extensive studies show the same trends (Owsley *et al.*, 1983). For disability glare, Vos (1995) and CIE (2002b) provide a modification of the disability glare formula to account for age (see Section 5.4.2.1 for the unmodified disability glare formula). The modified formula is

$$L_v = 10\,(1 + (A/70)^4)\,\Sigma(E_n\,\Theta_n^{-2})$$

where L_v is the equivalent veiling luminance (cd/m^2), A the age (years), E_n the illuminance at the eye from the nth glare source (lx), and Θ_n the angle between the line of sight and the nth glare source (degrees).

This formula shows that the equivalent veiling luminance increases with increasing age.

Figure 12.9 shows the average distribution of errors made on the Farnworth–Munsell 100 hue test as a function of illuminance and age

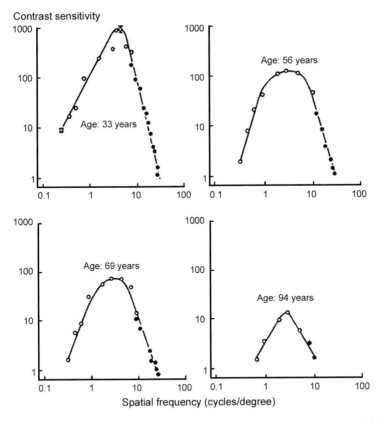

Figure 12.8 The contrast sensitivity function of four observers of different ages. Open circles are for near viewing distance. Filled circles are for far viewing distance (after McGrath and Morrison, 1981).

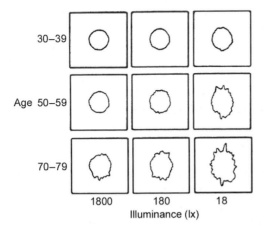

Figure 12.9 The average distribution of errors on the Farnsworth–Munsell 100 hue test as a function of illuminance and age (after Knoblauch *et al.*, 1987).

(Knoblauch *et al.*, 1987). The Farnsworth–Munsell 100 hue test is a test of hue discrimination that requires the subject to arrange a series of 85 colored disks of equal lightness and chroma but different hue into a consistent hue circle, i.e. into a circle in which the difference in hue between adjacent disks is a minimum. Performance on the test is scored by the magnitude of the misplacements of disks on the circle. In Figure 12.9, zero error is indicated by a smooth circle. As the number of errors increases, the circle becomes larger and more ragged. The distance from the center point in a given radial direction is a measure of the magnitude of the error in hue discrimination made for the particular hue represented by the radial direction. Examination of Figure 12.9 suggests that older people tend to make more errors in hue discrimination, particularly at low illuminances. Also, in old age, errors are more common for some hues than others. Specifically, the pattern of errors with hue suggest a reduced response from the short-wavelength cones because a similar pattern is obtained from people of any age with tritanopia, i.e. defective color vision associated with the absence of short-wavelength cones (Verriest, 1963).

While the trend of deterioration in visual function with age is evident in all these aspects of threshold performance, there are two other general conclusions that deserve emphasis. The first is that the deterioration in visual function with age starts in early adulthood and continues at a steady pace as the years pass. The second is that there are wide individual differences in visual function, sometimes wide enough to overcome the effects of age. This is most evident in Figure 2.29, which shows the threshold contrast for individuals of different ages. While the trend of increasing threshold contrast with increasing age is obvious, it is also clear that the individual differences are large enough for some people in their sixties to have lower threshold contrasts than some people in their twenties. Care should be taken before assuming that all older people are visually challenged. It is also important to appreciate that there are a few aspects of visual system performance that change little with normal aging. Specifically, vernier acuity, the ability to detect whether two lines are directly in line or are offset relative to each other, does not deteriorate with age (Enoch *et al.*, 1995), and neither do several aspects of color vision, such as the wavelength of unique hues (Werner and Kraft, 1995). The thing these two very different aspects of vision have in common, and what differentiates them from many other measures of visual function, is that they are the result of neural data processing of difference signals and difference signals are not sensitive to changes that affect both parts of the signal equally.

One group who can always be assumed to be visually challenged in some way are those with low vision. This group can be expected to show much worse threshold performance than people with normal sight. Indeed, as discussed in Section 12.4, a markedly poorer visual acuity is one of the criteria for classifying someone as having low vision. How dramatic the decline in threshold performance can be is shown in Figure 12.10 (Paulsson and

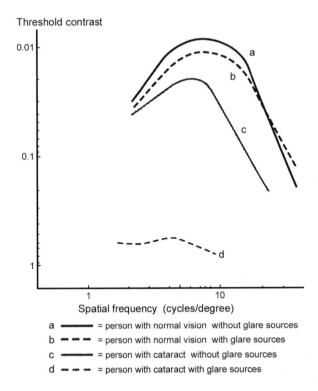

Threshold contrast

Spatial frequency (cycles/degree)

a ━━━━ = person with normal vision without glare sources
b ━ ━ ━ = person with normal vision with glare sources
c ━━━━ = person with cataract without glare sources
d ━ ━ ━ = person with cataract with glare sources

Figure 12.10 Threshold contrast plotted against spatial frequency for two people, one with normal vision and the other with cataract, with and without a glare source present (after Paulsson and Sjostrand, 1980).

Sjostrand, 1980). This shows the threshold contrast for a grating of different spatial frequencies, for two people, one with and the other without a cataract, with and without a high luminance surround. Clearly, the presence of a cataract increases threshold contrast even in the absence of a high luminance surround but when the surround is introduced, the difference between the two people increases greatly, because the light from the high luminance surround is scattered by the cataract over the part of the retinal image containing the threshold contrast target.

It is important to appreciate that the effect of low vision on threshold performance depends on the cause of the low vision and the nature of the task. The effect of the cause of low vision can be seen in the results of Julian (1983), who measured the contrast sensitivity of people with low vision at different illuminances. People with macular defects showed the greatest benefit from increases in illuminance. People with cataract showed little benefit from increased illuminance. People with glaucoma were little different from those with normal vision. Obviously, a very different rank order for the different causes of low vision would have been found if the task used had required peripheral vision rather than the foveal vision used when measuring contrast sensitivity.

The worsening in such threshold measures as absolute sensitivity to light, visual acuity, contrast sensitivity, and color discrimination with age has implications for the performance of many real tasks. Kosnik *et al.* (1988) confirmed this in surveys of several hundred people ranging from 18 to 100 years of age. The purpose of the survey was to determine what visual problems they experienced in daily life. Five types of visual problems that increased in difficulty with age were identified. They were: seeing in dim light, reading small print, distinguishing dark colors, reading moving information, and visual search. Morgan (1988), a former Dean of Optometry in the University of California, Berkeley, gives an account of his own experience of aging vision. He reports four situations which he finds difficult; driving at twilight, seeing in shadowed areas, entering and leaving dark areas, and visual search.

The concept that explains why the elderly have difficulty with such tasks is the separation of the stimuli provided by the task from their threshold values. This concept is evident in the work of Whittaker and Lovie-Kitchin (1993), who reviewed the literature on reading rate. They found that there were four factors that were important for improving the reading rate of those with low vision. The four factors were acuity reserve (print size relative to threshold visual acuity), contrast reserve (luminance contrast of print relative to threshold contrast), number of letters visible, and size of central field loss. Clearly, if visual acuity and threshold contrast worsen with age or the onset of low vision, then the acuity reserve and the contrast reserve are less, so reading rate will decrease. In general, the closer the stimulus provided by a task is to the threshold of the observer for that stimulus, the worse the task performance will be although, as discussed in Section 4.3.5, the relationship between task performance and visual stimuli is not a simple linear function. Rather, the importance of vision to the change in task performance with age, for a specific task, depends on two factors, the role of the visual component in the task, and how close to threshold the visual stimuli presented by the task are. If the visual component is insignificant then the change in task performance will be slight, even if visual capabilities are much diminished. If the task is dominated by vision then the changes in vision that occur with age will have an important effect on task performance. As for the proximity to threshold, the "plateau and escarpment" shape of visual performance (Boyce and Rea, 1987) implies that the effects of age on visual performance will be much more marked for tasks where the visual stimuli are close to threshold than when they are far above threshold. For example, Bailey *et al.* (1993) showed that reading speed improved as the acuity reserve increased, until the print size was about four times as big as the threshold size, after which no further improvement occurred. Given this understanding, the question that now needs to be addressed is what can be done to offset the decline in visual capabilities that occur with age.

12.7 What can be done to offset the effects of age?

There are four possible approaches to offsetting some or all of the effects of age on visual capabilities. They are: to change the optics of the eye so as to provide a sharp retinal image of the task; change the task so that the stimuli it presents are further from threshold; change the lighting to enhance the capabilities of the visual system or to move the stimuli presented by the task further from threshold; and to eliminate the need to do the task at all. Each will be discussed in turn.

12.7.1 Changing the optics

The first and most widely experienced effect of aging on the visual system is the recession of the near point caused by the increased rigidity of the lens. Eventually, the near point moves so far away that it is no longer possible to bring an object positioned at a normal distance to focus on the retina, e.g. it is no longer possible to read a newspaper, even when held at arms length. This problem can be overcome by the use of spectacles or contact lenses with the appropriate optical power. Figure 12.11 shows the effect of wearing spectacles on visual acuity, for near and distance vision, for people of different ages. Wearing spectacles to bring the retinal image into focus produces a marked improvement in visual acuity for older people, although it does not completely restore visual acuity to what is was when young

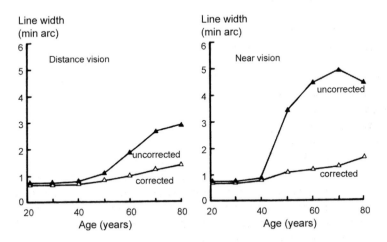

Figure 12.11 Subtended line widths of letters that can just be read by 50 percent of observers, for distant and near vision, with and without their usual spectacles, plotted against age, For distant vision, the test letters were 6 m from the observer while for near vision, they were 0.36 m from the observer (after US Department of Health, Education, and Welfare, 1964).

because simply wearing spectacles does nothing to offset the other optical and neural changes that occur in the eye with age.

The other reason why the optics of the eye are sometimes changed, a change that is becoming much more common in the elderly, is the development of cataract. Today it is routine to replace a cataracterous lens with a plastic lens of fixed optical power. It might be thought that removing a cataract and substituting a clear, intraocular lens has the potential to make the retinal image better than it was before the onset of the cataract, but this is not always so. Whether it does or not depends on the way in which the lens is removed and the optical characteristics of the intraocular lens (Nadler, 1990). Removing a cataract certainly reduces light absorption, light scatter, and lens fluorescence but may introduce new sources of light scatter and will do nothing for any retinal and neural degradation that has occurred. Nonetheless, for many people, removing a cataract will markedly improve visual acuity and contrast sensitivity and will enhance the saturation of colors (Asplund and Ejdervik Lindblad, 2002).

12.7.2   Changing the task

Another approach to offsetting the effects of age on task performance is to change the visual stimuli presented by the task. The Relative Visual Performance (RVP) model of visual performance (see Section 4.3.5) demonstrates that increasing the size or contrast of the task, and hence moving the task further from threshold, produces an improvement in visual performance for young people with normal vision. It is reasonable to assume that increasing size will have an even greater beneficial effect for the elderly and for people with low vision. Figure 12.12 shows the speed and accuracy of doing a high-contrast Landolt ring task (see Figure 4.5) of different sizes, for two age groups, one 18–28 years and the other 61–78 years (Boyce *et al.*, 2003b). Speed is measured as the number of Landolt rings examined in 20 s. Accuracy is measured as the number of Landolt rings of a specified gap orientation found as a percentage of the number of Landolt rings with the specified gap orientation examined in 20 s. As expected, Figure 12.12 shows that increasing the size of the gap in the Landolt ring leads to greater speed and higher accuracy until saturation occurs. This is true for both age groups. However, the effects of increasing size are more marked for the older age group, particularly for accuracy when the gap size is small. This is because smallest gap size is closer to threshold for the older age group; nine of the 38 older subjects being unable to do the task at the smallest size. Figure 12.12 also shows that enhancing the visual stimulus will not bring the level of performance of the older subjects to that of the young subjects. Even at the largest gap size, there is a difference in both speed and accuracy of performance between the two age groups. This is because increasing the size of the visual stimulus does nothing to address the increased absorption of light

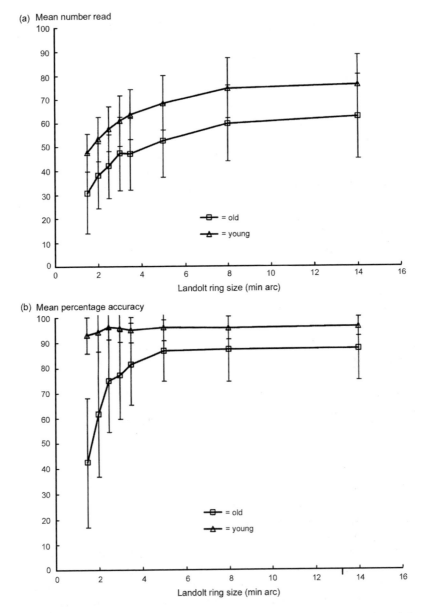

(a) Mean number read

= old
= young

Landolt ring size (min arc)

(b) Mean percentage accuracy

= old
= young

Landolt ring size (min arc)

Figure 12.12 Speed and accuracy of performance of a high-contrast Landolt ring task, plotted against Landolt ring gap size measured in angular subtense at the eye, for two age groups (18–28 and 61–78 years). Speed is measured as the mean number of Landolt rings examined in 20 s. Accuracy is measured as the mean number of Landolt rings of a specified gap orientation found as a percentage of the number of Landolt rings with the specified gap orientation examined, in 20 s. The error bars are standard deviations (after Boyce *et al.*, 2003b).

in the eye and the general slowing of cognitive function that occurs with advanced years.

The size of the retinal image of a task can be increased either by making the task bigger, e.g. large print books, or by bringing the task closer, although this may be limited by the need to keep the resulting image in focus on the retina; or by using some form of magnification. Magnification can be achieved either optically or electronically, but both forms need to be optimized for the individual and the task. This is because the greater is the magnification, the smaller is the field of view. If the task involves some form of scanning, e.g. reading text, then optimization of the magnification and field of view is essential.

One situation where magnification might be thought to be of great value is for people who have lost foveal vision, e.g. people with macular degeneration. Unfortunately, the benefit of magnification is not as great as expected. Legge *et al.* (1985) found that even with a very large character size, people who had lost their central vision could never read at a rate of greater than 70 words/min, whereas people with intact foveal vision could approach reading rates of 200–300 words/min. The improvement that does occur with magnification occurs because the effect is to enlarge the retinal image so that it extends over the near periphery of the retina, an area that is unaffected by macular degeneration. Unfortunately, visual acuity decreases with retinal eccentricity (Westheimer, 1987).

Size is just one dimension that can be used to make a task easier to do. Another is luminance contrast. Again, the RVP model of visual performance (see Section 4.3.5) shows that increasing luminance contrast will lead to better visual performance for young people. It is reasonable to assume that increasing luminance contrast will have even greater beneficial effects for the elderly and for people with low vision. Certainly, the idea of increasing luminance contrast is a feature of much advice on how to make life easier for people with low vision. For example, Sicurella (1977) recommends that people with low vision should have a sheet of black paper and a sheet of white paper on the kitchen wall. Then, the level of either a light or a dark liquid in a transparent container can be more easily seen by viewing the container against the opposite background. A similar approach can be used to allow people with cataract and other forms of low vision that result in a blurry retinal image to orientate themselves in a space. Specifically, a high luminance contrast between floor and walls, and between walls and door, and between door and door handle will help such a person find the door and open it. While high luminance contrasts of this type are undoubtedly useful, they should be attached only to salient aspects of the space. To enhance safe and confident movement about a space, the ideal to aim for is to create the impression of a line drawing of the scene, in which the high luminance contrast between salient parts of the scene represents the lines of the drawing. Too many different luminance contrasts produce a confusing picture for people with low vision to interpret.

Another factor that needs to be considered when seeking to maximize luminance contrast is the amount of scattered light produced in the eye. Scattered light will tend to reduce luminance contrast of the retinal image of the task. One simple means to reduce the amount of scatter is to reduce the luminance of the area immediately surrounding the task. Legge *et al.* (1985) found that people with cataract and other conditions that would lead to a large amount of scattered light, could read white letters against a black background much more easily than black letters against a white background. The reduction in scattered light from the background is what lies behind a device widely used by those with cataract to make reading easier. The device is a piece of black card with a slot cut in it. The slot is positioned over the page so that only one or two lines of print can be seen at a time. The low luminance of the immediate surround to the print minimizes the reduction of the luminance contrast of the print by scattered light.

There are several different means to reduce scattered light and straylight from a much larger visual field. These range from the wearing of an opaque visor or cap, which shields the eyes from the sun and sky outdoors and luminaires overhead indoors, to the wearing of photochromic, polarizing, spectrally selective sunglasses. Daylight outdoors can vary greatly in amount, is polarized, and constitutes the most common source of exposure to high-intensity UV radiation. The photochromic component of the sunglasses adjusts the transmittance of the glasses according to the amount of light available. The polarizing component removes vertically polarized light, which is produced as highlights after reflection from water and other specularly reflecting surfaces. These highlights reduce the luminance contrast of the object in two ways: at the object itself and by scattered light in the eye. Finally, the sunglasses are designed to transmit little light at wavelengths below about 550 nm. Thus, they cut out most of the incident radiation, both UV and short wavelength visible, that generates lens fluorescence and increase the luminance contrast between surfaces that are predominantly blue or green relative to those that are predominantly yellow or red. Glasses that stop short-wavelength radiation from reaching the eye have been shown to improve the vision of people with cataracts and macular degeneration (Tupper *et al.*, 1985; Rutkowsky, 1987; Zigman, 1992).

Another dimension which might be used to enhance visual performance is surface color. Color can be used to enhance visual performance in three different ways. The first is to identify objects. Wurm *et al.* (1993) found that color does improve the recognition of images of familiar foods by people with normal and low vision. The second is to make items more conspicuous. The value of color as an aid to conspicuity is discussed in Section 8.5. The third is as a substitute for luminance contrast. In the absence of luminance contrast, then a color difference between the task and its immediate background is the only way in which the task can be seen. However, this is a rather extreme situation. Color difference only becomes important when the luminance contrast is low. When luminance contrast is high, there is little to be gained in terms of visibility by enhancing color difference as shown by the fact that adding color to text produces no significant improvement in reading speed (Knoblauch *et al.*, 1991).

Finally, where what has to be seen can be presented on a self-luminous display, there is the possibility of using image enhancement. Peli and Peli (1984) suggest using an adaptive image enhancement technique in which the image is processed one pixel at a time based on its local characteristics. Specifically, the image is divided into its low and high spatial frequency components, which is analogous to the local contrast. The high frequency components are amplified while the low frequency components are shifted towards the mid-range. The effect is to enhance the contrast and sharpness of the elements of the image and hence make its details more visible to people with low vision.

12.7.3   Changing the lighting

The characteristics of lighting that can produce an improvement in visual performance are the amount of light, the spectrum of the light, and the spatial distribution of light. Each will be considered in turn.

The RVP model of visual performance (see Section 4.3.5) demonstrates that increasing the retinal illuminance will lead to an improvement in visual performance, although the magnitude of the improvement will vary with the separation of the size and contrast of the task from their respective threshold values, the greater the separation, the less the impact of increasing the retinal illuminance. The effect of age up to 65 years is taken into account in the RVP model by adjusting the retinal illuminance for the decreased pupil size and the increased absorption of light in the eye that occurs with age and by adjusting threshold contrast for light scattered in the eye (Rea and Ouellette, 1991).

Unfortunately, there is neither an equivalent to the RVP model for people with low vision, nor, given the large differences in the effects of the different forms of low vision and their wide inter-individual differences, is there ever likely to be. What can be said is that causes of low vision that reduce the retinal illumination without reducing the clarity of the retinal image are likely to benefit from increased illuminance. This view is supported by the observations of Sloan *et al.* (1973), who measured the ability of people with diseases of the macular to read under normal room lighting and under a high-intensity reading lamp. With the reading lamp, many of the patients were able to read continuous text without magnification or with much less magnification than was required under the normal room lighting. Eldred (1992) also reports faster reading speeds at dramatically higher illuminances by people with macular degeneration. Cornelissen *et al.* (1995) examined object perception in a simulated living room lit to illuminances in the range 1.6–5,000 lx. All the objects could be recognized by people with normal vision at 1.6 lx. All the subjects, who had several different forms of low vision, showed improvement in their ability to detect and recognize the objects as the illuminance was increased although there was considerable differences among the subjects with respect to whether, and at what illuminance, the improvement ceased. As might be expected, the ability to move through a space is enhanced for all people with low vision by an increase in illuminance (Kuyk *et al.*, 1996).

Given these changes in visual capabilities, it is not unreasonable to expect that different illuminances would be preferred by people with different types of low vision. Lindner *et al.* (1989) measured the preferred illuminance to read high contrast printing. Each subject could adjust the illuminance provided by a large array of ceiling-mounted fluorescent lamps over a wide range using a continuously variable dimming system. Table 12.4 shows the median illuminance selected and the 10th and 90th percentiles of groups of people with normal vision and various types of low vision, for three types of fluorescent lamp. The most obvious feature of these results,

Table 12.4 Median preferred illuminance and the 10th and 90th percentiles for reading high contrast printing of line width subtending 4.4 min arc at 30 cm, under three different types of fluorescent lamp (from Lindner *et al.*, 1989)

State of vision	Number of subjects	Fluorescent lamp type	Median preferred illuminance (lx)	10th and 90th percentile of illuminance (lx)
Emmetropic 20–30 years	50	White	900	329–2,072
		Warm-white	1,000	600–2,127
		Daylight	1,055	426–2,090
Emmetropic 40–79 years	50	White	268	75–817
		Warm-white	260	105–1,527
		Daylight	315	162–1,753
Cataract – pre-operative 40–80 years	75	White	325	98–1,800
		Warm-white	300	45–1,496
		Daylight	448	52–1,450
Cataract – post-operative with intra-ocular lens	50	White	121	70–1,162
		Warm-white	123	50–939
		Daylight	140	60–1,197
Cataract – post-operative with spectacle correction	25	White	119	75–439
		Warm-white	128	39–629
		Daylight	195	54–656
Glaucoma 40–82 years	50	White	596	100–1,071
		Warm-white	480	85–1,278
		Daylight	675	67–1,866

and of many others (CIE, 1997a), is the magnitude of the individual differences in preferred illuminance, within each group. The next most obvious feature is unexpected. It is that the median illuminances preferred by the young emmetropic group are much higher than for any other group. Given the clearer optic media of the younger group, it might be expected that they would prefer a lower illuminance than the others. The fact is they do not. This could be a matter of expectation based on the younger group's exposure to higher illuminances or simply because the older people in the other groups disliked higher illuminances because of the greater amount of scattered light and straylight produced in the eye. Three other aspects of these results deserve mention. The first is the tendency for pre-operative cataract patients to choose lower illuminances than glaucoma patients. This is to be expected because of the greater scattering of light in the eyes of some cataract patients. The second is the reduction in illuminance preferred by post-operative cataract patients. This is to be expected because of the increased light transmittance and reduced light absorption and scattering that occurs when the lens is removed. The third is the relatively small difference among the types of fluorescent lamp. This indicates that small differences in the lamp spectrum are unimportant.

One other effect of increasing the retinal illuminance is to improve the ability to discriminate colors. Figure 12.9 shows the effect of increased illuminance on the ability of people of different ages to discriminate the hue samples forming the Farnsworth–Munsell 100 hue test. It is clear that increasing the retinal illuminance enables finer hue discriminations to be made, particularly by older people. An alternative approach to enhancing the discrimination of colors is to change the light spectrum forming the illumination. It is much easier to discriminate colors that are widely separated in color space than those that are close together. The extent to which a light source will separate colors in color space is correlated to its CIE general CRI. Thus, the simplest advice for lighting places used by elderly people would be to use only lamps with a high CIE general CRI.

There is a caveat to this advice. Not all lamps with a high CIE general CRI are beneficial for color discrimination by the elderly. The problem lies in the wavelength dependence of lens fluorescence. Straylight generated by lens fluorescence will wash the whole retinal image with light of the wavelength emitted by the fluorophor. The effect is to diminish the distinctions between adjacent areas of different colors. These observations suggest that light sources that have significant spectral emissions below about 450 nm should be avoided for the very old, unless they are used in luminaires which filter out wavelengths below 450 nm. Light sources which have significant radiant flux below 450 nm are based on the mercury discharge, such as fluorescent lamps and metal halide discharge lamps. Incandescent lamps and lamps based on the sodium discharge have little energy below 450 nm, so they should be preferred, but sodium discharge lamps have poor color-rendering properties so they should be used only when luminous efficacy is of overwhelming importance. Tungsten–halogen lamps have significant emission in the ultraviolet (UV) and should be avoided unless they are used in luminaires that filter the light through a glass plate, which will remove most of the UV radiation.

Another aspect of lighting that can be important in determining the ability of the elderly and those with low vision to function is the distribution of light. This can be considered in two places: the surrounding space, either interior or exterior, and the task. In both cases, it is desirable that the light be uniformly distributed on all the relevant surfaces, without casting shadows (Julian, 1983). In a room, the primary factors that determine the illuminance uniformity of a lighting installation are the luminous intensity distribution of the luminaires, the spacing between luminaires, and the reflectances of the room surfaces. Different luminaires can have very different luminous intensity distributions so if the aim is to achieve uniform lighting the selection of an appropriate luminaire is important. Also, where uniform lighting is required, the maximum spacing between luminaires recommended by the manufacturer and based on the luminaire's luminous intensity distribution, should not be exceeded. As for surface reflectances, a uniform distribution of light is much easier to achieve when the room surface reflectances are high rather than low. Similar considerations apply to exteriors, although there the role of surface reflectance may be limited. Maintaining a uniform illuminance distribution is particularly important for people with low vision because of the problems they face with low light levels and because they may have difficulty in discriminating between a pattern of illuminance and a pattern of reflectance. Such conflicting patterns are likely to cause confusion, particularly where the pattern of illuminance differences produces higher luminance contrasts than the pattern of reflectances.

The distribution of light in the immediate task area is also important. Sanford (1996) examined the trade-off between the illuminance and the area illuminated for a group of people with macular degeneration, doing a reading task. The higher illuminance was preferred until the boundary of the area illuminated fell within the boundary of the area to be read. This is another example of the illuminance pattern conflicting with the reflectance pattern, and it emphasizes the desirability of uniformly lighting the area containing the task.

Light distribution is particularly important when self-luminous displays, such as television screens and computer monitors, are being used. The lighting in a room makes self-luminous displays less visible in two ways. First, ambient light reflected from the screen reduces the luminance contrast and desaturates the colors of the display. Second, light reflected from the front surface of the screen produces an image of the room, the screen acting as a low-reflectance mirror (Boyce, 1991; Lloyd *et al.*, 1996). If the room contains high-luminance luminaires or windows, two alternative views of the world can be seen: one generated by the display, and the other by the specular reflection. Again, separating these two views of the world will be difficult for people with low vision. In a small space, such as a private office or at home, reflections from the screen can be avoided by careful positioning. In large spaces containing many screens, luminaires specifically designed for use in such spaces, with restricted luminous intensity distributions, should be used (see Section 7.4.2.3).

An adverse aspect of light distribution that sometimes occurs is glare. Glare can take several forms (Vos, 1995).The two forms of concern here are discomfort and disability glare. The increased scattering of light in the eye that occurs with increasing age can be expected to produce increases in the level of discomfort and disability glare produced by a given lighting installation. Bennett (1977) quantified the effect of age on discomfort glare by the formula:

$$L_{BCD} = 85,750/A$$

where L_{BCD} is the luminance of the boundary between comfort and discomfort (cd/m^2) and A the age (years).

The stimulus used in this study was a 1° diameter circular target on a 5.5 cd/m^2 background, viewed directly. This experimental setting is obviously very different in size and background luminance from many real lighting installation but it is sufficient to demonstrate that older people are more sensitive to the range of luminances in a space than are young people.

As for disability glare, this is caused primarily by light scattered in the eye (Vos, 1984). The formula for predicting the magnitude of disability glare and the way it has been modified for the observer's age has been discussed in Section 12.6. The extent to which disability glare reduces visual capabilities depends on the luminance contrast of what is to be seen, its angular deviation from the glare source, and the luminance of the rest of the visual field. Elderly people generally have more light scatter and hence experience greater disability glare than the young, and people with such causes of low vision as cataract have even more (Storch and Bodis-Wollner, 1990; De Waard *et al.*, 1992). The formula for disability glare shows that the greater is the deviation of the glare source from the line of sight, the less is the magnitude of disability glare. The luminance contrast of the target and the luminance of the rest of the visual field are important because the luminance of the scattered light is superimposed on the luminance of the target and its background. The impact of scattered light is diminished the higher is the luminance contrast and the higher the luminance of the background.

The simplest approach to minimizing discomfort and disability glare for electric lighting is to use only luminaires in which there is no view of the light source, either directly or as a specularly reflected image, from common lines of sight; and to position the luminaires so that they are as far as possible from the common lines of sight. By restricting the view of the light source, the maximum luminance of the luminaire is reduced; by placing it far away from common lines of sight, the amount of light that is scattered onto the part of the retinal image representing what needs to be seen is also reduced. As for windows, the luminance of the window can be reduced by various types of blinds, some of which will preserve a diminished version of the view out. If the sun is directly visible through the window, however, there is no alternative but to use an opaque cover. Increasing the deviation from the line of sight, is usually a matter of moving what has to be seen away from the window.

Specific advice on lighting appropriate for different activities by the elderly and by people with low vision is given by both international and national lighting authorities (CIE, 1997a; IESNA, 1998), by organizations devoted to the welfare of the elderly (Figueiro, 2001), and by knowledgeable individuals (Lewis, 1992). Following this advice should lead to some improvement in visual capabilities for the elderly and for many people with low vision. Further, young people will not experience any loss in visual function following the provision of such lighting. The deterioration in the retinal and cortical processes that also occur with increasing age imply that any enhancement that occurs will most likely be limited. Nonetheless, for the elderly and those who have to live with low vision, any enhancement of their visual capabilities is welcome and may have a wider impact on their quality of life (Sorensen and Brunnstrom, 1995).

12.7.4 Eliminating the task

The final approach that can be used to offset the effects of age is to eliminate the need to do the task. This approach is evident in the common observation that elderly drivers give up driving at night while still feeling able to drive safely during the day. Being able to drive makes an important contribution to the independence and quality of life of the elderly (Jette and Branch, 1992). Many are reluctant to give up driving until forced to by circumstances beyond their control, medical problems related to vision loss being one of the most common circumstances (Campbell *et al.*, 1993a). Before this stage is reached, many of the elderly will recognize the stress of driving at night, in conditions of low luminance and in the presence of opposing headlights producing glare. The usual response is to time their journeys so that they can be completed before nightfall.

Deciding not to drive at night is a change in behavior in response to difficult visual conditions that cannot easily be changed. The other side of this coin is the possibility of maintaining behavior and changing the visual conditions to make them less difficult. An example of this is the use of transition zones between areas lit to very different illuminances. People with glaucoma, and other causes of low vision that affect peripheral vision, often experience delayed and diminished dark adaptation (CIE, 1997a). This makes it difficult for them to move safely from a brightly lit space to one that is dimly lit, e.g. from the interior of a building to the parking lot at night. Lighting can overcome this problem by eliminating the need for much dark adaptation. The features of the lighting that need attention if this approach is to be used are the range of adaptation luminances between the interior and the exterior, the grading of luminance between the interior and exterior so that a sudden change in luminance is avoided, and, of course, the control of glare.

12.8   Summary

As the visual system ages, a number of changes in its structure and capabilities occur. With increasing years the ability to focus close up is diminished, the amount of light reaching the retina is reduced, more of the light reaching the retina is scattered, the spectrum of the light reaching the retina is changed and more straylight is generated inside the eye. These changes start in early adulthood and continue at a steady rate with increasing age. The consequences of these changes with age for the capabilities of the visual system are many and varied. At the threshold level, old age is characterized by reduced absolute sensitivity to light, reduced visual acuity, reduced contrast sensitivity, reduced color discrimination, and greater sensitivity to glare. Outside the laboratory, the elderly have difficulty with seeing in dim light, moving from bright to dark conditions suddenly, reading small print and distinguishing dark colors.

These changes with age are the best that can be expected. With increasing age comes a greater likelihood of a pathological changes leading to low vision and eventual blindness. Low vision is a state which falls between normal vision and total blindness. The four most common causes of low vision in developed countries are cataract, macular degeneration, glaucoma, and diabetic retinopathy. These causes involve different parts of the eye and have different implications for how lighting might be used to help people with partial sight. Cataract is an opacity developing in the lens. The effect of cataract is to absorb and scatter more light as the light passes through the lens. This increased absorption results in reduced visual acuity and reduced contrast sensitivity over the entire visual field, as well as greater sensitivity to glare. Macular degeneration occurs when the macular, which covers the fovea, becomes opaque. An opacity immediately in front of the fovea implies a serious reduction in visual acuity and in contrast sensitivity at high spatial frequencies. Typically, these changes makes seeing detail difficult if not impossible. However, peripheral vision is unaffected so the ability to orient oneself in space and to find ones way around is little changed. Glaucoma is shown by a progressive narrowing of the visual field. Glaucoma is due to an increase in intra-ocular pressure which damages the blood vessels supplying the retina. Glaucoma will continue until complete blindness occurs unless the intra-ocular pressure is reduced. Diabetic retinopathy is a consequence of chronic diabetes mellitus and effectively destroys parts of the retina through the changes it produces in the vascular system that supplies the retina. The effect these changes have on visual capabilities depends on where on the retina the damage occurs and the rate at which it progresses.

These changes with age can be compensated, to some extent. The limited range of focus of the elderly can be overcome by the use of lenses. The tasks they have difficulty with can be redesigned to make them visually easier. This usually involves increasing the luminance contrast of the task details, making the task details bigger and using more saturated colors. Lighting can also be used to compensate for aging vision. The elderly benefit more from higher illuminances than do the young, but simply providing more light may not be enough. The light has to be provided in such a way that both disability and discomfort glare are carefully controlled and veiling reflections are avoided. Where elderly people are likely to be moving from a well-lit area to a dark area, a transition zone with a gradually reducing illuminance is desirable. Such a transition zone allows their visual system more time to make the necessary change in adaptation.

People with low vision may or may not benefit from such changes in lighting depending on the specific cause of the low vision. However, there is one approach which is generally useful. This approach is to simplify the visual environment and to make its salient details more visible by attaching high luminance contrast to those details, and only to those details.

Specific advice on lighting appropriate for different activities by the elderly and by people with low vision is given by both international and national lighting authorities (CIE, 1997a; IESNA, 1998), by organizations devoted to the welfare of the elderly (Figueiro, 2001), and by knowledgeable individuals (Lewis, 1992). Following this advice should lead to some improvement in visual capabilities for the elderly and for many people with low vision. Further, young people will not suffer any loss in visual capabilities from the provision of such lighting. However, the deteriorations in the retinal and cortical processes that also occur with increasing age imply that compensation will not be complete. Nonetheless, for the elderly and those who have to live with low vision, any enhancement of their visual capabilities is welcome and may have a wider impact on their quality of life.

13 Light and health

13.1 Introduction

Exposure to light is essential for the visual system to operate and desirable for entraining the circadian system. However, exposure to light can have both positive and negative impacts on human health, impacts that can become evident soon after exposure or only after many years. Unfortunately, health is an elastic term that can be stretched from the trivial to the fatal, from the individual to the population. Here, the impacts of lighting on health to be considered are limited in four ways. First, the impact is focused on the individual not on the population. Second, only impacts that have the potential to affect the health of many individuals are considered. Third, the aspects of health being considered are those for which an individual would be wise to seek out the services of a medical professional, although that professional's expertise might vary from ophthalmology, through dermatology and oncology to psychiatry. Fourth, the impacts are those where there is a plausible mechanism through which light can have its impact and/or, where light exposure has been used as a treatment for a condition, clinical trials have demonstrated the effectiveness of that treatment. In other words, this chapter is devoted to the proven effects of light exposure on the health of many individuals. Aspects of light and health that are matters of faith, such as color therapy, and plausible hypotheses based on science, such as the impact of exposure to light at night on breast cancer (Stevens *et al.*, 1997; Brainard *et al.*, 1999; Graham *et al.*, 2001), are not considered.

13.2 Light as radiation

People typically spend many hours of their lives bathed in electromagnetic radiation in the UV, visible, and IR wavelength ranges. This radiation can have an effect on human health simply as radiation, regardless of whether or not it stimulates the visual system or the circadian system.

13.2.1 Tissue damage

Body tissue can be damaged by many different means. The causes of damage can be broadly classified as mechanical, thermal, chemical, and biological.

The type of tissue damage of interest here is that caused by exposure to electromagnetic radiation in the UV, visible, and IR wavelength regions of the electromagnetic spectrum (see Figure 1.1). At first, the decision to include UV and IR irradiation in a book devoted to the effects of light may seem odd. It can be justified by the fact that many light sources produce UV and IR radiation as well as visible radiation and some light sources are deliberately designed to produce primarily UV or IR radiation, e.g. fluorescent lamps used in sunbeds and incandescent lamps used for industrial drying. Therefore, anyone who is using light sources should be aware of their potential for tissue damage and that means considering UV and IR radiation as well as visible radiation.

13.2.1.1 *Tissue damage by UV radiation*

The CIE has divided the UV components of the electromagnetic spectrum into three regions, UVA (400–315 nm), UVB (315–280 nm), and UVC (280–100 nm). Part of the UVA region (380–400 nm) stimulates the visual system, although according to this definition it is formally part of the UV radiation. Exposure to UV radiation affects both eye and skin. For the eye, exposure to UV radiation can produce photokeratitis of the cornea. This is a very unpleasant but temporary condition that can result in severe pain beginning several hours after exposure and persisting for 24 h or longer (Pitts and Tredici, 1971). The symptoms of photokeratitis are clouding of the cornea, reddening of the eye, tearing, photophobia, twitching of the eyelids, and a feeling of grit in the eye. Typically, all these symptoms clear up within about 48 h. Photokeratitis is an occupational hazard for electric arc welders (welders' flash) and polar explorers (snowblindness), the former because the electric arc produces copious amounts of UV radiation and the latter because snow reflects UV radiation very effectively. The factors that determine whether a person exposed to UV radiation will experience photokeratitis or not are the dose, i.e. the product of the irradiance on the eye and the time duration of the exposure, and the actual spectrum of the exposure. Figure 13.1 shows what is called the action spectrum of photokeratitis (Zuclich, 1998). An action spectrum displays the amount of a stimulus required to produce the same biological effect at different wavelengths (Coohill, 1998). The CIE standard observers used in the definition of light are derived from action spectra based on the perception of brightness (see Section 1.3). Figure 13.1 shows that irradiance in the wavelength range 200–400 nm is what causes photokeratitis, the sensitivity being the greatest for wavelengths around 270 nm.

Photokeratitis occurs because of a photochemical reaction to UV radiation at the cornea, but not all the UV radiation incident on the eye is absorbed at the cornea. Figure 13.2 shows the progress of different wavelengths in the UV region through the eye of adult humans. From Figure 13.2, it is clear that significant amounts of longer-wavelength UV radiation reach

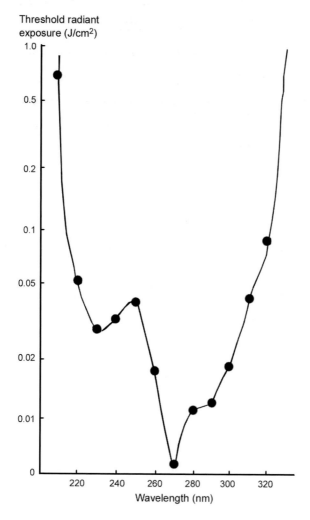

Figure 13.1 An action spectrum for photokeratitis presented as threshold
radiant exposure plotted against wavelength. Note that the vertical axis is a logarithmic scale. The data was obtained from rabbits.
Similar action spectra for photokeratitis have been found for
rabbit, monkey, and human (after Pitts, 1970).

and are absorbed in the lens. The effect of exposing the lens to UV radiation
is to produce a cataract, an opacity in the lens that absorbs and scatters
light, thereby severely degrading the retinal image. This cataract formation
can occur on two time scales, acute, i.e. with a few hours of exposure, and
chronic, i.e. after many years of exposure. Figure 13.3 shows the action
spectrum for acute cataract formation in the rabbit (Cullen, 1998). For this

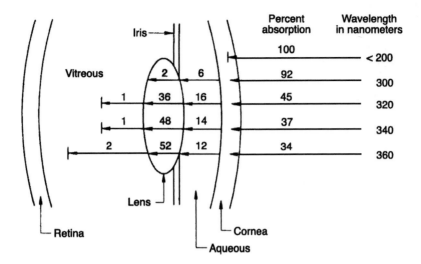

Figure 13.2 The percentage of UV radiation at specific wavelengths absorbed by various layers of the human eye (from IESNA, 2000a).

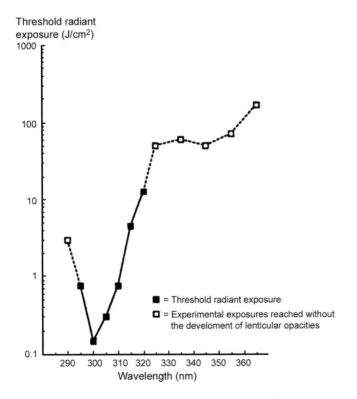

Figure 13.3 An action spectrum for acute cataract in the rabbit presented as threshold radiant exposure plotted against wavelength (from Pitts and Cullen, 1977; Pitts *et al.*, 1977).

animal, the peak sensitivity is at about 300 nm. There is no action spectrum available for acute cataract in humans. Acute cataract is evident by clouding of the lens several hours after exposure, a condition that is permanent for all but exposures close to threshold. As for chronic cataract, there is an ongoing argument about the extent to which UV exposure is the important factor, one side claiming that there is a statistically significant link between lens opacities and lifelong UV exposure in persons living and working in countries with high levels of sunlight (Taylor *et al.*, 1988), while others argue that this demonstrates a correlation and not a causation (Harding, 1995). This is an unsatisfactory situation that needs to be resolved but until it is, the cautious approach is to minimize exposure to UV radiation. Certainly there is plenty of evidence that prolonged exposure to UV radiation has the potential to cause serious damage to the eye.

Exposure to UV radiation also has an effect on the skin. Within a few hours of exposure, the skin reddens. This reddening is called erythema. Erythema reaches a maximum about 8–12 h after exposure and fades away after a few days. High-dose exposures may results in edema, pain, blistering and, after a few days, peeling of the skin, i.e. sunburn. Studies of the action spectra for erythema have a long history (Urbach, 1998). Figure 13.4 shows the action spectrum for minimal erythema derived from a statistical analysis of five different studies (McKinlay and Diffey, 1987). For minimal erythema, the peak sensitivity is around 290–300 nm. Repeated exposure to UV radiation produces a protective response in the skin. Specifically, with repeated exposure, pigment migration to the surface of the skin occurs and

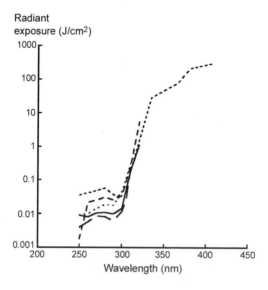

Figure 13.4 An action spectrum for minimal erythema of human skin presented as radiant exposure plotted against wavelength (after McKinlay and Diffey, 1987).

a new darker pigment is formed. Coincident with this, the outer layer of the skin thickens producing what used to be a socially acceptable tan. The effect of these changes is to decrease the sensitivity of the skin to UV radiation in the wavelength range below 290 nm (Farr and Diffey, 1985). It is just as well this screening process occurs because frequent and prolonged exposure of the skin to UV radiation is associated with skin aging and increases the risk of developing certain types of skin cancer (Freeman *et al.*, 1970). Skin cancer comes in three forms, basal cell, squamous cell, and malignant melanoma. The prevalences of both basal cell and squamous cell cancer show larger positive correlations with exposure to UV radiation from the sun than does malignant melanoma (McDonald, 1970). Basal and squamous cell cancers can often be cured if treated promptly but the cure rate for malignant melanoma is lower.

13.2.1.2 *Tissue damage by visible and near-IR radiation*

Electromagnetic radiation in the wavelength range 400–1,400 nm can damage the retina, because radiation in this wavelength range, unlike UV radiation, is transmitted through the ocular media and so reaches the retina. On arriving at the retina, some photons are absorbed in the photoreceptors, where they initiate the process of vision, while others are absorbed in the pigment epithelium, thereby increasing its temperature. Given enough energy, the temperature of the pigment epithelium can be elevated sufficiently to damage the tissue. This effect goes under the name of chorio-retinal injury. Such injuries have a long history, mostly derived from looking directly at the sun for a prolonged period. The main symptom of chorio-retinal injury is the presence of a "blind spot" or scotoma in the area where the absorption occurred. The location of the injury is important. If it occurs in the fovea, then it severely interferes with vision. If it is small and occurs in the far periphery, it may pass unnoticed. The scotoma can usually be seen under ophthalmic examination within 5 min of exposure and certainly within 24 h. Recovery from chorio-retinal injury is limited or non-existent.

The probability of chorio-retinal injury by exposure to visible and near-IR radiation basically depends on the retinal radiant exposure, weighted by the appropriate action spectrum. Figure 13.5 show the action spectrum for chorio-retinal injury, based on the energy measured at the cornea, within the pupil area, necessary to achieve a 50 percent probability of retinal alterations being evident 1 h after a 100 ms exposure, in the rhesus monkey (Lund, 1998). It is evident that the most sensitive wavelength region is from 400 to 1,000 nm. Of course, monkeys are not human but comparison studies have shown reasonable agreement between the retinal radiant exposures necessary to damage the retina in monkeys, rabbits, and humans (Geeraets and Nooney, 1973). Another factor that is important for chorio-retinal injury is the size of the retinal image. The relevance of retinal image size is simply that tissues in the retina can much more easily conduct heat away

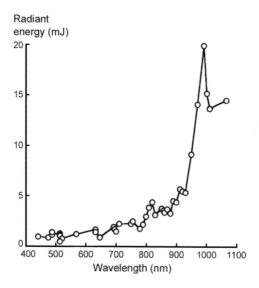

Figure 13.5 An action spectrum for chorio-retinal injury in the rhesus monkey presented as the radiant energy necessary for a 50 percent probability of producing an alteration in the appearance of the retina 1 h after a 100 ms exposure, plotted against wavelength (after Lund, 1998).

from the point of absorption for small retinal images, say less than 50 μm in diameter, than for large retinal images sizes, e.g. 1,000 μm. Therefore, large retinal images are much more likely to damage the retina than will a small area of the same retinal irradiance. Figure 13.6 shows a compilation of retinal injury data threshold data in terms of retinal radiant exposure plotted against exposure time (Sliney, 1972). Figure 13.6 shows that for chorio-retinal injury, radiant exposure is not the whole story because the threshold radiant exposure varies for different exposure times; the longer is the exposure time, the greater is the dose required for injury, presumably because the compensation by heat removal to surrounding tissue starts to operate. The curves in Figure 13.6 can be conveniently divided into two parts, longer and shorter than an exposure time of 150 ms. The time is of practical importance because it approximates to the time required for the operation of a simple mechanism used to protect the eye, the aversion response. The usual response to seeing a very bright light, which is what a high retinal irradiance in the wavelength range 400–760 nm will look like, is to blink and look away. These movements have a reaction time of 150–300 ms. For exposure times below 150 ms, no avoiding action is possible. Fortunately, very high retinal irradiances are required to produce a damaging radiant exposure in such short times. For example, for an exposure of 100 ms, retinal irradiances from about 50–1,000 W/cm², depending

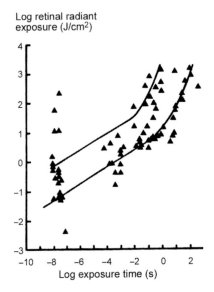

Figure 13.6 A compilation of retinal injury threshold data plotted as log retinal radiant exposure against log exposure time. The upper line is for small retinal image sizes (40–70 μm); the lower line is for larger retinal image sizes (800 μm) (after Geeraets and Nooney, 1973).

on the retinal image size, are necessary for injury to occur. For exposure times above 150 ms, lower retinal irradiances will cause injury but the probability that this will occur is reduced by the ability to take avoiding action. The most dangerous situation is if a source were to produce a lot of radiation in near-IR, i.e. the wavelength range 760–1,400 nm and very little in the visible. In this situation, there would be no high brightness cue to trigger the protective aversion response.

All the above discussion of chorio-retinal damage has been concerned with thermal damage to the retina. Unfortunately, there is also the possibility of rapid photochemical damage of the retina occurring following exposure to visible wavelengths. This is called photoretinitis. The exact nature of the chemical process by which photoretinitis occurs is not understood but what is known is that it can occur at radiant energy levels less than those required to cause threshold thermal damage and that the action spectrum is as shown in Figure 13.7 (Stuck, 1998). The greater sensitivity for photoretinitis in the visible wavelengths from 400 to 500 nm explains its original name of blue-light hazard. Photoretinitis is rare in practice because the normal aversion to very bright lights causes people to shield their eyes or to look away before damage can occur. However, if exposure is sufficient to cause photoretinitis, the damage will not usually become apparent until about 12 h later. Some recovery from the damage is possible.

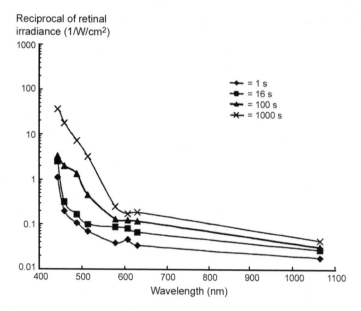

Reciprocal of retinal
irradiance (1/W/cm²)

● = 1 s
■ = 16 s
▲ = 100 s
✕ = 1000 s

Wavelength (nm)

Figure 13.7 Action spectra for threshold photoretinitis, defined as a minimally
visible retinal lesion 48 h after exposure, for exposure durations of
1–1,000 s. The data are from Ham *et al.* (1976) for non-human
primates (after Stuck, 1998).

13.2.1.3 *Tissue damage by IR radiation*

The CIE has treated the IR region of the electromagnetic spectrum in
the same way as the UV region, i.e. it has divided it into three parts: IRA
(780–1,400 nm), IRB (1,400–3,000 nm), and IRC (3,000–1,000,000 nm).
Measurements of the spectral transmittance of the ocular media have
shown that wavelengths up to 1,400 nm reach the retina, although an
increasing amount of radiation is absorbed in the lens with increasing wavelength. Between 1,400 and 1,900 nm, virtually all incident radiation is
absorbed in the cornea and aqueous humor. Above 1,900 nm, the cornea
is the sole absorber. The effect of energy in the IR-A region that reaches
the retina has already been considered in the discussion of chorio-retinal
damage. However, IR energy that is absorbed either in the ocular media
or in the cornea and lens also needs to be considered because it raises the
temperature of the tissue where it is absorbed and may, by conduction, raise
the temperature of adjacent areas. Fortunately, extremely high corneal irradiances, of the order of 100 W/cm², are necessary for changes in the lens to
occur within the time taken for the common aversive reaction to occur.
Further, only 10 W/cm² absorbed in the cornea will produce a powerful sensation of pain which should trigger the aversive response. It is generally

considered that the aversive reaction provides protection for the eye against thermal effects of IR radiation up to levels in excess of those that cause a flashburn of the skin.

So far, only the acute effect of IR radiation has been considered, but there are definitely adverse effects following prolonged exposure to IR radiation. Lydahl and Philipson (1984a,b) have shown an increased incidence of cataract amongst workers who have been exposed to molten glass or metal for many years. Unfortunately, little is known about how this effect occurs. There is argument over whether the damage is caused by thermal or photo-chemical mechanisms (Okuno, 1991; Wolbarsht, 1992) and there is no accepted action spectrum. This lack of knowledge has not stopped the development of recommendations for limiting exposure to IR radiation (see Section 13.2.2). From a practical point of view, the important point to note is that whenever exposure to a light source produces a marked sensation of warmth on the skin, the possibility of long-term IR radiation damage to the eye should be considered.

As for the skin, the effect of visible and IR radiation is simply to raise the temperature. If the temperature elevation is sufficient then burns will be produced. It is important to realize that the focusing process of the eye makes it much more sensitive than the skin to such injury for visible radiation and IRA radiation. However, the skin and eye are equally at risk from IRB and IRC radiations because the ocular media are virtually opaque for these wavelengths and the mechanism for acute damage is thermal. The efficiency with which a given irradiance raises the temperature of the skin depends on the exposed area, the reflectance of the skin, and the duration of exposure. The threshold irradiance for thermal injury of the skin is greater than 1 W/cm^2. Such irradiances are very unlikely to be produced by sunlight or conventional lighting of interiors so such sources are unlikely to produce any degree of thermal injury to the skin by radiation. In any case, for anything other than very short exposure times, considerations of heat stress become relevant before thermal damage can occur.

13.2.2 *Threshold limit values*

Given the potential for tissue damage by UV, visible, and IR radiation, it should not be too surprising that there are recommended limits to exposure to such radiation, and given the universality of the phenomena it should also come as no surprise that there are a number of bodies making such recommendations. The first organization to make recommendations limiting exposure was the American Conference of Governmental Industrial Hygienists (ACGIH). This is an independent, professional society dedicated to the advancement of occupational and environmental health. Its best-known contribution to protecting health has been the publication of threshold limit values for exposure to chemical and physical agents. The threshold limit values are levels of exposure and conditions under which it is believed, based on the best available scientific evidence, that nearly all healthy workers may be repeatedly exposed, day after day, without adverse health effects (Sliney and Bitran, 1998). The ACGIH publishes threshold limit values for exposure to UV radiation, to avoid photokertitis; for exposure to visible radiation, to avoid photoretinitis; and for visible and IR radiation to avoid cataract after prolonged exposure and chorio-retinal injury from low-luminance IR illumination sources. The threshold limiting values take various forms depending on the size of the source of radiation and the exposure time. For some situations, the threshold limit values are based on total irradiance at the eye, while for others they are based on the spectral irradiance at the eye or the spectral radiance of the source, multiplied by a weighting function based on the action spectrum of the damage being controlled. Figure 13.8 shows the weighting functions recommended by the ACGIH (ACGIH, 2001a). The recommendations of the ACGIH have been adopted by the Illuminating Engineering Society of North America (IESNA)and used to produce a recommended practice (IESNA, 1996b; Levin, 1998). They have also been adopted, with slight modifications by the International Committee on Non-Ionizing Radiation Protection (ICNIRP) (Hietanen, 1998). Following any of these recommendations will limit the likelihood of tissue damage by UV, visible, and IR radiation. Full details of

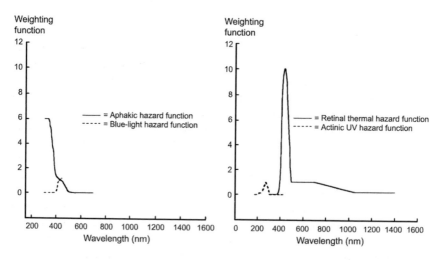

Figure 13.8 The ACGIH spectral weighting functions. The actinic UV hazard function is used to asses the risk of photokeratitis. The blue-light hazard
function is used to assess the risk of photoretinitis. The retinal thermal
hazard function is used to assess the risk of chorio-retinal injury. The
aphakic hazard function is used to assess the risk from UV and visible
radiation for people who have had their lens removed as treatment for
cataract (after ACGIH, 2001a).

the threshold limit values can be obtained from the publications of the organizations mentioned (INIRC/IRPA, 1991; ICNIRP, 1991, 1997; IESNA, 1996b; ACGIH, 2001a).

13.2.3   Hazardous light sources

The IESNA Recommended Practice 27 (IESNA, 1996b) not only adopts the ACGIH criteria for limiting tissue damage, it also gives details of how to make the necessary measurements and sets out a system for classifying light sources according to the level of potential risk they represent. This system has four classes; Exempt and Risk Groups 1, 2, and 3. Exempt light sources are those that do not pose a UV hazard for 8 h of exposure, nor a near-UV hazard, nor an IR cornea/lens hazard within 1,000 s; nor a retinal thermal hazard within 10 s, nor a blue-light hazard within 10,000 s. For light sources where sound assumptions about typical use can be made, the radiometric measurements necessary to evaluate the light source against these criteria are made at a location where the light source is producing 500 lx, or at 20 cm from the light source if the distance at which 500 lx is achieved is less than 20 cm. For light sources where sound assumptions about use cannot be made, the necessary radiometric measurements are made at a distance of 20 cm. Any light source that is assigned to Risk Groups 1, 2, or 3 must exceed one or more of the criteria used for the Exempt Group. The philosophical basis for Risk Group 1 (low risk) is that light sources in this group exceed the limits set for the Exempt Group, but do not pose a hazard due to normal behavioral limitations on exposure. The philosophical basis for Risk Group 2 (moderate risk) is that light sources in this group exceed the limits set for the Exempt Group and Risk Group 1, but do not pose a hazard due to the aversive response to very bright light or to thermal discomfort. Any light source in Risk Group 3 (high risk) is believed to pose a hazard, even for momentary exposures. The criteria defining Risk Groups 1, 2, and 3 are the same as those for the Exempt Group but the permitted exposure times are reduced. Lamps falling into any of the Risk Groups should carry a warning label, indicating the nature of the hazard and suggested precautions that should be taken.

Wood *et al.* (1998) report measurements of incandescent and fluorescent lamps commonly used for the lighting of residences, following the IESNA (1996b) procedure. They found that the incandescent and fluorescent lamps fell into the Exempt category and therefore are not a hazard for tissue damage in normal conditions of use. This comprehensive evaluation is consistent with the conclusions of other, more limited, studies of incandescent and fluorescent lamps for UV radiation (McKinlay *et al.*, 1988; Bergman *et al.*, 1995) and for blue-light hazard (Bullough, 2000). Table 13.1 shows the classification of a wider range of light sources used for general lighting, based on the IESNA (1996b) procedure (Kohmoto, 1999). Again, both

Table 13.1 IESNA 1966 classifications for a number of lamps used for general lighting (E = Exempt Group; RG1 = Risk Group 1; RG2 = Risk Group 2; RG3 = Risk Group 3). All lamps except the 500 W tungsten– halogen were measured at the distance at which they produced 500 lx. The 500 W tungsten-halogen lamp was measured at 20 cm (from Kohmoto, 1999)

Hazard	85 W tungsten– halogen	500 W tungsten– halogen	37 W linear fluorescent	36 W compact fluorescent	400 W mercury vapor	360 W high- pressure sodium	150 W compact metal halide
UV for eye and skin	E	RG3	E	E	E	E	RG3
UVA for eye	E	E	E	E	E	E	E
Chorio-retinal burn	E	E	E	E	E	E	E
Retinal blue light	E	RG1	E	E	RG1	RG1	RG1
IR eye hazard	E	RG2	E	E	E	E	E
IR eye hazard with weak visual stimulus	E	RG3	E	E	E	E	E
Thermal damage to skin	E	E	E	E	E	E	E

linear and compact fluorescent lamps fall into the Exempt Group for all criteria. The 85 W tungsten halogen was also in the Exempt Group for all criteria but the 500 W tungsten halogen was not, probably because the radiation was measured at only 20 cm from the lamp. This may seem unrealistic but as the lamp falls into Risk Group 3 on two criteria, the lamp does represent a hazard to people doing maintenance work on the luminaire. Other light sources commonly used for lighting industrial and commercial buildings, such as high-wattage high-pressure sodium, metal halide, and mercury discharge lamps, all fall into Risk Group 1 or 3 on one or more hazard criteria.

Of all the light sources available, the one to which most people are exposed and which represents the greatest potential for tissue damage is the sun. There is a strong correlation between exposure of the skin to the sun and the development of skin cancer (Moan and Dahlback, 1993). This is because, when overhead, the sun emits copious amounts of UV, visible, and IR radiation and easily falls into Risk Group 3. It is the realization of the hazard represented by exposure to optical radiation from the sun that has driven the development of more effective sunscreens to be applied to the skin (Forestier, 1998) and sunglasses to shield the eyes (Sliney, 1995; Mellerio, 1998).

It is important to appreciate that these observations about the potential for tissue damage posed by various light sources are generalizations. For electric light sources, the observations apply to lamps used for general lighting. They should not be taken to apply to all lamps of a given type. For example, while fluorescent lamps used for general lighting fall into the Exempt Group, there are fluorescent lamps used for sunbeds and for germicidal purposes that are designed to emit considerable UV radiation and that are not Exempt. Similarly, some mercury lamps are designed not as light sources, but rather as sources of UV radiation for industrial curing purposes. For the sun, the hazard posed depends on the path length through the atmosphere and the skin pigmentation of the individual. There is little hazard when the sun is low in the sky and the darker the skin pigmentation the less the risk at all sun elevations. The safest principle to follow when evaluating the potential for tissue damage from any specific light source is to assume the source is hazardous unless information suggesting otherwise is available. This is particularly true for new types of light source, such as LEDs. At the time of writing, LEDs do not produce sufficient radiant flux to be a hazard but great efforts are being made to increase their radiant flux and to make them suitable for general lighting. At some stage in this process, it will be necessary to evaluate LEDs for their potential for tissue damage.

13.2.4 Practical considerations

The key word in the above discussion of the hazards posed by different light sources is "potential." Whether the potential for tissue damage turns into actual damage depends on how the light source is used. The IESNA (1996b) classification of a light source assumes a bare lamp viewed directly for an defined time. Light sources are normally used in luminaires, and are rarely viewed directly for an extended period of time. Placing the light source in a luminaire may dramatically change the spectrum of the radiation received by the viewer. For example, the UV radiation emitted by tungsten–halogen lamps can be much reduced by using a glass cover. Dichroic reflectors can be used to transmit IR radiation while reflecting visible radiation. Different plastics and glasses have very different UV transmittances (McKinlay *et al.*, 1988; Lambrechts and Rothwell, 1996). Another factor that will change the spectrum of the radiation received by the viewer is what proportion of the radiation incident comes directly from the light source. The larger is the proportion of radiation received after reflection, the more likely it is that the spectral content will be changed, because there is no guarantee that the reflecting surface reflects UV, visible, and IR radiation equally. For example, snow reflects about 88 percent of UVB radiation (multiplied by the ACGIH actinic UV weighting function) while grass reflects less than

2 percent. What this variability implies is that where there is doubt about the risk of tissue damage by radiation from light sources, field measurements of the actual spectral radiance or irradiance are essential. If such measurements show that the hazard is actual rather potential, then action should be taken to reduce the hazard. Ideally, this would take the form of reducing the output from the light source to below that needed to create a hazard or reducing the exposure time. This is not always possible, either because the radiation is an inevitable product of the work being done, e.g. around a furnace, or is required to produce a particular effect on some component of the work, e.g. UV curing of dental fillings. In these circumstances, a degree of protection is required. This can take the form of screening the source with suitable materials, i.e. those opaque to the damaging radiation and/or personal protection in the form of eye filters, helmets, and clothing.

13.2.5 Special groups

All the methods for evaluating light sources for tissue damage are based on action spectra linked to the average adult human response to UV, visible, and IR radiation. Unfortunately, there are some groups who deviate markedly from that average sensitivity in the direction of making them much more sensitive to radiation in these wavelength ranges.

One such group consists of very premature babies, particularly those weighing less than 1,000 g at birth. These infants have eyes that are still developing and exposure to light is believed to be involved in the retinopathy of prematurity, a visual disorder that can permanently damage the retina of such babies. Proposals to limit the light exposure of babies in neonatal intensive care units have been made (Bullough and Rea, 1996). Even babies born after a normal gestation period have to be treated with care as regards light exposure because such infants have lenses with significant transmittance in the wavelength range from 300 to 350 nm, i.e. in the UVB and UVA regions (Barker and Brainard, 1991). This means care should be taken to limit the exposure of the eyes of newborns to light sources that emit a lot of UV radiation, such as the sun. Given the evidence that enhanced UV transmittance is still evident in young children (Sanford *et al.*, 1996) this care should be continued for several years.

Another population with a problem with exposure to light, but at the opposite end of life, are post-operative cataract patients who have had their lens removed, i.e. patients who are aphakic. Such patients are much more likely to suffer photochemical retinal damage due to short-wavelength visible and UV radiation exposure than are people with their biological lens intact, unless they are fitted with a UV-absorbing, intra-ocular lens (Werner and Hardenbergh, 1983; Werner *et al.*, 1990; CIE, 1997a). The ACGIH has recognized the hazard for aphakics by introducing a hazard weighting function specifically for this condition (see Figure 13.8).

Three other groups who need to take special care about exposure to UV radiation are those who have medical conditions that enhance photosensitivity, e.g. lupus erythematosus (Rihner and McGrath, 1992); those who are taking pharmaceuticals that increase photosensitivity; and those who are exposed to certain chemical agents in the environment, such as the whiteners used in some household products (Harber *et al.*, 1985). Unlike newborns and aphakics, where the hazard is confined to the retina, the effect of increased photosensitization primarily increases the hazard to the skin. How much the risk posed by exposure to UV radiation is increased will depend on the medical condition, or the specific pharmaceutical or chemical and the dose taken or level of exposure.

13.2.6   *Phototherapy*

So far, the effects of light as radiation on health have all been negative, but there are some positive effects. For example, exposure to UV radiation is important for the production of vitamin D in the skin. Vitamin D deficiency leads to bone softening diseases such as rickets in children and osteomalacia in adults. Most of the vitamin D requirements of children and adults are met by exposure to sunlight. Groups who cannot achieve sufficient exposure, such as the infirm, or those who live in areas where sunlight is limited for several months or those who have limited exposure and very dark skin, must depend on dietary sources and vitamin supplements to meet their vitamin D requirement (Holick, 1985).

There are also a number of other medical conditions where exposure to light as radiation has been shown to be helpful (Parrish *et al.*, 1985). Hyperbilirubinemia, commonly known as jaundice of the newborn, occurs frequently enough so that about 7–10 percent of babies born in the US require medical attention. Severe cases can lead to brain damage and death. The phototherapy for this condition involves exposing the naked baby to short-wavelength visible radiation, with the eyes shielded. UV radiation is also used in the treatment of skin diseases such as psoriasis and eczema. Patients are given multiple whole-body exposures to sub-erythemogenic doses of UVB radiation.

An alternative treatment for severe psoriasis, eczema, vitiligo, and some other skin disorders uses a combination of exposure to UVA radiation and a psolaren. This combined treatment is known as photochemotherapy. Chemotherapy operates by killing cells. The general problem of chemotherapy is how to limit this destruction to the desired cells. Psolaren has the potential to kill cells but it requires exposure to UVA to trigger the effect. Fortunately, UVA radiation penetrates the skin but does not reach internal organs so the combination of psolaren and UVA radiation limits the cytotoxic effects to the skin. This should not be taken to mean that photochemotherapy is without risk. Basal and squamous cell skin cancers have been found in patients who have been treated by photochemotherapy.

As in so many medical problems, the decision whether to use photo-chemotherapy or not is a matter of balancing one risk against another. Photochemotherapy can also be used to treat internal tumors. A chemical, which when injected into the blood stream binds to tumor cells, is triggered by exposure to visible radiation of 630 nm to kill tumor cells delivered via an endoscope. This process, which is also known as photodynamic therapy, has been shown to be effective against a wide range of tumors (Epstein, 1989). One other use of UV radiation is in the suppression of the immune system (Noonan and De Fabo, 1994). Such suppression may be helpful in cases of autoimmune diseases such as multiple sclerosis where hyperactivity of the immune system is a problem. Of course, it may also be dangerous for people who have already been immunosuppressed. Therapeutic exposure to UV should only be undertaken after consulting a qualified physician.

13.2.7 Aging effects

In addition to the hazards and benefits of exposure to UV, visible, and IR radiation discussed above, there are also possible effects of such exposure on the rate at which aging progresses. One example is the possibility of a link between the total light exposure over life and the likelihood of retinal damage. The proposed mechanism is that exposure to light causes damage to the retina. This damage can be repaired but the repair mechanisms become less effective with age, resulting in damage that accumulates more rapidly with greater retinal exposure to light (Marshall, 1981; Young, 1981). There is no doubt that the probability of retinal deterioration increases with age, and there are close similarities between the changes induced in the retina as a result of the aging process and those elicited by exposure to high levels of illumination (WHO, 1982), but whether it is really exposure to light that is responsible for the aging process in the retina is open to question. What is needed are comprehensive epidemiological studies examining the link between light exposure history and retinal deterioration with age. Until they are done, the effect of prolonged exposure to high levels of light on the rate of aging of the retina must remain unproven.

The other aging effect of prolonged exposure to radiation is well established and affects the skin. The most striking feature of severely photoaged skin is the presence of massive quantities of thickened, degraded elastic fibers which degenerate into amorphous masses. The result is a thicker skin resembling a crust, a condition reflected in the Australian term for exposing the body to the sun, "sun-baking." Photoaging is most commonly seen on the parts of the body that are not usually protected by clothing. The action spectrum for photoaging is not well defined but it is clear that the dominant radiation is in the UV region (Cesarini, 1998). Wearing a sunscreen while outdoors, particularly in regions where sunlight is copious, will provide some protection against the photoaging process.

13.3   Light operating through the visual system

The function of the visual system is to help us make sense of the visual environment around us. How well we can see may affect how we understand the visual environment and that in turn can affect our health.

13.3.1   Eyestrain

Light is a necessity for the visual system to operate but if used in the wrong way it can be injurious to health. The most common effect of lighting operating through the visual system on health is colloquially known as eyestrain. Eyestrain is the result of prolonged experience of lighting conditions that cause discomfort. What those conditions are is fully discussed in Chapter 5. As for eyestrain, there is some argument over whether it exists at all, some maintaining that the eye itself cannot be strained (Cogan, 1974), while others assert that the muscles of the eye can be strained and that, in any case, eyestrain is a useful portmanteau term in much the same way that backache is (Fry, 1974). What is beyond doubt is that the symptoms used to identify eyestrain are real and anyone who experiences them frequently can hardly be said to be enjoying the best of health. The symptoms of eyestrain are irritation of the eyes, evident as inflammation of the eyes and lids; breakdown of vision, evident as blurring or double vision; and referred effects, usually in the form of headaches, indigestion, giddiness, etc.

The symptoms of eyestrain are likely to appear whenever the visual system is faced with a difficult visual task, under or over stimulation, distraction, or perceptual confusion (see Section 5.3). These conditions can be brought about either by poor lighting, the inherent features of the task and its surroundings, inadequacies in the individual's visual system, or some combination of these factors. There are two mechanisms by which eyestrain can be caused, one physiological and one perceptual. The physiological is muscular strain occurring in the ocularmotor system, i.e. in the muscle systems that control the fixation, accommodation, convergence, and pupil size of the eyes. The perceptual is the stress that is felt when the visual system has difficulty in achieving its primary aim, to make sense of the world around us. Conditions that require the ocularmotor system to hold a fixed position for a long time or to make frequent changes of the same type are likely to produce eyestrain through muscular exhaustion. Conditions that make it difficult to see what needs to be seen or which distract attention from what needs to be seen are likely to produce eyestrain through stress. Lighting conditions which have been shown to lead to eyestrain are inadequate illuminance for the task (Simonson and Brozek, 1948), excessive luminance ratios between different elements of a task (Wibom and Carlsson, 1987), and lamp flicker, even when it is not visible (Wilkins *et al.*, 1989). Despite this list, it is important to appreciate that in conditions where the task is visually easy and free from distraction or perceptual confusion, the visual system can function for many hours without eyestrain. Carmichael and Dearborn (1947) measured the eye movement patterns of people continuously reading books printed in high contrast, 10-point print, for 6 h, at an illuminance of 160 lx, expecting to find signs of eyestrain. No such signs were found. Apparently the visual system is perfectly capable of prolonged activity without strain in the right conditions. Even when the conditions are not right, vision does not fail. Rather, it protests but will rapidly recover with rest.

13.3.2　*Migraine*

Everyone is likely to experience eyestrain in poor lighting conditions but there are some groups who are particularly sensitive to lighting conditions. One such group are those who suffer from photoepilepsy. Given fluctuating light of the right frequency, covering a large area and at a high percentage modulation, these individuals can be driven into a seizure. The frequency to which people with photoepilepsy are most sensitive is about 15 Hz, although about 50 percent still show signs of a photoconvulsive response at 50 Hz (Jeavons and Harding, 1975). Seizures start in the visual cortex and occur when normal physiological excitation involves more than a critical cortical area and are most likely when that cortical excitation is rhythmic (Wilkins, 1995).

A larger but related group who suffer adverse consequences from instability in light output are migraineurs. Migraine has been described as a neurovascular reaction to changes in the individual's internal or external environment. A migraine attack is much more than a severe headache. Nausea, vomiting, intolerance of smells, and photophobia are all part of a migraine attack. The exact cause of a migraine is not known, but Wilkins (1995) speculates that cortical hyperexcitability linked to the magnocellular pathway is responsible for triggering a migraine attack. What is known is that migraineurs are more sensitive to light than people who do not experience migraine, even when they are headache-free (Main *et al.*, 1997). This means migraineurs are much more likely to experience glare from luminaires and to complain about high light levels. In addition, migraineurs are likely to be hypersensitive to visual instability, no matter whether it is produced by fluctuations in light output from a light source, or by large area, regular patterns of very different reflectances (Marcus and Soso, 1989; Wilkins, 1995). Whether large area, high-contrast regular patterns are present in an environment is usually the responsibility of the architect or interior designer, but the presence of light output fluctuations are the responsibility of the lighting designer. One way to ensure that light output fluctuations do not cause trouble is either to use light sources that are inherently low in modulation, such as the incandescent lamp, or, if high modulation discharge light sources are to be used, to operate them from high-frequency control gear. Wilkins *et al.* (1989) carried out a field study in an office of the effect of replacing magnetic control gear operating from a 50 Hz electricity supply with electronic control gear operating at 32 kHz, on the frequency of headaches and eyestrain. The fluorescent lighting operating from the magnetic control gear had a modulation of about 45 percent at a fundamental frequency of 100 Hz. The same lamps operating from the electronic control gear had a modulation of less than 7 percent at 100 Hz. Figure 13.9 shows the percentage of the occupants experiencing various frequencies of headaches per week when working under the two types of fluorescent lighting. The distribution of headaches per week is strongly skewed. This implies that everybody in the office gets a headache now and again, for all sorts of reasons, but there a few people who experience

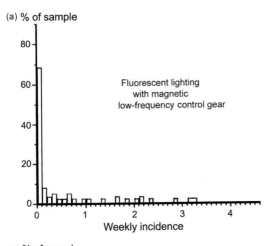

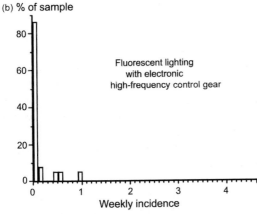

Figure 13.9 Percentage of a sample of office workers experiencing different frequencies of headaches per week, while working under fluorescent lighting operated on magnetic (50 Hz) control gear, and electronic (32 kHz) control gear (after Wilkins *et al.*, 1989).

headaches two or three times a week. Figure 13.9 demonstrates that changing from magnetic to electronic control gear does little for the mass of people but does help the people who frequently have headaches. With the electronic control gear nobody had a headache more frequently than 1.3 times per week. A similar change occurred in the distribution of the frequency of eyestrain per week. Kuller and Laike (1998) report a similar pattern in that individuals who had a high critical flicker frequency showed an increased arousal of the central nervous system when working under lighting controlled from conventional 50 Hz control gear.

13.3.3 *Autism*

Another group who can be expected to be sensitive to fluctuations in light output are the autistic. Autism is a neurological disorder that affects a child's ability to communicate, understand language, play, and relate to others. Symptoms are repetitive activities, stereotyped movements, resistance to changes in the environment and the daily routine, and unusual responses to sensory experiences. The level of arousal of autistic children is chronically high and repetitive behaviors are believed to be a way of regulating it (Hutt *et al.*, 1964). This implies that an increase in environmental stimulation will generate an increase in repetitive behavior and regular fluctuations in light output can be regarded as a form of environmental stimulation. Observations of autistic children have demonstrated that repetitive behavior does occur more frequently under fluorescent lighting than under incandescent lighting (Colman *et al.*, 1976; Fenton and Penney, 1985). This suggests that autistics too would benefit from the use of electronic control gear for fluorescent lamps. Care should also be taken to avoid lighting control systems that change light levels suddenly.

13.4 Light operating through the circadian system

The circadian system is fundamental to the functioning of many processes in life and a regular cycle of exposure to light and darkness is fundamental to the entrainment of the circadian system. Therefore, it should not be too surprising that light operating through the circadian system has a number of impacts on health.

13.4.1 *Sleep*

The sleep-wake cycle is one of the most obvious and important of the circadian rhythms. There are a number of common sleep disorders. Those susceptible to treatment with light are concerned with the timing and duration of sleep. Those associated with timing are delayed and advanced sleep phase disorders. Delayed phase sleep phase disorder is characterized by late sleep onset and late awakening, and is predominantly experienced by young people. Delayed sleep phase disorder need not necessarily cause a problem, provided sleep duration is normal and the individual can adjust his/her work and social schedules to his/her sleep pattern. However, if sleep duration is reduced and/or the timing of sleep is inconsistent with such societal requirements as being at work at a fixed time, then chronic sleep debt is likely. People with chronic sleep debt feel permanently tired and cannot be said to be enjoying the best of health.

Advanced phase sleep disorder is characterized by early sleep onset and early morning awakening and is predominantly experienced by the elderly. Again, advanced sleep phase disorder may not cause a problem as long as the duration of sleep is normal and the individual's lifestyle can be adjusted to accommodate it.

Exposure to light has been shown to be an effective treatment for these sleep disorders. Czeisler *et al.* (1988) have demonstrated that exposure to 10,000 lx at appropriate times results in significant phase advances for people with delayed sleep phase disorder and significant phase delay for those with advanced sleep phase disorder. As would be expected from the human phase response curve (see Section 3.5), the appropriate times are immediately on awakening for the delayed sleep phase disorder and in the evening for the advanced sleep phase disorder. Campbell *et al.* (1993b), in a study of elderly patients with advanced sleep phase disorder, showed not only a phase delay following exposure to 4,000 lx in the evening but also an improvement in sleep quality.

As for sleep duration disorders, the classic problems are sleep onset insomnia with normal awakening and normal sleep onset with sleep maintenance insomnia. Both these disorders are common in the elderly (Foley *et al.*, 1995). Campbell and Dawson (1991) and Lack and Schumacher (1993) have shown that exposure to bright light in the evening produces longer and better quality sleep for people who were experiencing sleep maintenance insomnia.

There can be little doubt that exposure to enough light at the right time is helpful in promoting sleep, but what is enough light, how long should exposure last and what spectrum should the light have? Unfortunately, there are no clear answers to these questions. A wide range of illuminances, from 2,500 to 10,000 lx, a range of times, from 15 min to 4 h, and a wide range of spectra, from fluorescent lamps to sunlight, have been shown to be effective in the treatment of sleep disorders (Terman *et al.*, 1995). Given that the assumed basis for the effectiveness of light exposure is the stronger stimulus given to the circadian system, it is only with the identification of the absolute sensitivity of that system that the minimum illuminance of a given spectrum delivered at the eye for a fixed time can be identified.

13.4.2 Seasonally affective disorder

Depression is one of the most common psychiatric conditions in patients visiting a doctor, with a lifetime prevalence of about 17 percent (Kessler *et al.*,

1994). Seasonally affective disorder (SAD) is a sub-type of major depression that is identified by a regular relationship between the onset of depression and the time of year; full remission of depression at another time of year; the pattern of onset and remission of depression at specific times of the year repeated over the last 2 years; no non-seasonal depression over the last 2 years; and episodes of seasonal depression substantially outnumbering non-seasonal depression over the individual's lifetime (American Psychiatric Association, 2000). Two forms of SAD have been identified, winter and summer SAD, the former being much more common than the latter. Winter SAD can be recognized by the increase in feelings of depression and a reduced interest in all or most activities, typical of depression, together with such atypical symptoms as increased sleep, increased irritability, and increased appetite with carbohydrate cravings and consequent weight gain. These symptoms disappear in summer. Summer SAD is also associated with an increase in feelings of depression and lack of interest in activities but in this case there is a decrease in sleep, poor appetite and weight loss (Wehr *et al.*, 1991). Winter SAD is experienced by about 5 percent of the population of the USA and about 10–20 percent have sub-syndromal symptoms, the percentages increasing with an increase in latitude (Wehr and Rosenthal, 1989; Kasper *et al.*, 1989b). Winter SAD is more common in females than males (Rosen *et al.*, 1990). Its prevalence increases with age until about the sixth decade, after which it declines dramatically.

The cause of winter SAD is unknown. Explanations based on disturbances to the circadian system and regulation of the hormone seratonin have been proposed but none have been proven. While the cause of winter SAD is unclear, what is clear is that exposure to bright light is often an effective treatment (Rosenthal *et al.*, 1985; Kasper *et al.*, 1989; Terman *et al.*, 1989; Tam *et al.*, 1995). What is meant by "bright light" is usually exposure to a light box that produces an illuminance at the eye of between 2,500 and 10,000 lx. Exposure durations range from 2 h for 2,500 lx to 30 min for 10,000 lx. The timing of the exposure to bright light is relatively unimportant (Wirz-Justice *et al.*, 1993). At these illuminances, the specific light spectrum is also not important. The fluorescent lamp is the light source most commonly used in light boxes, mainly because of its high luminous efficacy and large surface area. The latter makes it easier to provide the required illuminance from a large area source so that visual discomfort is less than would be the case for the same illuminance provided by a point source. A good light box will also have a filter to eliminate UV radiation from the light box.

Response to "bright light" can usually be expected with 2–4 days and a measurable improvement is often seen within 1 week, but symptoms will reappear if light treatment is discontinued. The symptoms that are atypical of depression in general are the ones that are most responsive to light treatment, i.e. hypersomnia, increased appetite, and carbohydrate cravings. As with most medical treatments, there are side-effects of prolonged exposure to the high illuminances of a light box. Typically they are mild disturbances of vision and headaches that subside with time. However, care should be taken with patients who have a tendency towards mania, and whose skin is photosensitive or who already have retinal damage and who have a medical condition that makes retinal damage likely (Levitt *et al.*, 1993; Gallin *et al.*, 1995; Kogan and Guilford, 1998). General guidance on the use of light in the treatment of SAD is available from a number of sources (Lam, 1998; Saeed and Bruce, 1998; Lam and Levitt, 1999).

13.4.3 *Alzheimer's disease*

Alzheimer's disease is a degenerative disease of the brain and is the most common cause of dementia. It first becomes evident to the external observer when the individual starts forgetting recent events or familiar tasks. As it develops, memory loss becomes more global, accompanied by personality change and reduced communication. It leads, eventually, to a complete unawareness of the world. The effect of these changes is to destroy the individual's personality and leave behind an empty shell. Lighting can influence the abilities and behavior of people with Alzheimer's disease, operating through both the visual system and the circadian system. Alzheimer's patients show a reduced contrast sensitivity function relative to healthy people of the same age (Gilmore and Whitehouse, 1995) (Figure 13.10). This pattern of change is consistent with the reports of cell loss at both retinal and cortical level in Alzheimer's disease, particularly for the magnocellular channel of vision (see Section 2.2.6) (Blanks *et al.*, 1991; Hof and Morrison, 1991; Kurylo *et al.*, 1991). It has been argued that such reduced

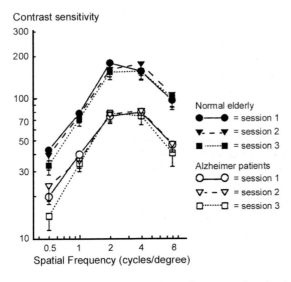

Figure 13.10 Contrast sensitivity functions for the healthy elderly and the elderly with Alzheimer's disease, measured at three different times in 1 year (after Gilmore and Whitehouse, 1995).

visual capabilities may exacerbate the effects of other cognitive losses in Alzheimer's patients, tending to increase confusion and social isolation (Mendez *et al.*, 1990; Uhlman *et al.*, 1991). This suggests that enhancing the luminance contrast of the stimulus would improve the functioning of Alzheimer's patients. Gilmore *et al.* (1996) have shown that increasing the luminance contrast does increase the speed of letter recognition by Alzheimer's patients (Figure 13.11). This finding, suggesting as it does that Alzheimer's patients are struggling to make sense of the world with diminished visual and cognitive capabilities, raises the intriguing possibility that lighting designed to enhance the capabilities of people with low vision might also be effective in helping people with Alzheimer's disease (see Section 12.7). This is a possibility that deserves investigation.

As for the circadian system, people with Alzheimer's disease and other forms of dementia often demonstrate fragmented rest/activity patterns throughout the day and night (Aharon-Peretz *et al.*, 1991; Van Someren *et al.*, 1996). This makes such patients difficult to care for and is one of the main reasons for having them institutionalized (Pollak and Perlick, 1991). The human circadian system is controlled by the suprachiasmatic nucleus, which in turn is entrained by exposure to alternate periods of light and dark

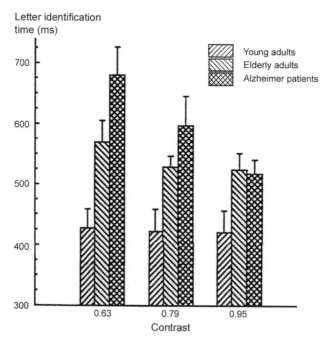

Figure 13.11 Mean time taken to identify individual letters by young, healthy adults, elderly healthy adults, and elderly adults with Alzheimer's disease, at letter luminance contrasts of 0.63, 0.79, and 0.95 (after Gilmore *et al.*, 1996).

(see Section 3.2). Degeneration is evident in the suprachiasmatic nucleus of people with Alzheimer's disease (Swaab *et al.*, 1985) and such patients are less likely to be exposed to bright light (Campbell *et al.*, 1988). This suggests that exposing Alzheimer's patients to bright light during the day and little light at night, thereby increasing the signal strength for entrainment, would help to make their rest activity patterns more stable. Studies using light boxes of the type used for the treatment of SAD have been used to demonstrate such benefits (Okawa *et al.*, 1989; Lovell *et al.*, 1995). Specifically, patients were placed in front of a light box producing 1,500–3,000 lx at the eyes for 2 h during the day. The result was reduced agitation and wandering at night and more stable rest/activity rhythms. Unfortunately, Alzheimer's patients are not the most compliant as regards instructions, so continuous supervision is necessary to keep the patient in front of the light box. A more practical alternative is to increase the general illuminance in rooms where patients spend their days to a high level. Van Someren *et al.* (1997a) tested this approach using 22 institutionalized patients with various forms of dementia. The average illuminance on the eyes of these patients when in their living rooms was increased from 436 to 1,136 lx by changing the lighting installation. After 4 weeks, the installation was returned to its original state, the resulting average illuminance being 372 lx (the values are different because the lighting includes a daylight component). Figure 13.12 shows the raw activity data for a patient with Alzheimer's disease and the average 24-h activity level for all patients, for the three lighting conditions. The decreased variability in activity and the generally low level of activity at night under the bright light conditions is obvious. Further, a careful regression analysis of the data showed that patients with severe visual impairment did not benefit from exposure to bright light, suggesting that the bright light is not acting as a placebo. There can be little doubt that lighting has a role to play in the management of Alzheimer's patients and maybe that of patients with other forms of dementia.

13.5 Light as a purifier

One other route whereby light can influence health deserves mention. It is the ability of UV radiation to destroy many types of viruses, bacteria, molds, and yeasts, most of which have the potential to damage human health. The mechanism of destruction is the absorption of UV radiation by the DNA molecule of the target organism. This absorption produces mutation or cell death, both of which stop the organism from multiplying. However, care has to be taken in the wavelength range used in irradiation because cell damage can be repaired by a photoreactive enzyme triggered by radiation in the wavelength range 320–410 nm (Harm *et al.*, 1971; Smith, 1978). The light source used to provide the UV radiation is an electric discharge passed though a low-pressure mercury vapor, the vapor being enclosed in either a special glass or a quartz tube that transmits

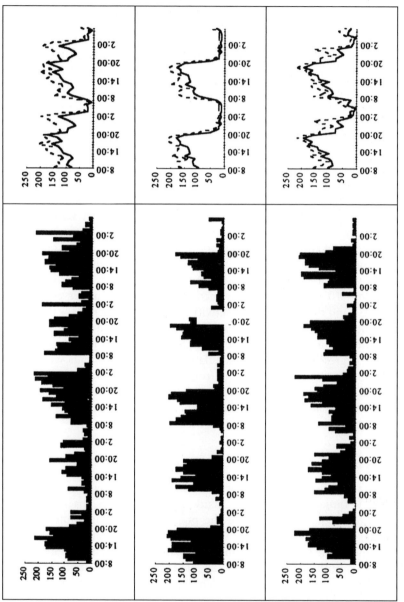

Clock time (h)

Figure 13.12 Raw hourly activity data of a patient with Alzheimer's disease over 5 days, before (upper panel), during (middle panel), and after (lower panel) bright light treatment. The right panels show the average activity levels over 48 h for 22 subjects with various forms of dementia for the same light exposure conditions (after Van Someren *et al.*, 1997a).

Table 13.2 Radiant exposure (J/m²) for 253.7 nm UV radiation to inhibit colony formation in different organisms (from IESNA, 2000a)

Organism	Radiant exposure (J/m²)
Bacillus anthracis	45.2
Salmonella enteritidis	40.0
Bacillus megatherium sp. (veg.)	37.5
Bacillus megatherium sp. (spores)	27.3
Bacillus paratyphosus	32.0
Bacillus subtilis	71.0
Bacillus subtilis spores	120.0
Corynebacterium diptheriae	33.7
Dysentry bacilli	22.0
Eberthella typhosa	21.4
Escherichia coli	30.0
Micrococcus candidus	60.5
Neisseria catarrhalis	44.0
Phytomonas tumefaciens	44.0
Proteus vulgaris	26.4
Pseudomonas aeruginosa	55.0
Pseudomonas fluorescens	35.0
Salmonella typhimurium	80.0
Sarcina lutea	197.0
Seratia marcescens	24.2
Shigella paradysenteriae	16.3
Spirillum rabrum	44.0
Staphylococcus albus	18.4
Staphylococcus aureus	26.0
Streptococcus hemolyticus	21.6
Streptococcus lactis	61.5
Streptococcus viridans	20.0
Yeasts	
Saccharomyces ellipsoideus	60.0
Saccharomyces sp.	80.0
Saccharomyces cerevisiae	60.0
Brewer's yeast	33.0
Baker's yeast	39.0
Common yeast cake	60.0
Mold spores	
Penicillium roqueforti	130.0
Penicillium digitatum	440.0
Aspergillus glaucus	440.0
Aspergillus flavus	600.0
Aspergillus niger	1,320.0
Rhizopus nigricans	1,110.0
Mucor racemosus	170.0
Oospora lactis	50.0

UV radiation. Ninety-five percent of the energy emitted by these germicidal lamps is at a wavelength of 253.7 nm and very little is in the range 320–410 nm. The effectiveness of this radiation in destroying microorganisms depends on many parameters, including the susceptibility of the specific organism which is related to the thickness of the cell wall, the spectrum of the radiation received and the radiant exposure. Table 13.2 shows the radiant exposure of 253.7 nm radiation required to inhibit colony formation, i.e. to achieve a 90 percent reduction in population, for a variety of microorganisms.

UV radiation has been used to purify air, liquids, and granular materials. Interest in air purification by UV radiation has grown in recent years with the emergence of drug-resistant strains of airborne disease, such as tuberculosis (Nardell, 1993; Riley, 1994). Germicidal lamps are usually placed in one of two locations. Where air conditioning is used, the lamps can be placed in the ductwork. For this location, exposure times are short because of the high air velocity in the duct, so a high irradiance is required. Alternatively, the germicidal lamps can be placed inside the occupied space. This poses a problem because, as discussed in Section 13.2.1.1, exposure to UV radiation can lead to photokeratitis and skin damage. There are two solutions to this problem. Where the ceiling height allows (>2.9 m) the lamps can be installed in luminaires that confine the UV radiation to the air volume above the occupants' heads, the surfaces directly irradiated having a reflectance at 253.7 nm of less than 0.05. Of course, this method relies on sufficient air circulation to move all the air in the occupied space through the irradiated volume frequently. If the ceiling height does not allow overhead purification, then some form of local air circulation that brings the air of the occupied space through an enclosed fitting containing the germicidal lamp is possible. When UV radiation is used to purify liquids, either a much higher radiant flux or a much longer exposure times are required than are used for air because of absorption by the liquid (Sommer *et al.*, 1997). Where very absorbing liquids are used, e.g. fruit juices, milk, the problem of absorption is overcome by spreading the liquid into a very thin film during exposure. As for granular material, such as sugar, the solution to exposing the material to the UV is to pass the granules along a vibrating conveyor illuminated by germicidal lamps. Clearly, UV radiation is useful for purifying air, liquids, and granular materials, but care is necessary wherever it is used. The system for delivering UV radiation to the material to be irradiated should always be designed so that the appropriate threshold limit values are not exceeded in the space occupied by people.

13.6   Summary

Exposure to light can have both positive and negative impacts on human health, impacts that can become evident soon after exposure or only after many years. This chapter is devoted to the proven effects of light exposure on health rather than beliefs and hypotheses about light and health.

The effects of light on health can be conveniently arranged in four classes. The first class is that of light treated as radiation. For this class the definition of light is stretched to include UV and IR as well as visible radiation, because many light sources produce all three types of radiation. In sufficient doses, exposure to light can cause damage to both the eye and skin, through both thermal and photochemical mechanisms. In the short term, UV radiation can cause photokeratitis of the eye and erythema of the skin. Prolonged exposure to UV radiation can lead to cataract in the lens as well as skin aging and skin cancer. Visible radiation can produce photoretinitis of the retina. Visible and short-wavelength IR radiation can cause thermal damage to the retina and burns to the skin. Prolonged exposure to IR radiation can lead to cataract and burns. Guidance setting out threshold limit values that should be observed to avoid these detrimental effects on health, and a lamp hazard classification system based on these threshold limit values, are available (IESNA, 1996b; ACGIH, 2001a). When evaluated according to these threshold limit values, most light sources used for general lighting pose no hazard to health, but a few do. The most hazardous light source to which most people are exposed is the sun.

The threshold limit values assume a normal response but there are some groups who are much more sensitive to light as radiation than the normal population. Among such groups are the newborn, aphakics, and people taking certain pharmaceuticals. All these effects of light as radiation are negative but light as radiation can also have positive effects on health. Specifically, controlled exposure to light can be used as a treatment for hyperbilirubinemia, some skin disorders, and some tumors.

The second class is light operating through the visual system. Lighting conditions that cause visual discomfort are likely to lead to eyestrain and anyone who frequently experiences eyestrain is not enjoying the best of health. The lighting conditions that cause visual discomfort are well known and easily avoided. There are also some people who have medical conditions that make them especially sensitive to lighting conditions. Migraineurs and autistics are two such groups. Both are sensitive to the temporal modulation of light.

The third class is light operating through the circadian system. The sleep-wake cycle is one of the most obvious circadian rhythms so it is hardly surprising that exposure to bright light at the right time can be used to treat some sleep disorders involving the timing and duration of sleep. Exposure to bright light is also a useful means of stabilizing the rest-activity cycle of people with Alzheimer's disease and of relieving the symptoms of seasonal affective disorder.

The fourth class is the use of UV radiation as a purifier of air, liquids, and granular materials. This role comes about because UV radiation has been shown to destroy many species of bacteria, molds, yeasts, and viruses, many of which have the potential to damage human health. Care is necessary when using UV radiation for these purposes to avoid damage to people.

Light is like fire, a good servant but a poor master. Exposure to light is essential for the visual system to operate, desirable for entraining the circadian system and valuable for the treatment of some medical conditions, but too much of the wrong wavelengths for too long and damage or injury may occur. It behooves anyone who is involved in the design and specification of lighting systems to be aware of these impacts of light on human health.

14 Codes and consequences

14.1 Introduction

Lighting research enters the lives of most people through documents specifying the amount of light to be provided for different applications and offering guidance on how to provide that light. Such documents, which are published by corporations, professional organizations, advocacy groups, quasi-governmental authorities, legislatures, and international bodies, influence the lighting of everything from cells to cinemas, from warehouses to workshops. This plethora of sometimes conflicting information has lead some to claim that there is too much advice available on lighting requirements. For example, Lam (1977) believes that, in interiors at least, the primary role of lighting should be to reveal the form of the architecture and thereby to fulfill the need for unambiguous perception. He asserts that there are many ways of doing this which are known to anyone with practical experience of lighting and that precise guidance is unnecessary. He also avers that one way of not revealing form in architecture is to follow slavishly recommendations which simply specify illuminances on a mythical horizontal plane. There is much truth in this criticism, as will be apparent to anyone who has compared the boring uniformity of many modern lighting schemes with what a talented lighting designer can achieve. However, there are some situations where the function of the lighting is too important to be left to the whims of designers; and there are some applications in which revealing the architectural form is very much a secondary consideration, assuming always that there is any architectural form worth revealing. An example of the first situation is escape lighting (see Chapter 9). Statutory requirements for the provision of escape lighting in buildings have existed for many years, for reasons of public safety. An example of the second situation is industrial lighting. For most industrial premises, there is no architecture worth revealing, and the primary focus of the lighting is to allow the workers to do their work quickly, accurately, and safely. Further, given the variety of positions, details, and types of tasks that can occur in a factory, uniform lighting across a hypothetical working plane, supplemented by localized lighting and task lighting as necessary, is often the most practical approach to achieving the necessary lighting conditions.

A further justification for the publication of advice on lighting is that much lighting is actually specified and/or installed by people who are not knowledgeable in lighting, who are not lighting designers or engineers, and who have very little interest in lighting as such. Very often lighting is one part of their job and it is a part to which they turn infrequently. For such people, simple guidance from an authoritative body on how much light should be used and how it should be provided is useful if poor quality lighting is to be avoided.

Thus, there is a place for advice about lighting, but that advice should never be used as a substitute for thought. Insisting that lighting codes and guides should always be followed, at all times and in all circumstances, is as silly as arguing they are totally irrelevant and should be resolutely ignored. Following guides to lighting practice published by authoritative bodies will usually ensure that poor quality lighting is avoided but will do little to ensure the good quality lighting is produced (see Section 5.6). For good-quality lighting, consideration of a wider range of factors than those listed in most lighting guides is essential (Loe and Rowlands, 1996). There is no substitute for the thoughtful application of knowledge, experience, and creativity to ensure that lighting is of good quality.

Given that there are many different forms of advice on lighting, it is important to be able to recognize where compliance is demanded and where it is voluntary. In the latter case, deciding whether to follow a guide or not requires that the reader be able to identify recommendations that are soundly based and those that can be safely ignored. To make such discriminations it is necessary to understand the various forms that such advice can take and how it is produced. The first part of this chapter is devoted to a discussion of these topics.

The operation of electric lighting systems has a number of consequences. The most obvious is the facilitation of the modern lifestyle. Before the widespread introduction of gas and then electric lighting, the lifestyles of many people were governed by the rising and setting of the sun. The widespread availability of copious amounts of light at the flick of a switch, at a modest cost, has broken that dependence and made it possible for people to work and play at any time, day or night. But the consequences of providing electric lighting do not stop there. For electric lighting to operate, electricity has to be generated and for that to happen fuel is usually burnt. This process certainly produces electricity, but it also produces air pollution. To produce light also requires the manufacture of lamps, control gear, and luminaires. This itself consumes resources and once the equipment fails, the lamps, control gear, and luminaires have to be disposed of, a process that may also generate pollution. Finally, the extensive and increasing use of exterior lighting has changed the appearance of the night sky in many areas, a phenomenon that is now known as light pollution. The second part of this chapter is concerned with these consequences of providing lighting.

14.2   Laws, regulations, codes, guides, and practices

Guidance on lighting requirements can be found under many different names. Regardless of the name, the first thing to be aware of when considering whether to follow such guidance is the legal status of the document. Laws that specify minimum lighting conditions have to be met within the area under the legislative authority's jurisdiction, no matter how inappropriate or downright stupid the specified conditions are. For example, in New York State, the automatic teller machine (ATM) Safety Act specifies a minimum illuminance of 2 footcandles (21.5 lx) to be provided on a horizontal plane, 5 ft (1.5 m) above ground level, at 50 ft (15 m) from an ATM set in an outside wall. This requirement is relevant to the illuminance on the top of the head but says little about the lighting of the face, which is what is of interest to anyone using an ATM machine and concerned with someone approaching. This legal requirement is enforced with inspections and fines for non-compliance.

Other forms of advice may have legal standing by reference. This is often the situation for regulations, where the law says something vague about adequate lighting being provided and regulations are used to define what that is. This route is often taken because it is usually easier, at least in democratic societies, to modify regulations by executive action rather than to revise a law. Yet other forms of guidance may have legal power even though they have no legal standing. This situation occurs when the guidance is generated by a body with no legal authority, such as a professional organization, but those recommendations become widely accepted as good practice. The power of such recommendations lies in the risk of litigation. Anyone who has ignored such recommendations is not able to base their defence on following best practice. This is the situation for many widely used, national lighting recommendations, such as the Chartered Institution of Building Service Engineers Code for Lighting (CIBSE, 2002a) and the Illuminating Engineering Society of North America's *Lighting Handbook* (IESNA, 2000a). Somewhat more remote but still with some legal power, although no legal standing, are the documents published by professional bodies to give guidance for specific applications, such as the *Lighting Guides* of the CIBSE and the *Recommended Practices* of the IESNA.

Having determined the legal status of the guidance, the next step is to examine the precision of any recommendations. This can vary widely from specific illuminances measured at specific positions to simple statements of desirable actions, such as that care should be taken to avoid veiling reflections. There is no simple relationship between level of precision and legal status. There are laws that have very precise lighting requirements and those that contain only a statement of intent. There are also guides and practices published by professional organizations that contain precise recommendations alongside vague exhortations. For documents with legal standing, the level of precision tends to be related to the consequences of lighting failing. The more serious are the consequences of failure, the more precise the recommendations in documents with legal standing tend to be. Thus, where poor lighting will place the safety or health of the public at risk, or will offend against other desirable public policy, such as reducing energy consumption, the recommendations tend to be quantified and measurable. Where the consequences of failure are slight or uncertain, statements of intent are usually considered sufficient.

It might be thought that documents with legal standing but imprecise recommendations are a bad thing, but a lack of enthusiasm for determining lighting practice by legislation is desirable, for two reasons. First, exact specification tends to reduce design flexibility and inhibit innovation. Second, there is the law of unintended consequences. This law refers to the observation that decisions can interact in unexpected ways, so a precise requirement in one area may have an unexpected impact in another. An example of this occurred when a 2 percent daylight factor was specified as the minimum daylight factor that should be provided anywhere in an area normally used as teaching accommodation in the UK (HMSO, 1959). To achieve this, schools were designed with large windows and roof lights. In consequence, many suffered from solar overheating in summer and wasted energy for heating in the winter. In this case, the precision of the recommendation was excessive. There was no danger to life and limb in using less daylight. Simple advice on how daylight could be used would have been sufficient, the decision on the extent to which it should be used being left to the designer. In fact, the statutory requirement for a 2 percent daylight factor distorted school design for many years. No doubt, the 2 percent daylight factor requirement was introduced for the best of intentions but, frequently the path to hell is paved with good intentions.

A more recent example of well-intentioned lighting recommendations having a disturbing outcome can be found in the field of office lighting. In 1996, detailed advice was published on lighting spaces containing computer monitors (CIBSE, 1996). The advice given was comprehensive and correct. The problem was that amongst the advice was a system for categorizing luminaires based on their average luminance when seen from different directions. The lighting industry jumped on this system to promote luminaires, implying that the choice of luminaire was all that was necessary to ensure a successful lighting installation. Unthinking specifiers and designers took them at their word and ignored all the rest of the advice. The result was a rash of gloomy and uninspiring office interiors, so much so that the luminaire category rating system was withdrawn in 2001, the concern being that the system was doing more harm than good (SLL, 2001). The moral of this sorry tale is twofold. For the writers of lighting recommendations, never underestimate the desire for a simple answer, preferably a single number answer. For the reader of lighting recommendations, think first, act afterwards, and remember that good lighting is rarely produced when only one aspect of the lighting requirements is considered.

One factor relevant to the likelihood of unexpected consequences occurring is the form the recommendations take. The two extremes are documents that specify what the lighting should do and those that specify what the lighting should be. The former codifies the ends, the latter the means. Clear examples of both these approaches can be found in the energy field. California specifies a maximum allowed power density (W/ft^2) for the lighting used in a large number of applications. This legal requirement does not tell the designer what light sources to use or what illuminances to provide. It simply says what the total maximum lighting power density should be. At the other extreme is the US Federal Energy Policy Act of 1992 which forbade the use of some inefficient light sources, including the 40 W Cool White fluorescent, a lamp that had been widely used in commercial buildings. With enforcement, this approach certainly removes energy-inefficient products from the market, but it also restricts the designer's options and that may lead to unexpected consequences.

Although this division between ends and means is useful for understanding, it should be realized that a complete division is the exception rather than the rule. Much lighting design in interiors and exteriors is done by people with limited experience of lighting. For this reason, most guidance on lighting contains recommended conditions to be achieved, gives simple design methods and summarizes the properties of lighting equipment through which the recommendations can be met. The reader is then left to choose whether to exercise his/her skills or to follow the simple design methods given.

Another point which is determined by the anticipated audience for the document is the extent of any explanation given for the recommendations. Obviously, there is a greater incentive for more explanation in documents which are to be followed voluntarily, but even here there is something to be said for keeping the explanation simple. The best approach is to give the advice unambiguously with references to other documents which provide some justification for the proposals. An extreme example of this approach is that taken by the American Conference of Governmental Industrial Hygienists (ACGIH) who publish threshold limit values for exposure to UV, visible, and IR radiation (ACGIH, 2001a) with a justification for those threshold limit values published separately (ACGIH, 2001b). However, the extent of the justification necessary for any advice is variable. For some forms of lighting, such as escape lighting, where the consequences of any failures are likely to be grave, some guidance is essential. In this situation, recommendations can be based on scientific evidence, experience and good judgement, and presented without extensive justification. Where recommendations are not essential, such as for avoiding veiling reflections, strong evidence is needed before precise recommendations can be justified. In a perfect world, every lighting recommendation would be completely justified, but this is not a perfect world. Lighting recommendations are made with varying degrees of backing, from prejudice to informed opinion to clear scientific evidence. It is one of the purposes of this book to indicate the limits of the evidence available as justification for many lighting recommendations.

14.3   Trends in lighting recommendations

Whatever their form, the role of all documents offering advice about lighting is to ensure that the lighting provided is suitable for its purpose. Different applications have different purposes and hence give different weights to different aspects of lighting. For industrial lighting, the priority is for the lighting to enable people to work quickly, easily, and accurately. For the lighting of a hotel foyer, the impression created is the first priority. Thus the recommendations for different applications can be expected to contain different criteria and different levels of the same criteria. This is made manifest in the latest edition of the Illuminating Engineering Society of North America *Lighting Handbook* (IESNA, 2000a) where there is a whole chapter devoted to the quality of the visual environment. In this chapter are a series of matrices indicating the relative importance of such factors as the appearance of the space and the luminaires, shadows, sparkle, and facial modeling, for many applications.

While there may be arguments about the value of such generic advice it is undeniable that its presence represents a recognition that lighting is about a lot more than just providing illuminance on a horizontal plane. Yet, this was where most lighting recommendations began. When the IES Code was fist published in the UK, in 1936, it consisted of little more than a list of illuminances. In 1961, a major change was made with the introduction of criteria to control discomfort glare in the form of the Glare Index system. In the latest edition (CIBSE, 2002a) the recommendations for each application are the illuminance on the task, the limiting Unified Glare Rating, the minimum Commission Internationale de l'Eclairage (CIE) General Color Rendering Index, and notes on aspects of lighting requiring care. In addition, criteria are given for the uniformity and diversity of illuminance; luminance ratios between task, immediate surround and background; surface reflectances and light distribution for good facial modeling.

Despite this increase in the number of criteria appearing in many current lighting recommendations, it remains a fact that for many people the first and sadly, sometimes, the only question asked about lighting is what illuminance should be provided. At first sight this is not unreasonable because, for many people, illuminance quantifies what a lighting installation is designed to deliver, light. However, a somewhat more sophisticated view is that lighting should be designed to deliver visibility, and while illuminance is often correlated with visibility, this is not always the case. A light source delivering disability glare certainly delivers light but little visibility. A much more sophisticated view is that lighting should be designed to deliver an impression. Illuminance will contribute to impression but it is not the only factor involved in creating an impression (see Chapter 6).

While illuminance undoubtedly has limitations as the one and only lighting criterion, it is unlikely that it will ever be displaced from its prime position. This is because it is easily understood, it is the basis of the simple lumen method of design, it can be easily calculated and measured, and it provides a simple means of deciding what you are getting for your money. Given the importance attached to illuminance criteria, it is interesting to consider how they have changed over the years. Mills and Borg (1999) have reviewed the illuminance recommendations made in 19 different countries. Figure 14.1 shows the changes in illuminance on a horizontal plane recommended for general offices at different times in the US, UK, and the Soviet Union/Russia. Clearly, there are wide differences between different countries in the illuminance recommended for the same application. From their review, Mills and Borg (1999) conclude that the historical pattern has been for recommended illuminances to increase over time by up to a factor of 10 until the early 1970s after which there was a stabilization or a decline. The current trend in illuminance recommendations is towards a convergence to similar values in different countries, values that are significantly lower than in earlier decades.

This variability in illuminance recommendations causes some conspiracy theorists to be suspicious of undue commercial influence. The question they ask is "If about 100 lx was sufficient for offices in 1936 why is 500 lx regarded as necessary today?" The answer to this question is that today is not 1936. The capabilities of the human visual system have not changed since 1936 but the nature of office work has changed, the means of providing light have changed, the furnishing of offices has changed and, most

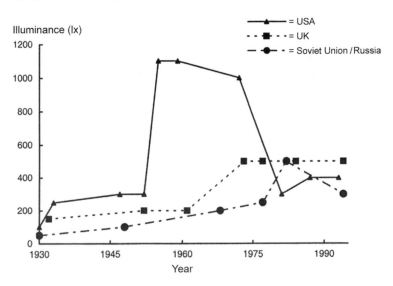

Figure 14.1 Illuminances recommended for general offices in the US, UK, and the Soviet Union/Russia since 1930 (after Mills and Borg, 1999).

importantly, people's expectations have changed. Variability comparable to that shown in Figure 14.1 can be found for other environmental criteria, e.g. temperatures recommended for thermal comfort and air quality.

The changes in lighting recommendations over time reveal a fact about all lighting recommendations that should always be remembered. Lighting recommendations are not immutable. They are not like the laws of physics, nor are they written on tablets of stone. Rather, they represent the best efforts of people to decide on reasonable lighting recommendations in the prevailing conditions. To reach this decision for any particular application, a number of factors have to be considered.

First, there are what might be called the aims of the lighting. For each application it is necessary to consider the relative weights to be given to task visibility, task performance, observer comfort, and perceived impression. Once the aims of the lighting have been specified the necessary lighting conditions can be derived from any available experimental evidence and from practical experience. Second, there is the extent to which the lighting desired can be achieved with available equipment. There is little point in recommending lighting that cannot be achieved. Third, the economics need to be assessed. What does it cost to produce the recommended lighting from the available equipment? There is little point in recommending lighting that is not economically viable. These three factors: the desired lighting, its technical possibility, and its cost to produce are all considered before making a decision about lighting criteria. The point to grasp from all this is that recommended illuminances, and all other quantitative lighting criteria, are matters of judgement, involving the balancing of several factors. Therefore, they inevitably represent a consensus view of what is reasonable for the conditions prevailing when they are written (Boyce, 1996). That consensus will be different in different countries, and different at different times in the same country, depending on the state of knowledge about lighting, the technical and economic situation, and the interests of the people contributing to the consensus.

By now it should be apparent that the writers of lighting recommendations do not have an easy task. They have to strike the right balance between a number of conflicting aims. They have to make recommendations that are precise and preferably quantitative but not so precise that they lose credibility. Equally, they have to avoid making recommendations that are so vague as to be meaningless. The recommendations have to be technically and economically feasible and simple enough to be implemented, although they should also reflect the complexity of the subject. Having achieved a balance between these factors appropriate for the application, and having written the document at a level suitable for the intended audience, they may be called upon to justify their recommendations. Often, this can only be done by an appeal to practical experience and a general consensus. An unenviable task indeed, yet the document writers' efforts are necessary. Advice is needed on appropriate lighting by people who buy lighting installations and by some who design them. To such people, it does not matter that every last word can be justified by experimental evidence. They do not have time to wait for science to catch up with technology. What does matter is if the recommendations made produce reasonable results in practice. It they do, then the recommendations will be accepted and the judgements of the writers vindicated. If they do not, then no amount of experimental evidence will convince anyone that the recommendations are correct. It is this test of practice that is the ultimate justification for many lighting standards codes and guides (Jay, 1973).

From the wide range of publications offering advice on lighting, it should be possible to find information for any specific application. This is not to say that all documents are the same. They all have different strengths and weaknesses, but they also have something in common. Ultimately, for any particular application, the recommendations come down to a consensus of opinion. For this reason, if for no other, it is always necessary to think about what lies behind any lighting recommendations before using them. Given this circumspect approach, lighting standards, codes, guides, and practices can be of great benefit to the art of lighting.

14.4 Consequences

The most obvious consequence of the use of lighting for the human race is the ability to function after the Sun has set. Indeed, modern life, with its trend toward globalization and the consequent provision of services throughout the 24 h, would be unthinkable without electric lighting. This book has been written with the aim of increasing the understanding of the impacts of lighting on human capabilities and perceptions. These impacts can range from enhancing the ability to work quickly, accurately, safely, and comfortably, at any time, in locations with and without access to daylight; to the ability to attract attention and modify mood and behavior. But the impacts of lighting do not stop at people. The provision of lighting also requires the consumption of natural resources and leads to pollution, both chemical and ethereal. It is these consequences that are of interest here.

14.4.1 Lighting and energy

Lighting is a major user of electricity. Estimates of the percentage of electricity generated that is used for lighting range from 5–19 percent in industrialized countries to 86 percent in underdeveloped countries. A recent model, based on estimates of electricity used for lighting in 38 countries, representing 3.7 billion people, predicts that globally, the amount of electricity used for lighting in 1997 was 2,016 TWh (Mills, 2000). Of this, 28 percent was used in the residential sector, 48 percent in the service sector, 16 percent in the industrial sector, and 8 percent for street lighting and other applications. Unfortunately, these values are essentially educated guesses. This is not for want of trying, or lack of expertise, but rather because the quality of the data available from different countries varies widely. The estimates from some countries are based on a simple residual analysis in which any electricity consumption that is left over after other end-uses have been identified is ascribed to lighting; others are based on extensive measurement and statistically validated surveys. This means that these global estimates should be treated as indicative rather than definitive. As for the allocation of consumption to different application areas, this will vary widely from country to country with the nature of the economy.

Given this variability, a better approach to considering the energy implications of lighting and what might be done to minimize the associated energy consumption is to focus on one country/state that has a well-established database on lighting usage and which has had a policy to reduce energy consumption for many years. Such a state is the State of California, the fifth biggest economy in the world, and for some, a model of the future. Based on a series of sample surveys carried out from 1992 to 1994, it has been estimated that lighting accounted for 23 percent of electricity use in California, 8 percent occurring in residences, 14 percent in commercial buildings, and 1 percent for road lighting. In absolute terms, this comes to 19,400 GWh used for residential lighting and 28,400 GWh used for commercial lighting, annually (California Energy Commission, 1999a). Figure 14.2 shows the electricity use for lighting in residences sub-divided into the percentages used in each room and the contribution of different light sources to that use. It is obvious from Figure 14.2 that the dominant light source in residences in California is the incandescent lamp. Fluorescent lamps only occur in significant numbers in the kitchen and the garage, and at the time of the survey these were usually linear fluorescent rather than compact fluorescent lamps. A rather different picture is evident for commercial buildings. Figure 14.3 shows the 14 percent of statewide electricity use that occurs in commercial buildings subdivided into the percentages used in different commercial building types and the contribution of different light sources to that use. For commercial buildings, the fluorescent lamp is the dominant light source, although there is significant use of incandescent and high intensity discharge lamps in some applications.

Over the last quarter of the twentieth century, there has been an increasing concern on the part of policymakers with reducing, or at least slowing the growth, of electricity consumption. There have been several reasons for this concern, reasons that have varied in their importance from year to year. Among the more common have been the idea that natural resources, such as oil, will eventually be completely consumed; the desire to reduce acid rain and other forms of pollution; the perceived need to reduce national dependence on fuels from other countries; the recognition that global warming is a threat; and the universal desire to reduce business costs. Lighting is an attractive candidate for action to people who want to reduce electricity use, for several reasons. First, the lighting industry has been diligent

% of statewide
residential energy use

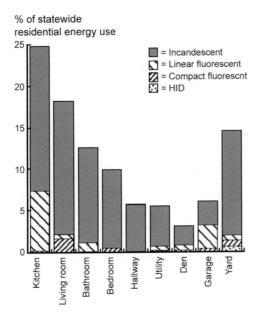

Figure 14.2 Statewide residential lighting energy use in California, classified by room and light source type, and expressed as a percentage (after California Energy Commission, 1999a).

% of statewide
commercial energy use

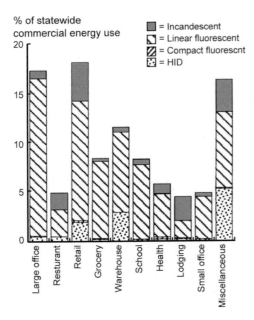

Figure 14.3 Statewide commercial lighting energy use in California, classified by application and light source type, and expressed as a percentage (after California Energy Commission, 1999a).

in producing more energy efficient light sources and luminaires. Second, the controls industry has produced a number of devices, such as occupancy sensors, that can be used to control lighting so that electricity is not wasted. Taken together, these innovations imply that the technology necessary to provide the same lighting conditions at lower energy consumption is available. Third, lighting installations have a much shorter life, typically 10–20 years, than a building. This means that lighting will be replaced several times during the life of a building, so the rate at which new lighting technology can be introduced throughout the building stock is faster for lighting than for other energy saving changes, such as improving insulation. Fourth, lighting in buildings is usually easily accessible and can be changed with much less disruption than some other proposed energy changes. The result of this evident suitability has been a series of actions, some legal and some promotional, aimed at reducing electricity use by lighting.

Probably the most successful legal action has been the introduction of maximum lighting power densities in buildings. These were first introduced in California in 1978 and have been revised regularly since then, the latest revision becoming effective in 2001. The beauty of this approach is that specifying the maximum allowed lighting power density does not restrict the designer unduly. Rather the lighting designer is free to allocate the amount and distribution of light and to choose the light sources to be used as desired, provided the lighting power density for the installation does not exceed the maximum. The maximum lighting power density can be applied to a whole building or, where different spaces in a building have very different visual requirements, on a space-by-space basis. The use of lighting controls to eliminate waste is also encouraged by the use of an effective lighting power density, the adjustment being made when specific lighting controls are incorporated into the design. The skill in setting the maximum lighting power densities is to make them stringent enough so that they encourage the development of more energy-efficient products and designs, yet not so stringent that they cannot be met with existing technology, without producing inadequate lighting. California's Title 24 Energy Efficiency Standards for Residential and Non-Residential Buildings have been largely successful in this respect. Figure 14.4 shows the distribution of the ratio of lighting power densities for a sample of new commercial buildings built in California between 1994 and 1998, to the relevant maximum allowed lighting power density. Ratios greater than unity do not comply with the 1995 Title 24 requirements. Ratios less than unity do. It is clear from Figure 14.4 that the majority of these new buildings do comply with the requirements, and that by making those requirements more stringent, there is an opportunity to reduce the energy consumption further. This tightening of the maximum allowed lighting power densities occurred in 2001. It is also evident that a few of these new buildings do not comply with the 1995 Title 24 requirements, a fact that suggests the need for stricter enforcement. Such legal requirements, provided they are backed by clear enforcement policies, can be very effective in

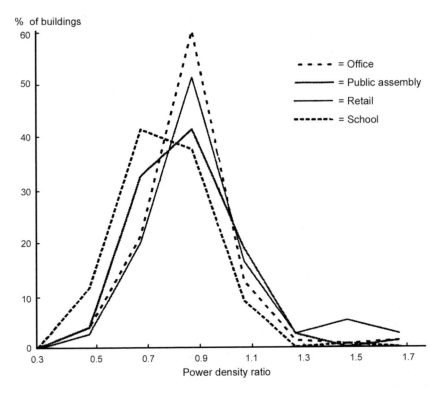

% of buildings

- - - - = Office
——— = Public assembly
——— = Retail
------- = School

Power density ratio

Figure 14.4 The distributions of the ratio of lighting power density for new commercial buildings to the relevant 1995 Title 24 maximum allowed lighting power density, in four types of building. Ratios greater than unity
indicate non-compliance with the Title 24 requirements. The distributions are based on a sample of 667 new buildings constructed in
California between 1994 and 1998 (after RLW Analytics, 1999).

reducing the use of electricity for lighting. Certainly, luminaire manufacturers
in the US have acknowledged Title 24 as the driving force behind the
increased production and marketing of energy efficient lighting technologies.

As for promotional activities, these have taken various forms at various
times in different countries. One approach has been to offer what is a subsidy for the use of energy efficient lighting technology. This has been widely
used by utilities, the subsidy coming in the form of either a reduced price
for the product or a payment based on estimated energy reduction after
installation (Mills, 1991). Another approach, widely used by government
agencies, is a combination of labeling and publicity. The idea is to make
energy efficient lighting products easy to identify and specify (Nirk, 1997;
Howarth *et al.*, 2000). Yet another approach is to undertake what is called
a technology procurement program (Stillesjo, 1991). This is a means of
accelerating the development and the introduction onto the market of

energy-efficient products, by subsidizing development costs and guaranteeing an initial market of a worthwhile size. These approaches, together with the good business practice of looking for reduced costs, have lead to the steady decrease in installed lighting power densities in commercial buildings.

Unfortunately, the same cannot be said for residential lighting where the incandescent lamp still reigns supreme. For more than a decade, many organizations in many countries have tried to bribe, cajole, or shame householders into using compact fluorescent lamps rather than incandescent lamps. The result of all this effort has been disappointing. In the European Union, the average household has 24 lamps, but only one of those is a compact fluorescent lamp (Palmer and Boardman, 1998). There are many reasons for this failure, but chief among them is the failure to recognize that the householder's decision to buy a compact fluorescent lamp is related to many factors, and that all the factors have to be successful for the majority of people to act (Boyce, 1998). These factors can be divided into those that determine whether a compact fluorescent represents an opportunity to use energy-efficient lighting, and those that determine whether compact fluorescent lighting gives the right impression, i.e. whether it is good or bad lighting. The opportunity factors are related to the technology and the costs, both first cost and operating cost. The impression factors are related to the values the householder has about costs, the environment and esthetics, their perception of the light source, and their expectations about what is good lighting. The present failure to make the compact fluorescent lamp the light source of choice for homes does not mean that attempts to promote energy efficient residential lighting should be abandoned. Rather, it first requires recognizing that the virtue of using energy efficiently is rarely enough to persuade householders to replace the cheap and effective incandescent lamp, even if the change would be accompanied by a distant economic benefit. Second, it means recognizing the multiple factors that influence householders' behavior and ensuring that any proposed energy efficient solution meets all those requirements. This is worthwhile because, as shown earlier, residential electricity use for lighting is large, incandescent lamps have a short life and so are changed frequently, and making the compact fluorescent lamp the lamp of choice for householders would reduce electricity use for lighting considerably (Palmer and Boardman, 1998). Further, it is not necessary to change all the lamps in a house. About 80 percent of the electricity consumed by lighting is taken by about 20 percent of the lamps in the house (California Energy Commission, 1999a). A number of publications have identified the barriers to widespread adoption of compact fluorescent lamps by householders (Palmer and Boardman, 1998; California Energy Commission, 1999b). What is needed is the will to tackle all of these barriers.

14.4.2 *Lighting and chemical pollution*

Lighting can generate chemical pollution either directly or indirectly. Direct chemical pollution comes about when lamps and control gear are scrapped.

Indirect chemical pollution occurs in the generation of the electricity consumed by the operation of the lamp.

Starting with direct chemical pollution, some lamps and control gear contain materials that are considered to be hazardous waste. Specifically, older electromagnetic control gear may contain polychlorinated biphenyls (PCB) and/or di(2-ethylhexyl)phthalate (DEHP). The appropriate method for disposing of such control gear, by landfill, recycling or incineration, depends on whether they are intact or leaking, and the local regulations. Modern control gear, both electromagnetic and electronic, do not contain such hazardous materials requiring special disposal procedures.

As for lamps, two toxic materials commonly found in lamps are lead and mercury. Lead occurs in solder and glass in virtually all lamp types, but lamps as a source are usually ignored because there are other more significant sources of lead. Much more attention has been given to mercury (Begley and Linderson, 1991; Clear and Berman, 1994). Mercury is used in many discharge lamps. In fluorescent and metal halide lamps, some mercury is essential because the mechanism for generating light is the creation of an electric discharge through a mercury atmosphere. Mercury is also used in high-pressure sodium discharge lamps as a starting aid and for voltage control. There is no prospect of completely eliminating mercury from fluorescent lamps but this may be possible for high-pressure sodium lamps. Despite the impossibility of eliminating mercury entirely, the lamp industry has responded to pressure from government and environmentalists by reducing the amount of mercury used in fluorescent lamps, from about 48 mg in a 4 ft fluorescent lamp in 1985 to 12 mg in 1999, and this process continues.

While any reduction in mercury content in lamps is welcome, it is important to appreciate that pursuing the objective of zero mercury in lamps may be self-defeating to the larger aim. This is because the amount of mercury released by lamp disposal is very small compared to the amounts of mercury released into the atmosphere naturally, through volcanos, and, more controllably, by the burning of fossil fuels to generate electricity (EPA, 1997). It would be possible to completely eliminate mercury pollution from lamps by using the incandescent lamp in preference to the compact fluorescent lamp, but then the amount of mercury emitted into the atmosphere would increase dramatically (Begley and Linderson, 1991). This perverse outcome is due to the low luminous efficacy of the incandescent lamp, causing more fuel to be burnt to generate the electricity to power the larger wattage of incandescent lamps necessary to provide the same lighting conditions. There are two conclusions that can be drawn from this argument, one general and one specific. The general conclusion is that when considering chemical pollution it is always necessary to consider all the various sources of the pollutant and the interactions between them. The specific conclusion is that the use compact fluorescent lamps, which do contain mercury, will lead to less total mercury pollution than the use of incandescent lamps, which do not (Gydesen and Maimann, 1991).

Table 14.1 Emissions of sulfur dioxide, nitrogen oxides, and carbon dioxide caused by electricity generation in the States of Oregon and Indiana, USA (from New Buildings Institute, 2001)

Location	Sulfur dioxide (lb/kW h)	Nitrogen oxides (lb/KW h)	Carbon dioxide (lb/KW h)
Oregon	0.000	0.001	0.183
Indiana	0.016	0.012	2.936

Turning now to indirect chemical pollution, the emissions of power plants generating the electricity used to operate light sources constitute pollution. Table 14.1 shows the pounds of sulfur dioxide, nitrogen oxides, and carbon dioxide produced per kilowatthour of electricity produced in two parts of the US. These chemicals are considered key indicators of pollution; sulfur dioxide, nitrogen oxides, and carbon dioxide being associated with air quality, acid rain, and global climate change, respectively. The two parts of the US are the states of Oregon and Indiana. These states have very different fuel mixes used in electricity generation. In Oregon, a significant proportion of electricity is generated by hydroelectric power. In Indiana, the dominant fuel used in electricity generation is coal. An examination of Table 14.1 shows that the amount of pollution produced per kilowatthour is very dependent on the fuel mix used. Despite this complicating factor, there can be little doubt that maximizing the efficiency of lighting installations and thereby reducing the electricity consumed to provide the desired lighting conditions, would have a beneficial effect on many aspects of pollution, everywhere.

14.4.3   Light pollution

Light, itself, can be considered a form of pollution. Complaints about light at night can be divided into two categories, light trespass and light pollution. Light trespass is local in that it is associated with complaints from individuals in a specific location. The classic case of light trespass is a complaint about light from a street lantern entering a bedroom window and keeping the occupant awake. Light trespass can be avoided by the careful selection, positioning, and aiming of luminaires with appropriate luminous intensity distributions. If that fails then some form of shielding can usually be devised. Advice on limiting light trespass, including guidance on appropriate maximum illuminances, is available (ILE, 1997; CIBSE, 1998; IESNA, 1999).

Light pollution is more diffuse than light trespass in that it can affect people over great distances and is more difficult to deal with. Complaints about light pollution originate from many people, ranging from those who have a professional interest in a dark sky, i.e. optical astronomers (McNally, 1994), to those who simply like to be able to see the stars at night. Light pollution is caused by the multiple scattering of light in the atmosphere, resulting in a diffuse distribution of luminance called sky glow. The problem this sky glow causes is that it reduces the luminance contrast of all the features of the night sky. A reduction in luminance contrast means that features that are naturally close to visual threshold will be taken below threshold by the addition of the sky glow. As a result, as sky glow increases the number of stars and other astronomical phenomena that can be seen is much reduced, so much so that in most cities it is difficult to see anything other than the moon and a few stars at night. To see the Milky Way, it is necessary to go to an area with a low population, many miles from any city.

Sky glow has two components, one natural and one due to human activity. Natural sky glow is light from the Sun, Moon, planets, and stars that is scattered by interplantery dust, and by air molecules, dust particles, water vapor, and aerosols in the Earth's atmosphere, and light produced by a chemical reaction of the upper atmosphere with UV radiation from the Sun. The luminance of the natural sky glow at zenith is of the order of $0.0002 \, cd/m^2$. The contribution of human activity is produced by light traversing the atmosphere and being scattered by dust and aerosols in the atmosphere. The magnitude of the contribution of city lights to sky glow at a specific location can be crudely estimated by Walker's law (Walker, 1977). This can be stated as

$$I = 0.01 \times P \times d^{-2.5}$$

where I is the proportional increase in sky luminance relative to the natural sky luminance, for viewing 45° above the horizon in the direction of the city (e.g. $I = 0.1 = 10$ percent increase), P the population of the city, and $d =$ distance to the city (km).

Obviously, this empirical formula assumes a certain use of light per head of population. Experience suggests the predictions are reasonable for cities where the number of lumens per person is between 500 and 1,000 lm. More sophisticated models based on the physics of light scatter have been used to generate light pollution maps, these models making allowances for the curvature of the earth and allowing predictions to be made for different altitudes and azimuths of viewing (Garstang, 1986; Albers and Duriscoe, 2001).

The problem in dealing with light pollution is not in measuring or predicting its effects on the visibility of the stars, but rather in agreeing what to do about it. The problem is that what constitutes the astronomer's pollution is often the business owner's commercial necessity and sometimes the citizen's preference. Residents of cities like their streets to be lit at night for the feeling of safety the lighting provides. Similarly, many roads are lit at night to enhance the safety of travel. Businesses use light to identify themselves at night and to attract customers. Further, the floodlighting of buildings and the

Figure 14.5 Lighting of a bridge in Cleveland, Ohio, that is attractive and dramatic
yet inevitably produces some light pollution (courtesy of Ross De Alessi).

lighting of landscapes are methods used to create an attractive environment
at night (Figure 14.5). The problem of light pollution is how to strike the
right balance between these conflicting desires. Different societies solve this
problem in different ways but something that is always useful is a quantitative understanding of the magnitude of the problem. Vos and Van
Bergem-Jansen (1995) determined this for a specific activity, the lighting of
greenhouses at night to promote plant growth. Such lighting is widely used
in The Netherlands from September to mid-May and the greenhouses are
commonly situated in the vicinity of residential areas. The illuminances on
the plants are typically in the range 3,000–4,000 lx and are provided by
high-pressure sodium lamps. To determine community reaction, a survey
was carried out in 10 areas around such lighted greenhouses, responses
being obtained from 391 residents. Light trespass was assessed by measuring the illuminance on the house facade and by asking if the resident was
annoyed by light from the greenhouses. The illuminance on the house
facade varied from 0.003 to 2 lx. The percentage of respondents who were
at least "a little annoyed" by the illumination of their rooms or garden by
light from the greenhouses was about 7 percent, while only about 3 percent
were "highly annoyed." There was no simple relationship between the level

of annoyance and the illuminance on the facade. As for sky glow, this was quantified by the luminance of the sky measured at an angle of 15° above the line of sight to the greenhouse. This luminance ranged from 0.09 to 0.67 cd/m². The percentage of respondents who were at least "a little annoyed" by the increased sky glow ranged from 15 to 45 percent. The percentage of respondents who were "very annoyed" by the increased sky glow ranged from 0 to 18 percent. The percentage annoyed in both categories increased with increasing luminance, although a closer relation between the level of annoyance and luminance was obtained by using the ratio of the luminance of the sky above the greenhouse to the luminance of a dark part of the sky not illuminated by the greenhouse, another illustration of the importance of contrast to visual perception (Figure 14.6). Of course, these data are for one situation in one country but they do indicate what can be done to determine whether complaints about light pollution are common in a community or the opinions of a few assertive individuals.

Given that complaints about light pollution are either widespread or are being made by a politically influential group, it may be necessary to address those complaints. One of the earliest attempts to do this involved the use of low-pressure sodium light sources for street lighting in cities adjacent to observatories. This approach was effective because astronomers could easily filter out the very limited range of wavelengths produced by the low-pressure sodium lamp. This approach has fallen by the wayside, for two reasons.

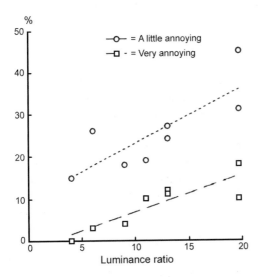

Figure 14.6 Percentage of survey respondents who rated the sky-glow above greenhouses illuminated at night at least as "a little annoying" or at least as "very annoying" plotted against the ratio of the luminance of the sky at an angle of 15° above the greenhouses and the luminance of a dark part of the sky not illuminated by the greenhouses (after Vos and Van Bergen-Jansen, 1995).

The first is that the non-existent color rendering properties of the low-pressure sodium lamp make it an unattractive prospect for city centers. The second is the growth in the use of light outdoors by commercial and residential property owners, using a wide range of light sources, has simultaneously increased the amount of light being emitted and undermined the effectiveness of the use of low-pressure sodium light sources by local authorities. Today, the most common advice given on how to reduce light pollution is to use what is called a full cut-off luminaire. A full cut-off luminaire is one that has zero luminous intensity emitted above the horizontal and whose luminous intensity per 1,000 lm at 80° above nadir is less than 100 cd (IESNA, 2000a). Following such advice is straightforward because luminaires that meet, or at least approach, the implied upward light output ratio limits are commercially available in a number of different forms (Pollard, 1994). Following this advice will certainly be effective in reducing light trespass, it may be less effective in reducing sky glow, for two reasons. The first is because restricting the upward light output from luminaires only limits the proportion of upward light and hence ignores the proportion of light scattered on the path to the surface to be illuminated as well as the light scattered after reflection from the illuminated surfaces. These sources of scattered light can be major factors in sky glow.

The second reason why simply using full cut-off luminaires may not be the best solution to the problem of light pollution is because the advice considers the luminaire in isolation and not as part of a lighting system. Keith (2000) calculated the total number of lumens going up into the sky from a roadway lighting installation, per unit area of road illuminated, including both light directly emitted upward and light reflected from the road and its surroundings (also known as upward luminous flux density). Figure 14.7 shows the upward luminous flux density for many different road lighting installations. The calculations were done for a collector road of 10 m width and of a specific reflectance, lit by lighting installations using a 250 W high-pressure sodium light source in full cut-off, cut-off, and semi-cut-off luminaires (IESNA, 2000a), arranged in a staggered pattern and spaced so as to meet average road surface luminance and luminance uniformity criteria. What is interesting in Figure 14.7 is that fact that the upward luminous flux density increases as the lighting power density increases. What this implies is that if the use of full cut-off luminaires demands a closer spacing of the luminaires to meet the road lighting criteria, the lighting power density will be increased and, consequently, the upward luminous flux density will increase. It is also suggested by Figure 14.7 that a semi-cut-off luminaire can be the best option, both for reducing installed power density and for reducing upward luminous flux density. However, this conclusion assumes that all directions of light emitted above the horizontal produce the same level of scattered light. Some have suggested that light emitted close to the horizontal produces more sky glow than light emitted vertically upward because of the greater path length through the atmosphere. Until the validity

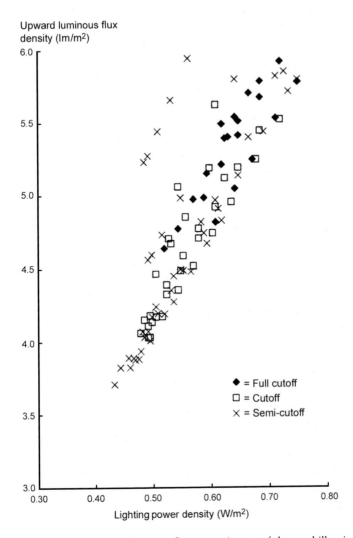

Upward luminous flux density (lm/m²)

Lighting power density (W/m²)

= Full cutoff
= Cutoff
= Semi-cutoff

Figure 14.7 Upward luminous flux per unit area of the road illuminated, plotted against lighting power density, for full cut-off, cut-off, and semi-cut-off luminaires. The values plotted are for a collector road of 10 m width, reflectance R3, lit by luminaires arranged in a staggered layout and using a 250 W high-pressure sodium lamp (after Keith, 2000).

of this argument is tested it would be premature to state that semi-cut-off luminaires are the best option. There are two other more general questions that arise from Figure 14.7. The first is whether the wide variation in upward luminous flux density for the same luminaire classification implies the need for some revision of the current luminaire classifications. The second concerns the validity of the luminance uniformity criterion used in roadway lighting, because if this criterion could be relaxed a different answer might be obtained.

It is important to recognize that the simple advice to always use a full cut-off luminaire does not consider the balance of interests in a specific location and the consequent interests of the people who live there. Fortunately, the CIE has published recommendations to limit sky glow that avoid the extremes of both environmental zealotry and just-do-it commercialism (CIE, 1997b). The approach adopted is first to classify an area into one of four major zones (Table 14.2). Then a maximum upward light output ratio for any luminaires used in each zone is given (see Table 14.3). Finally, a minimum distance between a reference point inside one zone and the borderline of adjacent zones is given, in kilometers. This minimum distance is designed to limit the contribution to sky glow from adjacent zones to a low level. This approach offers some flexibility in that it recognizes that areas of outstanding natural beauty are different from city centers and that it is ridiculous to give the same priority to sky glow limitation in the latter as in the former.

Of course, light output ratio is a metric of relative light distribution and does nothing to control the total amount of light emitted. To effectively

Table 14.2 The zoning system of the CIE (after CIE 1997b)

Zone	Zone description and examples of sub-zones
E1	Areas with intrinsically dark landscapes: National Parks, areas of outstanding natural beauty (where roads are usually unlit)
E2	Areas of "low district brightness": outer urban and rural residential areas (where roads are lit to residential road standard)
E3	Areas of "middle district brightness": generally urban residential areas (where roads are lit to traffic route standard)
E4	Areas of "high district brightness": generally, urban areas having mixed recreational and commercial land use with high night-time activity

Table 14.3 Maximum installed upward light output ratio; luminous flux emitted above the horizontal plane as a percentage of the total luminous flux emitted by the luminaire (after CIE, 1997b)

Zone	Maximum upward light output ratio (%)
E1	0
E2	5
E3	15
E4	25

limit the contribution of human activity to sky glow there are two complementary options. The first option is to limit the amount of light used at night, accepting that if taken too far such limits will increase the risk to life and limb through decreases in visibility and consequent increases in accidents and/or crime. The International Dark-Sky Association (IDA) has attempted to influence the amount of light used at night by providing a pattern code of practice that contains, along with much other useful advice, a maximum luminous flux density for exterior lighting, expressed in lumens per acre (Table 14.4). These maximum luminous flux densities are applicable to any new commercial or residential development, in each of the four CIE zones, plus one additional zone called Zone E1A and defined as a Dark-Sky Preserve (IDA, 2000). These maximum luminous flux densities imply the use of fully shielded luminaires, i.e. luminaires with no light output above the horizontal plane through the luminaire, the limits being much more stringent for unshielded luminaires. While such maximum luminous flux densities have the advantage of simplicity, they may be over-simple because they take no account of the fact that different activities in a zone may require different amounts of light to be performed safely. A better alternative is to adopt maximum lighting power densities for different activities in each zone, e.g. for roadways, for parks, and for petrol stations. Each maximum lighting power density can be based on knowledge of the illuminances considered appropriate for the given activity in the environmental zone, and the luminous efficacy of the light sources and the type of luminaires typically used to light such activities. Fortunately, there is plenty of advice on how much light is appropriate in specific exterior situations such as for road lighting (CIE, 1995c; IESNA, 2000b), for lighting landscapes (Moyer, 1992), and for perceptions of safety at night (Boyce *et al.*, 2000b). Legally enforceable maximum lighting power densities for exterior lighting are currently being developed in California.

Table 14.4 Maximum total luminous flux density, in lumens per acre, for commercial and industrial, and residential exterior lighting in different zones (after IDA, 2000)

Application	Zone E4	Zone E3	Zone E2	Zone E1	Zone E1A
Commercial and industrial – all luminaires	200,000	100,000	50,000	25,000	12,500
Commercial and industrial – unshielded luminaires only	10,000	10,000	4,000	2,000	1,000
Residential – all luminaires	20,000	10,000	10,000	10,000	5,000
Residential – unshielded luminaires only	5,000	5,000	1,000	1,000	0

The second option is to pay careful attention to the timing of the use of light. Light pollution is unlike chemical pollution in that when the light source is extinguished, the pollution goes away. This suggests that a curfew defining the times when lighting can and cannot be used could have a dramatic effect on sky glow. Of particular value would be the application of a curfew to the use of light for commercial purposes, other than for security. This is because one of the major forces leading to light pollution is the commercial need to be noticed. There is little point in using light at night to attract attention when there is no one about to notice. The power of this commercial imperative is evident in what has happened to the lighting of petrol stations in America. Over the last decade, a brightness war has broken out, the apparent aim being to have the brightest under-canopy lighting in the local market. If one petrol station in an area installs new higher-level lighting to create an impression of greater brightness, competing stations feel obliged to be brighter still. The result of this escalating process has been a steady increase in the illuminances produced by under-canopy lighting with the result that, today, it is not uncommon to find illuminances on the apron under the canopy that exceed 1,000 lx, more than double the illuminance used in offices. Until such excesses are reined in, success in limiting light pollution will itself be limited.

14.5 Summary

Lighting research enters the lives of most people through the medium of documents specifying the amount of light to be provided for different applications, and/or offering guidance on how to provide that light. Such advice is necessary because many lighting installations are specified or designed by people with limited knowledge of lighting. The advice can come in many different forms, ranging from laws and regulations which have to be complied with; to codes, guides, and recommended practices, which have no formal legal standing but which it may be wise to follow because they represent best practice and hence offer a defence in case of litigation.

The nature of published lighting advice has changed over the years, the number of criteria that have to be considered tending to increase and the values of those criteria changing, depending on technical and economic circumstances, and people's expectations. These changes occur because the preparation of lighting recommendations is inevitably a matter of consensus. What this variability implies is that lighting design should never be a thought-free process.

The provision of lighting has consequences beyond the ability to see. Specifically, the provision of lighting requires the consumption of electricity and that implies the generation of pollution. It has been estimated that in the State of California lighting uses 23 percent of the electricity generated; 14 percent in commercial buildings, 8 percent in residences, and 1 percent for road lighting. Over the last quarter century, governments have made attempts to reduce or at least slow the growth in electricity demand. Improving the efficiency of lighting installations is an attractive means to achieve this end, because more energy efficient energy lighting technology is available and the life of lighting installations is a matter of years rather than decades. Consequently, a number of governments have introduced laws either specifying a maximum lighting power density to be used in buildings or outlawing the use of specific energy-inefficient lighting equipment. The result has been a steady reduction in the lighting power densities in commercial buildings but little change for residences. For change to occur in residences, the incandescent lamp will have to be replaced as the light source of choice. For this to happen, attention needs to be given to all the factors that influence a householders decision to buy and not just the energy implications.

As regards pollution, this can take the forms of direct chemical pollution, indirect chemical pollution, and light pollution. Direct chemical pollution occurs when lamps and control gear are scrapped. Indirect chemical pollution occurs in the generation of the electricity required for the operation of the lamp. The direct chemical pollutant of concern at the moment is mercury. In response to pressure, the lamp industry has been steadily reducing the amount of mercury in fluorescent lamps and further reductions are being sought. However it is important to appreciate that mercury from lamp disposal is very small relative to the amount of mercury released by the burning of fossil fuels to generate electricity. Therefore reducing the mercury content of lamps will only be effective in reducing this form of chemical pollution as long as the use of the reduced-mercury lamps does not produce an increase in electricity consumption.

As for indirect chemical pollution, the generation of electricity produces increased levels of sulfur dioxide, nitrogen oxides, and carbon dioxide in the atmosphere, these chemicals being associated with poor air quality, acid rain, and global climate change, respectively. There can be little doubt that maximizing the efficiency of lighting installations and thereby reducing the electricity consumed to provide the desired lighting conditions, would have a beneficial effect on many aspects of chemical pollution.

Another consequence of the use of lighting is light pollution. Light pollution is evident as localized light trespass and a more general increased sky luminance, called sky glow. The effect of sky glow is to make many stars invisible because their luminance contrast is reduced to below threshold. There is a small natural contribution and a variable human contribution to sky glow. The magnitude of the human contribution is dependent on the amount of dust and aerosols in the atmosphere and the amount of light traversing the atmosphere. The only approaches to reducing light pollution that are likely to be successful long term are those of limiting the amount of light used outdoors at night and of limiting the hours for which such lighting is available. The latter is effective because light pollution, unlike chemical pollution, disappears the moment the light source is extinguished.

15 The way ahead

15.1 Introduction

What follows is unlike previous chapters in that it does not deal with definitions or quantities, or experimental results and their application. Rather it gives the author's opinions as to the actions that need to be taken if future lighting research is to be fruitful. Broadly these actions can be classified into: (a) avoiding the limitations of the past; (b) developing new approaches to the study of lighting and new measures of the effects of lighting; and (c) applying these approaches and measures to areas of practical importance. Each of these topics will be examined in turn.

15.2 The limitations of the past

A disinterested evaluation of past lighting research will reveal four aspects which limit the value of some of the work done. The first is that, by present day standards, some past experimentation has been of moderate methodological quality (Tiller, 1990; Veitch and Newsham, 1998b). Experiments can be found which are characterized by little control of experimental procedure, a small and/or unrepresentative sample of people and a limited statistical analysis of the data collected; although the control and description of the lighting conditions has often been good. Thankfully, such failings have become much rarer over the last two decades. This progress can be sustained if researchers adopt high standards of methodology and statistical analysis and the editors and referees of scientific journals insist on these standards for publication.

The second limiting aspect is the tendency to put lighting first and people second. This is most evident in the choice of the independent variables used in experiments. In the past the independent variables have frequently been such basic features of lighting as illuminance, different glare conditions, and color properties of lamps. These variables have been chosen because they are important to the lighting equipment manufacturer or are used in lighting design calculations, and not because they necessarily reflect factors that are important to the people using the lighting. It may be of some relevance to the designer's immediate problem if, for example, the color rendering index (CRI) of a lamp can be shown to relate to the ease with which different hues can be discriminated; but it is not getting at the root of the problem of the discrimination of colors. To do this, it is necessary to gain an understanding of how people discriminate between colors. Once this has been achieved, a suitable measure of the relevant lamp properties can be developed. To achieve a more fundamental understanding of how people respond to lighting it is necessary to put people first and the lighting conditions second. Further, the people considered have to be representative of actual users in all their variety and complexity. Unless this is done, future research will not be building on the successes of the past, merely repeating them.

The third limiting aspect, and a reason why the second is so prevalent, is the widespread absence of conceptual frameworks. Concepts are important for research. They form the unstated assumptions within which research is conceived. Concepts that are explicitly stated become theories and theories give rise to hypotheses that can be tested by experimentation. About the only area of lighting research which can be said to have followed this route through concept/theory/hypothesis/experimentation is the study of visibility. This is for two reasons. First, because visibility is an obvious impact of lighting and has been studied for many years. Second, because visibility is a unique effect of lighting. Other areas of interest to lighting research, such as discomfort and mood, are influenced by many factors in addition to lighting. In this situation, the need is for more general theories that span the effects of many different environmental and personal factors. Such theories exist and deserve investigation (e.g. Donovan and Rossiter, 1982; Gorn, 1982; Belcher and Kluczny, 1987; Kaplan, 1987).

The fourth limiting aspect is the concentration on the general, inherent in the neglect of context. Quite correctly, a lot of past lighting research has been concerned with establishing general rules for providing lighting that allows work to be done quickly and easily, without discomfort. These general rules can now be said to be established. What deserves attention in the future is the extent to which these general rules need to be modified for different contexts. The point is which lighting conditions are most suitable depends on context. Until the importance of context is acknowledged there is little likelihood of achieving a finer understanding of the effect of lighting in all its complexity.

15.3 Approaches and measures

Although putting people first, careful experimentation, developing conceptual frameworks, and an awareness of the importance of context are necessary for future research to be fruitful; alone they are not sufficient. For fruitful research to occur, the problems to be investigated have to be approached in the right way. In the broadest sense this means considering the direction in which research should aim to move. For the study of the effects of light operating through the visual system, there is plenty of knowledge on which to base very general rules but little that is applicable to specific tasks. Hence, the need in this field of study is for a move from the general to the specific. For the study of the effects of light operating through the circadian system, knowledge of how the circadian system works is growing rapidly. Unfortunately, understanding of how changes in the status of the circadian system impact everyday activities is sparse. For this field of study the present need is to move from the laboratory to the field. As for the study of the effect of lighting on impression and behavior, the direction of movement here should be from the specific to the general. At present there are some studies done in realistic conditions which have indicated how to create an impression with light. But these results only apply to the specific contexts in which they were obtained. What is needed in the future is for many more different contexts to be examined. If some consistency was then revealed, general rules about using lighting to create impression and direct behavior could be formulated.

The direction of any research is only one aspect of the approach adopted. Another aspect is the techniques used to study the problem. There are several different techniques that can be used to obtain information about the effect of lighting conditions. They can be summarized as follows:

Epidemiological approach: This approach is used to determine if two variables are correlated, e.g. if smoking cigarettes is related to the incidence of lung cancer. It is particularly useful as a method if there are many intervening factors that cannot be controlled, and/or the effect does not occur until long after exposure to the stimulus. The study of the presence of daylight provided through skylights on retail sales is typical of this approach (Heschong *et al.*, 2002). The overwhelming drawback of this approach is that it can only reveal whether two variables are correlated, not whether they are causally related. This means such studies are useful for determining if a relationship is worthy of further study, although such study should be undertaken only when a major effect is identified (Taubes, 1995). Practically, the main drawback of the epidemiological approach is that it requires extensive databases of all the relevant information, databases that often do not exist or are inaccessible.

Ecological approach: This approach is simply that of observation followed by interpretation, although it is sometimes possible to perturb the process by introducing a change in conditions. The study of Areni and Kim (1994) on the behavior of people in a wine store under "bright" and "soft" lighting is an example of this approach. This approach is most suitable where the context in which the study takes place is important and removing the activity from the context would destroy the phenomenon being studied. The main disadvantage of this approach is that it cannot provide an explanation of why effects occur. Explanations that are given when using this approach are post hoc rationalizations. However, for some studies, such as the effect of lighting conditions on behavior, there is little alternative, because this approach provides the minimum interference with the natural condition.

Stimulus/response approach: This is the approach conventionally used in human factors research, vision research, and psychophysics. In its simplest form, a stimulus is administered to the subject under controlled conditions and a response is measured. Experiments based on this approach require decisions about three classes of variables: independent, dependent, and intervening variables. Independent variables define the conditions being examined by the experiment. Dependent variables are the measures used to quantify the response to the independent variables. Intervening variables are all those factors that may influence the relationship the independent and dependent variables. There are two types of intervening variables: those that need to be controlled and those that need to be measured in order to identify the reason for any change in the dependent variables. Experimental design procedures allow for several independent variables and the interactions between them to be examined in one experiment. Provided care is taken with the selection, measurement, and control of independent, dependent, and intervening variables, and provided that statistical analysis of the collected data is thorough and appropriate, the stimulus/response approach can prove cause and effect. This is a great advantage over the epidemiological and ecological approaches. However, the stimulus/response approach does have one drawback; namely, that rigorous control of the intervening variables may destroy or modify the phenomenon being examined. This drawback does not confine the stimulus/response approach to the laboratory. Such experiments can be conducted in the field given the right conditions (Maniccia *et al.*, 1999). What the drawback does mean is that where the effect being studied may be influenced by the context, e.g. where behavior is being studied, the stimulus/response approach may not be appropriate because the very act of taking part in an experiment may change the response.

It would be a mistake to think these approaches are always mutually exclusive. Rather, different approaches are appropriate for answering different questions. It is very rare for a single experiment to provide a conclusive answer to a question. Usually, multiple experiments are required, with the results from different approaches providing mutual support. This ideal is called converging operations and is much like making a case for presentation in court. In the legal situation, the prosecutor has to prove that a crime occurred and that the accused had the means and the motive to carry out the crime. In scientific research, the researcher has to prove that lighting was responsible for the measured effect. To do that, the researcher has to prove that a change in response occurred and provide a proven mechanism through which lighting might act to produce that response.

As an aid in planning effective research, Wyon (1996) has introduced the idea of a linked mechanisms map. A linked mechanisms map sets out all the pathways between the independent variables and the dependent variables in a specific experiment. It is only when all the steps along one or more pathways have been proven that the effect of the independent variable on the dependent variable can be said to be established. Linked mechanisms maps provide a rational basis for answering the question, "Why do you expect your independent variable to affect your dependent variable?" This question needs to be addressed at the planning stage of an investigation. Without a rational answer to this question, any research project is reduced to a "fishing expedition."

Although choosing an appropriate approach to a problem is important, new ways of measuring the effects of lighting are also required. For examining the effects of light on work, the most urgent requirement is for some measure of how easy a target is to see, as opposed to how well it is lit. Ideally, a measure of how easy a target is to see would include both lighting and target characteristics in one metric. One possibility is based on minimum presentation time. The idea here would be to steadily reduce the time for which the target was available until the details of interest were no longer visible. This method would certainly have the variability associated with all threshold measurements but it would have the advantage that time intervals can be easily and accurately measured and varied; and it would be easy to introduce an element of visual search into the scene, if that was called for. At the moment this is only a possibility which might not work but there can be little doubt that some simple measure of how easy an object is to see would be of value. It would certainly be useful in establishing a taxonomy of visual task difficulty. At present, the only visual task classification system that is widely available is based on the physical measures of size and contrast of critical detail, rather than on a direct visual measure of how easy something is to see. A measure of ease of seeing is a topic well worth pursuing.

But a measure of ease of seeing is not the only requirement. One of the points considered in Section 4.2 was that for the vast majority of tasks the visual component is only part of the complete task. The visual component can vary from very large, as in driving in a strange town at night, to very small, as in audio typing. If a more precise estimate of the effect of lighting on work is to be gained, then a measure of the magnitude of the visual component in the work is necessary. This might be achieved by a two-step process using eye movement recording and restricted viewing time. Basically, the idea would be to use eye movement recording first to identify where the worker was looking while doing the task. This would enable the features of the visual scene of importance for the performance of the task to be identified. Second, the times for which the elements of the visual scene identified as important were available could be reduced until task performance started to decline. From such data, it should be possible to estimate how much of the total time was spent obtaining the visual information, the greater the time, the larger the visual component. Such a task analysis system would be valuable in suggesting to what tasks light could most effectively be applied to increase productivity.

So far the emphasis has been on new measures and new approaches for use with conventional experiments. But it should be realized that there are other ways of gaining knowledge apart from experimentation. One way is scholarship. This is rather a grand name for reading the literature. Scientific understanding is a cumulative, self-correcting process recorded in the literature. Familiarity with the literature is essential if effort is not to be wasted and progress is to be made. Another approach to gaining knowledge which has been out of fashion for many years but which may still be of value is introspection. By considering one's own reactions to the lighting conditions it is possible to produce some information on the factors that create certain impressions, facilitate performance, or cause discomfort. Finally, the heretical course of listening to artists, stage lighting experts and photographers, can be advocated. Such people know a lot about creating an impression with light. Introspection, and discussions with artists are unlikely to be sufficient alone to enable firm conclusions about lighting to be drawn because they are so open to bias. However, they can introduce observations, ideas and concepts which can then be tested by carefully designed experiments. This is their real function – like scholarship, they tend to give insight so that a useful hypothesis can be constructed and then tested.

15.4   Areas of application

Having considered the possible methods of investigation, it is appropriate to discuss the problems to which they might be applied. There are numerous specific problems that need investigation but here consideration will be given only to the major areas that deserve study. In what might be called the pure research area three topics deserve attention. The first is mesopic vision. This state of vision is poorly understood. There is no established relative spectral senstivity function for mesopic vision and visual capabilities are subject to large changes over small luminance ranges. A better understanding of mesopic vision and how to make the best use of its capabilties would be useful for exterior lighting practice. The second topic is the effect of lighting conditions on peripheral vision. Currently the effects of lighting on vision are almost invariably considered in terms of foveal vision, and yet many everyday activities and perceptions involve peripheral vision. This practical relevance, together with what is known about the way in which increases in task difficulty restrict the efficiency of peripheral vision, suggest it would be a fruitful field of study. The third topic is the the impact of light through the circadian system on everyday activities. The circadian system operates at a very basic level of human physiology, so it has the potential to influence human behavior in many ways. Until the reach and nature of those influences are understood, we do not know if the abolition of the night by the wide-spread availability of electric lighting has hidden costs. Also, it is only after understanding the full impact of the circadian system that it will be possible to optimise lighting to serve both the visual system and the circadian system.

As for applied research, there are two areas deserving attention. The first is industrial lighting. Much of existing lighting research has been concerned with offices. Unfortunately, factories are not like offices in that they often involve three-dimensional, as opposed to two-dimensional tasks, and the lighting is frequently subject to severe obstruction. Further, many industrial tasks involve the perception of form rather than the meaning of symbols and hence may call for different lighting conditions. At present most industrial lighting is based on data obtained with office work in mind. It may well be that different approaches to the lighting of factories need to be developed.

The second area for investigation is the influence of lighting on behavior. Such an understanding would be very relevant for retail lighting, a field that is characterized more by trial and error than systematic study. It is undeniable that lighting conditions can influence behavior. At the simplest level, people will change their behavior to compensate for poor lighting conditions (Rea *et al.*, 1985a). But the effects of lighting on behavior are more extensive than simply overcoming problems of poor visibility. Lighting can be used to guide movement through a space (Taylor and Socov, 1974) and has been found to influence speech levels and the topic of conversations (Gifford, 1988; Veitch and Kaye, 1988) Probably the most interesting aspect of the effect of lighting conditions on behavior from the commercial viewpoint is whether lighting conditions can be used to increase sales of a product. Certainly, lighting can be used to draw attention to a product (LaGiusa and Perney, 1974) and can change the attractiveness of a product (Mangum, 1998) but it cannot guarantee a sale. An understanding of how lighting might be used in retail situations would be of considerable interest.

15.5   Why bother?

All the preceding comments may be interesting but are they necessary? After all, existing lighting seems to be adequate. Work gets done and if interiors are not always pleasant they are not usually uncomfortable. In which case why is it necessary to continue lighting research at all, let alone attempt to improve it? There are two answers. The first is simply that the general standard of lighting could be better, particularly exterior lighting. Greater knowledge of how people respond to lighting will aid and encourage better practice.

The second answer is that for most people lighting is important, although its value is often not appreciated until it is elminated by a power cut. However, for most people lighting is not an end in itself; it is a means to an end. The most widely recognized "ends" are increasing productivity, reducing energy consumption, enhancing health, and improving the quality of life.

The need for increased productivity arises from the globalization of business. The developed countries, where lighting is most widely used, are all high-wage economies. The globalization of business has meant that high-wage economies have to be more productive and more intelligent than low-wage economies to keep their populations employed. Providing appropriate lighting has been recognized as an aid to productivity for many years.

The need to reduce energy consumption arises from the threat of global warming and the consequent climate change, and the ultimately limited global supply of fuel, a limit that is likely to be reached sooner as some countries with large populations strive to increase their standard of living. Lighting is a major user of electricity. This suggests that in the future there will be consistent pressure to minimize the use of electric light. When responding to such pressure, it would be as well to understand what the consequences might be in terms of human response.

As for the desire to enhance health, in developed countries many forms of previously fatal medical conditions have been largely eliminated but there is increasing concern about non-life-threatening health issues. This tendency is exacerbated by the presence of an aging population, many of whom can be expected to live longer than any previous generation. They may also have to work longer, because the number of young people available to replace retiring older workers is inadequate. Lighting has a role to play in enabling older workers to perform their work effectively and in promoting some aspects of health for people of all ages.

As for the quality of life, this means different things to different people. For much of the world population, improving the quality of life means having enough to eat, returning to a peaceful existence, and escaping poverty. For many citizens of developed countries, it means improving the education system and making the streets safe. For others, particularly older citizens, it means maintaining their independence in their own homes; while for yet others, it means having an attractive environment around them. Lighting can have an impact on all these aims.

As long as these "ends" remain relevant, lighting research directed towards a better understanding of how lighting can be used to fulfill them will be worthwhile.

15.6 Summary

This chapter is addressed to those who are, or will be concerned with lighting research. It discusses the actions that need to be taken if future research is to be fruitful. On a philosophical level, these actions involve putting people before lighting and accepting the importance of context. On a technical level, they involve developing new approaches to the study of lighting, and new measures to quantify the magnitude of the visual component and how easy things are to see. On a practical level, they involve applying these approaches and measures to areas of interest, a number of which are suggested.

References

Adrian, W. (1976) Method of calculating the required luminances in tunnel entrances, *Light. Res. Technol.*, 8, 103–6.

Adrian, W. (1982) Investigations on the required luminance in tunnel entrances, *Light. Res. Technol.*, 14, 151–9.

Adrian, W. (1987) Adaptation luminance when approaching a tunnel in daytime, *Light. Res. Technol.*, 19, 73–9.

Adrian, W. (1989) Visibility of targets: model for calculation, *Light. Res. Technol.*, 21, 181–8.

Adrian, W. (1991) Comparison between the CBE and CIE glare mark formula and earlier discomfort glare descriptions, *First International Symposium on Glare*, New York: The Lighting Research Institute.

Adrian, W. (1995) Change of visual acuity with age, in W. Adrian (ed.) *Lighting for Aging Vision and Health*, New York: Lighting Research Institute.

Adrian, W. and Eberbach, K. (1969) On the relationship between the visual threshold and the size of the surrounding field, *Light. Res. Technol.*, 1, 251–4.

Adrian, W. and Gibbons, R. (1994) Visual performance and its metric, *Light Eng.*, 2, 1–34.

Adrian, W. and Schreuder, D.A. (1970) A simple method for the appraisal of glare in street lighting, *Light. Res. Technol.*, 2, 61–73.

Adrian, W.K., Gibbons, R.G., and Thomas, L. (1993) Amendments in calculating STV: influence of light reflected from the road surface on the target luminance, *Proceedings of the 2nd International Symposium on Visibility and Luminance in Roadway Lighting*, New York: Lighting Research Institute.

Aharon-Peretz, J., Masiah, A., Pillar, T., Epstein, T., Tzichinsky, O., and Lavie, P. (1991) Sleep–wake cycles in multi-infarct dementia and dementia of the Alzheimer type, *Neurology*, 41, 1616–19.

Aizlewood, C.E. and Webber, G.M.B. (1995) Escape route lighting: comparison of human performance with traditional lighting and wayfinding systems, *Light. Res. Technol.*, 27, 133–43.

Akashi, Y. and Rea, M.S. (2001a) Peripheral detection while driving under a mesopic light level, *Proceedings of the IESNA Annual Conference*, Ottawa, New York: IESNA.

Akashi, Y. and Rea, M.S. (2001b) The effect of oncoming headlight glare on peripheral detection under mesopic light levels, *Proceedings of the Symposium on Progress in Automobile Lighting*, Darmstadt, Germany: Herbert Utz Verlag GmbH.

Akashi, Y., Akashi, I., Tanabe, Y., and Kanaya, S. (1995) The sparkle effect of luminaires on the sensation of brightness, *Proceedings of the CIE, 23rd Session, New Dehli*, Vienna: CIE.

Akashi, Y., Muramatsu, R., and Kanaya, S. (1996) Unified glare rating (UGR) and subjective appraisal of discomfort glare, *Light. Res. Technol.*, 28, 199–206.

Akerstedt, T. (1985) Adjustment of physiological circadian rhythms and the sleep–wake cycle to shiftwork, in S. Folkard and T.H. Monk (eds) *Hours of Work*, New York: John Wiley and Sons.

Albers, S. and Duriscoe, D. (2001) Modeling light pollution from population data and implications for National Park Service lands, Available online at http://laps.fsl.noaa.gov/albers/lp/gwpaper/lppaper.html (accessed, December 10, 2001).

Alferdinck, J.W.A.M. (1996) Traffic safety aspects of high-intensity discharge headlamps; discomfort glare and direction indicator conspicuity, in A.G. Gale, I.D. Brown, C.M. Haslegrave, and S.P. Taylor (eds) *Vision in Vehicles – V*, Amsterdam: North-Holland.

Alferdinck, J.W.A.M. and Padmos, P. (1988) Car headlamps: influence of dirt, age and poor aim on glare and illumination intensities, *Light. Res. Technol.*, 20, 195–8.

Allen, T.M., Dyer, F.N., Smith, G.M., and Janson, M.H. (1967) Luminance requirements for illuminated signs, *Highway Res. Rec.*, 179, 16–37.

American Association of State Highway and Transportation Officials (AASHTO) (1990) *A Policy on Geometric Design of Highways and Streets*, Washington, DC: AASHTO.

American Conference of Governmental Industrial Hygienists (ACGIH) (2001a) *TLVs and BEIs Threshold Limit Values for Chemical Substances and Physical Agents, Biological Exposure Indices*, Cincinnati, OH: ACGIH.

American Conference of Governmental Industrial Hygienists (ACGIH) (2001b) *Documentation of Physical Agents Threshold Limit Values*, Cincinnati, OH: ACGIH.

American National Standards Institution (ANSI) (1998) *Safety Color Code*, ANSI Z535.1-1998, New York: ANSI.

American Psychiatric Association (APA) (2000) *Diagnostic and Statistical Manual of Mental Disorders*, DSM-IV-TR, Washington, DC: APA.

American Society for Testing and Materials (ASTM) (1996a) *Standard Practice for Lighting Cotton Classing Rooms for Color Grading*, D1684-96, Philadelphia, PA: ASTM.

American Society for Testing and Materials (ASTM) (1996b) *Standard Practice for Visual Appraisal of Colors and Color Differences of Diffusely-Illuminated Opaque Materials*, D1729-96, Philadelphia, PA: ASTM.

American Society for Testing and Materials (ASTM) (1996c) *Standard Practice for Specifying Colors by the Munsell System*, D1535-1596, Philadelphia, PA: ASTM.

Anderson, N.H. (1981) *Foundations of Information Integration Theory*, New York: Academic Press.

Anon, (1977) Recommendations on uniform color spaces, color difference equations and metric color terms, *Color Res. Appl.*, 2, 5–6.

Anon, (1983) Twelve die in fire at Westchase Hilton Hotel, *Fire J.*, January, 11–15, 20–3, 54–6.

Anon, (1998) *Proceedings of the First International Symposium on Human Behaviour in Fire*, Belfast, UK: University of Ulster.

Areni, C.S. and Kim, D. (1994) The influence of in-store lighting on consumer's examination of merchandise in a wine store, *Int. J. Res. in Mktg.*, 11, 117–25.

Arndt, W., Bodmann, H.W., and Muck, E. (1959) Untersuchung uber die psycholgische Blendung durch mehrere Lichtquellen, *Proceedings of the CIE 14th Session*, Brussels, Paris: CIE.

Aschoff, J. (1969) Desynchronization and resynchronization of human circadian rhythms, *Aerospace Med.*, 40, 844–9.

Aschoff, J., Fatranska, M., Hiedke, H., Doerr, P., Stamm, D., and Wisser, H. (1971) Human circadian rhythms in continuous darkness: entrainment by social cues, *Science*, 171, 213–15.

Aschoff, J., Hoffmann, K., Pohl, H., and Wever, R. (1975) Re-entrainment of circadian rhythms after phase shifts of the Zeitgeber, *Chronobiologica*, 2, 23–78.

Asplund, R. and Ejdervik Lindblad, B. (2002) The development of sleep in persons undergoing cataract surgery, *Archives of Gerontology and Geriatrics*, 35, 179–87.

Assum, T., Bjornskau, T., Fosser, S., and Sagberg, F. (1999) Risk compensation – the case of road lighting, *Accid. Anal. Prev.*, 31, 545–53.

Aston, S.M. and Bellchambers, H.E. (1969) Illumination, color rendering and visual clarity, *Light. Res. Technol.*, 1, 259–61.

Atkins, S., Husain, S., and Storey, A. (1991) *The Influence of Street Lighting on Crime and the Fear of Crime*, Paper 28, London: Home Office Crime Prevention Unit, Home Office.

Avery, D.H., Bolte, M.A., Wolfson, J.K., and Kazaras, A.L. (1994) Dawn simulation compared with a dim red signal in the treatment of winter depression, *Biol. Psychiatry*, 36, 180–8.

Badia, P., Myers, B., Boecker, M., and Culpeper, J. (1991) Bright light effects on body temperature, alertness, EEG and behavior, *Physiol. Behav.*, 50, 583–8.

Bailey, I., Clear, R., and Berman, S. (1993) Size as a determinant of reading speed, *J. Illumin. Eng. Soc.*, 22, 102–17.

Bajorski, P., Dhar, S., and Sandhu, D. (1996) Forward-lighting configurations for snowplows, *Transport. Res. Rec.*, 1533, 59–66.

Balazsi, A.G., Rootman, J., Drance, S.M., Schulze, M., and Douglas, G.R. (1984) The effect of age on the nerve fiber population of the human optic nerve, *Am. J. Ophthalmol.*, 97, 760–6.

Barker, F.M. and Brainard, G.C. (1991) *The Direct Spectral Transmittance of the Excised Human Lens as a Function of Age*, FDA 785345 0090 RA, Washington, DC: US Food and Drug Administration.

Baron, R.A., Rea, M.S., and Daniels, S.G. (1992) Effects of indoor lighting (illuminance and spectral distribution) on the performance of cognitive tasks and interpersonal behaviors: the potential mediating role of positive affect, *Motivation Emotion*, 16, 1–33.

Barr, R. and Lawes, H. (1991) *Towards a Brighter Monsall: Street Lighting as a Factor in Community Safety – the Manchester Experience*, Manchester, UK: Manchester University.

Bartlett, F.C. (1953) Psychological criteria of fatigue, in W.F. Floyd and A.T. Welford (eds) *Symposium on Fatigue*, London: H.K. Lewis and Co.

Bartley, S.H. and Chute, E. (1947) *Fatigue and Impairment in Man*, New York: McGraw Hill.

Bartness, T.J. and Goldman, B.D. (1989) Mammalian pineal melatonin: a clock for all seasons, *Experientia*, 45, 939–45.

Batchelor, B.G., Hill, D.A., and Hodgson, D.C. (1985) *Automated Visual Inspection*, Bedford, UK and Amsterdam: IFS (Publications) Ltd. and North Holland.

Bates, B.T., Osternig, L.R., and James, S.L. (1977) Fatigue effects in running, *J. Motor Behav.*, 9, 203–7.

Bean, A.R. and Bell, R.I. (1992) The CSP Index: a practical measure of office lighting quality, *Light. Res. Technol.*, 24, 215–25.

Beckstead, J.W. and Boyce, P.R. (1992) Structural equation modeling in lighting research: application to residential acceptance of new lighting, *Light. Res. Technol.*, 24, 189–201.

Bedocs, L., Hart, A., Lynes, J.A., and Page, R.K. (1982) The return of the uplighter, *Proceedings of the CIBS National Lighting Conference*, Warwick, UK, London: Chartered Institution of Building Services.

Beersma, D.G.M., Spoelstra, K., and Daan, S. (1999) Accuracy of human circadian entrainment under natural light conditions: model simulations, *J. Biol. Rhythms*, 14, 523–31.

Begemann, S.H.A., Tenner, A., and Aarts, M. (1994) Daylight, artificial light and people, *Proceedings, IES Lighting Convention*, Sydney, Australia: Illuminating Engineering Societies of Australia.

Begemann, S.H.A., van den Beld, G.J., and Tenner, A.D. (1995) Daylight, artificial light and people, Part 2, *Proceedings of the CIE 23rd Session*, New Dehli, India, Vienna: CIE.

Begley, K. and Linderson, T. (1991) Management of mercury in lighting products, *Proceedings of the 1st European Conference on Energy-Efficient Lighting*, Stockholm, Sweden: Swedish National Board for Industrial and Technical Development.

Belcher, M.C. and Kluczny, R. (1987) A model for the possible effects of light on decision-making, *Light. Des. Appl.*, 17, 19–21; 54–5.

Bell, J.R. (1979) Fifteen residents die in mental hospital fire, *Fire J.*, July, 68–76.

Bellchambers, H.E. and Godby, A.C. (1972) Illumination, color rendering and visual clarity, *Light. Res. Technol.*, 4, 104–6.

Bennett, C.A. (1977) The demographic variables of discomfort glare, *Light. Des. Appl.*, 7, 22–4.

Bennett, T. and Wright, R. (1984) *Burglars on Burglary*, Hants, UK: Gower.

Bennett, C.A., Chitlangia, A., and Pengrekar, A. (1977) Illumination levels and performance of practical visual tasks, *Proceedings of the Human Factors Society, 21st Annual Meeting*, San Francisco, CA.

Bergman, R.S., Parham, T.G., and McGowan, T.K. (1995) UV emission from general lighting lamps, *J. Illumin. Eng. Soc.*, 24, 13–24.

Berman, S.M. (1992) Energy efficiency consequences of scotopic sensitivity, *J. Illumin. Eng. Soc.*, 21, 3–14.

Berman, S.M., Jewett, D.L., Fein, G., Saika, G., and Ashford, F. (1990) Photopic luminance does not always predict perceived room brightness, *Light. Res. Technol.*, 22, 37–42.

Berman, S.M., Greenhouse, D.A., Bailey, I.D., Clear, R.D., and Raasch, T.W. (1991) Human electroretinogram responses to video displays, fluorescent lighting and other high frequency sources, *Opt. Vis. Sci.*, 68, 645–62.

Berman, S.M., Fein, G., Jewett, D.L., Saika, G., and Ashford, F. (1992) Spectral determinants of steady-state pupil size with a full field of view, *J. Illumin. Eng. Soc.*, 21, 3–13.

Berman, S.M., Fein, G., Jewett, D.L., and Ashford, F. (1993) Luminance-controlled pupil size affects Landolt C task performance, *J. Illumin. Eng. Soc.*, 22, 150–65.

Berman, S.M., Fein, G., Jewett, D.L., and Ashford, F. (1994) Landolt-C recognition in elderly subjects is affected by scotopic intensity of surround illuminants, *J. Illumin. Eng. Soc.*, 23, 123–30.

Bernecker, C.A. and Mier, J.M. (1985) The effect of source luminance on the perception of environmental brightness, *J. Illumin. Eng. Soc.*, 14, 253–68.

Bertelson, P. and Joffe, R. (1963) Blocking in prolonged serial responding, *Ergonomics*, 6, 109–16.

Best, R.L. (1977) *Reconstruction of a Tragedy: the Beverly Hills Supper Club Fire, Southgate, KY*, May 28, NFPA No LS-2, Quincy, MA: National Fire Protection Association.

Beutell, A.W. (1934) An analytical basis for a lighting code, *Illumin. Eng. (London)*, 27, 5–11.

Bierman, A. and Conway, K.M. (2000) Characterizing daylight photosensor system performance to help overcome market barriers, *J. Illumin. Eng. Soc.*, 29, 101–15.

Billmeyer, F.W., Jr. (1987) Survey of color order systems, *Color Res. Appl.*, 12, 173–86.

Bills, A.G. (1931) A new principle of mental fatigue, *Am. J. Psychol.*, 43, 230.

Bjorset, H.H. and Frederiksen, E.A. (1979) A proposal for recommendations for the limitation of the contrast reduction in office lighting, *Proceedings of the CIE 19th Session*, Kyoto, Japan, Paris: CIE.

Bjorvatn, B., Kecklund, G., and Akerstedt, T. (1999) Bright light treatment used for adaptation to night work and re-adaptation back to day life. A field study at an oil platform in the North Sea, *J. Sleep Res.*, 8, 105–12.

Blackwell, H.R. (1946) Contrast thresholds of the human eye, *J. Opt. Soc. Am.*, 36, 624–43.

Blackwell, H.R. (1959) Development and use of a quantitative method for specification of interior illumination levels on the basis of performance data, *Illumin. Eng.*, 54, 317–53.

Blackwell, H.R. and Blackwell, O.M. (1971) Visual performance data for 156 normal observers of various ages, *J. Illumin. Eng. Soc.*, 1, 3–13.

Blackwell, H.R. and Blackwell, O.M. (1980) Population data for 140 normal 20–30 year olds for use in assessing some effects of lighting upon visual performance, *J. Illumin. Eng. Soc.*, 9, 158–74.

Blackwell, H.R. and Moldauer, A.B. (1958) *Detection Thresholds for Point Sources in the Near Periphery* (EPRI Project 2455) Ann Arbor, MI: Engineering Research Institute, University of Michigan.

Blackwell, H.R., Schwab, R.N., and Pritchard, B.S. (1964) Visibility and illumination variables in roadway visual tasks, *Illumin. Eng.*, 59, 277–308.

Blanks, J.C., Torigoe, Y., Hinton, D.R., and Blanks, R.H.I. (1991) Retinal degeneration in the macula of patients with Alzheimer's disease, *Ann. NY Acad. Sci.*, 640, 44–6.

Bloomfield, J.R. (1975a) Theoretical approaches to visual search, in C.G. Drury and J.G. Fox (eds) *Human Reliability in Quality Control*, London: Taylor and Francis.

Bloomfield, J.R. (1975b) Studies in visual search, in C.G. Drury and J.G. Fox (eds) *Human Reliability in Quality Control*, London: Taylor and Francis.

Bodmann, H.W. (1992) Elements of photometry, brightness and visibility, *Light. Res. Technol.*, 24, 29–42.

Bodmann, H.W. and La Toison, M. (1994) Predicted brightness – luminance phenomena, *Light. Res. Technol.*, 26, 135–43.

Bodmann, H.W., Sollner, G., and Voit, E. (1963) Evaluation of lighting level with various kinds of light, *Proceedings of the CIE, 15th Session*, Vienna, Paris: CIE.

Bodmann, H.W., Sollner, G., and Senger, E. (1966) A simple glare evaluation system, *Illumin. Eng.*, 61, 347–52.

Boettner, E.A. and Wolter, J.R. (1962) Transmission of the ocular media, *Invest. Ophthalmol.*, 1, 776–83.

Boff, K.R. and Lincoln, J.E. (1988) *Engineering Data Compendium: Human Perception and Performance*, Wright-Patterson AFB, OH: Harry G. Armstrong Aerospace Medical Research Laboratory.

Boivin, D.B. and James, F.O. (2002) Phase-dependent effect of room light exposure in a 5-h advance of the sleep-wake cycle: implications for jet lag, *J. Biol. Rhythms*, 17, 266–76.

Bornstein, M.H., Kessen, W., and Weiskopf, S. (1976) The categories of hue in infancy, *Science*, 191, 201.

Boulos, Z., Campbell, S.S., Lewy, A.J., Terman, M., Dijk, D.-J., and Eastman, C.I., (1995) Light treatment for sleep disorders: consensus report, VII. Jet lag, *J. Biol. Rhythms*, 10, 167–76.

Bourdy, C., Chiron, A., Cottin, F., and Monot, A. (1988) Visibility at a tunnel entrance: effect of temporal luminance variation, *Light. Res. Technol.*, 20, 199–200.

Boyce, P.R. (1973) Age, illuminance, visual performance and preference, *Light. Res. Technol.*, 5, 125–44.

Boyce, P.R. (1974) Illuminance, difficulty, complexity and visual performance, *Light. Res. Technol.*, 6, 222–6.

Boyce, P.R. (1977) Investigations of the subjective balance between illuminance and lamp colour properties, *Light. Res. Technol.*, 9, 11–24.

Boyce, P.R. (1978) Variability of contrast rendering factor in lighting installations, *Light. Res. Technol.*, 10, 94–105.

Boyce, P.R. (1979a) Users' attitudes to some types of local lighting, *Light. Res. Technol.*, 11, 158–64.

Boyce, P.R. (1979b) The effect of fence luminance on the detection of potential intruders, *Light. Res. Technol.*, 11, 78–84.

Boyce, P.R. (1980) Observations of the manual switching of lighting, *Light. Res. Technol.*, 12, 195–205.

Boyce, P.R. (1985) Movement under emergency lighting: the effect of illuminance, *Light. Res. Technol.*, 17, 51–71.

Boyce, P.R. (1986) Movement under emergency lighting: the effects of changeover from normal lighting, *Light. Res. Technol.*, 18, 1–18.

Boyce, P.R. (1991) Lighting and lighting conditions, in J.A.J. Roufs (ed.) *The Man–Machine Interface, Volume 15 of Visual and Visual Dysfunction*, London: Macmillan.

Boyce, P.R. (1996) Illuminance selection based on visual performance – and other fairy stories, *J. Illumin. Eng. Soc.*, 25, 41–9.

Boyce, P.R. (1998) Promoting energy-efficient lighting: the need for parallel processing, *Light Eng.*, 6, 1–6.

Boyce, P.R. and Bruno, L.D. (1999) An evaluation of high pressure sodium and metal halide light sources for parking lot lighting, *J. Illumin. Eng. Soc.*, 28, 16–32.

Boyce, P.R. and Cuttle, C. (1990) Effect of correlated colour temperature on the perception of interiors and color discrimination performance, *Light. Res. Technol.*, 22, 19–36.

Boyce, P.R. and Cuttle, C. (1998) Discussion of Veitch, J.A. and Newsham, G.R., Determinants of lighting quality 1: state of the science, *J. Illumin. Eng. Soc.*, 27, 92–106.

Boyce, P.R. and Eklund, N.H. (1997) *Evaluations of Four Solar 1000 Sulphur Lamp Installations*, Troy, NY: Lighting Research Center.

Boyce, P.R. and Eklund, N.H. (1998) Simple tools for evaluating lighting, *Proceedings of the CIBSE National Lighting Conference*, Lancaster, UK, London: CIBSE.

Boyce, P.R. and Mulder, M.M.C. (1995) Effective directional indictors for exit signs, *J. Illumin. Eng. Soc.*, 24, 64–72.

Boyce, P.R. and Rea, M.S. (1987) Plateau and escarpment: the shape of visual performance, *Proceedings of the CIE, 21st Session*, Venice, Vienna: CIE.

Boyce, P.R. and Rea, M.S. (1990) Security lighting: the effects of illuminance and light source on the capabilities of guards and intruders, *Light. Res. Technol.*, 22, 57–79.

Boyce, P.R. and Van Derlofske, J. (2002) *Pedestrian Crosswalk Safety: Evaluating In-Pavement Flashing Warning Lights*, Troy, NY: Lighting Research Center.

Boyce, P.R., Eklund, N., Mangum, S., Saalfield, C., and Tang, L., (1995) Minimum acceptable transmittance of glazing, *Light. Res. Technol.*, 27, 145–52.

Boyce, P.R., Beckstead, J.W., Eklund, N.H., Strobel, R.W., and Rea, M.S. (1997) Lighting the graveyard shift: the influence of a daylight-simulating skylight on the task performance and mood of night-shift workers, *Light. Res. Technol.*, 29, 105–42.

Boyce, P.R., Eklund, N.H., and Simpson, S.N. (2000a) Individual lighting control: task performance, mood and illuminance, *J. Illumin. Eng. Soc.*, 29, 131–42.

Boyce, P.R., Eklund, N.H., Hamilton, B.J., and Bruno, L.D. (2000b) Perceptions of safety at night in different lighting conditions, *Light. Res. Technol.*, 32, 79–91.

Boyce, P.R., Bierman, A., Carter, B., Hunter, C., Bullough, J., Figueiro, M., and Conway, K. (2000c) *The Color Identification of Traffic Signals*, Troy, NY: Lighting Research Center.

Boyce, P.R., Hunter, C.M., and Carter, C.B. (2002) Perceptions of full-spectrum, polarized lighting, *J. Illumin. Eng. Soc.*, 31, 119–35.

Boyce, P.R., Hunter, C.M., and Inclan, C. (2003a) Overhead glare and visual discomfort, *J. Illumin. Eng. Soc.*, 32, 73–88.

Boyce, P.R., Akashi, Y., Hunter, C.M., and Bullough, J.D. (2003b) The impact of spectral power distribution on the performance of an achromatic visual task, *Light. Res. Technol.*, 35, in press.

Boynton, R.M. (1987) Categorical color perception and color rendering of light sources, *Proceedings of the CIE, 21st Session*, Venice, Vienna: CIE.

Boynton, R.M. and Clarke, F.J.J. (1964) Sources of entoptic scatter in the human eye. *J. Opt. Soc. Am.*, 54, 110–19.

Boynton, R.M. and Gordon, J. (1965) Bezold–Brucke hue shift measured by a color naming technique, *J. Opt. Soc. Am.*, 55, 78–86.

Boynton, R. and Olson, C. (1987) Locating basic colors in the OSA space, *Color Res. Appl.*, 12, 94–105.

Boynton, R.M. and Purl, K.F. (1989) Categorical colour perception under low pressure sodium lighting with small amounts of added incandescent illumination, *Light. Res. Technol.*, 21, 23–7.

Brainard, G.C., Lewy, A.J., Menaker, M., Miller, L.S., Fredrickson, R.H., Weleber, R.G., Cassone, V., and Hudson, D. (1988) Dose response relationship between light irradiance and the suppression of melatonin in human volunteers, *Brain Res.*, 454, 212–18.

Brainard, G.C., Rollag, M.D., and Hanifin, J.P. (1997) Photic regulation of melatonin in humans: ocular and neural signal transduction, *J. Biol. Rhythms*, 12, 537–46.

Brainard, G.C., Kavet, R., and Kheifets, L.I. (1999) The relationship between electromagnetic field and light exposures to melatonin and breast cancer: a review of the relevant literature, *J. Pineal Res.*, 26, 65–100.

Brainard, G.C., Hanifin, J.P., Rollag, M.D., Greeson, J., Byrne, B., Glickman, G., Gerner, E., and Sanford, B. (2001a) Human melatonin regulation is not mediated by the three cone photopic visual system, *J. Clin. Endocrinol. Metab.*, 86, 433–6.

Brainard, G.C., Hanifin, J.P., Greeson, J.M., Byrne, B., Glickman, G., Gerner, E., and Rollag, M.D. (2001b) Action spectrum for melatonin regulation in humans: evidence for a novel circadian photoreceptor, *J. Neurosci.*, 21, 6405–12.

Brandston, H.M., Peterson, A.J., Jr., Simonson, E.K., and Boyce, P.R. (2000) A white-LED post-top luminaire for rural applications, *Proceedings of the Illuminating Engineering Society of North America Annual Conference*, Washington, DC, New York: IESNA.

British Standards Institution (BSI) (1998) *BS 5266-2: 1998 Emergency Lighting. Code of Practice for Electrical Low-Mounted Way Guidance Systems for Emergency Use*, London: BSI.

British Standards Institution (BSI) (1999) *BS 5266-6: 1999 Emergency Lighting. Code of Practice for Non-Electrical Low-Mounted Way Guidance Systems for Emergency Use, Photoluminescent Systems*, London: BSI.

Brock, M.A. (1991) Chronobiology and aging, *J. Am. Geriatr. Soc.*, 39, 74–91.

Bronson, F.H. (1995) Seasonal variation in human reproduction: environmental factors, *Q. Rev. Biol.*, 70, 141–64.

Brown, I.D. (1994) Driver fatigue, *Hum. Factors*, 36, 298–314.

Brown, I.D., Tickner, A.H., and Simmons, D.C. (1970) Effect of prolonged driving on overtaking criteria, *Ergonomics*, 13, 239–42.

Brown, J.L., Graham, C.H., Leibowitz, H., and Ranken, H.B. (1953) Luminance thresholds for the resolution of visual detail during dark adaptation, *J. Opt. Soc. Am.*, 43, 197–202.

Bryan, J.L. (1977) *Smoke as a Determinant of Human Behavior in Fire Situations*, Report from Fire Protection Curriculum, College of Engineering, University of Maryland.

Buchanan, T.L., Barker, K.N., Gibson, J.T., Jiang, B.C., and Pearson, R.E. (1991) Illumination and errors in dispensing, *Am. J. Hosp. Pharm.*, 48, 2137–45.

Buck, J.A., McGowan, T.K., and McNelis, J.F. (1975) Roadway visibility as a function of light source color, *J. Illumin. Eng. Soc.*, 5, 20–5.

Bullough, J.D. (2000) The blue-light hazard: a review, *J. Illumin. Eng. Soc.*, 29, 6–14.

Bullough, J. and Rea, M.S. (1996) Lighting for neonatal intensive care units: some critical information for design, *Light. Res. Technol.*, 28, 189–98.

Bullough, J. and Rea, M.S. (1997) A simple model of forward visibility for snow plow operators through snow and fog at night, *Transport. Res. Rec.*, 1585, 19–24.

Bullough, J. and Rea, M.S. (2000) Simulated driving performance and peripheral detection at mesopic and low photopic light levels, *Light. Res. Technol.*, 32, 194–8.

Bullough, J.D. and Rea, M.S. (2001) Driving in snow: effects of headlamp color at mesopic and photopic light levels, *Lighting Technology Developments for Automobiles*, Warrendale, PA: Society of Automotive Engineers.

Bullough, J.D., Boyce, P.R., Bierman, A., Conway, K.M., Huang, K., O'Rourke, C.P., Hunter, C.M., and Nakata, A. (2000) Response to simulated traffic signals using light emitting diode and incandescent sources, *Transport. Res. Rec.*, 1724, 39–46.

Bullough, J.D., Boyce, P.R., Bierman, A., Hunter, C.M., Conway, K.M., Nakata, A., and Figueiro, M.G. (2001a) Traffic signal luminance and visual discomfort at night, *Transport. Res. Rec.*, 1754, 42–7.

Bullough, J.D., Rea, M.S., Pysar, R.M., Nakhla, H.K., and Amsler, D.E. (2001b) Rear lighting configurations for winter maintenance vehicles, *Proceedings of the IESNA Annual Conference*, Ottawa, New York: IESNA.

Bunning, E. (1936) Die endogene Tagesrhythmik als Grundlage der photoperiodischen Reaktion, *Ber. Dtsch. Bot. Ges.*, 54, 590–607.

Bunning, E. and Stern, K. (1930) Uber die tagesperiodischen Bewegungen der Primarblatter von *Phaseolus multiflorus*. II. Die Bewegungen bei Thermokonstanz, *Ber. Dtsch. Bot. Ges.* 48, 227–52.

Burden, T. and Murphy, L. (1991) *Street Lighting, Community Safety and the Local Environment*, Leeds, UK: Leeds Polytechnic.

Burg, A. (1967) *The Relationship between Vision Test Scores and Driving Records: General Findings*, Report 67-24, Los Angeles, CA: Department of Engineering, University of California.

Bursill, A.E. (1958) The restriction of peripheral vision during exposure to hot and humid conditions, *Q. J. Exp. Psychol.*, 10, 113–29.

Buswell, G.T. (1937) *How Adults Read*, Supplementary Adult Monograph 45, Chicago, IL: University of Chicago.

Butler, D.L. and Biner, P.M. (1989) Effects of setting on window preferences and factors associated with those preferences, *Environ. Behav.*, 21, 17–31.

Cagnacci, A., Krauchi, K., Wirz-Justice, A., and Volpe, A. (1997a) Homeostatic versus circadian effects of melatonin on core body temperature in humans, *J. Biol. Rhythms*, 12, 509–17.

Cagnacci, A., Soldani, R., and Yen, S.S.C. (1997b) Contemporaneous melatonin administration modifies the circadian response to nocturnal bright light stimuli, *Am. J. Physiol.*, 272, R482–6.

Caird, J.K. and Hancock, P.A. (2002) Left turn and gap acceptance crashes, in R.E. Dewar and P.L. Olson (eds) *Human Factors in Traffic Safety*, Tucson, AZ: Lawyers and Judges Publishing Company.

Cairney, P. and Catchpole, J. (1996) Patterns of perceptual failures at intersections of arterial roads and local streets, in A.G. Gale, I.D. Brown, C.M. Haslegrave, and S.P. Taylor (eds) *Vision in Vehicles V*, Amsterdam: North-Holland.

Cajochen, C., Khalsa, S.B.S., Wyatt, J.K., Czeisler, C.A., and Dijk, D.-J. (1999) EEG and ocular correlates of circadian melatonin phase and human performance decrements during sleep loss, *Am. J. Physiol.*, 277, R640–9.

Cajochen, C., Zeitzer, J.M., Czeisler, C.A., and Dijk, D.-J. (2000) Dose–response relationship for light intensity and ocular and electroencephalographic correlates of human alertness, *Behav. Brain Res.*, 115, 75–83.

California Energy Commission (CEC) (1999a) *Lighting Efficiency Technology Report, Volume 1: California Baseline*, Sacramento, CA: CEC.

California Energy Commission (CEC) (1999b) *Lighting Efficiency Technology Report, Volume 4: Recommendations Report*, Sacramento, CA: CEC.

Cameron, C. (1974) A theory of fatigue, in A.T. Welford (ed.) *Man Under Stress*, London: Taylor and Francis.

Caminada, J.F. and van Bommel, W.J.M. (1980) New lighting considerations for residential areas, *Int. Light. Rev.*, 3, 69–75.

Campbell, S.S. (1995) Effects of timed bright-light exposure on shift work adaptation in middle-aged subjects, *Sleep*, 18, 408–16.

Campbell, S.S. and Dawson, D. (1991) Bright light treatment of sleep disturbance in older subjects, *Sleep Res.*, 20, 448.

Campbell, F.W. and Green, D.G. (1965) Optical and retinal factors affecting visual resolution, *J. Physiol.*, 181, 576–93.

Campbell, S.S., Kripke, D.F., Gillin, J.C., Hrubovcak, J.C. (1988) Exposure to light in healthy elderly subjects and Alzheimer's patients, *Physiol. Behav.*, 42, 141–4.

Campbell, M.K., Bush, T.L., and Hale, W.E. (1993a) Medical conditions associated with driving cessation in community dwelling, ambulatory elders, *J. Gerontol.*, 48, S230–4.

Campbell, S.S., Dawson, D., and Anderson, M.W. (1993b) Alleviation of sleep maintenance insomnia with timed exposure to bright light, *J. Am. Geriartr. Soc.*, 41, 829–36.

Campbell, S.S., Dijk, D.J., Boulos, Z., Eastman, C.I., Lewy, A.J., and Terman, M. (1995) Light treatment for sleep disorders: consensus report III. Alerting and activating effects, *J. Biol. Rhythms*, 10, 129–32.

Canter, D. (ed.) (1980) *Fires and Human Behaviour*, Chichester, UK: John Wiley and Sons.

Canter, D., Breaux, J., and Sime, J. (1980) Domestic, multiple occupancy and hospital fires, in D. Canter (ed.) *Fires and Human Behaviour*, Chichester, UK: John Wiley and Sons.

Carlton, J.W. (1982) Effective use of lighting, in D.C. Pritchard (ed.) *Developments in Lighting – 2*, London: Applied Science Publishers.

Carmichael, L. and Dearborn, W.F. (1947) *Reading and Visual Fatigue*, Westport, CT: Greenwood Press.

Carnevale, A.P. and Rose, S.J. (1998) *Education for What? The New Office Economy*, Princeton, NJ: Educational Testing Service.

Carter, D.J., Slater, A.I., and Moore, T. (1999) A study of occupier controlled lighting system, *Proceedings of the CIE, 24th Session, Warsaw*, Vienna: CIE.

Cesarini, J.-P. (1998) UV skin aging, in R. Matthes and D. Sliney (eds) *Measurements of Optical Radiation Hazards*, Oberschleißheim, Germany: International Commission on Non-Ionizing Radiation Protection.

Chandler, D. (1949) *The Rise of the Gas Industry in Britain*, London: Kelly and Kelly.

Chartered Institution of Building Services Engineers (CIBSE) (1979) *Lighting Guide: Shipbuilding and Ship Repair*, London: CIBSE.

Chartered Institution of Building Services Engineers (CIBSE) (1983) *Application Guide: Lighting for Hostile and Hazardous Environments*, London: CIBSE.

Chartered Institution of Building Services Engineers (CIBSE) (1985) *Technical Memorandum 10: The Calculation of Glare Indices*, London: CIBSE.

Chartered Institution of Building Services Engineers (CIBSE) (1992) *Lighting Guide: The Outdoor Environment*, London: CIBSE.

Chartered Institution of Building Services Engineers (CIBSE) (1994) *Code for Interior Lighting*, London: CIBSE.

Chartered Institution of Building Services (CIBSE) (1996) *Lighting Guide LG3: The Visual Environment for Display Screen Use*, London: CIBSE.

Chartered Institution of Building Services Engineers (CIBSE) (1998) *Environmental Considerations for Exterior Lighting*, Lighting Division Factfile, No. 7, London: CIBSE.

Chartered Institution of Building Services Engineers (CIBSE) (1999a) Environmental factors affecting office worker performance: a review of evidence, *CIBSE Technical Memorandum TM24*, London: CIBSE.

Chartered Institution of Building Services Engineers (CIBSE) (1999b) *Daylighting and Window Design, CIBSE Lighting Guide LG10*, London: CIBSE.

Chartered Institution of Building Services Engineers (CIBSE) (2002a) *Code for Lighting*, London, CIBSE.

Chartered Institution of Building Services Engineers (CIBSE) (2002b) *Lighting Guide 1: The Industrial Environment*, London: CIBSE.

Chesterfield, B.P., Rasmussen, P.G., and Dillon, R.D. (1981) *Emergency Cabin Lighting Installations: an Analysis of Ceiling vs. Lower Cabin Mounted Lighting during Evacuation Trials*, Federal Aviation Administration, Report FAA-AM-81-7, Washington, DC: US Department of Transportation.

Chittum, C.B. and Rasmussen, P.G. (1989) An evaluation of several light sources in a smoke filled environment, paper presented to the *Society of Automotive Engineers A20C Committee, Nashville, TN*, Warrendale, PA: Society of Automotive Engineers.

Chylack, L.T. (2000) Age-related cataract, in B. Silverstone, M.A. Lang, B.P. Rosenthal, and E.E. Faye (eds) *The Lighthouse Handbook on Vision Impairment and Vision Rehabilitation*, New York: Oxford University Press.

Clarke, R.V. (1995) Situational crime prevention, in M. Tonry and D.P. Farrington (eds) *Building a Safer Society: Strategic Approaches to Crime Prevention*, Chicago, IL: University of Chicago Press.

Clear, R. and Berman, S. (1990) Speed, accuracy and VL, *J. Illumin. Eng. Soc.*, 19, 124–31.

Clear, R. and Berman, S. (1994) Environmental and health aspects of lighting: Mercury, *J. Illumin. Eng. Soc.*, 23, 138–56.

Coaton, J.R. and Marsden, A.M. (1996) *Lamps and Lighting*, 4th Edition, London: Butterworth-Heinemann.

Cobb, J. (1989) *Roadside Survey of Vehicle Lighting 1989*, Research Report 290, Crowthorne, UK: Transport and Road Research Laboratory.

Cockram, A.H., Collins, J.B., and Langdon, F.J. (1970) A study of user preference for fluorescent lamp colours for daytime and night-time lighting, *Light. Res. Technol.*, 2, 249–56.

Cogan, D. (1974) Lighting and health hazards, in *NIOSH Symposium, The Occupational Safety and Health Effects Associated with Reduced Levels of Illumination*, Washington, DC: US Department of Health, Education and Welfare.

Cole, B.L. and Brown, B. (1966) Optimum intensity of red road-traffic signal lights for normal and protanopic observers, *J. Opt. Soc. Am.*, 56, 516–22.

Cole, R.J. and Kripke, D.F. (1989) Amelioration of jet lag by bright light treatment: effects on sleep consolidation, *Sleep Res.*, 18, 411.

Collins, W.M. (1962) The determination of the minimum identifiable glare sensation interval, *Trans. Illumin. Eng. Soc. (London)*, 27, 27–34.

Collins, B.L. (1991) Visibility of exit signs and directional indicators, *J. Illumin. Eng. Soc.*, 20, 117–33.

Collins, J.J. and Hall, R.D. (1992) Legibility and readability of light reflecting matrix variable message road signs, *Light. Res. Technol.*, 24, 143–8.

Collins, B.L. and Lerner, N.D. (1983) *An Evaluation of Exit Symbol Visibility*, Washington, DC: US Department of Commerce.

Collins, B.L. and Worthey, J.A. (1985) Lighting for meat and poultry inspection, *J. Illumin. Eng. Soc.*, 15, 21–8.

Collins, B.L., Kuo, B.Y., Mayerson, S.E., Worthey, J.A., and Howett, G.L. (1986) *Safety Color Appearance under Selected Light Sources*, NBS IR86-3403, Washington, DC: US Department of Commerce.

Collins, B.L., Dahir, M.S., and Madrzykowski, D. (1990) *Evaluation of Exit Signs in Clear and Smoke Conditions*, NISTIR 4399, Washington, DC: US Department of Commerce.

Colman, R.S., Frankel, F., Ritvo, E., and Freeman, B.J. (1976) The effects of fluorescent and incandescent illumination upon repetitive behaviors in autistic children, *J. Autism Child. Schizophr.*, 6, 157–62.

Colombo, E.M., Kirschbaum, C.F., and Raitelli, M. (1987) Legibility of texts: the influence of blur, *Light. Res. Technol.*, 19, 61–71.

Comerford, J.P. and Kaiser, P.K. (1975) Luminous efficiency functions determined by heterochromatic brightness matching, *J. Opt. Soc. Am.*, 64, 466–8.

Commission Internationale de l'Eclairage (CIE) (1972) *A Unified Framework of Methods for EvaluatingVisual Performance Aspects of Lighting*, CIE Publication 19, Paris: CIE.

Commission Internationale de l'Eclairage (CIE) (1973) *Standardization of Luminance Distribution of Clear Skies*, CIE Publication 22, Vienna: CIE.

Commission Internationale de l'Eclairage (CIE) (1978) *Light as a True Visual Quantity*, CIE Publication 41, Vienna: CIE.

Commission Internationale de l'Eclairage (1979) *Road Lighting for Wet Conditions*, CIE Publication 47, Paris: CIE.

Commission Internationale de l'Eclairage (1981) *An Analytic Model for Describing the Influence of Lighting Parameters upon Visual Performance*, CIE Publication 19/2, Paris: CIE.

Commission Internationale de l'Eclairage (CIE) (1983a) *The Basis of Physical Photometry*, CIE Publication 18.2, Vienna: CIE.

Commission Internationale de l'Eclairage (CIE) (1983b) *Discomfort Glare in the Interior Working Environment*, CIE Publication 55, Paris: CIE.

Commission Internationale de l'Eclairage (CIE) (1986) *Colorimetry*, CIE Publication 15.2, Vienna: CIE.

Commission Internationale de l'Eclairage (CIE) (1989) *Mesopic Photometry: History, Special Problems and Practical Solutions*, CIE Publication No. 81, Vienna: CIE.

Commission Internationale de l'Eclairage (CIE) (1990a) *CIE 1988 2° Spectral Luminous Efficiency Function for Photopic Vision*, CIE Publication No. 86, Vienna: CIE.

Commission Internationale de l'Eclairage (CIE) (1990b) *Guide for the Lighting of Road Tunnels and Underpasses*, CIE Publication No. 88, Vienna: CIE.

Commission Internationale de l'Eclairage (CIE) (1992a) *Fundamentals of the Visual Task of Night Driving*, CIE Publication 100, Vienna: CIE.

Commission Internationale de l'Eclairage (CIE) (1992b) *Road Lighting as an Accident Countermeasure*, CIE Technical Report 93, Vienna: CIE.

Commission Internationale de l'Eclairage (CIE) (1994) *Review of the Official Recommendations of the CIE for the Colors of Signal Lights*, CIE Technical Report 107, Vienna: CIE.

Commission Internationale de l'Eclairage (CIE) (1995a) *Method of Measuring and Specifying Color Rendering Properties of Light Sources*, CIE Publication 13.3, Vienna: CIE.

Commission Internationale de l'Eclairage (CIE) (1995b) *Discomfort Glare in Interior Lighting*, CIE Publication 117, Vienna: CIE.

Commission Internationale de l'Eclairage (CIE) (1995c) *Recommendations for the Lighting of Roads for Motor and Pedestrian Traffic*, CIE Technical Report 115, Vienna: CIE.

Commission Internationale de l'Eclairage (CIE) (1997a) *Low Vision: Lighting Needs for the Partially Sighted*, CIE Technical Report 123, Vienna: CIE.

Commission Internationale de l'Eclairage (CIE) (1997b) *Guidelines for Minimizing Sky Glow*, CIE Publication 126-1997, Vienna: CIE.

Commission Internationale de l'Eclairage (CIE) (1998) *The CIE 1997 Interim Colour Appearance Model CIECAM97*, CIE Publication 131-1998, Vienna: CIE.

Commission Internationale de l'Eclairage (CIE) (2002a) *The Correlation of Models of Vision and Visual Performance*, CIE Publication 145-2002, Vienna: CIE.

Commission Internationale de l'Eclairage (CIE) (2002b) *CIE Collection on Glare*, CIE Publication 146-2002, Vienna: CIE.

Coohill, T.P. (1998) Photobiological action spectra – what do they mean, in R. Matthes and D. Sliney (eds) *Measurements of Optical Radiation Hazards*, Oberschleißheim, Germany: International Commission on Non-Ionizing Radiation Protection.

Cook, G.K., Wright, M.S., Webber, G.M.B., and Bright, K.T. (1999) Emergency lighting and wayfinding provision systems for visually impaired people: phase 2 of a study, *Light. Res. Technol.*, 31, 43–8.

Cooper, J.R., Wiltshire, T., and Warren, B. (1973) Glazing considerations for buildings, *Glass in Buildings*.

Copinschi, G. and van Cauter, E. (1995) Effects of ageing on modulation of hormonal secretions by sleep and circadian rhythmicity, *Horm. Res.*, 43, 20–4.

Cornelissen, F.W., Bootsma, A., and Kooijman, A.C. (1995) Object perception by visually impaired people at different light levels, *Vision Res.*, 35, 161–8.

Corso, J.F. (1967) *The Experimental Psychology of Sensory Behavior*, New York: Holt, Rinehart and Winston.

Craig, A. and Cooper, R.E. (1992) Symptoms of acute and chronic fatigue, *Handbook of Human Performance*, New York: Academic Press.

Crawford, B.H. (1949) The scotopic visibility function, *Phys. Soc. Proc.*, 62, 321.

Crawford, B.H. (1972) The Stiles–Crawford effects and their significance in vision, in D. Jameson and L.M. Hurvich (eds) *Handbook of Sensory Physiology, Vol. VII/4 Visual Psychophysics*, Berlin: Springer-Verlag.

Cridland, W. (1995) *The Impact of Street Lighting Improvements on Crime, Fear of Crime and Quality of Life. The Larkhill Estate, Stockport*, Billericay, UK: Personal and Professional Management Services.

Crisp, V.H.C. (1978) The light switch in buildings, *Light. Res. Technol.*, 10, 69–82.

Crisp, V.H.C. and Henderson, G. (1982) The energy management of artificial lighting use, *Light. Res. Technol.*, 14, 193–206.

Croft, T.A. (1971) Failure of visual estimation of motion under strobe, *Nature*, 231, 397.

Cullen, A.P. (1998) The lens – ultraviolet and infrared action spectra for cataract, acute in vivo studies, in R. Matthes and D. Sliney (eds) *Measurements of Optical Radiation Hazards*, Oberschleißheim, Germany: International Commission on Non-Ionizing Radiation Protection.

Cuttle, C. (1979) Subjective assessments of the appearance of special performance glazing in offices, *Light. Res. Technol.*, 11, 140–9.

Cuttle, C. (1983) People and windows in workplaces, *Proceedings of the People and Physical Environment Research Conference*, Wellington, New Zealand.

Cuttle, C. (2003) *Lighting by Design*, London: Architectural Press.

Czeisler, C.A. and Allen, J.S. (1987) Acute circadian phase reversal in man via bright light exposure: application to jet-lag, *Sleep Res.*, 16, 605.

Czeisler, C.A., Richardson, G.S., Zimmerman, J.C., Moore-Ede, M.C., and Weitzman, E.D. (1981) Entrainment of human circadian rhythms: a reassessment, *Photochem. Photobiol.* 34, 239–47.

Czeisler, C.A., Rios, C.D., Sanchez, R., Brown, E.N., Richardson, G.S., Ronda, J.M., and Rogacz, S. (1986) Phase advance and reduction in amplitude of the endogenous circadian oscillator correspond with systematic changes in sleep/wake habits and daytime functioning in the elderly, *Sleep Res.*, 15, 268.

Czeisler, C.A., Kronauer, R.E., Johnson, M.P., Allen, J.S., and Dumont, M. (1988) Action of light on the human circadian pacemaker: treatment of patients with circadian rhythm sleep disorders, in J. Horn (ed.) *Sleep '88.* Stuttgart, Germany: Verlag.

Czeisler, C.A., Johnson, M.P., Duffy, J.F., Brown, E.N., Ronda, J.M., and Kronauer, R.E. (1990) Exposure to bright light and darkness to treat physiologic maladaptation to night work, *N. Engl. J. Med.*, 322, 1253–9.

Czeisler, C.A., Chiasera, A.J., and Duffy, J.F. (1991) Research on sleep, circadian rhythms and aging: application to manned spaceflight, *Environ. Gerontol.*, 26, 217–32.

Czeisler, C.A., Shanahan, T.L., Klerman, E.B., Martens, H., Brotman, D.J., Emens, J.S., Klein, T., and Rizzo, J.F., III, (1995) Suppression of melatonin secretion in some blind patients by exposure to bright light, *N. Engl. J. Med.*, 332, 6–11.

Davidson, P.A. (1978) The role of drivers vision in road safety, *Light. Res. Technol.*, 10, 125–39.

Davidson, N. and Goodey, J. (1991) *Street Lighting and Crime: The Hull Project*, Hull, UK: University of Hull.

Davis, R.G. and Ginthner, D.N. (1990) Correlated color temperature, illuminance level and the Kruithof curve, *J. Illumin. Eng. Soc.*, 19, 27–38.

Dawson, D. and Campbell, S.S. (1990) Bright light treatment: are we keeping our subjects in the dark, *Sleep*, 13, 267–71.

De Boer, J.B. (1951) Fundamental experiments on visibility and admissible glare in road lighting, *Proceedings of the CIE, 12th Session, Stockholm*, Paris: CIE.

De Boer, J.B. (1967) *Public Lighting*, Eindhoven, The Netherlands: Philips Technical Library.

De Boer, J.B. (1974) Modern light sources for highways, *J. Illumin. Eng. Soc.*, 3, 142–52.

De Boer, J.B. (1977) Performance and comfort in the presence of veiling reflections, *Light. Res. Technol.*, 9, 169–76.

De Boer, J.B., Burghout, F., and van Heemskerk-Veekens, F.T. (1960) Appraisal of quality of public lighting based on road surface luminance and glare, *Proceedings of the CIE, 14th Session, Brussels*, Paris: CIE.

De Waard, P.W.T., IJspeert, J.K., van den Berg, T.J.T.P., and de Jong, P.T.V.M. (1992) Intraocular light scattering in age-related cataracts, *Invest. Ophthalmol. Vision Sci.*, 33, 618–25.

Desimone, R. (1991) Face-selective cells in the temporal cortex of monkeys, *J. Cogn. Neurosci.*, 3, 1–8.

Devaney, K.O. and Johnson, H.A. (1980) Neuron loss in the ageing visual cortex of man, *J. Gerontol.*, 35, 836–41.

Diaper, G. (1990) The Hawthorne Effect: a fresh examination, *Educ. Stud.*, 16, 261–7.

Dijk, D-J., Boulos, Z., Eastman, C.I., Lewy, A.J., Campbell, S.S., and Terman, M. (1995) Light treatment for sleep disorders: consensus report II. Basic properties of circadian physiology and sleep regulation, *J. Biol. Rhythms*, 10, 113–25.

Ditton, J. and Nair, G. (1994) Throwing light on crime: a case study of the relationship between street lighting and crime prevention, *Security J.*, 5, 125–32.

Domey, R.G. and McFarland, R.A. (1961) Dark adaptation as a function of age: individual prediction, *Am. J. Ophthalmol.*, 51, 1262–8.

Donovan, R.J. and Rossiter, J.R. (1982) Store atmosphere: an environmental psychology approach, *J. Retail.*, 58, 34–57.

Drasdo, N. (1977) The neural representation of visual space, *Nature*, 266, 554–6.

Drew, G.C. (1940) An experimental study of mental fatigue, *Flying Personnel Research Committee Report 227*, London: His Majesty's Stationery Office.

Drury, C.G. (1975) Inspection of sheet material – model and data, *Hum. Factors*, 17, 257–65.

Duke-Elder, W.S. (1944) *Textbook of Ophthalmology, Vol 1.*, St Louis, M.: C.V. Mosby & Co.

Dunbar, C. (1938) Necessary values of brightness contrast in artificially lighted streets, *Trans. Illumin. Eng. Soc. (London)*, 3, 187–95.

Dureman, E.I. and Boden, C. (1972) Fatigue in simulated car driving, *Ergonomics*, 15, 299–308.

Eastman, C.I. (1990) Circadian rhythms and bright light recommendations for shift work, *Work Stress*, 4, 245–60.

Eastman, A.A. and McNelis, J.F. (1963) An evaluation of sodium, mercury and filament lighting for roadways, *Illumin. Eng.*, 58, 28–34.

Eastman, C.I., Stewart, K.T., Mahoney, M.P., Liu, L., and Fogg, L.F. (1994) Dark goggles and bright light improve circadian rhythm adaptation to night shift work, *Sleep*, 17, 535–43.

Eastman, C.I., Boulos, Z., Terman, M., Campbell, S.S., Dijk, D.J., and Lewy, A.J. (1995) Light treatment for sleep disorders: consensus report VI. Shift work, *J. Biol. Rhythms*, 10, 157–64.

Einhorn, H.D. (1991) Discomfort glare from small and large sources, *Proceedings of the First International Symposium on Glare*, New York: Lighting Research Institute.

Eklund, N.H. (1999) Exit sign recognition for color normal and color deficient observers, *J. Illumin. Eng. Soc.*, 28, 71–81.

Eklund, N.H. and Boyce, P.R. (1996) The development of a reliable, valid, and simple office lighting survey, *J. Illumin. Eng. Soc.*, 25, 25–40.

Eklund, N.H., Boyce, P.R., and Simpson, S.N. (2000) Lighting and sustained performance, *J. Illumin. Eng. Soc.*, 29, 116–30.

Eklund, N.H., Boyce, P.R., and Simpson, S.N. (2001) Lighting and sustained performance: modeling data-entry task performance, *J. Illumin. Eng. Soc.*, 30, 126–41.

Eldred, K.B. (1992) Optimal illumination for reading in patients with age-related maculopathy, *Optometry Vision Sci.*, 69, 46–50.

Eloholma, M., Halonen, L., and Setala, K. (1999) The effects of light spectrum on visual acuity in mesopic lighting levels, *Proceedings: Vision at Low Light Levels, EPRI/LRO Fourth International Lighting Research Symposium*, Palo Alto, CA: Electric Power Research Institute.

Elton, P.M. (1920) A study of output in silk weaving during winter months, *Industrial Fatigue Research Board Report No. 9*, London: His Majesty's Stationery Office.

Elvik, R. (1995) Meta-analysis of evaluations of public lighting as accident countermeasure, *Transport. Res. Rec.*, 1485, 112–23.

Engel, F.L. (1971) Visual conspicuity, directed attention and retinal locus, *Vision Res.*, 11, 563–76.

Engel, F.L. (1977) Visual conspicuity, visual search, and fixation tendencies of the eye, *Vision Res.*, 17, 95–108.

Enoch, J.M., Rynders, M., Lakshminarayanan, V., Vilar, E.Y., Giraldez-Fernandez, M.J., Grosvenor, T., Knowles, R., and Srinivasan, R. (1995) Two vision response functions which vary very little with age, in W. Adrian (ed.) *Lighting for Aging Vision and Health*, New York: Lighting Research Institute.

Environmental Protection Agency (EPA) (1997) *Mercury Study Report to Congress*, Washington, DC: EPA.

Environmental Protection Agency (EPA) (2001) *Criteria for ENERGY STAR labeled exit signs*, from www.epa.gov/nrgystar/purchasing/6j_exit signs.html#specs_exitsigns, accessed March 12, 2001.

Epstein, J.H. (1989) Photomedicene, in K.C. Smith (ed.) *The Science of Photobiology*, New York: Plenum.

European Committee for Standardization (CEN) (1996) *Lighting Applications – Lighting of Work Places*, prEN 12464, Brussels: CEN.

European Committee for Standardization (CEN) (1998) *European Standard: Traffic Control Equipment – Signal Heads (Draft)*, prEN 12368, Brussels: CEN.

European Committee for Standardization (CEN) (1999) *EN 1838 Lighting Applications – Emergency Lighting*, Brussels: CEN.

Farr, P.M. and Diffey, B.L. (1985) The erythemal response of human skin to ultraviolet radiation, *Br. J. Dermatol.*, 113, 65–76.

Faulkner, T.W. and Murphy, T.J. (1973) Lighting for difficult visual tasks, *Hum. Factors*, 15, 149–62.

Feeny-Burns, L., Burns, R.P., and Gao, C.L. (1990) Age-related macular changes in humans over 90 years old, *Am. J. Ophthalmol.*, 109, 265–78.

Fechner, G.T. (1860) *Elemente der Psychophysik*, translated by H.E. Alder, *Elements of Psychophysics*, New York: Holt.

Federal Highways Administration (FHA) (2000) *Manual of Uniform Traffic Control Devices*, Washington, DC: US Department of Transportation.

Fenton, D.M. and Penney, R. (1985) The effects of fluorescent and incandescent lighting on the repetitive behaviours of autistic and intellectually handicapped children, *Aust. NZ J. Dev. Disabil.*, 11, 137–41.

Ferguson, S.A., Preusser, D.F., Lund, A.K., Zador, P.L., and Ulmer, R.G. (1995) Daylight saving time and motor vehicle crashes: The reduction in pedestrian and vehicle occupant fatalities, *Am. J. Public Health*, 85, 92–5.

Ferris, F.L. (1993) Diabetic retinopathy, *Diabetes Care*, 16, (Supplement 1), 322–5.

Figueiro, M.G. (2001) *Lighting the Way: A Key to Independence*, Washington, DC and Troy, NY: AARP Andrus Foundation and the Lighting Research Center.

Fischer, D. (1972) The European glare limiting method, *Light. Res. Technol.*, 4, 97–100.

Fischer, D. (1991) Discomfort glare in interiors, *Proceedings of the First International Symposium on Glare*, New York: Lighting Research Institute.

Fisher, A.J. and Hall, R.R. (1976) Road luminances based on detection of change of visual angle, *Light. Res. Technol.*, 8, 187–94.

Fisher, B.S. and Nasar, J.L. (1992) Fear of crime in relation to three exterior site features; prospect, refuge and escape, *Environ. Behav.*, 24, 35–65.

Flannagan, M.J. (1999) *Subjective and Objective Aspects of Headlamp Glare: Effects of Size and Spectral Power Distribution*, Report UMTRI-99-36, Ann Arbor, MI: University of Michigan.

Flannagan, M.J., Sivak, M., Gellatly, A.W., and Louma, J. (1992) *A Field Study of Discomfort Glare from High Intensity Discharge Headlamps*, Report UMTRI-92-16, Ann Arbor, MI: University of Michigan.

Floyd, W.F. and Welford, A.T. (1953) *Fatigue*, London: H.K. Lewis & Co. Ltd.

Flynn, J.E. (1977) A study of subjective responses to low energy and non-uniform lighting systems, *Light. Des. Appl.*, 7, 6–15.

Flynn, J.E., Hendrick, C., Spencer, T.J., and Martyniuk, O. (1973) Interim study of procedures for investigating the effect of light on impression and behavior, *J. Illumin. Eng. Soc.*, 3, 87–94.

Flynn, J.E., Spencer, T.J., Martyniuk, O., and Hendrick, C. (1975) The effect of light on human judgement and behavior, *IERI Project 92, Interim Report to the Illuminating Engineering Research Institute*, New York: IERI.

Flynn, J.E., Hendrick, C., Spencer, T.J., and Martyniuk, O. (1979) A guide to methodology procedures for measuring subjective impressions in lighting, *J. Illumin. Eng. Soc.*, 8, 95–110.

Foley, D.J., Monjan, A.A., Brown, S.L., Simonsick, E.M., Wallace, R.B., and Blazer, D.G. (1995) Sleep complaints among elderly persons: an epidemiologic study of three communities, *Sleep*, 18, 425–32.

Folkard, S. and Monk, T.H. (1979) Shiftwork and performance, *Hum. Factors*, 21, 483–92.

Forbes, T.W. (1972) Visibility and legibility of highway signs, in T.W. Forbes (ed.) *Human Factors in Highway Safety Traffic Research*, New York: Wiley Interscience.

Forestier, S. (1998) Sunscreens, in-vivo versus in-vitro testing: pros and cons, in R. Matthes and D. Sliney (eds) *Measurements of Optical Radiation Hazards*, Oberschleißheim, Germany: International Commission on Non-Ionizing Radiation Protection.

Forger, D.B., Jewett, M.E., and Kronauer, R.E. (1999) A simpler model of the human circadian clock, *J. Biol. Rhythms*, 14, 532–7.

Foster, R.G., Provencio, I., Hudson, D., Fiske, S., De Grip, W., and Menaker, M. (1991) Circadian photoreception in the retinally degenerate mouse (rd/rd), *J. Comp. Physiol. A*, 169, 39–50.

Fotios, S.A. (2001) Lamp color properties and apparent brightness: a review, *Light. Res. Technol.*, 33, 163–81.

Fotios, S.A. and Levermore, G.J. (1997) The perception of electric light sources of different color properties, *Light. Res. Technol.*, 29, 161–71.

Fotios, S.A. and Levermore, G.J. (1998a) Chromatic effect on apparent brightness in interior spaces I: introduction and color gamut models, *Light. Res. Technol.*, 30, 97–102.

Fotios, S.A. and Levermore, G.J. (1998b) Chromatic effect on apparent brightness in interior spaces III: chromatic brightness models, *Light. Res. Technol.*, 30, 97–102.

Fotios, S.A. and Levermore, G.J. (1998c) Chromatic effect on apparent brightness in interior spaces II: SWS lumens model, *Light. Res. Technol.*, 30, 97–102.

Frederiksen, E. and Sorensen, K. (1976) Reflection classification of dry and wet road surfaces, *Light. Res. Technol.*, 8, 175–86.

Freeman, R.G., Hudson, H.T., and Carnes, R. (1970) Ultra-violet wavelength factors in solar radiation and skin cancer, *Int. J. Dermatol.*, 9, 232–5.

French, J., Hannon, P., and Brainard, G.C. (1990) Effects of bright illuminance on body temperature and human performance, *Annu. Rev. Chronopharmacol.*, 7, 37–40.

Froberg, J. (1985) Sleep deprivation and prolonged working hours, in S. Folkard and T.H. Monk (eds) *Hours of Work*, New York: John Wiley and Sons.

Fry, G.A. (1974) Ocular discomfort and other symptoms of eyestrain at low levels of illumination, *NIOSH Symposium: The Occupational Safety and Health Effects Associated with Reduced Levels of Illumination*, Cincinnati, OH: National Institute of Occupational Safety and Health.

Fry, G.A. and King, V.M. (1975) The pupillary response and discomfort glare, *J. Illumin. Eng. Soc.*, 4, 307–24.

Gabor, T. (1990) Crime displacement and situational prevention: toward the development of some principles, *Can. J. Criminol.*, 32, 41–74.

Gallagher, V.P., Janoff, M.S., and Farber, E. (1974) Interaction between fixed and vehicular illumination systems on city streets, *J. Illumin. Eng. Soc.*, 4, 3–10.

Gallin, P.F., Terman, M., Reme, C.E., Rafferty, B., Terman, J.S., and Burde, E.M. (1995) Ophthalmologic examination of patients with seasonal affective disorder, before and after light therapy, *Am. J. Ophthalmol.*, 119, 202–10.

Garlick, G.F.J. (1949) *Luminescent Materials*, Oxford: Clarenden.

Garstang, R.H. (1986) Model for artificial night sky illumination, *Publ. Astron. Soc. Pacific.*, 98, 364–75.

Geeraets, W.J. and Nooney, T.W. (1973) Observations following high intensity white light exposure to the retina, *Am. J. Optom. Arch. Am. Acad. Optom.*, 50, 405–15.

Geyer, T.A.W., Bellamy, L.J., Max-Lino, P.I., Bahraini, Z., and Modha, B. (1988) An evaluation of the effectiveness of the components of informative fire warning systems, in J. Sime (ed.) *Safety in the Built Environment*, London: E. & F.N. Spon.

Gibson, K.S. and Tyndell, E.P.T. (1923) Visibility of radiant energy, *Bull. Bureau Standards*, 19: 131–91.

Gifford, R. (1988) Light, decor, arousal, comfort and communication, *J. Environ. Psychol.*, 8, 177–89.

Gilmore, G.C. and Whitehouse, P.J. (1995) Contrast sensitivity in Alzheimer's disease: a 1-year longitudinal analysis, *Optometry Vision Sci.*, 72, 83–91.

Gilmore, G.C., Thomas, C.W., Klitz, T., Persanyi, M.W., and Tomsak, R. (1996) Contrast enhancement eliminates letter identification speed deficits in Alzheimer's disease, *J. Clin. Geropsychol.*, 2, 307–20.

Girasole, T., Roze, C., Maheu, B., Grehan, G., and Menard, J. (1998) Visibility distances in a foggy atmosphere: comparison between lighting installations by Monte Carlo simulation, *Light. Res. Technol.*, 30, 29–36.

Glasgow Crime Survey Team (1991) *Street Lighting and Crime: The Strathclyde Twin Site Study*, Glasgow, UK: Criminology Research Unit, Glasgow University.

Gorn, G.J. (1982) The effects of music in advertising on choice behavior: a classical conditioning approach, *J. Mktg.*, 46, 94–101.

Graeber, R.C., Sing, H.C., and Cuthbert, B.N. (1981) The impact of transmeridian flight on deploying soldiers, in L.C. Johnson, D.I. Tepas, W.P. Colquhoun, and M.J. Colligan (eds) *The Twenty-four Hour Workday: Proceedings of a Symposium on Variations in Work–Sleep Schedules*, Cincinnati, OH: US Department of Health and Human Services (NIOSH).

Graham, C.G., Cook, M.R., Gerkovich, M.M., and Sastre, A. (2001) Examination of the melatonin hypothesis in women exposed at night to EMF and bright light, *Environ. Health Perspect.*, 109, 501–7.

Grandjean, E. (1969) *Fitting the Task to the Man: An Ergonomic Approach*, London: Taylor and Francis.

Green, J. and Hargroves, R.A. (1979) A mobile laboratory for dynamic road lighting measurement, *Light. Res. Technol.*, 11, 197–203.

Griswold, D.B. (1984) Crime prevention and commercial burglary: a time series analysis, *J. Criminal Justice*, 7, 493–501.

Groos, G.A. and Mason, R. (1980) The visual properties of rat and cat suprachiasmatic neurones, *J. Comp. Physiol. A*, 135, 349–56.

Groos, G.A. and Meijer, J.A. (1985) Effects of illumination on suprachiasmatic nucleus electrical discharge, in R.J. Wurtman, M.J. Baum, and J.T. Potts, Jr. (eds) *The Medical and Biological Effects of Light*, New York: New York Academy of Sciences.

Gross, H.G. (1986) Koch emergency egress lighting systems for adverse optical conditions for military and commercial aircraft and other applications, *Proceedings of the System for Automated Flight Efficiency 24th Annual Symposium*, San Antonio, TX.

Gross, H.G. (1988) Wayfinding lighting breakthroughs for smoke in buildings as fallout from aircraft/ship programs, *Proceedings of the National Fire Protection Association, 93rd Annual Meeting*, Washington, DC.

Groupe de Travail Bruxelles (GTB) (1999) *Rationale of Harmonized Dipped (Low) Beam Pattern*, Report No. C.E.-3160, Geneva, Switzerland: GTB.

Guth, S.K. (1963) A method for the evaluation of discomfort glare, *Illumin. Eng.*, 57, 351–64.

Gwinner, E. (1975) Circadian and circannual rhythms in birds, in D. Farner and J. King (eds) *Avian Biology*, New York: Academic Press.

Gydesen, A. and Maimann, D. (1991) Life cycle analyses of integral compact fluorescent lamps versus incandescent lamps, *Proceedings of the 1st European Conference on Energy-Efficient Lighting*, Stockholm, Sweden: Swedish National Board for Industrial and Technical Development.

Hadjisophocleous, G.V., Proulx, G., and Liu, Q. (1997) *Occupant Evacuation Model for Apartment and Office Buildings*, Institute for Research in Construction Report 741, Ottawa, Canada: National Research Council Canada.

Hall, E.T. (1966) *The Hidden Dimension*, New York: Anchor Books/Doubleday.

Hallet, P.E. (1963) Spatial summation, *Vision Res.*, 3, 9–24.

Ham, W.T., Jr., Mueller, H.A., and Sliney, D.H. (1976) Retinal sensitivity to damage from short wavelength light, *Nature*, 260, 153–5.

Hamm, M. and Rosenhahn, E.O. (2001) System strategies and technology for improved safety and comfort with adaptive headlamps, *Lighting Technology Developments for Automobiles*, Society of Automotive Engineers Report SP1595, Warrendale, PA: SAE.

Han, S. (2002) *Effect of Illuminance, CCT and Decor on the Perception of Lighting*, MS Thesis, Troy, New York: Rensselaer Polytechnic Institute.

Hanscom, F.R. and Pain, R.F. (1990) *Service Vehicle Lighting and Traffic Control Systems for Short-Term and Moving Operations*, NCHRP Report 337, Washington, DC: National Cooperative Highway Research Program.

Harber, L.C., Whitman, G.B., Armstrong, R.B., and Deleo, V.A. (1985) Photosensitivity diseases related to interior lighting, in R.J. Wurtman, M.J. Baum, and J.T. Potts, Jr. (eds) *The Medical and Biological Effects of Light*, New York: New York Academy of Sciences.

Harding, J.J. (1995) The untenability of the sunlight hypothesis of caractogenesis, *Documenta Ophthalmol.*, 88, 345–9.

Hargroves, R.A., Hugill, J.R., and Thomas, S.R. (1996) Security, surveillance and lighting, *Proceedings of the CIBSE National Lighting Conference*, London: Chartered Institution of Building Services Engineers.

Hargroves, R.A. and Scott, P.P. (1979) Measurements of road lighting and accidents – the results, *Public Light.*, 44, 213–21.

Harm, W., Rupert, C.S., and Harm, H. (1971) The study of photoenzymatic repair of UV lesions in DNA by flash photolysis, in A.C. Giese (ed.) *Photophysiology: Current Topics in Photobiology and Photochemistry*, New York: Academic Press.

Harrington, R.E. (1954) Effect of color temperature on apparent brightness, *J. Opt. Soc. Am.*, 44, 113–16.

Hartmann, E., Finsterwalder, J., and Muller, M. (1986) Kinetic luminance measurement and assessment of road tunnels, *Light. Res. Technol.*, 18, 28–36.

Hartmann, E. and Moser, E.A. (1968) The law of physiological glare at very small glare angles, *Lichttechnik*, 20, 67A–69A.

Hartnett, O.M. and Murrell, K.F.H. (1973) Some problems of field research, *Appl. Ergonom.*, 4, 219–21.

Harvey, L.O., DiLaura, D.L., and Mistrick, R.J. (1984) Quantifying reactions of visual display unit operators to indirect lighting, *J. Illumin. Eng. Soc.*, 14, 515–46.

Hasson, P., Lutkevich, P., Ananthanarayanan, B., Watson, P., and Knoblauch, R., (2002) Field test for lighting to improve safety at pedestrian crosswalks, in

Proceedings of the 16th Biennial Symposium on Visibility and Simulation, Washington, DC, Transportation Research Board.

Haubner, P. (1977) *Zur Helligkeitsbewertung quasi-achromatischer Reize*, Dissertation, Universitat Karslruhe, Germany.

Hawkes, R.J., Loe, D.L., and Rowlands, E. (1979) A note towards the understanding of lighting quality, *J. Illumin. Eng. Soc.*, 8, 111–20.

He, Y., Rea, M.S., Bierman, A. and Bullough, J. (1997) Evaluating light source efficacy under mesopic conditions using reaction times, *J. Illumin. Eng. Soc.*, 26, 125–38.

He, Y., Bierman, A., and Rea, M.S. (1998) A system of mesopic photometry, *Light. Res. Technol.*, 30, 175–81.

Hebert, M., Dumont, M., and Paquet, J. (1998) Seasonal and diurnal patterns of human illumination under natural conditions, *Chronobiol. Int.*, 15, 59–70.

Hecht, S. and Smith, E.L. (1936) Intermittent stimulation by light, VI. Area and the relation between critical frequency and seeing, *J. Gen. Physiol.*, 19, 979–89.

Hedge, A. (1994) Reactions of computer users to three different lighting systems in windowed and windowless offices, *Work with Display Units '94*, B54–6.

Hedge, A., Sims, W.R., Jr., and Becker, F.D. (1995) Effects of lensed-indirect and parabolic lighting on the satisfaction, visual health and productivity of office workers, *Ergonomics*, 38, 260–80.

Heerwagen, J.H. (1990) The psychological aspects of windows and window design, *Environ. Des. Res. Assoc.*, 269–79.

Heerwagen, J.H. and Orians, G.H. (1986) Adaptations to windowlessness: a study of the use of visual decor in windowed and windowless offices, *Environ. Behav.*, 5, 623–39.

Helander, M.G. and Zhang, L. (1997) Field studies of comfort and discomfort in sitting, *Ergonomics*, 40, 895–915.

Helmers, G. and Rumar, K. (1975) High beam intensity and obstacle visibility, *Light. Res. Technol.*, 7, 38–42.

Her Majesty's Stationery Office (HMSO) (1959) *Statutory Instrument 890: The Standards for School Premises Regulations*, London: HMSO.

Herbert, D. and Moore, L. (1991) *Street Lighting and Crime: The Cardiff Project*, Swansea, UK: University College of Swansea.

Heschong, L., Mahone, D., Rubinstein, F., and McHugh, J. (1998) *Skylighting Guidelines*, Sacramento, CA: Heschong-Mahone Group.

Heschong, L., Wright, R.L., and Okura, S. (2002) Daylighting impacts on retail sales performance, *J. Illumin. Eng. Soc.*, 31, 21–5.

Hietanen, M. (1998) ICNIRP action spectra and guidelines, in R. Matthes and D. Sliney (eds) *Measurements of Optical Radiation Hazards*, Oberschleißheim, Germany: International Commission on Non-Ionizing Radiation Protection.

Hills, B.L. (1975a) Visibility under night driving conditions. Part 1. Laboratory background and theoretical considerations, *Light. Res. Technol.*, 7, 179–84.

Hills, B.L. (1975b) Visibility under night driving conditions. Part 2. Field measurements using disc obstacles and a pedestrian dummy, *Light. Res. Technol.*, 7, 251–8.

Hills, B.L. (1976) Visibility under night driving conditions: Derivation of $(\Delta L, A)$ characteristics and factors in their application, *Light. Res. Technol.*, 8, 11–26.

Hills, B.L. (1980) Vision, visibility and perception in driving, *Perception*, 9, 183–216.

Hilz, R. and Cavonius, C.R. (1974) Functional organization of the peripheral retina: sensitivity to periodic stimuli, *Vision Res.*, 14, 1333–7.

Hockey, G.R.J. (1970) Signal probability and spatial location as possible bases for increased selectivity in noise, *Q. J. Exp. Psychol.*, 22, 37–42.

Hockey, G.R.J. (1983) *Stress and Fatigue in Human Performance*, Chichester, UK: John Wiley and Sons.

Hof, P.R. and Morrison, J.H. (1991) Quantitative analysis of a vulnerable subset of pyramidal neurons in Alzheimer's disease: II. primary and secondary visual cortex, *J. Comp. Neurol.*, 301, 55–64.

Hofner, H. and Williams, D.R. (2002) The eye's mechanisms for autocalibration, *Optics Photonics News*, January, 34–49.

Holding, D., Loeb, M., and Baker, M. (1983) Effects and aftereffects of continuous noise and computation work on risk and effort choices, *Motivation Emotion*, 7 (4), 331–44.

Hole, G.J. and Tyrrell, L. (1996) Possible penalties of motorcyclist's' daytime headlight use: an experimental investigation, in A.G. Gale, I.D. Brown, C.M. Haslegrave, and S.P. Taylor (eds) *Vision in Vehicles V*, Amsterdam: North-Holland.

Holick, M.F. (1985) The photobiology of vitamin D and its consequences for humans, in R.J. Wurtman, M.J. Baum, and J.T. Potts, Jr. (eds) *The Medical and Biological Effects of Light*, New York: New York Academy of Sciences.

Holladay, L.L. (1926) The fundamentals of glare and visibility, *J. Opt .Soc. Am.*, 12, 271–319.

Holz, M. and Weidel, E. (1998) Night vision enhancement system using diode laser headlights, in *Electronics for Trucks and Buses*, Report SP-1401, Warrendale, PA: Society of Automotive Engineers.

Hopkinson, R.G. (1940) Discomfort glare in lighted streets, *Trans. Illumin. Eng. Soc. (London)*, 5, 1–29.

Hopkinson, R.G. (1963) *Architectural Physics: Lighting*, London: Her Majesty's Stationery Office.

Hopkinson, R.G. and Collins, J.B. (1970) *The Ergonomics of Lighting*, London: McDonald & Co. (Publishers) Ltd.

Houpt T.A., Bolus, Z., and Moore-Ede, M.C. (1996) Midnight sun: software for determining light exposure and phase-shifting schedules during global travel, *Physiol. Behav.*, 59, 561–8.

Howarth, C.I. and Bloomfield, J.R. (1969) A rational equation for predicting search times in simple inspection tasks, *Psychonom. Sci.*, 17, 225–6.

Howarth, R. B., Haddad, B.M., and Paton, B. (2000) The economics of energy efficiency: insights from voluntary participation programs, *Energy Policy*, 28, 477–86.

Hunt, D.R.G. (1979) The use of artificial lighting in relation to daylight levels and occupancy, *Building Environ.*, 14, 21–33.

Hunt, D.R.G. (1980) Predicting artificial lighting use – a method based upon observed patterns of behavior, *Light. Res. Technol.*, 12, 7–14.

Hunt, R.W.G. (1982) A model of color vision for predicting color appearance, *Color Res. Appl.*, 7, 95–112.

Hunt, R.W.G. (1987) A model of color vision for predicting color appearance in various viewing conditions, *Color Res. Appl.*, 12, 297–314.

Hunt, R.W.G. (1991) Revised color appearance model for related and unrelated colors, *Color Res. Appl.*, 16, 146–65.

Hutt, D.L., Bissonnette, L.R., St Germain, D., and Oman, J. (1992) Extinction of visible and infrared beams by falling snow, *Appl. Optics*, 31, 5121–32.

Hutt, C., Hutt, S., Lee, D., and Ounsted, C. (1964) Arousal and childhood autism, *Nature*, 204, 908–9.

IES Industrial Lighting Committee (1949) Lighting for flour mills, *Illumin. Eng.*, 44, 691–6.

IES Industrial Lighting Committee (1950a) Lighting for bakeries, *Illumin. Eng.*, 45, 387–97.

IES Industrial Lighting Committee (1950b) Lighting for canneries, *Illumin. Eng.*, 45, 45–65.

IES Industrial Lighting Committee (1952) Lighting for steel mills. Part 1: open Hearth, *Illumin. Eng.*, 47, 165–71.

IES Industrial Lighting Committee (1953) Lighting for foundries, *Illumin. Eng.*, 48, 279–90.

IES Industrial Lighting Committee (1962a) Lighting for the manufacturing of men's clothing, *Illumin. Eng.*, 57, 379–86.

IES Industrial Lighting Committee (1962b) Railroad yard lighting, *Illumin. Eng.*, 57, 239–51.

IES Industrial Lighting Committee (1971) Lighting for cotton gins, *J. Illumin. Eng. Soc.*, 1, 68–72.

IES Industrial Lighting Committee (1975) Lighting for the aircraft/airline industries – manufacturing and maintenance, *J. Illumin. Eng. Soc.*, 4, 207–19.

IES Industrial Lighting Committee (1976a) Lighting for the aircraft/airline industries – airframe maintenance, *Light. Des. Appl.*, 8, 41–7.

IES Industrial Lighting Committee (1976b) Nuclear power plant lighting, *J. Illumin. Eng. Soc.*, 5, 107–16.

IES Industrial Lighting Committee (1977) Lighting for petroleum and chemical plants, *J. Illumin. Eng. Soc.*, 6, 184–92.

IES Industrial Lighting Committee (1990) Lighting for the logging and sawmill industries, *J. Illumin. Eng. Soc.*, 19, 147–64.

Illuminating Engineering Society of North America (IESNA) (1980) Roadway Lighting Subcommittee, *Visual Comfort: The CBE Recommendation*, New York: IESNA.

Illuminating Engineering Society of North America (IESNA) (1996a) *IESNA Recommended Practice for Tunnel Lighting*, RP-22-96, New York: IESNA.

Illuminating Engineering Society of North America (IESNA) (1996b) *ANSI/IESNA RP-27-96, Recommended Practice for Photobiological Safety for Lamps and Lamp Systems*, New York: IESNA.

Illuminating Engineering Society of North America (IESNA) (1998) *IESNA Recommended Practice RP-28-98 Lighting and the Visual Environment for Senior Living*, New York: IESNA.

Illuminating Engineering Society of North America (IESNA) (1999) *IESNA Recommended Practice RP-33-99 Lighting for Exterior Environments*, New York: IESNA.

Illuminating Engineering Society of North America (IESNA) (2000a) *The IESNA Lighting Handbook, 9th Edition*, New York: IESNA.

Illuminating Engineering Society of North America (IESNA) (2000b) *Recommended Practice 8-00: Roadway Lighting*, New York: IESNA.

Illuminating Engineering Society of North America (IESNA) (2001) RP-7-01 *Recommended Practice for Lighting Industrial Facilities*, New York: IESNA.

Inditsky, B., Bodmann, H.W., and Fleck, H.J. (1982) Elements of visual performance: contrast metric – visibility lobes – eye movements, *Light. Res. Technol.*, 14, 218–31.

Institute of Transportation Engineers (ITE) (1985) *Vehicle Traffic Control Signal Heads: A Standard of the Institute of Transportation Engineers*, Washington, DC: ITE.

Institute of Transportation Engineers (ITE) (1998) *Vehicle Traffic Control Signal Heads – Part 2: Light Emitting Diode Traffic Control Signal Modules*, Washington, DC: ITE.

Institution of Lighting Engineers (ILE) (1997) *Guidance Notes for the Reduction of Light Pollution*, Rugby, UK: ILE.

International Commission on Non-Ionizing Radiation Protection (ICNIRP) (1991) Statement: guidelines on UV radiation exposure limits, *Health Phys.*, 71, 978.

International Commission on Non-Ionizing Radiation Protection (ICNIRP) (1997) Guidelines on limits of exposure to broad-band incoherent optical radiation (0.38 to 3 μm), *Health Phys.*, 77, 539–55.

International Dark-Sky Association (IDA) (2000), *Outdoor Lighting Code Handbook*, Version 1.11, http://www.darksky.org, accessed January 31, 2002.

International Maritime Organization (IMO) (1993) *Guidelines on Evaluation, Testing and Application of Low-Location Lighting on Passenger Ships, MSC62/WP.17 Annex*, Washington, DC: IMO.

International Non-Ionizing Radiation Committee of the International Radiation Protection Association (INIRC/IRPA) (1991) Guidelines on limits of exposure to ultraviolet radiation of wavelength between 180 nm and 400 nm, in A.S. Duchene, J.R.A. Lakey, and M.H. Repacholi (eds) *IRPA Guidelines on Protection against Non-Ionizing Radiation*, Oxford, UK: Pergamon Press.

International Standardization Organization (ISO) (1984) *Safety Colours and Safety Signs*, ISO 3864, Geneva, Switzerland: ISO.

Isen, A.M. and Baron, R.A. (1991) Affect as a factor in organizational behavior, in B.M. Staw and L.L. Cummings (eds) *Research in Organizational Behavior*, Greenwich, CT: JAI Press.

Ishida, T. and Ogiuchi, Y. (2002) Psychological determinants of brightness of a space – perceived strength of light source and amount of light in the space, *J. Light Vis. Environ.*, 26, 29–35.

Jacobs, R. and Krohn, D.L. (1976) Variations in fluorescence characteristics of intact human crystalline lens segments as a function of age, *J. Gerontol.*, 38, 641–3.

Janoff, M.S. (1990) The effect of visibility on driver performance: a dynamic experiment, *J. Illumin. Eng. Soc.*, 19, 57–63.

Janoff, M.S. (1992) The relationship between visibility level and subjective ratings of visibility, *J. Illumin. Eng. Soc.*, 21, 98–107.

Janoff, M.S., Freedman, M., and Koth, B. (1977) Driver and pedestrian behavior – the effect of specialized crosswalk illumination, *J. Illumin. Eng. Soc.*, 6, 202–8.

Japanese Standards Association (1992) *Japanese Industrial Standard: Recommended Levels of Illumination: JIS-Z-9110-1979*, Tokyo: Japanese Standards Association.

Japuntich, D.A. (2001) Polarized task lighting to reduce reflective glare in open-plan office cubicles, *Appl. Ergonom.*, 32, 485–99.

Jaschinski, W. (1982) Conditions of emergency lighting, *Ergonomics*, 25, 363–72.

Jay, P.A. (1967) Scales of luminance and apparent brightness, *Light Lighting*, 60, 42–5.

Jay, P.A. (1971) Lighting and visual perception, *Light. Res. Technol.*, 3, 133–46.

Jay, P.A. (1973) The theory of practice in lighting engineering, *Light Lighting*, 66, 303–6.

Jeavons, P.M. and Harding, G.F.A. (1975) *Photosensitive Epilepsy*, London: Heinneman.

Jerome, C.W. (1977) The rendering of ANSI safety colors, *Light. Des. Appl.*, 6 (3), 180–3.

Jette, A.M. and Branch, L.G. (1992) A ten-year follow-up of driving patterns among community-dwelling elderly, *Hum. Factors*, 34, 25–31.

Jewett, M.E., Kronauer, R.E., and Czeisler, C.A. (1991) Light induced suppression of endogenous circadian amplitude in humans, *Nature*, 350, 59–62.

Jewett, M.E., Rimmer, D.W., Duffy, J.F., Klerman, E.B., Kronauer, R.E., and Czeisler, C.A. (1997) Human circadian pacemaker is sensitive to light throughout subjective day without evidence of transients, *Am. J. Physiol.*, 273, R1800–R1809.

Jin, T. (1978) Visibility through fire smoke, *J. Fire Flammabil.*, 9, 135–55.

Jin, T., Kawai, S., Takahashi, S., and Tanabe, R. (1985) *Evaluation on Visibility and Conspicuity of Exit Signs*, Tokyo: Fire Research Institute.

Jin, T., Takahashi, S., Kawai, S., Takeuchi, Y., and Tanabe, R. (1987) Experimental study on visibility and conspicuousness of an exit sign, *Proceedings of the CIE, 21st Session*, Venice, Vienna: CIE.

Jones, H.V. and Heimstra, N.W. (1964) Ability of drivers to make critical passing judgements, *J. Eng. Psychol.*, 3, 117–22.

Johnson, C.A. and Casson, E.J. (1995) Effects of luminance, contrast and blur on visual acuity, *Optometry Visual Sci.*, 72, 864–9.

Judd, D.B. (1951), Report of the US Secretariat Committee on Colorimetry and Artificial Daylight, *Proceeding of the CIE, 12th Session*, Stockholm, Vienna: CIE.

Judd, D.B. and Wyszecki, G.W. (1963) *Colour in Business, Science and Industry*, New York: John Wiley and Sons.

Judd, D.B., MacAdam, D.L., and Wyszecki, G.W. (1964) Spectral distribution of typical daylight as function of correlated color temperature, *J. Opt. Soc. Am.*, 54, 1031–8.

Julian, W.G. (1983) The use of light to improve the visual performance of people with low vision, *Proceedings of the CIE, 20th Session*, Amsterdam, Paris: CIE.

Kaiser, P.K. (1981) Photopic and mesopic photometry: yesterday, today and tomorrow, in *Golden Jubilee of Colour in the CIE*, Bradford, UK: The Society of Dyers and Colourists.

Kaiser, P.K. (1986) Models of heterochromatic brightness matching, *CIE J.*, 5, 57–9.

Kaiser, P.K. and Boynton, R.M. (1996) *Human Color Vision*, Washington, DC: Optical Society of America.

Kaplan, S. (1987) Aesthetics, affect, and cognition: environmental preference from an evolutionary perspective, *Environ. Behavi.*, 19, 2–32.

Kasper, S., Rogers, S.L.B., Yancey, A., Schulz, P.M., Skwerer, R.G., and Rosenthal, N.E. (1989a) Phototherapy in individuals with and without subsyndromal seasonal affective disorder, *Arch. Gen. Psychiatry*, 46, 837–44.

Kasper, S., Wehr, T.A., Bartko, J.J., Gaist, P.A., and Rosenthal N.E. (1989b) Epidemiological finding of seasonal changes in mood and behavior, *Arch. Gen. Psychiatry*, 46, 823–33.

Keighly, E.C. (1973a) Visual requirements and reduced fenestration in offices – a study of multiple apertures and window area, *Building Sci.*, 8, 32.

Keighly, E.C. (1973b) Visual requirements and reduced fenestration in office buildings – a study of window shape, *Building Sci.*, 8, 311.

Keith, D.M. (2000) Roadway lighting design for optimization of UPD, STV and uplight, *J. Illumin. Eng. Soc.*, 29, 15–23.

Kelly D.H. (1961) Visual response to time-dependent stimuli, 1. Amplitude sensitivity measurements, *J. Opt. Soc. Am.*, 51, 422–9.

Kelly, K.L. and Judd, D.B. (1965) *The ISCC-NBS Centroid Color Charts*, Washington, DC: National Bureau of Standards.

Kessler, R.C., McGonagle, K.A., Zhao, S., Nelson, C.B., Hughes, M., and Eshleman, S. (1994) Lifetime and 12-month prevalence of DSM-III-R psychiatric disorders in the United States, *Arch. Gen. Psychiatry*, 51, 8–19.

Ketvirtis, A. (1977) *Road Illumination and Traffic Safety*, Ottawa: Transport Canada.

Khek, J. and Krivohlavy, J. (1967) Evaluation of the criterion to measure the suitability of visual conditions, *Proceedings of the CIE, 16th Session, Washington*, Paris: CIE.

Klein, K.E. and Wegmann, H.M. (1974) The resynchronization of human circadian rhythms after transmeridian flights as a result of flight direction and mode of activity, in L.E. Scheving, F. Halberg, and J.E. Pauly (eds) *Chronobiology*, Tokyo; Igaku Shoin.

Klein, D.C., Moore, R.Y., and Reppert, S.M. (1991) *Suprachiasmatic Nucleus: The Mind's Clock*, Oxford: Oxford University Press.

Klein, T., Martens, H., Dijk, D.-J., Kronauer, R., Seely, E.W., and Czeisler, C.A. (1993) Circadian sleep regulation in the absence of light perception: chronic non-24-hour circadian rhythm sleep disorder in a blind man with a regular 24-h sleep–wake schedule, *Sleep*, 16, 333–43.

Klein, R., Klein, B.E.K., Jensen, S.L., *et al.* (1997) The five-year incidence and progression of age-related maculopathy: the Beaver Dam Study, *Ophthalmology*, 104, 7–21.

Kleinhoonte, A. (1929) Uber die durch das Licht regulierten autonomen Bewegungen der Canavalia-Blatter, *Arch. neerl Sci. ex et nat.*, 5, 1–110.

Knoblauch, K, Arditi, A., and Szlyk, J. (1991) Effects of chromatic and luminance contrast on reading, *J. Opt. Soc. Am.*, A8, 428–39.

Knoblauch, K., Saunders, F., Kusada, M., Hynes, R., Podgor, M., Higgins, K.E., and de Monasterio, F.M. (1987) Age and illuminance effects in the Farnsworth–Munsell 100 hue test scores, *Appl. Optics*, 26, 1441–8.

Kogan, A.O. and Guilford, P.M. (1998) Side effects of short-term 10,000-lux light therapy, *Am. J. Psychiatry*, 155, 293–4.

Kohmoto, K. (1999) Evaluation of actual light sources with proposed photobiological lamp safety standard and its applicability to guide on lighted environment, *Proceedings of the CIE, 24th Session*, Warsaw, Vienna: CIE.

Kokoschka, S. and Bodmann, H.W. (1986) Visual inspection of sealing rings – a case study on lighting and visibility, *Light. Res. Technol.*, 18, 98–101.

Kosmatka, W.J. (1995) Obstacle detection rationale for vehicle headlamps, *J. Illumin. Eng. Soc.*, 24, 36–40.

Kosnik, W., Winslow, L., Kline, D., Rasinski, K. and Sekular, R. (1988) Visual changes throughout adulthood, *J. Gerontol.: Psychol. Sci.*, 43, 63–70.

Kraemer, Sieverts and Partners (1977) *Open-Plan Offices*, London: McGraw-Hill.

Krause, P.B. (1977) The impact of high intensity street lighting on nighttime business burglary, *Hum. Factors*, 19, 235–9.

Krochmann, J. and Seidl, M. (1974) Quantitative data on daylight for illuminating engineering, *Light. Res. Technol.*, 6, 165–71.

Kronauer, R.E. (1990) A quantitative model for the effects of light on the amplitude and phase of the deep circadian pacemaker, based on human data, in J. Horne (ed.) *Sleep 90, Proceedings of the Tenth European Congress on Sleep Research*, pp. 306–9, Dusseldorf: Pontenagel.

Kronauer, R.E., Forger, D.B., and Jewett, M.E. (1999) Quantifying human circadian pacemaker response to brief, extended and repeated light episodes over the photopic range, *J. Biol. Rhythms*, 14, 500–15.

Kruithof, A.A. (1941) Tubular luminescence lamps for general illumination, *Philips Tech. Rev.*, 6, 65–96.

Kuller, R. and Laike, T. (1998) The impact of flicker from fluorescent lighting on well-being, performance and physiological arousal, *Ergonomics*, 41, 433–47.

Kurtin, W.E. and Zuclich, J.A. (1978) Action spectrum of oxygen-dependent near-ultraviolet induced corneal damage, *Photochem. Photobiol.*, 27, 329–33.

Kurylo, D.D., Corkin, S., Schiller, P.H., Golan, R.P., and Growdon, J.H. (1991) Disassociating two visual systems in Alzheimer's disease, *Invest. Ophthalmol. Vis. Sci.*, 32, 1283.

Kuyk, T., Elliot, J.L., Biehl, J., Fuhr, P.S. (1996) Environmental variables and mobility performance in adults with low vision, *J. Am. Optomet. Assoc.*, 67, 403–9.

Kwong, R.C., Michaiski, L., Nugent, M., Rajan, K., Ngo, T., Brown, J.J., Lamansky, S., Djurovich, P., Murphy, D., Abdel-Razzaq, F., Brooks, J., Thompson, M.E., Adachi, C., Baldo, M., and Forrest, S.R. (2001) Recent advances in organic light emitting diodes, *Proceedings of the 9th International Symposium on the Science and Technology of Light Sources*, Ithaca, NY: Cornell University Press.

LaGuisa, F. and Perney, L.R. (1974) Further studies on the effects of brightness variations on attention span in a learning environment, *J. Illumin. Eng. Soc.*, 3, 249–52.

Lack, L. and Schumacher, K. (1993) Evening light treatment of early morning insomnia, *Sleep Res.*, 22, 225.

Lam, R.W. (1998) *Seasonal Affective Disorder and Beyond: Light Treatment for SAD and Non-SAD conditions*, Washington, DC: American Psychiatric Press.

Lam, W.M.C. (1977) *Perception and Lighting as Formgivers for Architecture*, New York: McGraw-Hill.

Lam, R.W. and Levitt, A.J. (1999) *Canadian Consensus Guidelines for the Treatment of Seasonal Affective Disorder*, Vancouver, BC: Clinical and Academic Publishing.

Lamar, E.S. (1960) Operational background and physical conditions relative to visual search problems, in A. Morris and E.P. Horne (eds) *Visual Search Techniques*, Washington, DC: US National Academy of Sciences.

Lambrechts, S.M. and Rothwell, H.L., Jr., (1996) A study on UV protection in lighting, *J. Illumin. Eng. Soc.*, 25, 104–12.

Land, E.H. (1948) The polarized headlight system, *Highway Res. Board Bull.*, 11, 1–19.

Landsberger, H.A. (1958) *Hawthorne Revisited: "Management and the Worker," its Critics and Developments in Human Relations in Industry*, Ithaca, NY: Cornell University.

Larson, C.T. (1973) *The Effect of Windowless Classrooms on Elementary Schoolchildre*n, Architectural Research Laboratory, University of Michigan.

Lathrop, J.K. (1975) The Summerland fire: 50 die on Isle of Man, *Fire J.*, 69, 5–12.

Lebensohn, J.E. (1951) Photophobia: mechanisms and implications, *Am. J. Ophthalmol.*, 34, 1294–1300.

Lecocq, J. (1994) Visibility and lighting of wet road surfaces, *Light. Res. Technol.*, 26, 75–87.

Legge, G.E., Rubin, G.S., Pelli, D.G., and Schleske, M.M. (1985) Psychophysics of reading, II. Low vision, *Vision Res.*, 25, 253–66.

Leibig, J. and Roll, R.F. (1983) Acceptable luminances reflected on VDU screens in relation to the level of contrast and illumination, *Proceedings of the CIE, 20th Session, Amsterdam*, Paris: CIE.

Leibowitz, H.W. and Owens, D.A. (1975) Night myopia and the intermediate dark focus of accommodation, *J. Opt. Soc. Am.*, 65, 1121–8.

Lennie, P. and D'Zmura, M. (1988) Mechanisms of color vision, *CRC Crit. Rev. Neurobiol.*, 3, 333–400.

Leonard, B. and Charles, S. (2000) Diabetic retinopathy, in B. Silverstone, M.A. Lang, B.P. Rosenthal, and E.E. Faye (eds) *The Lighthouse Handbook on Vision Impairment and Vision Rehabilitation*, New York: Oxford University Press.

Leslie, R.P. and Rodgers, P.A. (1996) *The Outdoor Lighting Pattern Book*, New York: McGraw-Hill.

Lestina, D.C., Miller, T.R., Knoblauch, R., and Nitzburg, M. (1999) Benefits and costs of ultraviolet fluorescent lighting, *Proceedings of the Association for the Advancement of Automotive Medicine*, Barcelona, Spain.

Levin, R.E. (1998) Photobiological safety and risk ANSI/IESNA RP-27 series, *J. Illumin. Eng. Soc.*, 27, 136–43.

Levitt, A.J., Joffe, R.T., Moul, D.E., Lam, R.W., Teicher, M.H., and Lebegue, F. (1993) Side effects of light therapy in seasonal affective disorder, *Am. J. Psychiatry*, 150, 650–2.

Lewis, A.L. (1992) Lighting considerations for the low vision patient, *Problems Optometry*, 4, 20–33.

Lewis, A.L. (1999) Visual performance as a function of spectral power distribution of light sources at luminances used for general outdoor lighting, *J. Illumin. Eng. Soc.*, 28, 37–42.

Lewis, E.B. and Sullivan T.T. (1979) Controlling crime and citizen attitudes: a study of the corresponding reality, *Criminal Justice*, 7, 71–9.

Lighting Research Center (LRC) (1994) *Specifier Reports: Exit Signs*, Troy, NY: LRC.

Lighting Research Center (LRC) (1995) *Specifier Reports Supplements: Exit Signs*, Troy, NY: LRC.

Lindner, H., Hubnert, K., Schlote, H.W., and Rohl, F. (1989) Subjective lighting needs of the old and the pathological eye. *Light. Res. Technol.*, 21, 1–10.

Lion, J.S., Richardson, E., and Browne, R.C. (1968) A study of the performance of industrial inspection under two levels of lighting, *Ergonomics*, 11, 23–4.

Lipinski, M.E. and Shelby, B.L. (1993) Visibility measures of realistic roadway tasks, *J. Illumin. Eng. Soc.*, 22, 94–101.

Lippincott, H.W. and Stark, H. (1982) Optical–digital detection of dents and scratches on specular metal surfaces, *Appl. Optics*, 21, 2875–81.

Littlefair, P.J. (1981) The luminance distribution of an average sky, *Light. Res. Technol.*, 13, 192–8.

Littlefair, P.J. (1990) Innovative daylighting: review of systems and evaluation methods, *Light. Res. Technol.*, 22, 1–17.

Littlefair, P.J., Aizlewood, M.E., and Birtles, A.B. (1994) The performance of innovative daylighting systems, *Renewable Energy*, 5, 920–34.

Livingston, M.S. and Hubel, D.H. (1981) Effects of sleep and arousal on the processing of visual information in the cat. *Nature*, 291, 554–61.

Lloyd, C.J., Boyce, P.R., Ferzacca, N., Eklund, N.H., and He, Y. (1999) Paint inspection lighting optimization of lamp width and spacing, *J. Illumin. Eng. Soc.*, 28, 92–102.

Lloyd, C.J., Mizukami, M., and Boyce, P.R. (1996) A preliminary model of lighting–display interaction, *J. Illumin. Eng. Soc.*, 25, 59–69.

Lobban, M.C. (1961) The entrainment of circadian rhythms in man, *Cold Spring Harb. Symp. Quant. Biol.*, 25, 325–332.

Loe, D.L., Mansfield, K.P., and Rowlands, E. (1994) Appearance of lit environment and its relevance in lighting design: experimental study, *Light. Res. Technol.*, 26, 119–33.

Loe, D.L. and Rowlands, E. (1996) The art and science of lighting: a strategy for lighting design, *Light. Res. Technol.*, 28, 153–64.

Loomis, D., Marshall, S.W., Wolf, S.H., Runyan, C.W., and Butts, J.D. (2002) Effectiveness of safety measures recommended for prevention of workplace homicide, *J. Am. Med. Assoc.*, 287, 1011–17.

Love, J.A. (1998) Manual switching patterns in private offices, *Light. Res. Technol.*, 30, 45–50.

Lovell, B.B., Ancoli-Isreal, S., and Gevirtz, R. (1995) Effect of bright light treatment on agitated behavior in institutionalized elderly subjects, *Psychiatry Res.*, 57, 7–12.

Ludlow, A.M. (1976) The functions of windows in buildings, *Light. Res. Technol.*, 8, 57–68.

Lund, D.J. (1998) Action spectrum for retinal thermal injury, in R. Matthes and D. Sliney (eds) *Measurements of Optical Radiation Hazards*, Oberschleißheim, Germany: International Commission on Non-Ionizing Radiation Protection.

Lydahl, E. and Philipson, B. (1984a) Infrared radiation and cataract. I. Epidemiologic investigation of iron- and steel-workers, *Acta Ophthalmol.*, 62, 961–75.

Lydahl, E. and Philipson, B. (1984b) Infrared radiation and cataract. II. Epidemiologic investigation of glass workers, *Acta Ophthalmol.*, 62, 976–92.

Lynes, J.A. (1971) Lightness, colour and constancy in lighting design, *Light. Res. Technol.*, 3, 24–42.

Lynes, J.A. (1977) Discomfort glare and visual distraction, *Light. Res. Technol.*, 9, 51–2.

Lynes, J.A. and Littlefair, P.J. (1990) Lighting energy savings for daylight: estimation at the sketch design stage, *Light. Res. Technol.*, 22, 129–37.

Lyons, S.L. (1980) *Exterior Lighting for Industry and Security*, London: Applied Science Publishers.

Lyons, S.L. (1981) *Handbook of Industrial Lighting*, London: Butterworth.

Lythgoe, R.J. (1932) *The Measure of Visual Acuity*, MRC Special Report, No. 173, London: His Majesty's Stationery Office.

MacAdam, D.L. (1942) Visual sensitivity to color differences in daylight, *J. Opt. Soc. Am.*, 32, 247–74.

McCloughan, C.L.B., Aspinall, P.A., and Webb, R.S. (1999) The impact of lighting on mood, *Light. Res. Technol.*, 31, 81–8.

McColl, S.L. and Veitch, J.A. (2001) Full-spectrum fluorescent lighting: a critical review of its effects on physical and mental health, *Psychol. Med.*, 31, 949–64.

McDonald, E.J. (1970) The epidemiology of skin cancer, *J. Invest. Dermatol.*, 32, 379.

McGrath, C. and Morrison, J.D. (1981) The effects of age on spatial frequency perception in human subjects, *Q. J. Exp. Physiol.*, 66, 253–61.

McGuiness, P.J. and Boyce, P.R. (1984) The effect of illuminance on the performance of domestic kitchen work by two age groups, *Light. Res. Technol.*, 16, 131–6.

McIntyre, D.A. (2002) Colour blindness: causes and effects, Chester UK: Dalton Publishing.

McIntyre, I.A., Norman, T.R., Burrows, G.D., and Armstrong, S.M. (1989) Quantal melatonin suppression by exposure to low intensity light in man, *Life Sci.*, 45, 327–32.

McKinlay, A.F. and Diffey, B.L. (1987) A reference action spectrum for ultraviolet induced erythema in human skin, *CIE J.*, 6, 17–22.

McKinlay, A.F., Harlen, F., and Whillock, M.J. (1988) *Hazards of Optical Radiation*, Bristol, UK: Adam Hilger.

McNally, D. (1994) *The Vanishing Universe*, Cambridge, UK: Cambridge University Press.

McNelis, J.F. (1973) Human performance – a pilot study, *J. Illumin. Eng. Soc.*, 2, 190–6.

Mace, D., Garvey, P., Porter, R.J., Schwab, R., and Adrian, W. (2001) *Countermeasures for Reducing the Effects of Headlight Glare*, Washington, DC: AAA Foundation for Traffic Safety.

Mackworth, N.H. (1948) The breakdown of vigilance during prolonged visual search, *Q. J. Exp. Psychol.*, 1, 6–21.

Main, A., Dowson, A., and Gross, M. (1997) Photophobia and phonophobia in migraineurs between attacks, *Headache*, 376, 492–5.

Malven, F.C. (1986) *Directional Continuity in Escape Lighting*, New York: Lighting Research Institute.

Manabe, H. (1976) The assessment of discomfort glare in practical lighting installations, *Oteman Economics Studies, No.9*, Osaka, Japan: Oteman Gakuin University.

Mandlebaum, J. and Sloan, L.L. (1947) Peripheral visual acuity, *Am. J. Ophthalmol.*, 30, 581–8.

Mangum, S.R. (1998) Effective constrained illumination of three-dimensional, light-sensitive objects, *J. Illumin. Eng. Soc.*, 27, 115–31.

Maniccia, D., Rutledge, B., Rea, M.S., and Morrow, W. (1999) Occupant use of manual lighting controls in private offices, *J. Illumin. Eng. Soc.*, 28, 42–56.

Marcus, D.A. and Soso, M.J. (1989) Migraine and stripe-induced visual discomfort, *Arch. Neurol.*, 46, 1129–32.

Markus, T.A. (1967) The significance of sunshine and view for office workers, in R.G. Hopkinson (ed.) *Sunlight in Buildings*, Rotterdam, The Netherlands: Boewcentrum International.

Marsden, A.M. (1969) Brightness – a review of current knowledge, *Light. Res. Technol.*, 1, 171–81.

Marsden, A.M. (1970) Brightness–luminance relationships in an interior, *Light. Res. Technol.*, 2, 10–16.

Marshall, J. (1981) Light damage and the practice of ophthalmology, in E. Rosen, E. Arnott, and W. Haining (eds) *Intraocular Lens Implantation*, London: Mosby-Yearbook.

Marshall, J., Grindle, J., Ansell, P.L., and Borwein, B. (1979) Convolution in human rods: an aging process, *Br. J. Ophthalmol.*, 63, 181–7.

Megaw, E.D. (1979) Factors affecting visual inspection accuracy, *Appl. Ergonom.*, 10, 27–32.

Megaw, E.D. and Richardson, J. (1979) Eye movements and industrial inspection, *Appl. Ergonom.*, 10, 145–54.

Meijer, J.H. and Rietveld, W.J. (1989) Neurophysiology of the suprachiasmatic circadian pacemaker in rodents, *Physiol. Rev.*, 69, 671–707.

Mellor, E.F. (1986) Shift work and flexitime: how prevalent are they? *Monthly Labor Rev.*, 109, 14–21.

Mellerio, J. (1987) Yellowing of the human lens: nuclear and cortical contributions, *Vision Res.*, 27, 1581–7.

Mellerio, J. (1998) The design of effective ocular protection for solar radiation, in R. Matthes and D. Sliney (eds) *Measurements of Optical Radiation Hazards*, Oberschleißheim, Germany: International Commission on Non-Ionizing Radiation Protection.

Menaker, M. (1997) Commentary: what does melatonin do and how does it do it, *J. Biol. Rhythms*, 12, 532–4.

Menaker, M. and Tosini, G. (1996) The evolution of vertebrate circadian systems, in K. Honma and S. Honma (eds) *Circadian and Oscillatory Coupling: Proceedings of the Sixth Sapporo Symposium*, Sapparo, Japan: Hokkaido University Press.

Mendez, M.F., Tomsak, R.L., and Remler, B. (1990) Disorders of the visual system in Alzheimer's disease, *J. Clin. Neuro-Ophthalmol.*, 10, 62–9.

Middleton, W.E.K. (1952) *Vision Through the Atmosphere*, Toronto: University Press.

Miles, L.E.M., Raynal, D.M., and Wilson, M.A. (1977) Blind man living in normal society has circadian rhythms of 24.9 hours, *Science*, 198, 421–3.

Miller, J.W. and Ludvigh, E. (1962) The effects of relative motion on visual acuity, *Survey Ophthalmol.*, 7, 83–116.

Miller, N.J. (2000) Lighting for VDT offices: an improved metric for judging the acceptability of direct luminaires in offices using computer screens, MS Thesis, Troy, NY: Rensselaer Polytechnic Institute.

Miller, N.J., McKay, H., and Boyce, P.R. (1995) An approach to the measurement of lighting quality, *Proceedings of the IESNA Annual Conference*, New York, New York: IESNA.

Miller, N.J., Boyce, P.R., and Ngai, P.Y. (2001) A metric for judging acceptability of direct luminaires in computer offices, *J. Illumin. Eng. Soc.*, 30, 12–29.

Mills, E. (1991) Using financial incentives to promote compact fluorescent lamps in Europe: cost effectiveness and consumer response in 10 countries, *Proceedings of the 1st European Conference on Energy-Efficient Lighting*, Stockholm, Sweden: Swedish National Board for Industrial and Technical Development.

Mills, E. (2000) Global lighting: 1000 power plants, *International Association of Energy Efficient Lighting (IAEEL) Newsletter*, 1–2/00.

Mills, E. and Borg, N. (1999) Trends in recommended illuminance levels: an international comparison, *J. Illumin. Eng. Soc.*, 28, 155–63.

Minors, D.S. and Waterhouse, J.M. (1981) *Circadian Rhythms and the Human*, Bristol, UK: Wright.

Mistrick, R.G. and Choi, A. (1999) A comparison of the visual comfort probability and unified glare rating systems, *J. Illumin. Eng. Soc.*, 28, 94–101.

Miyamoto, Y. and Sancar, A. (1998) Vitamin B_2-based blue-light photoreceptors in the retinalhypothalamic tract as the photoactive pigments for setting the circadian clock in mammals, *Proc. Natl. Acad. Sci. USA*, 95, 6097–102.

Mizokami, Y., Ikeda, M., and Shinoda, H. (2000) Color property of the recognized visual space of illumination controlled by interior color as the initial visual information, *Opt. Rev.*, 7, 358–63.

Moan, J. and Dahlback, A. (1993) Ultraviolet radiation and skin cancer: epidemiological data from Scandinavia, in A.R. Young, L.O. Bjorn, J. Moan and W. Nultsch (eds) *Environmental UV Photobiology*, New York: Plenum.

Mollon, J.D. (1989) "Tho' she kneel'd in that place where they grew … " The uses and origins of primate colour vision, *J. Exp. Biol.*, 146, 21–38.

Monk, T.H. and Folkard, S. (1985) Shiftwork and performance, in S. Folkard and T.H. Monk (eds) *Hours of Work*, Chichester, UK: John Wiley and Sons.

Monk, T.H., Knauth, P., Folkard, S., and Rutenfranz, J. (1978) Memory based performance measures in studies of shiftwork, *Ergonomics*, 21, 819–26.

Moore, R.L. (1952) *Rear Lights of Motor Vehicles and Pedal Cycles*, Road Research Technical Paper 25, London: His Majesty's Stationery Office.

Moore, R.Y., Speh, J.C., and Card, J.P. (1995) The retino-hypothalamic tract originates from a distinct subset of retinal ganglion cells, *J. Comp. Neurol.* 352, 351–66.

Morgan, M. (1988) Vision through my aging eyes, *J. Am. Optomet. Assoc.*, 59, 278–80.

Mortimer, R.G. and Becker, J.M. (1973) *Development of a Computer Simulation to Predict the Visibility Distances Provided by Headlamp Beams*, Report UM-HSRI-IAF-73-15, Ann Arbor, MI: University of Michigan.

Mortimer, R.G. and Jorgeson, C.M. (1974) *Eye Fixations of Drivers in Night Driving with Three Headlight Beams*, Report UM-HSRI-74-17, Ann Arbor, MI: University of Michigan.

Mourant, R.R. and Rockwell, T.H. (1972) Strategies of visual search by novice and experienced drivers, *Hum. Factors*, 14, 325–35.

Moyer, J.L. (1992) *The Landscape Lighting Book*, New York: John Wiley and Sons.

Muck, E. and Bodmann, H.W. (1961) Die bedeutung des beleuchtungsniveaus bei praktische sehtatigkeit, *Lichttechnik*, 13, 502–7.

Murata, Y. (1987) Light absorption characteristics of the lens capsule, *Ophthalmic Res.*, 19, 107–12.

Nadler, D.J. (1990) Glare and contrast sensitivity in cataracts and pseudophakia, in M.P. Nadler, D. Miller, and D.J. Nadler (eds) *Glare and Contrast Sensitivity for Clinicians*, New York: Springer-Verlag.

Nair, G. and Ditton, J. (1994) "In the dark, a taper is better then nothing", *Light. J.*, 59, 25–6.

Nair, G., Ditton, J., and Phillips, S. (1993) Environmental improvements and the fear of crime, the sad case of the "Pond" area in Glasgow, *Br. J. Criminol.*, 33, 555–61.

Nakano, Y., Yamada, K., Suehara, K., and Yano, T. (1999) A simple formula to calculate brightness equivalent luminance, *Proceedings of the CIE, 24th Session*, Warsaw, Vienna: CIE.

Nardell, E.A. (1993) Environmental control of tuberculosis, *Med. Clin. North Am.*, 77, 1315–34.

Narendran, N. and Bullough, J.D. (2001) Light emitting diodes as light sources, *Proceedings of the 9th International Symposium on the Science and Technology of Light Sources*, Ithaca, NY: Cornell University Press.

Narendran, N., Vasconez, S., Boyce, P., and Eklund, N. (2000) Just-perceivable color difference between similar light sources in display lighting applications, *J. Illumin. Eng. Soc.*, 29, 78–82.

Narendran, N., Bullough, J.D., Maliyagoda, N., and Bierman, A. (2001) What is useful life for white light LEDs?, *J. Illumin. Eng. Soc.*, 30, 57–67.

Narisada, K. (1971) Influence of non-uniformity in road surface luminance of public lighting installations upon perception of objects on the road surface by car drivers, *Proceedings of the CIE, 17th Session,* Barcelona, Paris: CIE.

Narisada, K. and Yoshikawa, K. (1974) Tunnel entrance lighting – effect of fixation point and other factors on the determination of requirements, *Light. Res. Technol.*, 6, 9–18.

Nathan, J., Henry, G., and Cole, B. (1964) Recognition of colored traffic light signals by normal and color vision defective observers, *J. Opt. Soc. Am.*, 54, 1041–45.

National Bureau of Standards (1976) *Color: Universal Language and Dictionary of Names*, Special Publication 440, Washington, DC: National Bureau of Standards.

National Fire Protection Association (NFPA) (1997) *NFPA 101 Life Safety Code*, Quincy, MA: NFPA.

National Highway Traffic Safety Administration (NHTSA) (1992) *Fatality Analysis Reporting System*, Washington, DC: US Department of Transportation.

National Highway Traffic Safety Administration (NHTSA) (2000) *Traffic Safety Facts 1999: A Compilation of Motor Vehicle Crash Data from the Fatality Analysis Reporting System and the General Estimates System*, Washington, DC: US Department of Transportation.

National Police Agency (Japan) (1986) *Specification for Metal Vehicle Traffic Control Signals*, Tokyo: The National Police Agency.

Nayatani, Y., Sobagaki, H., and Hashimoto, K. (1994) Existence of two kinds of representations of the Helmholtz–Kohlrausch effect II. The models, *Color Res. Appl.*, 19, 262–72.

Ne'eman, E. and Hopkinson, R.G. (1970) Critical minimum acceptable window size, a study of window design and provision of view, *Light. Res. Technol.*, 2, 17–27.

Ne'eman, E., Craddock, E., and Hopkinson, R.G. (1976) Sunlight requirements in Buildings – 1. Social survey, *Building Environ.*, 11, 217.

Neitz, J., Carroll, J., and Neitz, M. (2001) Color vision: almost reason enough for having eyes, *Optics Photonics News*, January, 26–33.

Neubauer, O., Harrer, S., Marre, M., and Verriest, G. (1978) Colour vision and traffic, in G. Verriest (ed.) *Modern Problems in Ophthalmology*, Basel, Switzerland: Karger.

New Buildings Institute (2001) *Advanced Lighting Guidelines*, White Salmon, WA: New Buildings Institute.

Newman, J.S. and Kahn, M.M. (1984) *Standard Test Criteria for Evaluation of Underground Fire Detection Systems*, Washington, DC: Bureau of Mines, US Department of the Interior.

Newsham, G. and Veitch, J. (2001) Lighting quality recommendations for VDT offices: a new method of derivation, *Light. Res. Technol.*, 33, 97–116.

Ngai, P.Y. and Boyce, P.R. (2000) The effect of overhead glare on visual discomfort, *J. Illumin. Eng. Soc.*, 29, 29–38.

Nickerson, D. (1948) The illuminant in textile color matching, *Illumin Eng.*, 43, 416–67.

Nickerson, D. (1957) Horticultural color chart names with a Munsell key, *J. Opt. Soc. Am.*, 47, 619–21.

Nikitin, V.D. (1973) Minimum required level of illumination intensity for emergency illumination in evacuation of persons, *Svetoteknika*, 6, 9–10.

Nilsson, L. and Alm, H. (1996) Effects of a vision enhancement system on drivers' ability to drive safely in fog, in A.G. Gale, I.D. Brown, C.M. Haslegrave, and S.P. Taylor (eds) *Vision in Vehicles V*, Amsterdam: North-Holland.

Nirk, L. (1997) The ENERGY STAR residential lighting program, *Proceedings of Right Light 4*, Frederiksberg, Denmark: DEF Congress Service.

Noonan, F.P. and De Fabo, E.C. (1994) UV-induced immunosuppression, in A.R. Young, L.O. Bjorn, J. Moan, and W. Nultsch (eds) *Environmental UV Photobiology*, New York: Plenum.

Norris, D. and Tillett, L. (1997) Daylight and productivity: is there a causal link?, *Proceedings, Glass Processing Days Conference*, Tampere, Finland.

Novellas, F. and Perrier, J. (1985) New lighting method for road tunnels, *CIE J.*, 4, 58–70.

O'Dea, W.T. (1958) *The Social History of Lighting*, New York: Macmillan.

Ogle, K.V. (1961) Foveal contrast thresholds with blurring of the retinal image and increasing size of test stimulus, *J. Opt. Soc. Am.*, 51, 862–7.

Okawa, M., *et al.* (1989) Sleep–wake rhythm disorders and their phototherapy in elderly patients with dementia, *Jpn. J. Psychiatry Neurol.*, 43, 293–5.

Okuno, T. (1991) Thermal effect of infra-red radiation on the eye: a study based on a model, *Ann. Occup. Hyg.*, 35, 1–12.

Olson, P.L. and Sivak, M. (1983) Comparison of headlamp visibility distance and stopping distance, *Perceptual Motor Skills*, 57, 1177–8.

Olson, P.L., Battle, D.S., and Aoki, T. (1989) *The Detection Distance of Highway Signs as a Function of Color and Photometric Properties*, Technical Report UMTRI-89-36, Ann Arbor, MI: University of Michigan.

Olson, P.L., Aoki, T., Battle, D.S., and Flannagan, M.J. (1990) *Development of a Headlight System Performance Evaluation Tool*, Report UMTRI-90–41, Ann Arbor, MI: University of Michigan.

Opstelten, J.J. (1983) The establishment of a representative set of test colors for the specification of the color rendering properties of light sources. *Proceedings of the CIE, 20th Session*, Amsterdam, Vienna: CIE.

Ouellette, M.J. (1988) Exit signs in smoke: design parameters for greater visibility, *Light. Res. Technol.*, 20, 155–60.

Ouellette, M.J. and Rea, M.S. (1989) Illuminance requirements for emergency lighting, *J. Illumin. Eng. Soc.*, 18, 37–42.

Ouellette, M.J., Tansley, B.W., and Pasini, I. (1993) The dilemma of emergency lighting: theory vs. reality, *J. Illumin. Eng. Soc.*, 22, 113–21.

Owsley, C., Sekular, R., and Siemsen, D. (1983) Contrast sensitivity throughout adulthood, *Vision Res.*, 23, 689–99.

Oya, H., Mitsuhashi, K. and Ando, K. (2000) A study on visibility at the fusion of road lighting and headlamps, *Proceedings of the Transportation Research Board, 79th Annual Meeting*, Washington, DC: Transportation Research Board.

Padgham, C.A. and Saunders, J.E. (1975) *The Perception of Light and Color*, London: G. Bell and Sons Ltd.

Padmos, P., Van den Brink, T.D.J., Alferdinck, J.W.A.M., and Folles, E. (1988) Matrix signs for motorways: system design and optimum photometric features, *Light. Res. Technol.*, 20, 55–60.

Painter, K. (1988) *Lighting and Crime Prevention: The Edmonton Project*, Hatfield, UK: Middlesex Polytechnic.

Painter, K. (1989) *Lighting and Crime Prevention for Community Safety; The Tower Hamlets Study, 1st Report*, Hatfield, UK: Middlesex Polytechnic.

Painter, K. (1991a) An evaluation of public lighting as a crime prevention strategy: the West Park surveys, *Light. J.*, 56, 228–32.

Painter, K. (1991b) *An Evaluation of Public Lighting as a Crime Prevention Strategy with Special Focus on Women and Elderly People*, Hatfield, UK: Middlesex Polytechnic.

Painter, K. (1994) The impact of street lighting on crime, fear, and pedestrian street use, *Security J.*, 5, 116–24.

Painter, K. (1996) Street lighting, crime and fear of crime: a summary of research, in T.H. Bennett (ed.) *Preventing Crime and Disorder: Targeting Strategies and Responsibilities, 22nd Cropwood Round Table Conference*, Cambridge, UK: University of Cambridge.

Painter, K. (1999) The social history of street lighting (part 1). *Light. J.*, 64, 14–24.

Painter, K. (2000) The social history of street lighting (part 2). *Light. J.*, 65, 24–30.

Painter, K. and Farrington, D.P. (1997) The crime reducing effect of improved street lighting: the Dudley project, in R.V. Clarke (ed.) *Situational Crime Prevention: Successful Case Studies*, Albany, NY: Harrow and Heston.

Painter, K. and Farrington, D.P. (1999) Street lighting and crime: diffusion of benefits in the Stoke-on-Trent project, in K. Painter and N. Tilley (eds) *Crime Prevention Studies*, Monsey, NY: Criminal Justice Press.

Painter, K.A. and Farrington, D.P. (2001a) Evaluating situational crime prevention using a young people's survey, *Br. J. Criminol.*, 41, 266–84.

Painter, K.A. and Farrington, D.P. (2001b) The financial benefits of improved street lighting based on crime reduction, *Light. Res. Technol.*, 33, 3–12.

Palmer, J. and Boardman, B. (1998) *DELight: Domestic Efficient Lighting*, Oxford, UK: Environment Change Unit, University of Oxford.

Parrish, J.A., Rosen, C.F., and Gange, R.W. (1985) Therapeutic uses of light, in R.J. Wurtman, M.J. Baum, and J.T. Potts, Jr. (eds) *The Medical and Biological Effects of Light*, New York: New York Academy of Sciences.

Parsons, H.M. (1974) What happened at Hawthorne?, *Science*, 183, 922–32.

Pasini, R. and Proulx, G. (1988) Building access and safety for the visually impaired person, in J.D. Sime (ed.) *Safety in the Built Environment*, London: E & FN Spon.

Paul, B.M. and Einhorn, H.D. (1999) Discomfort glare from small light sources, *Light. Res. Technol.*, 31, 139–44.

Paulmier, G., Brusque, C., Carta, V., and Nguyen, V. (2001) The influence of visual complexity on the detection of targets investigated by computer generated images, *Light. Res. Technol.*, 33, 197–207.

Pauls, J.L. (1980) Building evacuation: research findings and recommendations, in D. Canter (ed.) *Fires and Human Behaviour*, Chichester, UK: John Wiley and Sons.

Pauls, J.L. (1988) Egress time criteria related to design rules in codes and standards, in J. Sime (ed.) *Safety in the Built Environment*, London: E & FN Spon.

Paulsen, T. (1994) The effect of escape route information on mobility and wayfinding under smoke logged conditions, *Proceedings of the Fourth International Symposium, Fire Safety Sci.*, 7, 693–704.

Paulsson, L. and Sjostrand, J. (1980) Contrast sensitivity in the presence of a glare light. *Invest. Ophthalmol.*, 19, 401–6.

Pease, K. (1997) Crime prevention, *Oxford Handbook of Criminology*, Oxford, UK: Clarendon.

Pease, K. (1999) A review of street lighting evaluations: crime reduction effects, in K. Painter and N. Tilley (eds) *Crime Prevention Studies*, Monsey, NY: Criminal Justice Press.

Peli, E. (1990) Contrast in complex images, *J. Opt. Soc. Am. A*, 7, 2032–40.

Peli, E. and Peli, T. (1984) Image enhancement for the visually impaired, *Opt. Eng.*, 23, 47–51.

Petherbridge, P. and Hopkinson, R.G. (1950) Discomfort glare and the lighting of buildings, *Trans. Illumin. Eng. Soc. (London)*, 15, 39–79.

Philips Lighting (1999) *International Lighting Review 993*, Eindhoven, The Netherlands: Philips Lighting.

Phillips, C. (1999) A review of CCTV evaluations: crime reduction effects and attitudes towards its use, in K. Painter and N. Tilley (eds) *Crime Prevention Studies*, Monsey, NY: Criminal Justice Press.

Photoluminescent Safety Products Association (PSPA) (1997) *Standard 002 Emergency Way-finding Guidance Systems, Part 1: Code of Practice for the Installation of Emergency Way-finding Guidance (LLL) Systems Produced from Photoluminescence for Use in Public, Industrial and Commercial Buildings*, West Sussex, UK: PSPA.

Pittendrigh, C.S. (1981) Circadian systems: entrainment, in J. Aschoff (ed.) *Handbook of Behavioral Neurobiology (Biological Rhythms)*, New York: Plenum Press.

Pittendrigh, C.S. and Daan, S. (1976) A functional analysis of circadian pacemakers in nocturnal rodents: V. pacemaker structure – a clock for all seasons, *J. Comp. Physiol. A*, 106, 333–55.

Pitts, D.G. (1970) A comparative study of the effects of ultraviolet radiation on the eye, *Am. J. Optom. Arch. Am. Acad. Optom.*, 50, 535–46.

Pitts, D.G. and Cullen, A.P. (1977) *Ocular Effects of Ultraviolet Radiation from 295 nm to 400 nm in the Rabbit Eye*, NIOSH Publication No 77–175, Washington, DC: US Department of Health, Education and Welfare.

Pitts, D.G. and Tredici, T.J. (1971) The effects of ultraviolet on the eye, *Am. Ind. Hyg. Assoc.*, 32, 235–46.

Pitts, D.G., Cullen, A.P., and Hacker, P.D. (1977) Ocular effects of ultraviolet radiation from 295 nm to 365 nm, *Invest. Opthalmol. Vis. Sci.*, 16, 932–9.

Poffenberger, A.T. (1928) The effects of continuous work upon output and feelings, *J. Appl. Psychol.*, 12, 459–67.

Pointer, M.R. (1986) Measuring color rendering – a new approach, *Light. Res. Technol.*, 18, 175–83.

Pollak, C.P. and Perlick, D. (1991) Sleep problems and institutionalization of the elderly, *J. Geriatr. Psychiatry Neurol.*, 4, 204–10.

Pollard, N.E. (1994) Sky-glow conscious lighting design, *Light. Res. Technol.*, 26, 151–6.

Poulton, E.C. (1977) Quantitative subjective assessments are almost always biased, sometimes completely misleading, *Br. J. Psychol.*, 68, 409.

Poyner, B. (1991) Situational prevention in two car parks, *Security J.*, 2, 96–101.

Pritchard, D. (1964) Industrial lighting in windowless buildings, *Light Lighting*, 63, 292–6.

Pritchard, R.M., Heron, W., and Hebb, D.O. (1960) Visual perception approached by the method of stabilized images, *Can. J. Psychol.*, 14, 67–77.

Proulx, G. (1998) The impact of voice communication messages during a residential highrise fire, in J. Shields (ed.) *Proceedings of the First International Symposium on Human Behavior in Fire*, Belfast, UK: University of Ulster.

Proulx, G. (1999) Occupant response during a residential highrise fire, *Fire and Materials*, 23, 317–23.

Proulx, G. (2000a) *Why Building Occupants Ignore Fire Alarms*, Construction Technology Update No. 42, Ottawa, Canada: National Research Council Canada.

Proulx, G. (2000b) *Strategies for Ensuring Appropriate Occupant Response to Fire Alarm Signals*, Construction Technology Update, No. 43, Ottawa, Canada: National Research Council Canada.

Proulx, G. and Koroluk, W. (1997) Fires mean people need fast, accurate information, *CABA Home Building Automat. Q.*, Summer, 17–19.

Proulx, G., Creak, J., and Kyle, B. (2000) A field study of photoluminescent signage used to guide building occupants to exit in complete darkness, *Proceedings of the Human Factors and Ergonomics Society 2000 Congress*, San Diego, CA.

Pu, M. (2000) Physiological response properties of the cat retinal ganglion cells projecting to the suprachiasmatic nucleus, *J. Biol. Rhythms*, 15, 31–6.

Rahmani, B., Tielsch, J.M., Katz, J., Gottsch, J., Quigley, H.A., Javitt, J., and Sommer, A. (1996) The cause-specific prevalence of visual impairment in an urban population: The Baltimore Eye Survey, *Ophthalmology*, 103, 1721–6.

Rea, M.S. (1981) Visual performance with realistic methods of changing contrast, *J. Illumin. Eng. Soc.*, 10, 164–77.

Rea, M.S. (1982) Calibration of subjective scaling responses, *Light. Res. Technol.*, 14, 121–9.

Rea, M.S. (1984) Window blind occlusion: a pilot study, *Building Environ.*, 19, 133–7.

Rea, M.S. (1986) Toward a model of visual performance: foundations and data, *J. Illumin. Eng. Soc.*, 15, 41–58.

Rea, M.S. (1987) Toward a model of visual performance: a review of methodologies, *J. Illumin. Eng. Soc.*, 16, 128–42.

Rea, M.S. (1991) Solving the problem of VDT reflections, *Progressive Architect.*, 10, 35–40.

Rea, M.S. (1996) Essay by invitation, *Light. Des. Appl.*, October, 15–16.

Rea, M.S. (2001) The road not taken, *Light. J.*, 66, 18–25.

Rea, M.S. and Boyce, P.R. (1999) Different sources for different courses under mesopic light levels, *Vision at Low Light Levels, Proceedings of the 4th International Lighting Research Symposium*, Orlando, FL, EPRI Technical Report TR 110738.

Rea, M.S. and Ouellette, M.J. (1988) Visual performance using reaction times, *Light. Res. Technol.*, 20, 139–53.

Rea, M.S. and Ouellette, M.J. (1991) Relative visual performance: a basis for application, *Light. Res. Technol.*, 23, 135–44.

Rea, M.S., Ouellette, M.J., and Kennedy, M.E. (1985a) Lighting and task parameters affecting posture, performance and subjective ratings, *J. Illumin. Eng. Soc.*, 15, 231–8.

Rea, M.S., Clark, F.R.S., and Ouellette, M.J. (1985b) *Photometric and psychophysical Measurements of Exit Signs through Smoke*, DBR Paper 1291, Ottawa: National Research Council Canada.

Rea, M.S., Ouellette, M.J., and Tiller, D.K. (1990a) The effects of luminous surroundings on visual performance, pupil size and human preference, *J. Illumin. Eng. Soc.*, 19, 45–58.

Rea, M.S., Robertson, A.R., and Petrusic, W.M. (1990b) Color rendering of skin under fluorescent lamp illumination, *Color Res. Appl.*, 15, 80–91.

Rea, M.S., Bullough, J.D., and Figueiro, M.G. (2002) Phototransduction for human melatonin suppression, *J. Pineal Res.*, 32, 1–5.

Reading, V. (1966) Yellow and white headlamp glare and age, *Trans. Illumin. Eng. Soc. (London)*, 31, 108–21.

Reitmaier, J. (1979) Some effects of veiling reflections in papers, *Light. Res. Technol.*, 11, 204–9.

Renfrew, J.W., Pettigrew, K.D., and Rapoport, S.I. (1987) Motor activity and sleep duration as function of age in healthy men, *Physiol. Behav.*, 41, 627–34.

Reynolds. R.L., Karpala, F., Clarke, D.A., and Hagenlers, O.L. (1993) Theory and applications of a surface inspection technique using double pass retroreflection, *Opt. Eng.*, 32, 2122–9.

Richter, M. and Witt, K. (1986) The story of the DIN color system, *Color Res. Appl.*, 11, 138–45.

Rihner, M. and McGrath, H., Jr. (1992) Fluorescent light photosensitivity in patients with systemic lupus erythematosus, *Arthritis Rheum.*, 35, 949–52.

Riley, R.L. (1994) Ultraviolet air disinfection: rationale for whole building irradiation, *Infect. Control Hosp. Epidemiol.*, 15, 324–8.

Ritch, R. (2000) Glaucoma, in B. Silverstone, M.A. Lang, B.P. Rosenthal, and E.E. Faye (eds) *The Lighthouse Handbook on Vision Impairment and Vision Rehabilitation*, New York: Oxford University Press.

RLW Analytics Inc., (1999) *Non-Residential/New Construction Baselines Study*, Report for the California Board for Energy Efficiency of the California Public Utilities Commission.

Robbins, C.L. (1986) *Daylighting Design and Analysis*, New York: Van Norstrand Reinhold.

Roberts, B. (1997) *The Quest for Comfort*, London: Chartered Institution of Building Services Engineers.

Robertson, A.R. (1977) The CIE 1976 color difference formulae, *Color Res. Appl.* 2, 7–11.

Roenneberg, T. and Aschoff, J. (1990a) Annual rhythm of human reproduction: I. Biology, sociology or both?, *J. Biol. Rhythms*, 5, 217–39.

Roenneberg, T. and Aschoff, J. (1990b) Annual rhythm of human reproduction: II. Environmental considerations, *J. Biol. Rhythms*, 5, 217–39.

Roenneberg, T. and Foster, R.G. (1997) Twilight times: light and the circadian system, *Photochem. Photobiol.*, 66, 549–61.

Roethlisberger, F.J. and Dickson, W.J. (1939) *Management and the Worker*, Cambridge, MA: Harvard University Press.

Roll, R.F. (1987) Recent results on the illumination of VDU and CAD workstations, in B. Knave and P.G. Wideback (eds) *Work with Display Units*, 86, Amsterdam: North-Holland.

Rombauts, P., Vandewyngaerde, H., and Maggetto, G. (1989) Minimum semi-cylindrical illuminance and modeling in residential lighting, *Light. Res. Technol.*, 21, 49–55.

Roper, V.J. and Howard, E.A. (1938) Seeing with motor car headlamps, *Trans. Illumin. Engng. Soc. (London)*, 33, 417–38.

Rosen, L.N., Targum, S.D., Terman, M., Bryant, M.J., Hoffman, H., Kasper, S.F., Hamovit, J.R., Docherty, J.P., Welch, B., and Rosenthal, N.E. (1990) Prevalence of seasonal affective disorder at four latitudes, *Psychiatry Res.*, 31, 131–44.

Rosenhahn, E.O. and Hamm, M. (2001) Measurements and ratings of HID head-lamp impact on traffic safety aspects, in *Lighting Technology Developments for Automobiles, Society of Automotive Engineers Report, SP1595*, Warrendale, PA: SAE.

Rosenthal, N.E., Sack, D.A., James, S.P., Parry, B.L., Mendelson, W.B., Tamarkin, L., and Wehr, T.A. (1985) Seasonal affective disorder and phototherapy, in R.J. Wurtman, M.J. Baum, and J.T Potts, Jr. (eds) *The Medical and Biological Effects of Light*, New York: New York Academy of Sciences.

Roufs, J.A.J., (1972) Dynamic aspects of vision 1. Experimental relations between flicker and flash threshold, *Vision Res.*, 12, 261–78.

Roufs, J.A.J., (1991) The man–machine interface, in J. Cronley-Dillon (ed.) *Vision and Visual Dysfunction*, vol. 15, London: Macmillan Press.

Roufs, J.A.J. and Boschman, M.C. (1991) Visual comfort and performance, in J.A.J. Roufs (ed.) *The Man–Machine Interface*, London: MacMillan Press.

Rowlands, E., Loe, D.L., Waters, I.M., and Hopkinson, R.G., (1971) Visual performance in illuminance of different spectral quality, *Proceedings of the CIE, 17th Session*, Barcelona, Paris: CIE.

Ruberg, F.L., Skene, D.J., Hanifin, J.P., Rollag, M.D., English, J., Arendt, J., and Brainard, G.C. (1996) Melatonin regulation in humans with color vision defi-ciencies, *J. Clin. Endocrinol. Metab.*, 81, 2980–5.

Rubin, A.I., Collins, B.L., and Tibbott, R.L. (1978) *Window Blinds as a Potential Energy Saver – A Case Study*, NBS Building Science Series 112, Gaithersburg, MD: National Bureau of Standards.

Rubinstein, F., Avery, D., Jennings, J., and Blanc, S. (1997) On the calibration and commissioning of lighting controls, *Proceedings of Right Light 4*, Frederiksberg, Denmark: DEF Congress Service.

Rubinstein, F., Jennings, J., Avery, D., and Blanc, S. (1999) Preliminary results from an advanced lighting controls testbed, *J. Illumin. Eng. Soc.*, 28, 130–41.

Rutenfranz, J., Haider, M., and Koller, M. (1985) Occupational health measures for nightworkers and shiftworkers, in S. Folkard and T.H. Monk (eds) *Hours of Work*, Chichester, UK: John Wiley and Sons.

Ruth, W., Carlsson, L., Wibom, R., and Knave, B. (1979) Workplace lighting in foundries, *Light. Des. Appl.*, 9 (11), 22–9.

Rutkowsky, W. (1987) Light filtering lenses as an alternative to cataract surgery, *J. Am. Optomet. Assoc.*, 58, 640–1.

Ruys, T. (1970) *Windowless Offices*, MA Thesis, University of Washington.

Saalfield, C. (1995) *The Effect of Lamp Spectra and Illuminance on Color Identification*, MS Thesis, Troy, NY: Rensselaer Polytechnic Institute.

Sabey, B.E. and Staughton, E.C. (1975) Interacting roles of road environment, vehicle and road user in accidents, *Proceedings of the 5th Conference of the International Association for Accident and Traffic Medicine*.

Sack, R.L., Blood, M.L., and Lewy, A.J. (1992a) Melatonin rhythms in night shift workers, *Sleep*, 15, 434–41.

Sack, R.L., Lewy, A.J., Blood, M.L., Keith, L.D., and Nakagawa, H. (1992b) Circadian rhythm abnormalities in totally blind people: phase advances and entrainment, *J. Biol. Rhythms*, 6, 249–61.

Saeed, S.A. and Bruce, T.J. (1998) Seasonal affective disorders, *Am. Fam. Physician*, 57, 1340–6, 1351–2.

Sagawa, K. and Takahashi, Y. (2001) Spectral luminous efficiency as a function of age, *J. Opt. Soc. Am.*, 18, 2659–67.

Sanford, L.J. (1996) Visual environment for the partially sighted, *Proceedings of the IESNA Annual Conference*, Cleveland, OH. New York: Illuminating Engineering Society of North America.

Sanford, B.E., Neacham, S., Hanifin, J.P., Hannon, P., Streletz, L., Sliney, D., and Brainard, G.C. (1996) The effects of ultraviolet-A radiation on visual evoked potentials in the young human eye, *Acta Ophthalmol. Scand.*, 74, 553–7.

Sasaki, M., Kurosaki, Y., Onda, M., Yamaguchi, O., Nishimura, H., Kashimura, K., and Graeber, R.C. (1989) Effects of bright light on circadian rhythmicity and sleep after transmeridian flight, *Sleep Res.*, 18, 442.

Sato, M., Inui, M., Nakamura, Y., and Takeuchi, Y. (1989) Visual environment of a control room, *Light. Res. Technol.*, 21, 99–106.

Saunders, J.E. (1969) The role of the level and diversity of horizontal illumination in an appraisal of a simple office task, *Light. Res. Technol.*, 1, 37–46.

Schivelbusch, W. (1988) *Disenchanted Night: The Industrialization of Light in the Nineteenth Century*, Oxford, UK: Berg Publishing.

Schmidt-Clausen, H.J. and Bindels, J.H. (1974) Assessment of discomfort glare in motor vehicle lighting, *Light. Res. Technol.*, 6, 79–88.

Schonpflug, W. (1983) Coping efficiency and situational demands, in G.R.J. Hockey (ed.) *Stress and Fatigue in Human Performance*, Chichester, UK: John Wiley and Sons.

Schooley, L.C. and Reagan, J.A. (1980a) Visibility and legibility of exit signs, Part 1: analytical predictions, *J. Illumin. Eng. Soc.*, 10, 24–8.

Schooley, L.C. and Reagan, J.A. (1980b) Visibility and legibility of exit signs. Part 2: experimental results, *J. Illumin. Eng. Soc.*, 10, 29–32.

Schreuder, D.A. (1964) *The Lighting of Vehicular Traffic Tunnels*, Thesis, Centrex: Techische Hochschule Eindhoven.

Schreuder, D.A. (1976) *White or Yellow Light for Vehicle Head-Lamps*, Voorburg, The Netherlands: Institute for Road Safety Research.

Schreuder, D.A. (1993) Energy saving in tunnel entrance lighting, *Proceedings of Right Light, the Second European Conference on Energy-Efficient Lighting*, Arnhem, The Netherlands: Netherlands Institute of Illuminating Engineering.

Schreuder , D.A. (1998) *Road Lighting for Safety* , London: Thomas Telford.

Schwab, R.N. and Mace, D.J. (1987) Luminance measurements for signs with complex backgrounds, *Proceedings of the CIE, 21st Session*, Venice, Vienna: CIE.

Schwartz, S.D. (2000) Age-related maculopathy and age-related macular degeneration, in B. Silverstone, M.A. Lang, B.P. Rosenthal, and E.E. Faye (eds) *The Lighthouse Handbook on Vision Impairment and Vision Rehabilitation*, New York: Oxford University Press.

Scott, P.P. (1980) *The Relationship between Road Lighting Quality and Accident Frequency*, TRRL Laboratory Report 929, Crowthorne, UK: Transport and Road Research Laboratory.

Sekular, R. and Blake, R. (1994) *Perception*, New York: McGraw-Hill.

Sharp, G.W.G. (1960) The effect of light on diurnal leucocyte variations, *J. Endocrinol.* 21, 213–23.

Shao, L., Elmualim, A.A., and Yohannes, I. (1998) Mirror lightpipes: daylighting performance in real buildings, *Light. Res. Technol.*, 30, 37–44.

Shepherd, A.J., Julian, W.G., and Purcell, A.T. (1989) Gloom as a psychophysical phenomenon, *Light. Res. Technol.*, 21, 89–97.

Shepherd, A.J., Julian, W.G., and Purcell, A.T. (1992) Measuring appearance: parameters indicated from gloom studies, *Light. Res. Technol.*, 24, 203–14.

Shields, M.B., Ritch, R., and Krupin, T.K. (1996) Classification and mechanisms of the glaucomas, in R. Rich, M.B. Shields, and T. Krupin (eds) *The Glaucomas*, St Louis, MO: Mosby.

Shingledecker, C.A. and Holding, D.H. (1974) Risk and effort measures of fatigue, *J. Motor Behav.*, 6, 17–25.

Shlaer, S. (1937) The relation between visual acuity and illumination, *J. Gen. Physiol.*, 21, 165–8.

Shlaer, S., Smith, E.L., and Chase, A.M. (1941) Visual acuity and illumination in different spectral regions, *J. Gen. Physiol.*, 25, 553–69.

Sicurella, V.J. (1977) Colour contrast as an aid for visually impaired persons, *Visual Impairment Blindness*, 71, 252–7.

Sime, J. (1980) The concept of "panic," in D. Canter (ed.) *Fires and Human Behaviour*, Chichester, UK: John Wiley and Sons.

Sime, J. and Kimura, M. (1988) The timing of escape: exit choice behavior in fires and building evacuations, in J. Sime (ed.) *Safety in the Built Environment*, London: E & F.N. Spon.

Simmons, R.C. (1975) Illuminance, diversity and disability glare in emergency lighting, *Light. Res. Technol.*, 7, 125–32.

Simons, R.H. and Bean, A.R. (2000) *Lighting Engineering*, London: Butterworth-Heinemann.

Simons, R.H., Hargroves, R.A., Pollard, N.E., and Simpson, M.D. (1987) Lighting criteria for residential roads and areas, *Proceedings of the CIE, 21st Session*, Venice, Vienna: CIE.

Simonson, E. and Brozek, J. (1948) Effects of illumination level on visual performance and fatigue, *J. Opt. Soc. Am.*, 38, 384–7.

Sivak, M. and Flannagan, M.J. (1993) A fast rise brake light as a collision-prevention device, *Ergonomics*, 36, 391–5.

Sivak, M., Simmons, C.J., and Flannagan, M. (1990) Effect of headlamp area on discomfort glare, *Light. Res. Technol.*, 22, 49–52.

Sivak, M., Flannagan, M., and Gellatly, W. (1993) Influence of truck driver eye position on effectiveness of retroreflective traffic signs, *Light. Res. Technol.*, 25, 31–6.

Sivak, M., Sato, T., and Flannagan, M.J. (1994a) Effects of voltage drop on the rise time and light output of incandescent brake lights on trucks, *Light. Res. Technol.*, 26, 89–90.

Sivak, M., Flannagan, M.J., Traube, E.C., Battle, D.S., and Sato, T. (1994b) Evaluations of in-traffic performance of high-intensity discharge headlamps, *Light. Res. Technol.*, 26, 181–8.

Sivak, M., Flannagan, M.J., Traube, E.C., and Kojima, S. (1998) Automobile rear signal lamps: effects of realistic levels of dirt on light output, *Light. Res. Technol.*, 30, 24–8.

Sivak, M., Flannagan, M.J., Traube, E.C., and Kojima, S. (1999) The influence of stimulus duration on discomfort glare for persons with and without visual correction, *Transport. Hum. Factors*, 1, 147–58.

Sivak, M., Flannagan, M.J., and Miyokawa, T. (2001) A first look at visually aimable and internationally harmonized low-beam headlamps, *J. Illumin. Eng. Soc.*, 30, 26–33.

Slater, A.I. and Boyce, P.R. (1990) Illuminance uniformity on desks: where is the limit?, *Light. Res. Technol.*, 22, 165–74.

Slater, A.L., Perry, M.J., and Crisp, V.H.C. (1983) The applicability of the CIE visual performance model to lighting design, *Proceedings of the CIE, 20th Session*, Amsterdam, Paris: CIE.

Slater, A.I., Perry, M.J., and Carter, D.J. (1993) Illuminance differences between desks: Limits of acceptability, *Light. Res. Technol.*, 25, 91–103.

Sliney, D.H. (1972) Non-ionising radiation, in L.V. Cralley (ed.) *Industrial Environmental Health*, London: Academic Press.

Sliney, D.H. (1995) Ultraviolet radiation and its effect on the aging eye, in W. Adrian, D. Sliney, and J. Werner (eds) *Lighting for Aging Vision and Health*, New York: Lighting Research Institute.

Sliney, D.H. and Bitran, M. (1998) The ACGIH action spectra for hazard assessment: The TLV's, in R. Matthes and D. Sliney (eds) *Measurements of Optical Radiation Hazards*, Oberschleißheim, Germany: International Commission on Non-Ionizing Radiation Protection.

Sliney, D., Fast, P., and Ricksand, A. (1995) Optical radiation hazards analysis of ultraviolet headlamps, *Appl. Optics*, 34, 4912–22.

Sloan, L.L., Habel, A., and Feiock, K. (1973) High illumination as an auxiliary reading aid in diseases of the macula, *Am. J. Ophthalmol.*, 76, 745–57.

Smith, S.W. (1976) Performance of complex tasks under different levels of illumination. Part 1 – needle probe task, *J. Illumin. Eng. Soc.*, 5, 235–42.

Smith, K.C. (1978) Multiple pathways of DNA repair in bacteria and their roles in mutagenesis, *Photochem. Photobiol.*, 28, 121–9.

Smith, S.W. and Rea, M.S. (1978) Proofreading under different levels of illumination, *J. Illumin. Eng. Soc.*, 8, 47–52.

Smith, S.W. and Rea, M.S. (1979) Relationships between office task performance and ratings of feelings and task evaluations under different light sources and levels, *Proceedings of the CIE, 19th Session*, Kyoto, Japan, Paris: CIE.

Smith, S.W. and Rea, M.S. (1982) Performance of a reading test under different levels of illumination, *J. Illumin. Eng. Soc.*, 12, 29–33.

Smith, S.W. and Rea, M.S. (1987) Check value verification under different levels of illumination, *J. Illumin. Eng. Soc.*, 16, 143–9.

Snow, C.E. (1927) Research on industrial illumination, *Tech. Eng. News*, 8, 257–82.

Society of Automotive Engineers (SAE) (1995a) *Harmonized Vehicle Headlamp Performance Requirements, SAE J1735*, Warrendale, PA: SAE.

Society of Automotive Engineers (SAE) (1995b) *Discharge Forward Lighting, SAE J2009*, Warrendale, PA: SAE.

Society of Automotive Engineers (SAE) (1995c) *Color Specification, SAE J578*, Warrendale, PA: SAE.

Society of Automotive Engineers (SAE) (2001) *Ground Vehicle Lighting Standards Manual, SAE HS-34*, Warrendale, PA: SAE.

Society of Light and Lighting (SLL) (2001) *Lighting Guide 3: Addendum 2001, The Visual Environment for Display Screen Use. A New Standard of Performance*, London: SLL.

Sollner, G. (1972) Glare from luminous ceilings, *Lichttechnik*, 24, 557–60.

Sollner, G. (1974) Subjective appraisal of discomfort glare in high halls, illuminated by luminaires for high intensity discharge lamps, *Lichttechnik*, 26, 169–72.

Sommer, R., Cabaj, A., Pribil, W., and Haider, T. (1997) Influence of lamp intensity and water transmittance on the uv disinfection of water, *Water Sci. Technol.*, 35, 113–18.

Sommer, A., Tielsch, J.M., Katz, J., Quigley, H.A., Gottsch, J.D., Javitt, J.C., Martone, J.F., Royall, R.M., Witt, K., and Ezrine, S. (1991) Racial differences in the cause-specific prevalence of blindness in east Baltimore, *New Eng. J. Med.*, 325, 1412–17.

Sorensen, K. (1987) Comparison of glare index definitions, *Research Note of the Danish Illuminating Engineering Laboratory, DK 2800*, Lyngby, Denmark: DIES.

Sorensen, K. (1991) Practical aspects of discomfort glare evaluation: interior lighting, *Proceedings of the First International Symposium on Glare*, New York: Lighting Research Institute.

Sorensen, S. and Brunnstrom, G. (1995) Quality of light and quality of life: an intervention study among older people. *Light. Res. Technol.*, 27, 113–18.

Stenzel, A.G. (1962) Experience with 1000 lx in a leather factory, *Lichttechnik*, 14, 16–18.

Stenzel, A.G. and Sommer, J. (1969) The effect of illumination on tasks which are largely independent of vision, *Lichttechnik*, 21, 143–6.

Stevens, S.S. (1961) The psychophysics of sensory function, in W.A. Rosenblith (ed.) *Sensory Communication*, Cambridge, MA: MIT Press.

Stevens, R.G., Wilson, B.W., and Anderson, L.E. (1997) *The Melatonin Hypothesis: Breast Cancer and the Use of Electric Power*, Columbus, OH: Battelle Press.

Steward, J.M. and Cole, B.L. (1989) What do colour defectives say about everyday tasks, *Optom. Vis. Sci.*, 66, 288–95.

Stiles, W.S. (1930) The scattering theory of the effect of glare on the brightness difference threshold, *Proc. Roy. Soc. London*, 105B, 131–46.

Stiles, W.S. and Crawford, B.H. (1937) The effects of a glaring light source on extrafoveal vision, *Proc. Roy. Soc. London*, 122B, 255–80.

Stillesjo, S. (1991) Using innovative procurement mechanisms to help commercialize new energy-efficient lighting and ventilation products, *Proceedings of the 1st European Conference on Energy-Efficient Lighting*, Stockholm, Sweden: Swedish National Board for Industrial and Technical Development.

Stone, P.T. and Harker, S.D.P. (1973) Individual and group differences in discomfort glare response, *Light. Res. Technol.*, 5, 41–9.

Stone, N. and Irvine, J. (1991) Performance, mood, satisfaction and task type in various work environments: a preliminary study, *J. Gen. Psychol.*, 120, 489–97.

Storch, R.L. and Bodis-Wollner, I. (1990) Overview of contrast sensitivity and neuro-ophthalmic disease, in M.P. Nadler, D. Miller, and D.J. Nadler (eds) *Glare and Contrast Sensitivity for Clinicians*, New York: Springer-Verlag.

Stuck, B.E. (1998) The retina and action spectrum for photoretinitis ("blue light hazard"), in R. Matthes and D. Sliney (eds) *Measurements of Optical Radiation Hazards*, Oberschleißheim, Germany: International Commission on Non-Ionizing Radiation Protection.

Subisak, G.J. and Bernecker, C.A. (1993) Psychological preferences for industrial lighting, *Light. Res. Technol.*, 25, 171–7.

Sullivan, J.M. and Flannagan, M.J. (1999) *Assessing the Potential Benefit of Adaptive Headlighting using Crash Databases, University of Michigan Transportation Research Institute Report UMTRI-99-21*, Ann Arbor, MI: University of Michigan.

Sumner, R., Baguley, C., and Burton, J. (1977) *Driving in Fog on the M4*. Supplementary Report 281, Crowthorne, UK: Transport Research Laboratory.

Swaab, D.F., Fliers, E., and Partiman, T.S. (1985) The suprachiasmatic nucleus of the human brain in relation to sex, age and senile dementia, *Brain Res.*, 342, 37–44.

Tam, E.M., Lam, R.W., and Levitt, A.J. (1995) Treatment of seasonal affective disorder: a review, *Can. J. Psychiatry*, 40, 457–66.

Tanner, J.C. and Harris, A.J. (1956) Comparison of accidents in daylight and in darkness, *Int. Road Safety Traffic Rev.*, 4, 11–14, 39.

Tansley, B.W. and Boynton, R.M. (1978) Chromatic border perception: the role of red- and green-sensitive cones, *Vision Res.*, 18, 683–97.

Taubes, G. (1995) Epidemiology faces its limits, *Science*, 269, 164–9.

Taylor, R.B. and Gottfredson, S. (1986) Environmental design, crime and crime reduction: an examination of community dynamics, in A.J. Reiss and M. Tonry (eds) *Communities and Crime*, Chicago, IL: University of Chicago Press.

Taylor, L.H. and Socov, E.W. (1974) The movement of people towards lights, *J. Illumin. Eng. Soc.*, 3, 237–41.

Taylor, H.R., West, S.K., Rosenthal, F.S., Munoz, B., Newland, H.S., Abbey, H., and Emmett, E.A. (1988) Effect of ultraviolet radiation on cataract formation, *New Engl. J. Med.*, 319, 1429–33.

Teichner, W. and Krebs, M. (1972) The laws of simple reaction time, *Psychol. Rev.*, 79, 344–58.

Tenkink, E. (1988) Lane keeping and speed choice with restricted sight, in T. Rothengatter and R. de Bruin (eds) *Road User Behaviour: Theory and Research*, Assen/Maastricht, The Netherlands: Van Gorcum.

Terman, M. (1989) On the question of mechanism in phototherapy for seasonal affective disorder: considerations of clinical efficacy and epidemiology, in N.E. Rosenthal and M.C. Blehar (eds) *Seasonal Affective Disorders and Phototherapy*, New York: Guilford.

Terman, M., Terman, J.S., Quitkin, F.M., McGrath, P.J., Stewart, J.W., and Rafferty, B. (1989) Light therapy for seasonal affective disorder: A review of efficacy, *Neuropsychopharmacology*, 2, 1–22.

Terman, M., Lewy, A.J., Dijk, D.-J., Boulos, Z., Eastman, C.I., and Campbell, S.S. (1995) Light treatment for sleep disorders: consensus report IV. Sleep phase and duration disturbances, *J. Biol. Rhythms*, 10, 135–47.

Thapan, K., Arendt, J., and Skene, D.J. (2001) An action spectrum for melatonin suppression: Evidence for a novel non-rod, non-cone photoreceptor system in humans, *J. Physiol.*, 535, 261–7.

Theeuwes, J. and Alferdinck, J.W.A.M. (1996) *The Relation between Discomfort Glare and Driver Behavior, Report DOT HS 808 452*, Washington, DC: US Department of Transportation.

Thylefors, B., Negrel, A.D., Pararajasegaram, R., and Dadzie, K.Y. (1995) Global data on blindness, *Bull. WHO*, 73, 115–21.

Tielsch, J.M. (2000) The epidemiology of vision impairment, in B. Silverstone, M.A. Lang, B.P. Rosenthal, and E.E. Faye (eds) *The Lighthouse Handbook on Vision Impairment and Vision Rehabilitation*, New York: Oxford University Press.

Tielsch, J.M., Sommer, A., Witt, K., Katz, J., and Royall, R.M., (1990) Blindness and visual impairment in an American urban population, *Arch. Ophthalmol.*, 108, 286–90.

Tien, J.M., O'Donnell, V.F., Barnett, A., and Mirchandani, P.B. (1979) *Street Lighting Projects National Evaluation Program: Phase 1 Report*, Washington, DC: US Department of Justice.

Tiller, D.K. (1990) Towards a deeper understanding of psychological effects of lighting, *J. Illumin. Eng. Soc.*, 19, 59–65.

Tiller, D.K. and Rea, M.S. (1992) Semantic differential scaling: prospects for lighting research, *Light. Res. Technol.*, 24, 43–52.

Tilley, A.J., Wilkinson, R.T., Warren, P.S.G., Watson, B., and Drud, M. (1982) The sleep and performance of shift workers, *Hum. Factors*, 24, 629–41.

Timmers, H. (1978) An effect of contrast on legibility of printed text, *Institute of Perception Annual Report*, Eindhoven, Netherlands: IPO.

Tong, D. and Canter, D. (1985) Informative warnings: in situ evaluations of fire alarms, *Fire Safety J.*, 9, 267–79.

Touw, L.M.C. (1951) Preferred brightness ratio of task and its immediate surround, *Proceedings of the CIE, 12th Session,* Stockholm, Paris: CIE.

Tregenza, P. and Loe, D. (1998) *Design of Lighting*, London: E & FN Spon.

Tupper, B., Miller, D., and Miller, R. (1985) The effect of a 550 nm cutoff filter on the vision of cataract patients, *Ann. Ophthalmol.*, 17, 67–72.

Turner, B.P., Ury, M.G., Leng, Y., and Love, W.G. (1997) Sulfur lamps – progress in their development, *J. Illumin. Eng. Soc.*, 26, 10–16.

Turner, D., Nitzburg, M., and Knoblauch, R. (1998) Ultraviolet headlamp technology for nighttime enhancement of roadway markings, *Transport. Res. Rec.*, Paper No. 98–1187, 124–31.

Ueki, K. Ohkubo, N., and Fujimura, H. (1992) Motorway accidents in tunnels in relation to lighting, in *Road Lighting as an Accident Counter-Measure*, CIE Publication No. 93,Vienna: CIE.

Uhlman, R.F., Larson, E.B., Koepsell, T.D., Rees, T.S., and Duckert, L.G. (1991) Visual impairment and cognitive dysfunction in Alzheimer's disease, *J. Gen. Intern. Med.*, 6, 126–32.

Underwriters Laboratories Inc. (UL) (1997) *Low Level Path Marking and Lighting Systems*, UL-1994, Northbrook, IL: UL.

United States Department of Health, Education and Welfare (USDHEW) (1964) *Binocular Visual Acuity of Adults – US 1960–1962*, Washington, DC: USDHEW.

United States Department of Justice (USDOJ) (1998) *Crime in the United States – 1998: Uniform Crime Reports*, Washington, DC: USDOJ.

Urbach, F. (1998) The ultraviolet action spectrum for erythema – history, in R. Matthes and D. Sliney (eds) *Measurements of Optical Radiation Hazards*, Oberschleißheim, Germany: International Commission on Non-Ionizing Radiation Protection.

Urwick, L. and Brech, E.F.L. (1965) *The Making of Scientific Management: Vol. 3, The Hawthorne Investigations*, London: Pitmans.

Van den Berg, T.J.T.P. (1993) Quantal and visual efficiency of fluorescence in the lens of the human eye. *Invest. Ophthalmol. Vision Sci.*, 34, 3566–73.

Van den Berg, T.J.T.P., IJspeert, J.K., and de Waard, P.W.T. (1991) Dependence of intraocular straylight on pigmentation and transmission through the ocular wall, *Vision Res.*, 31, 1361–7.

Van Ierland, J.F.A.A. (1967) *Two Thousand Dutch Office Workers Evaluate Lighting*, Publication 283, Delft, The Netherlands: Research Institute for Environmental Hygiene, TNO.

Van Kemenade, J.T.C. and van der Burgt, P.J.M. (1988) Light sources and colour rendering: Additional information for the R_a index, *Proceedings of the CIBSE National Lighting Conference*, Cambridge, London: CIBSE.

Van Lierop, F.H., Rojas, C.A., Nelson, G.J., Dielis, H., and Suijker, J.L.G. (2000) 4000 K low wattage metal halide lamps with ceramic envelopes: a breakthrough in color quality, *J. Illumin. Eng. Soc.*, 29, 83–8.

Van Nes, F.L. and Bouman, M.A. (1967) Spatial modulation transfer in the human eye, *J. Opt. Soc. Am.*, 47, 401–6.

Van Reeth, O., Sturis, J., Byrne, M.M., Blackman, J.D., L'Hermite-Balriaux, M., Leproult, R., Oliner, C., Retetoff, S., Turek, F.W., and Van Cauter, E. (1994) Nocturnal exercise phase delays circadian rhythms of melatonin and thyrotropin secretion in normal men, *Am. J. Physiol.*, 266, E964–E974.

Van Someren, E.J.W., Hagebeuk, E.E.O., Lijzenga, C., Schellens, P., Rooij, S., Eja, de, Jonker, C., Pot, M.A., Mirmiran, M., and Swaab, D. (1996) Circadian rest–activity rhythm disturbances in Alzheimer's disease, *Biol. Psychiatry*, 40, 259–70.

Van Someren, E.J.W., Kessler, A., Mirmiran, M., and Swaab, D.F. (1997a) Indirect bright light improves circadian rest–activity rhythm disturbances in demented patients, *Biol. Psychiatry*, 41, 955–63.

Van Someren, E.J., Lijzenga, C., Mirmiran, M., and Swaab, D.F. (1997b) Long-term fitness training improves the circadian rest–activity rhythm in healthy elderly males, *J. Biol. Rhythms*, 12, 146–56.

Veitch, J.A. (2001a) Lighting quality considerations from biophysical processes, *J. Illumin. Eng. Soc.*, 30, 3–16.

Veitch, J.A. (2001b) Psychological processes influencing lighting quality, *J. Illumin. Eng. Soc.*, 30, 124–40

Veitch, J.A. and Kaye, S.M. (1988) Illumination effects on conversational sound levels and job-candidate evaluation, *J. Environ. Psychol.*, 8, 223–33.

Veitch, J.A. and McColl, S.L. (2001) A critical examination of the perceptual and cognitive effects attributed to full-spectrum fluorescent lighting, *Ergonomics*, 44, 255–79.

Veitch, J.A. and Newsham, G.R. (1996) Experts quantitative and qualitative assessments of lighting quality, *Proceedings of the Annual Conference of the IESNA*, Cleveland, OH, New York: IESNA.

Veitch, J.A. and Newsham, G.R. (1998a) Lighting quality and energy efficiency effects on task performance, mood, health, satisfaction and comfort, *J. Illumin. Eng. Soc.*, 27, 107–29.

Veitch, J.A. and Newsham, G.R. (1998b) Determinants of lighting quality 1. State of the science, *J. Illumin. Eng. Soc.*, 27, 92–106.

Verriest, G. (1963) Further studies on acquired deficiency of color discrimination, *J. Opt. Soc. Am.*, 53, 185–95.

Vos, J.J. (1984) Disability glare – a state of the art report, *CIE J.*, 3, 39–53.

Vos, J.J. (1995) Age dependence of glare effects and their significance in terms of visual ergonomics, in W. Adrian (ed.) *Lighting for Aging Vision and Health*, New York: Lighting Research Institute.

Vos, J.J. (1999) Glare today in historical perspective: towards a new CIE glare observer and a new glare nomenclature, *Proceedings of the CIE, 24th Session*, Warsaw, Vienna: CIE.

Vos, J.J. and van Bergem-Jansen, P.M. (1995) Greenhouse lighting side-effects: community reaction, *Light. Res. Technol.*, 27, 45–51.

Vos, J.J. and Boogaard, J. (1963) Contribution of the cornea to entoptic scatter, *J. Opt. Soc. Am.*, 53, 869–73.

Vos, J.J. and Padmos, P. (1983) Straylight, contrast sensitivity and the critical object in relation to tunnel lighting, *Proceedings of the CIE, 20th Session, Amsterdam*, Vienna: CIE.

Wald, G. (1945) Human vision and the spectrum, *Science*, 101, 653–8.

Waldhauser, F. and Dietzel, M. (1985) Daily and annual rhythms in human melatonin secretion: role in puberty control, in R.J. Wurtman, M.J. Baum, and J.T. Potts, Jr. (eds) *The Medical and Biological Effects of Light, Ann.* New York: New York Academy of Sciences.

Walker, M.F. (1977) The effects of urban lighting on the brightness of the night sky, *Publ. Astron. Soc. Pacific*, 89, 405–9.

Walker, J. (1985) Social problems of shiftwork, in S. Folkard and T.H. Monk (eds) *Hours of Work*, Chichester, UK: John Wiley and Sons.

Walraven, P.L. (1974) A closer look at the tritanopic convergence point, *Vision Res.*, 14, 1339–43.

Ware, C. and Cowan, W.B. (1983) *National Research Council Technical Report No. 26055 Specification of Heterochromatic Brightness Matches: A Conversion Factor for Calculating Luminances of Small Stimuli which are Equal in Brightness*, Ottawa, Canada: National Research Council Canada.

Warren, N. and Clark, B. (1936) Blocking in mental and motor tasks during a 65 hour vigil, *Psychol. Bull.*, 33, 814–15.

Watanabe, Y., Nayuki, K., and Torisaki, K. (1973) *Actions of Firemen in Smoke*, Fire Research Institute of Japan, Report 37, Tokyo: Fire Research Institute of Japan.

Waters, I. and Loe, D.L. (1973) Visual performance in illumination of differing spectral quality, *Visual Performance Research Project, University College Environmental Research Group*, London: University College.

Waters, C.E., Mistrick, R.G., and Bernecker, C.A. (1995) Discomfort glare from sources of nonuniform luminance, *J. Illumin. Eng. Soc.*, 24, 73–85.

Weale, R.A. (1982) *A Biography of the Eye: Development, Growth, Age*, London: H.K. Lewis.

Weale, R.A. (1985) Human lenticular fluorescence and transmissivity and their effects on vision, *Experimental Eye Research*, 41, 457–73.

Weale, R.A. (1988) Age and the transmittance of the human crystalline lens, *J. Physiol.*, 395, 577–87.

Weale, R.A. (1990) Evolution, age and ocular focus, *Mech. Aging. Dev.*, 53, 85–9.

Weale, R.A. (1991) The lenticular nucleus, light and the retina, *Exp. Eye Res.*, 52, 213–18.

Weale, R.A. (1992) *The Senescence of Human Vision*, Oxford, UK: Oxford University Press.

Weaver, F.M. and Carroll, J.S. (1985) Crime perceptions in a natural setting by expert and novice shoplifters, *Social Psychol. Q.*, 48, 349–59.

Webber, G.M.B. and Aizlewood, C.E. (1993a) *Emergency Wayfinding Lighting*, Building Research Establishment (BRE) Information Paper IP1/93, Garston, Watford, UK: BRE.

Webber, G.M.B. and Aizlewood, C.E. (1993b) Investigation of emergency wayfinding lighting systems, *Proceedings of Lux-Europa 1993*, London: CIBSE.

Webber, G.M.B. and Aizlewood, C.E. (1994) Emergency lighting and wayfinding systems in smoke, *Proceedings of the CIBSE National Lighting Conference, Cambridge*, London: CIBSE.

Webber, G.M.B. and Hallman, P.J. (1989) *Photoluminescent Markings for Escape*, Building Research Establishment (BRE) Information Paper IP17/89, Garston, Watford, UK: BRE.

Webber, G.M.B., Hallman, P.J., and Salvidge, A.C. (1988) Movement under emergency lighting: comparison between standard provisions and photo-luminescent markings, *Light. Res. Technol.*, 20, 167–75.

Wehr, T.A. (1996) A "clock for all seasons" in the human brain, in R.M. Buijs, A. Kalsbeek, J.H. Romihn, C.M.A. Pennartz, and M. Mirmiran (eds) *Prog. Brain Res.*, 111, 319–40, Amsterdam: Elsevier.

Wehr, T.A. (1997) Melatonin and seasonal rhythms, *J. Biol. Rhythms*, 12, 518–27.

Wehr, T.A. (2001) Photoperiodism in humans and other primates: evidence and implications, *J. Biol. Rhythms*, 16, 348–64.

Wehr, T.A. and Rosenthal, N.E. (1989) Seasonality and affective illness, *Am. J. Psychiatry*, 146, 829–39.

Wehr, T.A., Giesen, H.A., Schulz, P. M., Anderson, J.L., Joseph-Vanderpool, J.R., Kell, K. (1991) Contrasts between symptoms of summer depression and winter depression, *J. Affect. Disord.*, 23, 178–83.

Wehr, T.A., Giesen, H.A., Moul, D.E., Turner, E.H., and Schwatrz, P.J. (1995) Suppression of human responses to seasonal changes in day-length by modern artificial lighting, *Am. J. Physiol*, 269, R173–R178.

Welford, A.T., Brown, R.A., and Gabb, J.E. (1950) Two experiments on fatigue as affecting skilled performance in civilian aircrew, *Br. J. Psychol.*, 40, 195–211.

Werner, J. and Hardenbergh, F.E. (1983) Spectral sensitivity of the pseudophakic eye, *Arch. Ophthalmol.*, 101, 758–60.

Werner, J.S. and Kraft, J.M. (1995) Color vision senescence: implications for lighting design, in W. Adrian (ed.) *Lighting for Aging Vision and Health*, New York: Lighting Research Institute.

Werner, J.S., Peterzell, D.H., and Scheetz, A.J. (1990) Light, vision and aging. *Optomet. Vis. Sci.*, 67, 214–29.

Westheimer, G. (1987) Visual acuity and hyperacuity: resolution, localization, form, *Am. J. Optomet. Physiol. Optics*, 64, 567–74.

Weston, H.C. (1922) A study of the efficiency in fine linen-weaving, *Industrial Fatigue Research Board Report No. 20*, London: His Majesty's Stationery Office.

Weston, H.C. (1935) The relation between illumination and visual efficiency: the effect of size of work, *Industrial Health Research Board and the Medical Research Council*, London: His Majesty's Stationery Office.

Weston, H.C. (1945) The relation between illumination and visual efficiency: the effect of brightness contrast, *Industrial Health Research Board, Report No. 87*, London: His Majesty's Stationery Office.

Weston, H.C. and Taylor, S.K. (1926) The relation between illumination and efficiency in fine work (typesetting by hand), *Final Report of the Industrial Fatigue Research Board and the Illumination Research Committee (DSIR)*, London: His Majesty's Stationery Office.

Wetterberg, L. (1993) *Light and Biological Rhythms in Man*, New York: Pergamon Press.

Wever, R.A. (1979) *The Circadian System of Man: Results of Experiments under Temporal Isolation*, New York: Springer-Verlag.

White M.E. and Jeffrey D.J. (1980) *Some Aspects of Motorway Traffic Behavior in Fog*, Report LR 958, Crowthorne, UK: Transport Research Laboratory.

Whitlock and Weinberger Transportation (1998) *An Evaluation of a Crosswalk Warning System Utilizing In-Pavement Flashing Lights*, Santa Rosa, CA: Whitlock and Weinberger Transportation.

Whittaker, J. (1996) An investigation in to the effects of British summer time on road traffic accident casualties in Cheshire, *J. Accid. Emerg. Med.*, 13, 189–92.

Whittaker, S.G. and Lovie-Kitchin, J. (1993) Visual requirements for reading, *Optomet. Vision Sci.*, 70, 54–65.

Wibom, R.I. and Carlsson, W. (1987) Work at visual display terminals among office employees: visual ergonomics and lighting, in B. Knave and P.G. Wideback (eds) *Work with Display Units, 86*, Amsterdam: North-Holland.

Wierda, M. (1996) Beyond the eye: cognitive factors in drivers' visual perception, in A.G. Gale, I.D. Brown, C.M. Haslegrave, and S.P. Taylor (eds) *Vision in Vehicles V*, Amsterdam: North-Holland.

Wiggle, R., Gregory, W., and Lloyd, C.J. (1997) Paint inspection lighting, Society of Automotive Engineers (SAE) Paper 982315, *Proceedings of the International Body Engineering Conference*, Detroit, MI: SAE.

Wilkins, A.J. (1995) *Visual Stress*, Oxford: Oxford University Press.

Wilkins, A.J., Nimmo-Smith, I., Slater, A.J., and Bedocs, L. (1989) Fluorescent lighting, headaches and eyestrain, *Light. Res. Technol.*, 21, 11–18.

Wilkinson, R.T. (1969) Some factors influencing the effect of environmental stress on performance, *Psychol. Bull.*, 72, 260–72.

Willey, A.E. (1971) Unsafe exiting conditions! Apartment house fire, Boston, Massachusetts, *Fire J.*, July, 16–23.

Williams, L.G. (1966) The effect of target specification on objects fixated during visual search, *Percept. Psychophys.*, 1, 315–18.

Williams, T.D. (1983) Aging and central visual field area, *Am. J. Optomet. Physiol. Opt.* 60, 888–91.

Wirz-Justice A., Graw, P., Krauchi, K., Gisin, B., and Jochum, A. (1993) Light therapy in seasonal affective disorder is independent of time of day or circadian phase, *Arch. Gen. Psychiatry*, 50, 929–37.

Wolbarsht, M.L. (1992) Cataract from infrared lasers: evidence for photochemical mechanisms, *Lasers Light Ophthalmol.*, 4, 91–6.

Wolf, E. and Gardiner, J.S. (1965) Studies on the scatter of light in the dioptric media of the eye as a basis of visual glare, *Arch. Ophthalmol.*, 74, 338–45.

Wood, P.G. (1980) A survey of behavior in fires, in D. Canter (ed.) *Fires and Human Behaviour*, Chichester, UK: John Wiley and Sons.

Wood, R.L. Jr., Franks, J.K., and Sliney, D.H. (1998) Measurements of representative lamps for the ANSI/IESNA RP-27.3–96 photobiological safety standard for lamps, in R. Matthes and D. Sliney (eds) *Measurements of Optical Radiation Hazards*, Oberschleißheim, Germany: International Commission on Non-Ionizing Radiation Protection.

World Health Organization (WHO) (1977) *Manual of the International Classification of Diseases, Injuries and Causes of Death*, Geneva, Switzerland: WHO.

World Health Organization (WHO) (1982) *Lasers and Optical Radiation*, Environmental Health Criteria Document 23, Geneva, Switzerland: WHO.

Worthey, J.A. (1985) An analytical visual clarity experiment, *J. Illumin. Eng. Soc.*, 15, 239–51.

Wright, M.S., Cook, G.K., and Webber, G.M.B. (1999) Emergency lighting and wayfinding provision systems for visually impaired people: Phase 1 of a study, *Light. Res. Technol.*, 31, 35–42.

Wright, R.M., Keilweil, P., Pelletier, P., and Dickinson, K. (1974) *The Impact of Street Lighting on Crime, Part 1*, Washington, DC: National Institute of Law Enforcement and Criminal Justice.

Wurm, L.H., Legge, G.E., Isenberg, L.M., and Luebker, A. (1993) Colour improves object recognition in normal and low vision, *J. Exp. Psychol.: Hum. Percept. Perform.*, 19, 899–911.

Wyatt, S. and Langdon, J.N. (1932) *Inspection Processes in Industry*, Medical Research Council Industrial Health Research Board, Report 63, London: His Majesty's Stationery Office.

Wyon, D.P. (1996) Indoor environmental effects on productivity, *Proceedings of the Indoor Air Quality 96 Conference*, Atlanta, GA: ASHRAE.

Wyszecki, G. (1981) Uniform color spaces, in *Golden Jubilee of Colour in the CIE*, Bradford, UK: The Society of Dyers and Colourists.

Wyszecki, G. and Stiles, W.S. (1982) *Color Science: Concepts and Methods, Quantitative Data and Formulas*, New York: John Wiley and Sons.

Yerrell, J.S. (1971) *Headlamp Intensities in Europe and Britain, LR 383*, Crowthorne, UK: Road Research Laboratory.

Yerrell, J.S. (1976) Vehicle headlights, *Light. Res. Technol.*, 8, 69–79.

Young, R.W. (1981) A theory of central retinal disease, in M.L. Sears (ed.) *New Directions in Ophthalmic Research*, New Haven, CT: Yale University Press.

Zeitzer, J.M., Dijk, D.-J., Kronauer, R.E., Brown, E.N., and Czeisler, C.A. (2000) Sensitivity of the human circadian pacemaker to nocturnal light: melatonin phase resetting and suppression, *J. Physiol.*, 526, 695–702.

Zhang, L., Helander, M., and Drury, C.G. (1996) Identifying factors of comfort and discomfort in seating, *Hum. Factors*, 38, 377–89.

Zigman, S. (1992) Light filters to improve vision, *Optomet. Vision Sci.*, 69, 325–8.

Zuclich, J.A. (1998) The corneal ultraviolet action spectrum for photokeratitis, in R. Matthes and D. Sliney (eds) *Measurements of Optical Radiation Hazards*, Oberschleißheim, Germany: International Commission on Non-Ionizing Radiation Protection.

Index